Handbook of Experimental Pharmacology

Volume 234

More information about this series at http://www.springer.com/series/164

Tobias Langenhan • Torsten Schöneberg
Editors

Adhesion G Protein-coupled Receptors

Molecular, Physiological and Pharmacological Principles in Health and Disease

Springer

Editors
Tobias Langenhan
Institute of Physiology
Department of Neurophysiology
University of Würzburg
Würzburg, Germany

Torsten Schöneberg
Institute of Biochemistry
Medical Faculty
University of Leipzig
Leipzig, Germany

ISSN 0171-2004 ISSN 1865-0325 (electronic)
Handbook of Experimental Pharmacology
ISBN 978-3-319-41521-5 ISBN 978-3-319-41523-9 (eBook)
DOI 10.1007/978-3-319-41523-9

Library of Congress Control Number: 2016955552

Printed on acid-free paper

This Springer imprint is published by Springer Nature
The registered company is Springer International Publishing AG
The registered company address is: Gewerbestrasse 11, 6330 Cham, Switzerland

Preface

The very first monograph that was dedicated to a general overview on the adhesion family of G protein-coupled receptors (aGPCRs) was published just in 2010 (Adhesion-GPCRs: Structure to Function. Yona and Stacey, Ed., Landes Bioscience and Springer). It was the earliest attempt by a small group of researchers to cast the scarce information on these enigmatic molecules into a general concept on what they do and how they do it.

The absence of such public face for the biology of aGPCRs was painfully felt by all colleagues who were actively researching aGPCRs in these days. Scepticism was high from many neighbouring fields why aGPCRs rise to such grotesque dimensions with thousands of residues dedicated to their extracellular tails alone. Also how their exotic functions during the development of organs could be accounted for by their peculiar bipartite adhesive/receptive structure was a constant source of doubt (and motivation for further investigation). Not least, whether aGPCRs are 'true' GPCRs and can thus be attacked by the immense technological armoury that has accumulated during the decades of research on other members of the GPCR superfamily was possibly the most pressing question we were confronted with. Next to the simple matter: what do these receptors sense, after all?

While many of these points could not be satisfactorily answered yet back then, the 2010 book project brought them on the map for the first time in a collective effort. Therefore, this venture from a group of adhesion GPCR aficionados was an incontestable sign of a growing community of researchers that had formed to pursue the inherent questions on aGPCRs with seriousness and persistence.

The roster of colleagues that have contributed their expertise, time and dedication to the current monograph bears testimony to that spirit, and we are immensely grateful for their support. We also wish to thank the Editorial Board of the *Handbook of Experimental Pharmacology* for allotting us an entire volume of this eminent book series to document our knowledge on aGPCRs. We are indebted to Susanne Dathe, Wilma McHugh, Rahila Nahid and Sumathy Thanigaivelu from *Springer Nature* for excellent editorial and technical support, and for generous funding from the Deutsche Forschungsgemeinschaft (DFG) to several chapter authors through a Research Unit Grant (FOR 2149), a first award of its kind to a coordinated scientific initiative dedicated to the study of aGPCRs.

The chapters of this volume are authored by renowned experts in the aGPCR field and chart the current state of aGPCR research. Following their contributions, the reader will learn that some of the pressing molecular issues of 2010 have begun to find answers:

- aGPCRs can signal via canonical signaling outlets and a credible mechanism on how they get activated has been recently devised (Liebscher et al., *Tethered agonism: a common activation mechanism of adhesion GPCRs*; Kishore et al., *Versatile signaling activity of adhesion GPCRs*).
- In some cases, the receptors' structural peculiarities have been experimentally matched with highly intriguing biochemical and biological phenomena such as in the case of the GAIN domain and other extracellular protein folds (Araç et al., *Understanding the structural basis of adhesion GPCR functions*).
- One such class of events regards the extensive proteolytic processing of aGPCRs and is discussed by Nieberler et al. (*Control of adhesion GPCR function through proteolytic processing*). Knapp et al. explore the central position of *adhesion GPCRs-related protein networks*, roles that are mainly relayed through their intracellular domains.
- Other vital components of their architecture such as the structure of the heptahelical transmembrane domain of aGPCRs have remained locked to our efforts, but it is clear that in the near future the focus will shift evermore into their direction and offer new vantage points to interfere with their activity. Nijmeijer et al. explored these possibilities in their chapter on *7TM domain structure of adhesion GPCRs*.
- Kovacs et al. review *the relevance of genomic signatures at adhesion GPCR loci* (specifically of human homologs), informing us about their role in phenotypic variation and disease aetiology, an overdue endeavour in the omics era that is aided by the novel harmonised nomenclature and classification system of the aGPCR family introduced by Krishnan et al. (*Classification, nomenclature and structural aspects of adhesion GPCRs*).

The second part of this book is dedicated to physiological and pathological aspects of aGPCRs:

- Scholz et al. describe the emerging concept of *adhesion GPCRs as a putative class of metabotropic mechanosensors*, which distinguishes them from the rest of the GPCR superfamily.
- Several chapters relate to this discovery with specialist focus on its implications in the nervous system (Harty et al., *Adhesion GPCRs as novel actors in neural and glial cell functions: from synaptogenesis to myelination*), in skeletal muscle (White et al., *Control of skeletal muscle cell growth and size through adhesion GPCRs*) and lung physiology (Ludwig et al., *Adhesion GPCR function in pulmonary development and disease*) and in the immune system (Hamann et al., *Adhesion GPCRs as modulators of immune cell function*). Musa et al. describe that *heart development, angiogenesis and blood-brain barrier*

function are modulated by adhesion GPCRs, adding further organ systems that require those receptors for their respective setups and daily operations.
- Finally, Strutt et al. discuss that *adhesion GPCRs govern polarity of epithelia and cell migration*, while the chapter of Aust et al. review the current state of knowledge on *adhesion GPCRs in tumorigenesis.*

We are certain that the research described in this book marks several milestones in the maturation of our understanding on how aGPCRs impact biology. It is to be hoped that the concepts on several aspects of aGPCRs unveiled in the last years are stepping stones to grasp their roles in human disease and therapeutic intervention. We are much looking forward to witness and participate in the exciting developments of this thriving area of biomedical research.

The book will start though with look back at the *History of the adhesion GPCR field* (Hamann and Petrenko), to record the path of our community and remark its scientific course throughout the last 20 years.

Würzburg, Germany — Tobias Langenhan
Leipzig, Germany — Torsten Schöneberg

Contents

List of Contributors

Demet Araç Department of Biochemistry and Molecular Biology, University of Chicago, Chicago, IL, USA

Gabriela Aust* Department of Surgery, Research Laboratories, University of Leipzig, Leipzig, Germany

James P. Bridges Department of Pediatrics, Section of Pulmonary Biology, Perinatal Institute, Cincinnati Children's Hospital Medical Center, Cincinnati, OH, USA

Division of Pulmonary Biology, Perinatal Institute, Cincinnati Children's Hospital Medical Center, Cincinnati, OH, USA

Chris de Graaf Department of Medicinal Chemistry, Amsterdam Institute for Molecules, Medicines and Systems, Vrije Universiteit Amsterdam, Amsterdam, The Netherlands

Felix B. Engel* Experimental Renal and Cardiovascular Research, Department of Nephropathology, Institute of Pathology, Friedrich-Alexander-Universität Erlangen-Nürnberg, Erlangen, Germany

Oliver P. Ernst Department of Biochemistry, University of Toronto, Toronto, ON, Canada

Department of Molecular Genetics, University of Toronto, Toronto, ON, Canada

Franziska Fiedler* Medical Faculty, Institute of Biochemistry, Leipzig University, Leipzig, Germany

Randy A. Hall Department of Pharmacology, Emory University School of Medicine, Atlanta, GA, USA

Jörg Hamann* Department of Experimental Immunology, K0-144, Academic Medical Center, University of Amsterdam, Amsterdam, The Netherlands

Breanne L. Harty Department of Developmental Biology, Washington University School of Medicine, St. Louis, MO, USA

Cheng-Chih Hsiao Department of Experimental Immunology, K0-144, Academic Medical Center, University of Amsterdam, Amsterdam, The Netherlands

Ayush Kishore Department of Pharmacology, Emory University School of Medicine, Atlanta, GA, USA

Robert J. Kittel* Department of Neurophysiology, Institute of Physiology, University of Würzburg, Würzburg, Germany

Barbara Knapp* Cell and Matrix Biology, Institute of Zoology, Johannes Gutenberg University of Mainz, Mainz, Germany

Peter Kovacs Integrated Research and Treatment Center (IFB) AdiposityDiseases, University of Leipzig, Leipzig, Germany

Arunkumar Krishnan Department of Neuroscience, Functional Pharmacology, Uppsala University, Biomedical Center, Uppsala, Sweden

Tobias Langenhan* Department of Neurophysiology, Institute of Physiology, University of Würzburg, Würzburg, Germany

Chang Sup Lee Department of Microbiology, Immunology, and Cancer Biology, University of Virginia, Charlottesville, VA, USA

Ines Liebscher* Medical Faculty, Institute of Biochemistry, University Leipzig, Leipzig, Germany

Hsi-Hsien Lin Department of Microbiology and Immunology, College of Medicine, Chang Gung University, Tao-Yuan, Taiwan

Chang Gung Immunology Consortium, Chang Gung Memorial Hospital and Chang Gung University, Tao-Yuan, Taiwan

Marie-Gabrielle Ludwig Novartis Institutes for Biomedical Research, Basel, Switzerland

Kelly R. Monk Department of Developmental Biology, Washington University School of Medicine, St. Louis, MO, USA

Gentian Musa* Experimental Renal and Cardiovascular Research, Department of Nephropathology, Institute of Pathology, Friedrich-Alexander-Universität Erlangen-Nürnberg, Erlangen, Germany

Colin Niaudet Department of Immunology, Genetics and Pathology, Rudbeck Laboratory, Uppsala University, Uppsala, Sweden

Matthias Nieberler* Department of Neurophysiology, Institute of Physiology, University of Würzburg, Würzburg, Germany

Saskia Nijmeijer Department of Medicinal Chemistry, Amsterdam Institute for Molecules, Medicines and Systems, Vrije Universiteit Amsterdam, Amsterdam, The Netherlands

Alexander G. Petrenko Laboratory of Receptor Cell Biology, Shemyakin-Ovchinnikov Institute of Bioorganic Chemistry, Moscow, Russia

Xianhua Piao Division of Newborn Medicine, Boston Children's Hospital, Harvard Medical School, Boston, MA, USA

Simone Prömel* Medical Faculty, Institute of Biochemistry, Leipzig University, Leipzig, Germany

Kodi S. Ravichandran Department of Microbiology, Immunology, and Cancer Biology, University of Virginia, Charlottesville, VA, USA

Torsten Schöneberg* Medical Faculty, Rudolf Schönheimer Institute of Biochemistry, University of Leipzig, Leipzig, Germany

Helgi B. Schiöth Department of Neuroscience, Functional Pharmacology, Uppsala University, Biomedical Center, Uppsala, Sweden

Ralf Schnabel Institute of Genetics, TU Braunschweig, Braunschweig, Germany

Nicole Scholz* Department of Neurophysiology, Institute of Physiology, University of Würzburg, Würzburg, Germany

Elena Seiradake Department of Biochemistry, University of Oxford, Oxford, UK

Fekrije Selimi Center for Interdisciplinary Research in Biology (CIRB), Collège de France, Paris, France

CNRS UMR 7241 and INSERM U1050, Paris, France

Klaus Seuwen Novartis Institutes for Biomedical Research, Basel, Switzerland

Séverine M. Sigoillot Center for Interdisciplinary Research in Biology (CIRB), Collège de France, Paris, France

CNRS UMR 7241 and INSERM U1050, Paris, France

Norbert Sträter* Institute for Bioanalytical Chemistry, Leipzig University, Leipzig, Germany

David Strutt Bateson Centre and Department of Biomedical Science, University of Sheffield, Sheffield, UK

Erwin G. Van Meir Department of Neurosurgery and Hematology & Medical Oncology, School of Medicine and Winship Cancer Institute, Emory University, Atlanta, GA, USA

James P. White Duke University Medical Center, Durham, NC, USA

Steffen Wolf Department of Biophysics, CAS-MPG Partner Institute for Computational Biology, Key Laboratory of Computational Biology, Shanghai Institutes for Biological Sciences, Chinese Academy of Sciences, Shanghai, P.R. China

Institute of Physics, Albert Ludwigs University, Freiburg, Germany

Uwe Wolfrum* Cell and Matrix Biology, Institute of Zoology, Johannes Gutenberg University of Mainz, Mainz, Germany

Lei Xu Department of Biomedical Genetics, University of Rochester Medical Center, Rochester, NY, USA

Dan Zhu Department of Neurosurgery and Hematology & Medical Oncology, School of Medicine and Winship Cancer Institute, Emory University, Atlanta, GA, USA

*Member of the DFG Research Unit FOR2149 *Elucidation of Adhesion GPCR Signaling*

Introduction: History of the Adhesion GPCR Field

Jörg Hamann and Alexander G. Petrenko

Graphical Abstract

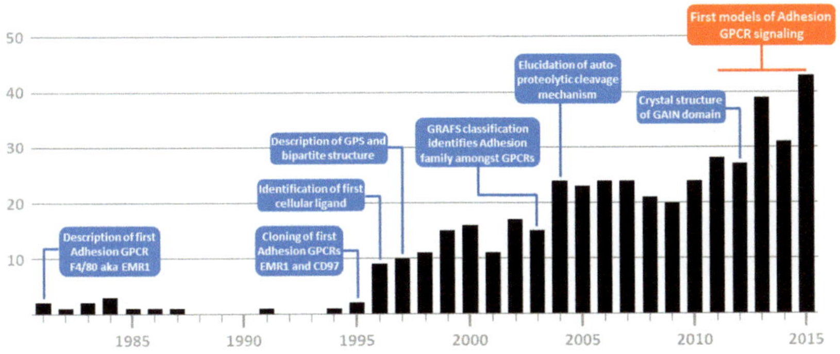

Development of the aGPCR scientific field based on PubMed-listed research articles and selected key findings

Contents

J. Hamann (✉)
Department of Experimental Immunology, K0-144, Academic Medical Center, University of Amsterdam, Meibergdreef 9, 1105AZ Amsterdam, The Netherlands
e-mail: j.hamann@amc.uva.nl

A.G. Petrenko (✉)
Laboratory of Receptor Cell Biology, Shemyakin-Ovchinnikov Institute of Bioorganic Chemistry, Russian Academy of Sciences, ul. Miklukho-Maklaya 16/10, 117997 Moscow, Russia
e-mail: petrenkoag@gmail.com

T. Langenhan, T. Schöneberg (eds.), *Adhesion G Protein-coupled Receptors*, Handbook of Experimental Pharmacology 234, DOI 10.1007/978-3-319-41523-9_1

Abstract

Since the discovery of adhesion G-protein-coupled receptors (aGPCRs) 20 years ago, reverse genetics approaches have dominated the elucidation of their function and work mechanisms. Seminal findings in this field comprise the description of aGPCRs as seven-transmembrane (7TM) molecules with an extended extracellular region, the identification of matricellular ligands that bind to distinct protein folds at the N-terminus, the clarification of an autoproteolytic cleavage event at a juxtamembranous GPCR proteolysis site (GPS), the elucidation of the crystal structure of the GPCR autoproteolysis-inducing (GAIN) domain that embeds the GPS and connects the receptor fragments, the demonstration that a short N-terminal sequence of the seven-transmembrane (7TM) region can serve as a tethered agonist, and, recently, the notification that aGPCRs can serve as mechanosensors. We here discuss how these discoveries have moved forward aGPCR research and, finally, linked the field to the GPCR field. We argue that crucial questions remain to be addressed before we can fully appreciate the biological nature of these fascinating receptors.

Keywords

Adhesion GPCRs • History • Biology • Structure • Signaling • Pharmacology

1 A Novel Type of Seven-Transmembrane Receptors

After the discovery of hormones as "first messenger" and cyclic adenosine monophosphate (cAMP) as a "second messenger" in the twentieth century, the search for molecules that transduce the "message" through the cell membrane led to the discovery of the first G proteins and then G-protein-coupled receptors (GPCRs). Forward biological approaches, searching for membrane-bound cognate receptors for biocative molecules, subsequently resulted in the identification of many GPCRs and the finding that seven-transmembrane (7TM) receptors possess the largest receptor family in nature. Yet, in contrast to rhodopsin, secretin, glutamate, and Frizzled GPCRs, members of the fifth GPCR family, the adhesion (a) GPCRs, were not discovered via their ligand molecules. Their identification about 20 years ago was the result of genetic approaches that became available through the development of cDNA cloning techniques in the late 1980s (Fig. 1). In 1995, the primary structure of the leukocyte surface molecules CD97 and EMR1 (EGF module-containing, mucin-like hormone receptor 1; in the mouse known as F4/80) was described [1, 2]. The mature proteins were found to comprise a 7TM region, the hallmark of all GPCRs. Most notable was the extended extracellular part, possessing several tandem epidermal growth factor (EGF)-like domains at the N-terminus. With reference to the binary molecule structure, the name EGF-TM7 was coined for these novel receptors [3].

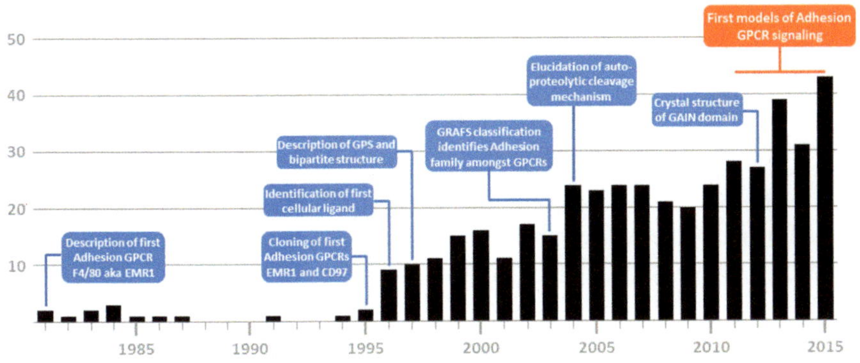

Fig. 1 Development of the aGPCR scientific field. PubMed-listed research articles reporting on aGPCRs indicated per year throughout the last 35 years. Selected key findings are highlighted

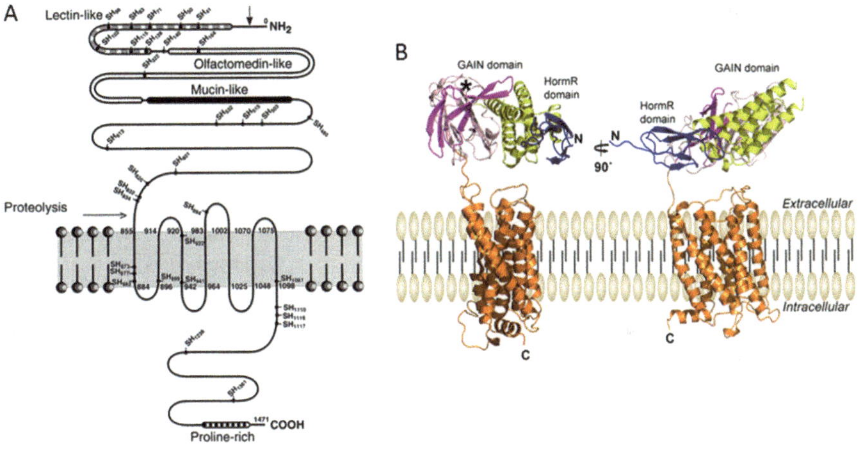

Fig. 2 Progress in the structural understanding of aGPCRs. (**a**) Protein structure of CIRL-1/latrophilin 1 predicted in 1997 from the deciphered amino acid sequence of the mature polypeptide, with indicated the 7TM region, the extended extracellular region with several protein domains, and the juxtamembranous proteolysis site (reproduced from [4]). (**b**) Model of CIRL-1/latrophilin 1 suggested in 2012 based on crystal structures of the GAIN and hormone receptor domain and modeling of the 7TM moiety (reproduced from [6])

Soon after, CIRL-1 (calcium-independent receptor of α-latrotoxin 1)/latrophilin 1, a neuronal receptor for the black widow spider poison α-latrotoxin, was shown to possess a similar structure and strong homology to the 7TM cores of EGF-TM7 receptors [4, 5] (Fig. 2). However, instead of repetitive EGF-like domains, CIRL-1/latrophilin 1 contains singular lectin-like, olfactomedin, and hormone receptor motif domains in its extracellular part. Subsequent description of other homologous 7TM receptors, including latrophilins, EMRs, CELSRs (cadherin EGF LAG seven-pass G-type receptors), BAIs (brain-specific angiogenesis inhibitors), HE6 (human epididymal 6), and VLGR1 (very large GPCR 1), confirmed the existence of a novel type of GPCR with a large extracellular part, differently composed of structural modules that are typically found in cell adhesion proteins, suggesting their role in coupling cell-to-cell interaction to intracellular signaling.

Right at the beginning, it became clear that these chimeric GPCRs undergo intensive posttranslational modifications and that CD97, CIRL-1/latrophilin 1, and several of their relatives consist of two noncovalently attached fragments that arise from cleavage of the full-length precursor molecules at a juxtamembranous GPCR proteolysis site (GPS) [4, 7, 8]. By pulse-and-chase labeling, it was shown that the cleavage at the GPS site takes place in the endoplasmic reticulum, rendering it fundamentally different from other proteolytic steps, such as furin processing, which occurs in the Golgi apparatus. In 2004, Hsi-Hsien Lin and colleagues showed that the cleavage is an autocatalytic event commonly employed by N-terminal nucleophile hydrolases [9]. The GPS motif appeared to be highly conserved in the aGPCRs family, representing essentially the eighth region of homology within the family. N-terminal to the GPS, larger regions with no sequence homology were found, linking the cell adhesion-like domains. Why the N-terminal protein adhesion-like domains in many aGPCRs are separated from the 7TM part by a large spacer sequence remained unclear for many more years.

Deciphering of the human genome finally disclosed the existence of 33 related receptors that, based on phylogenetic comparison of the 7TM part, assemble a distinct family of GPCRs. The original interest in these proteins was based primarily on their potential of linking cell-to-cell interactions to intracellular signaling. Helgi Schiöth and coworkers thus called them aGPCRs and subdivided them into nine subfamilies [10]. Their unique molecular design clearly sets the aGPCRs apart from other GPCR families, including the secretin GPCRs [11]. It is of note that in spite of their intricate structure, aGPCRs seem to be ancestral to most other GPCR families and have been found even in the most ancient metazoan phyla [12].

As aGPCRs increasingly received attention from a wide spectrum of biomedical fields, the Adhesion GPCR Consortium, together with the International Union of Basic and Clinical Pharmacology Committee on Receptor Nomenclature and Drug Classification (NC-IUPHAR), recently proposed a unified nomenclature [13]. The new names carry ADGR as a common dominator, followed by a letter and a number to denote each subfamily and subtype, respectively.

2 Receptor Biology Convenes a New Research Field

In line with the discovery of aGPCRs through genomic approaches, an interest in these molecules developed concurrently in different biomedical areas. In particular, immunologists, neuroscientists, and developmental biologists were among the first who studied these intriguing receptors. Far before any molecular structures were disclosed, Jon Austyn and Siamon Gordon had described in 1981 a monoclonal antibody directed against an antigen on mouse macrophages, called F4/80 [14]. This antigen, currently known as EMR1 (ADGRE1), has become widely used as a macrophage marker, expressed during development and throughout adult life in a range of inflammatory, infectious, tumor, and other disease models. Other ADGREs (EMRs) are expressed in specific granulocyte populations [15]. More recently, also ADGRBs (BAIs) and ADGRGs have been identified in immune cells, and BAI1 (ADGRB1) attracted interest as a macrophage receptor for danger-associated molecular patterns [16, 17].

A link between aGPCRs and neuronal function was first established by the finding that α-latroxin evokes massive neurotransmitter release and hormone secretion upon binding to CIRL-1/latrophilin 1 (ADGRL1) [5, 18]. More recently, involvement of several ADGRLs (latrophilins) in high-affinity transsynaptic interactions has been reported, suggesting involvement in synaptic functions [19–21]. Another highly intriguing observation was the discovery that defects in the ADGRG subfamily member GPR56 (ADGRG1) cause a cortical malformation, known as bilateral frontoparietal polymicrogyria (BFPP) [22]. GPR56-associated BFPP, also studied in mouse models, has become a prime example for a monogenic disorder arising from aGPCR dysfunction. More recently, additional roles for GPR56 in gyral patterning and in neocortex evolution as well as in oligodendrocyte development have been described [23–25]. Furthermore, elegant studies in zebrafish and mice have linked GPR126 (ADGRG6) and GPR56 on Schwann cells and oligodendrocytes to myelination of peripheral and central nervous axons, respectively [24, 26, 27].

Investigation of invertebrate aGPCRs has helped to understand fundamental developmental processes in health and disease. The Drosophila CELSR (ADGRC) homolog Flamingo/Starry night governs planar cell polarity (PCP) through facilitating the asymmetric distribution of Frizzled and Disheveled [28–30], and chicken CELSR1 (ADGRC1) facilitates core-PCP signaling-mediated closure of the neural tube [31]. In a similar way, the latrophilin homolog LAT-1 organizes cell division planes across the anterior–posterior axis of the *C. elegans* embryo, acting in parallel with noncanonical Wnt/Frizzled signaling [32]. Moreover, CELSR homologs in *C. elegans* and mice regulate axon guidance and neural circuit development [33, 34]. Finally, CELSR1 and VLGR1 (ADGRV1) are required for the development of sensory epithelia; mutations in the latter are associated with the human Usher syndrome, a severe sensory-neuronal disorder that affects vision and hearing [35].

Next to developmental effects in several organ systems, including the reproductive tract, the role of aGPCRs in tumorigenesis evoked interest in the clinical implication of the receptors. Gabriela Aust was the first who showed that expression of CD97 (ADGRE5) correlates with dedifferentiation and invasiveness in various carcinomas [36, 37]. Inversely, GPR56 controls melanoma growth and metastasis [38].

The examples of aGPCR research provided here are far from complete. However, they illustrate a research field that developed in parallel and fairly separated within different biomedical areas, resulting in a steadily growing number of publications (Fig. 1). It was Siamon Gordon who organized in 2002 a 1-day workshop for immunologists and tumor biologists working on EGF-TM7 receptors

Table 1 Biennial adhesion GPCR workshops

Date	Place	Organizers	Talks	Scientific highlights
April 4, 2002	Oxford	Siamon Gordon	16	Identification of cellular ligands
March 19, 2004	Leipzig	Gabriela Aust	19	Autoproteolytic cleavage at the GPS
March 24, 2006	Amsterdam	Jörg Hamann	22	Interaction between receptor fragments
March 29, 2008	Oxford	Martin Stacey	15	Adhesion GPCRs in development
May 1, 2010	Leipzig	Gabriela Aust	19	In vivo models for Adhesion GPCRs
September 6–8, 2012	Würzburg	Tobias Langenhan	23	Crystal structure of the GAIN domain; Autonomous signaling by the CTF; Tethered vs inverse agonist models
June 5–7, 2014	Boston	Xianhua Piao	33	Stachel mechanism of receptor activation; Receptor triggering by mechanosensation
June 2–4, 2016	Leipzig	Torsten Schöneberg; Tobias Langenhan	38	TBD

CTF C-terminal fragment, *GAIN* GPCR autoproteolysis-inducing, *GPCR* G-protein-coupled receptor, *GPS* GPCR proteolysis site, *TBD* to be determined

in Oxford (Table 1). During the following events in Leipzig (2004), Amsterdam (2006), Oxford (2008), and Leipzig (2010), aGPCRs expressed outside the immune system, such as CIRL-1/latrophilin 1, GPR64 (ADGRG2), and VLGR1, slowly entered the stage. The more recent events in Würzburg (2012) and Boston (2014) dealt with all aspects of aGPCR biology and saw a strongly expanding audience [39, 40]. Yet, despite the transformation into 3-day events, the aGPCR Workshops are still informal gatherings, at which novel, unpublished work is presented, and open questions are discussed in an intimate setting. By catalyzing cross talk and collaboration between aGPCR researchers with a different scientific background, the aGPCR Workshops had a tremendous impact on the field. Currently, the community is looking forward to the next event in Leipzig in 2016.

A decisive step in the development of the field was the founding of the aGPCR Consortium (AGC; www.adhesiongpcr.org) in 2012. As an international, open network of academic and nonacademic laboratories interested in aGPCRs, the AGC currently connects more than 60 scientists from 15 countries. The AGC has become a meeting place for everyone interested in aGPCRs and organizes, currently, the biennial workshops. Moreover, the AGC provides information and visibility for the aGPCR community, works on terminology and nomenclature issues, and serves as a starting ground for collaborative research initiatives. The latter has led to the establishment of the Research Unit 2149—*Elucidation of*

Adhesion GPCR signaling—in 2015, which is supported by the Deutsche Forschungsgemeinschaft (www.adhesiongpcr.de).

3 From Molecular Structure to Pharmacology

While work on aGPCRs transcended different biomedical areas, the central question on the mechanism by which these receptors signal remained hard to answer for a long time. The existence of numerous protein domains implied that aGPCRs might engage in cell–cell interactions. A similar function of receptor tyrosine kinases has been very well described, with its importance in cancer biology and development. In 1996, Jörg Hamann demonstrated that CD97 binds decay-accelerating factor/CD55, a molecule associated with regulation of the complement cascade [41]. Since then, interacting partners, often matricellular molecules, have been identified for about ten aGPCRs. However, no comprehensive picture arose that would fit the concept of agonistic ligands as these have identified for other GPCR families [42].

Recently, the juxtamembrane part of the aGPCRs containing the GPS motif has been implicated in the receptor signaling. Demet Araç showed that the GPS is an integral part of a much larger domain that was termed GPCR autoproteolysis-inducing (GAIN) domain [6]. Crystal structures of GAIN domains from CIRL-1/latrophilin 1 and BAI3 (ADGRB3) revealed a conserved, novel fold that fine-tunes the chemical environment at the GPS to catalyze peptide bond hydrolysis (Fig. 2). Another key finding by the groups of Randy Hall and Lei Xu was the observation that the C-terminal fragment (CTF) of some aGPCRs shows intense metabotropic and biological activity, implying that the N-terminal fragment (NTF) controls receptor signaling [43, 44], and that the ectodomain of aGPCRs may act as a tethered ligand for their 7TM domain [45].

Building forth on these studies, the laboratories of Ines Liebscher, Torsten Schöneberg, and Gregory Tall have proposed a tethered agonist mechanism according to which displacement of the NTF exposes a short N-terminal sequence of the 7TM domain, designated *Stachel* (German for stinger), that is hidden within the GAIN domain [46, 47]. Synthetic peptides, comprising these *Stachel* sequences, have been shown to potently trigger various aGPCRs, in vitro and also in vivo. Finally, work from the groups of Bruce Spiegelman, Kelly Monk, and Tobias Langenhan uncovered that mechanical cues trigger the activity of aGPCRs under physiological conditions, adding mechanosensation to the sensory canon of the GPCR superfamily [27, 48, 49].

The ability to activate aGPCRs enabled studies aiming at identifying downstream signaling modes. The demonstration that the receptors can couple to all subclasses of G proteins [46, 50] led to the recognition as bona fide GPCRs [55]. Consequently, established GPCR conferences currently discuss developments in aGPCR research. Yet, uncertainties remain (Table 2). Information concerning the ability of aGPCR binding partners to trigger G proteins is very scarce so far [27, 51], and we do not know whether the functioning of the receptors is confined to

Table 2 Certainties and uncertainties about GPCRs

What do we know	What we are not sure of
• GPCRs are bipartite molecules with a large extracellular region that is connected through a GAIN domain to a 7TM moiety	• A functional link between adhesive capacity and receptor signaling remains to be established
• The majority of aGPCRs undergo autocatalytic processing at a GPS embedded within the GAIN domain	• Lack of cleavage of some aGPCRs suggests that a bipartite structure is not a prerequisite for receptor function
• aGPCRs can be activated through a tethered agonist (*Stachel* sequence)	• Mechanisms allowing exposure of the *Stachel* need to be determined, in particular for solid tissues and non-cleavable receptors
• aGPCRs are widely distributed and cause distinct biological phenotypes	• It is not clear whether aGPCRs display cell type-specific or general cellular functions

7TM seven-transmembrane, *aGPCR* adhesion GPCR, *GAIN* GPCR autoproteolysis-inducing, *GPCR* G-protein-coupled receptor, *GPS* GPCR proteolysis site

G-protein signaling. Early studies on PCP in Drosophila showed that the CELSR homolog Flamingo arranges in *trans* and in *cis* with other transmembrane molecules to execute its functions [28–30], possibly presenting another major working mechanism. In addition, CD97 has been shown to heterodimerize with the lysophosphatidic acid (LPA) receptor to amplify LPA-initiated Rho-dependent signaling and invasion in prostate cancer cells [52].

Another interesting possibility exists that the large NTF of aGPCRs may serve autonomously by interacting as a ligand with other receptors. In the original discovery of BAI1, its ectodomain soluble fragment had a role in the inhibition of brain-specific angiogenesis [53]. Also, the presence of a specifically cleaved soluble fragment of CIRL-1/latrophilin 1 was detected in the brain, comprising about 5 % of the total amount of the receptor expressed [54]. Finally, a genetic study uncovered a specific role of the N-terminal fragment of GPR126 in axon sorting [27].

Based on the anecdotal identification of most of its members, the aGPCR cohort is one of the prime examples of genome-sequencing effort-driven identification and definition of an entire molecule class. As a consequence, research on aGPCRs grows out of a "molecule-centered" rather than "biology-centered" history since almost two decades. This situation has recently changed with the advent of molecular models on the signaling paradigm of aGPCRs and their physiological mode of activation. The field has now entered a highly intriguing stage, and we predict that the pharmacological insight and tools that are currently developed will boost novel attempts to understand the biological functions of aGPCRs in health and disease.

Acknowledgments The writing of this manuscript was supported by the Deutsche Forschungsgemeinschaft (Research Unit 2149) and by grants of the Thyssen Foundation (2015-00387) to JH and the Russian Science Foundation (14-14-01195) to AGP.

Competing Financial Interests The authors declare no competing financial interests.

References

1. Baud V, Chissoe SL, Viegas-Péquignot E, Diriong S, N'Guyen VC, Roe BA et al (1995) EMR1, an unusual member in the family of hormone receptors with seven transmembrane segments. Genomics 26:334–344
2. Hamann J, Eichler W, Hamann D, Kerstens HM, Poddighe PJ, Hoovers JM et al (1995) Expression cloning and chromosomal mapping of the leukocyte activation antigen CD97, a new seven-span transmembrane molecule of the secretion receptor superfamily with an unusual extracellular domain. J Immunol 155:1942–1950
3. McKnight AJ, Gordon S (1996) EGF-TM7: a novel subfamily of seven-transmembrane-region leukocyte cell-surface molecules. Immunol Today 17:283–287
4. Krasnoperov VG, Bittner MA, Beavis R, Kuang Y, Salnikow KV, Chepurny OG et al (1997) alpha-Latrotoxin stimulates exocytosis by the interaction with a neuronal G-protein-coupled receptor. Neuron 18:925–937
5. Lelianova VG, Davletov BA, Sterling A, Rahman MA, Grishin EV, Totty NF et al (1997) Alpha-latrotoxin receptor, latrophilin, is a novel member of the secretin family of G protein-coupled receptors. J Biol Chem 272:21504–21508
6. Araç D, Boucard AA, Bolliger MF, Nguyen J, Soltis SM, Südhof TC et al (2012) A novel evolutionarily conserved domain of cell-adhesion GPCRs mediates autoproteolysis. EMBO J 31:1364–1378
7. Gray JX, Haino M, Roth MJ, Maguire JE, Jensen PN, Yarme A et al (1996) CD97 is a processed, seven-transmembrane, heterodimeric receptor associated with inflammation. J Immunol 157:5438–5447
8. Krasnoperov V, Lu Y, Buryanovsky L, Neubert TA, Ichtchenko K, Petrenko AG (2002) Post-translational proteolytic processing of the calcium-independent receptor of alpha-latrotoxin (CIRL), a natural chimera of the cell adhesion protein and the G protein-coupled receptor. Role of the G protein-coupled receptor proteolysis site (GPS) motif. J Biol Chem 277:46518–46526
9. Lin HH, Chang GW, Davies JQ, Stacey M, Harris J, Gordon S (2004) Autocatalytic cleavage of the EMR2 receptor occurs at a conserved G protein-coupled receptor proteolytic site motif. J Biol Chem 279:31823–31832
10. Bjarnadóttir TK, Fredriksson R, Höglund PJ, Gloriam DE, Lagerström MC, Schiöth HB (2004) The human and mouse repertoire of the adhesion family of G-protein-coupled receptors. Genomics 84:23–33
11. Lagerström MC, Schiöth HB (2008) Structural diversity of G protein-coupled receptors and significance for drug discovery. Nat Rev Drug Discov 7:339–357
12. Nordström KJ, Lagerström MC, Wallér LM, Fredriksson R, Schiöth HB (2009) The Secretin GPCRs descended from the family of Adhesion GPCRs. Mol Biol Evol 26:71–84
13. Hamann J, Aust G, Araç D, Engel FB, Formstone C, Fredriksson R et al (2015) International Union of Basic and Clinical Pharmacology. XCIV. Adhesion G protein-coupled receptors. Pharmacol Rev 67:338–367
14. Austyn JM, Gordon S (1981) F4/80, a monoclonal antibody directed specifically against the mouse macrophage. Eur J Immunol 11:805–815
15. Gordon S, Hamann J, Lin HH, Stacey M (2011) F4/80 and the related adhesion-GPCRs. Eur J Immunol 41:2472–2476
16. Park D, Tosello-Trampont AC, Elliott MR, Lu M, Haney LB, Ma Z et al (2007) BAI1 is an engulfment receptor for apoptotic cells upstream of the ELMO/Dock180/Rac module. Nature 450:430–434
17. Das S, Owen KA, Ly KT, Park D, Black SG, Wilson JM et al (2011) Brain angiogenesis inhibitor 1 (BAI1) is a pattern recognition receptor that mediates macrophage binding and engulfment of Gram-negative bacteria. Proc Natl Acad Sci U S A 108:2136–2141
18. Lang J, Ushkaryov Y, Grasso A, Wollheim CB (1998) Ca2+-independent insulin exocytosis induced by alpha-latrotoxin requires latrophilin, a G protein-coupled receptor. EMBO J 17:648–657

19. Silva JP, Lelianova VG, Ermolyuk YS, Vysokov N, Hitchen PG, Berninghausen O et al (2011) Latrophilin 1 and its endogenous ligand Lasso/teneurin-2 form a high-affinity transsynaptic receptor pair with signaling capabilities. Proc Natl Acad Sci U S A 108:12113–12118
20. Boucard AA, Maxeiner S, Südhof TC (2014) Latrophilins function as heterophilic cell-adhesion molecules by binding to teneurins: regulation by alternative splicing. J Biol Chem 289:387–402
21. O'Sullivan ML, de Wit J, Savas JN, Comoletti D, Otto-Hitt S, Yates JR III et al (2012) FLRT proteins are endogenous latrophilin ligands and regulate excitatory synapse development. Neuron 73:903–910
22. Piao X, Hill RS, Bodell A, Chang BS, Basel-Vanagaite L, Straussberg R et al (2004) G protein-coupled receptor-dependent development of human frontal cortex. Science 303:2033–2036
23. Bae BI, Tietjen I, Atabay KD, Evrony GD, Johnson MB, Asare E et al (2014) Evolutionarily dynamic alternative splicing of GPR56 regulates regional cerebral cortical patterning. Science 343:764–768
24. Giera S, Deng Y, Luo R, Ackerman SD, Mogha A, Monk KR et al (2015) The adhesion G protein-coupled receptor GPR56 is a cell-autonomous regulator of oligodendrocyte development. Nat Commun 6:6121
25. Ackerman SD, Garcia C, Piao X, Gutmann DH, Monk KR (2015) The adhesion GPCR Gpr56 regulates oligodendrocyte development via interactions with Gα12/13 and RhoA. Nat Commun 21:6122
26. Monk KR, Naylor SG, Glenn TD, Mercurio S, Perlin JR, Dominguez C et al (2009) A G protein-coupled receptor is essential for Schwann cells to initiate myelination. Science 325:1402–1405
27. Petersen SC, Luo R, Liebscher I, Giera S, Jeong SJ, Mogha A et al (2015) The adhesion GPCR GPR126 has distinct, domain-dependent functions in Schwann cell development mediated by interaction with laminin-211. Neuron 85:755–769
28. Usui T, Shima Y, Shimada Y, Hirano S, Burgess RW, Schwarz TL et al (1999) Flamingo, a seven-pass transmembrane cadherin, regulates planar cell polarity under the control of Frizzled. Cell 98:585–595
29. Shimada Y, Usui T, Yanagawa S, Takeichi M, Uemura T (2001) Asymmetric colocalization of Flamingo, a seven-pass transmembrane cadherin, and Dishevelled in planar cell polarization. Curr Biol 11:859–863
30. Chen WS, Antic D, Matis M, Logan CY, Povelones M, Anderson GA et al (2008) Asymmetric homotypic interactions of the atypical cadherin flamingo mediate intercellular polarity signaling. Cell 133:1093–1105
31. Nishimura T, Honda H, Takeichi M (2012) Planar cell polarity links axes of spatial dynamics in neural-tube closure. Cell 149:1084–1097
32. Langenhan T, Prömel S, Mestek L, Esmaeili B, Waller-Evans H, Hennig C et al (2009) Latrophilin signaling links anterior-posterior tissue polarity and oriented cell divisions in the C. elegans embryo. Dev Cell 17:494–504
33. Steimel A, Wong L, Najarro EH, Ackley BD, Garriga G, Hutter H (2010) The Flamingo ortholog FMI-1 controls pioneer-dependent navigation of follower axons in C. elegans. Development 137:3663–3673
34. Tissir F, Goffinet AM (2013) Shaping the nervous system: role of the core planar cell polarity genes. Nat Rev Neurosci 14:525–535
35. Weston MD, Luijendijk MW, Humphrey KD, Möller C, Kimberling WJ (2004) Mutations in the VLGR1 gene implicate G-protein signaling in the pathogenesis of Usher syndrome type II. Am J Hum Genet 74:357–366
36. Aust G, Eichler W, Laue S, Lehmann I, Heldin NE, Lotz O et al (1997) CD97: a dedifferentiation marker in human thyroid carcinomas. Cancer Res 57:1798–1806
37. Steinert M, Wobus M, Boltze C, Schütz A, Wahlbuhl M, Hamann J et al (2002) Expression and regulation of CD97 in colorectal carcinoma cell lines and tumor tissues. Am J Pathol 161:1657–1667

38. Xu L, Begum S, Hearn JD, Hynes RO (2006) GPR56, an atypical G protein-coupled receptor, binds tissue transglutaminase, TG2, and inhibits melanoma tumor growth and metastasis. Proc Natl Acad Sci U S A 103:9023–9028
39. Araç D, Aust G, Calebiro D, Engel FB, Formstone C, Goffinet A, Hamann J et al (2012) Dissecting signaling and functions of adhesion G protein-coupled receptors. Ann N Y Acad Sci 1276:1–25
40. Liebscher I, Ackley B, Araç D, Ariestanti DM, Aust G, Bae BI et al (2014) New functions and signaling mechanisms for the class of adhesion G protein-coupled receptors. Ann N Y Acad Sci 1333:43–64
41. Hamann J, Vogel B, van Schijndel GM, van Lier RA (1996) The seven-span transmembrane receptor CD97 has a cellular ligand (CD55, DAF). J Exp Med 184:1185–1189
42. Langenhan T, Aust G, Hamann J (2013) Sticky signaling–adhesion class G protein-coupled receptors take the stage. Sci Signal 6:re3
43. Paavola KJ, Stephenson JR, Ritter SL, Alter SP, Hall RA (2011) The N terminus of the adhesion G protein-coupled receptor GPR56 controls receptor signaling activity. J Biol Chem 286:28914–28921
44. Yang L, Chen G, Mohanty S, Scott G, Fazal F, Rahman A et al (2011) GPR56 Regulates VEGF production and angiogenesis during melanoma progression. Cancer Res 71:5558–5568
45. Prömel S, Frickenhaus M, Hughes S, Mestek L, Staunton D, Woollard A et al (2012) The GPS motif is a molecular switch for bimodal activities of adhesion class G protein-coupled receptors. Cell Rep 2:321–331
46. Liebscher I, Schön J, Petersen SC, Fischer L, Auerbach N, Demberg LM et al (2014) A tethered agonist within the ectodomain activates the adhesion G protein-coupled receptors GPR126 and GPR133. Cell Rep 9:2018–2026
47. Stoveken HM, Hajduczok AG, Xu L, Tall GG (2015) Adhesion G protein-coupled receptors are activated by exposure of a cryptic tethered agonist. Proc Natl Acad Sci U S A 112:6194–6199
48. White JP, Wrann CD, Rao RR, Nair SK, Jedrychowski MP, You JS et al (2014) G protein-coupled receptor 56 regulates mechanical overload-induced muscle hypertrophy. Proc Natl Acad Sci U S A 111:15756–15761
49. Scholz N, Gehring J, Guan C, Ljaschenko D, Fischer R, Lakshmanan V et al (2015) The adhesion GPCR latrophilin/CIRL shapes mechanosensation. Cell Rep 11:866–874
50. Gupte J, Swaminath G, Danao J, Tian H, Li Y, Wu X (2012) Signaling property study of adhesion G-protein-coupled receptors. FEBS Lett 586:1214–1219
51. Paavola KJ, Sidik H, Zuchero JB, Eckart M, Talbot WS (2014) Type IV collagen is an activating ligand for the adhesion G protein-coupled receptor GPR126. Sci Signal 7:ra76
52. Ward Y, Lake R, Yin JJ, Heger CD, Raffeld M, Goldsmith PK et al (2011) LPA receptor heterodimerizes with CD97 to amplify LPA-initiated RHO-dependent signaling and invasion in prostate cancer cells. Cancer Res 71:7301–7311
53. Nishimori H, Shiratsuchi T, Urano T, Kimura Y, Kiyono K, Tatsumi K et al (1997) A novel brain-specific p53-target gene, BAI1, containing thrombospondin type 1 repeats inhibits experimental angiogenesis. Oncogene 15:2145–2150
54. Krasnoperov V, Deyev IE, Serova OV, Xu C, Lu Y, Buryanovsky L et al (2009) Dissociation of the subunits of the calcium-independent receptor of alpha-latrotoxin as a result of two-step proteolysis. Biochemistry 48:3230–3238
55. Monk KR, Hamann J, Langenhan T, Nijmeijer S, Schöneberg T, Liebscher I (2015) Adhesion G protein-coupled receptors: from in vitro pharmacology to in vivo mechanisms. Mol Pharmacol 88:617–623

Part I

Molecular and Pharmacological Properties of Adhesion GPCRs

Classification, Nomenclature, and Structural Aspects of Adhesion GPCRs

Arunkumar Krishnan, Saskia Nijmeijer, Chris de Graaf, and Helgi B. Schiöth

Contents

Arunkumar Krishnan and Saskia Nijmeijer are equally contributed.

A. Krishnan • H.B. Schiöth (✉)
Department of Neuroscience, Functional Pharmacology, Uppsala University, Biomedical Center, Box 593, Uppsala 75 124, Sweden
e-mail: arunkumar.krishnan@neuro.uu.se; helgi.schioth@neuro.uu.se

S. Nijmeijer • C. de Graaf
Department of Medicinal Chemistry, Amsterdam Institute for Molecules, Medicines and Systems, Vrije Universiteit Amsterdam, Amsterdam 1081HV, The Netherlands
e-mail: s.nijmeijer@vu.nl; c.de.graaf@vu.nl

T. Langenhan, T. Schöneberg (eds.), *Adhesion G Protein-coupled Receptors*,
Handbook of Experimental Pharmacology 234, DOI 10.1007/978-3-319-41523-9_2

Graphical Abstract

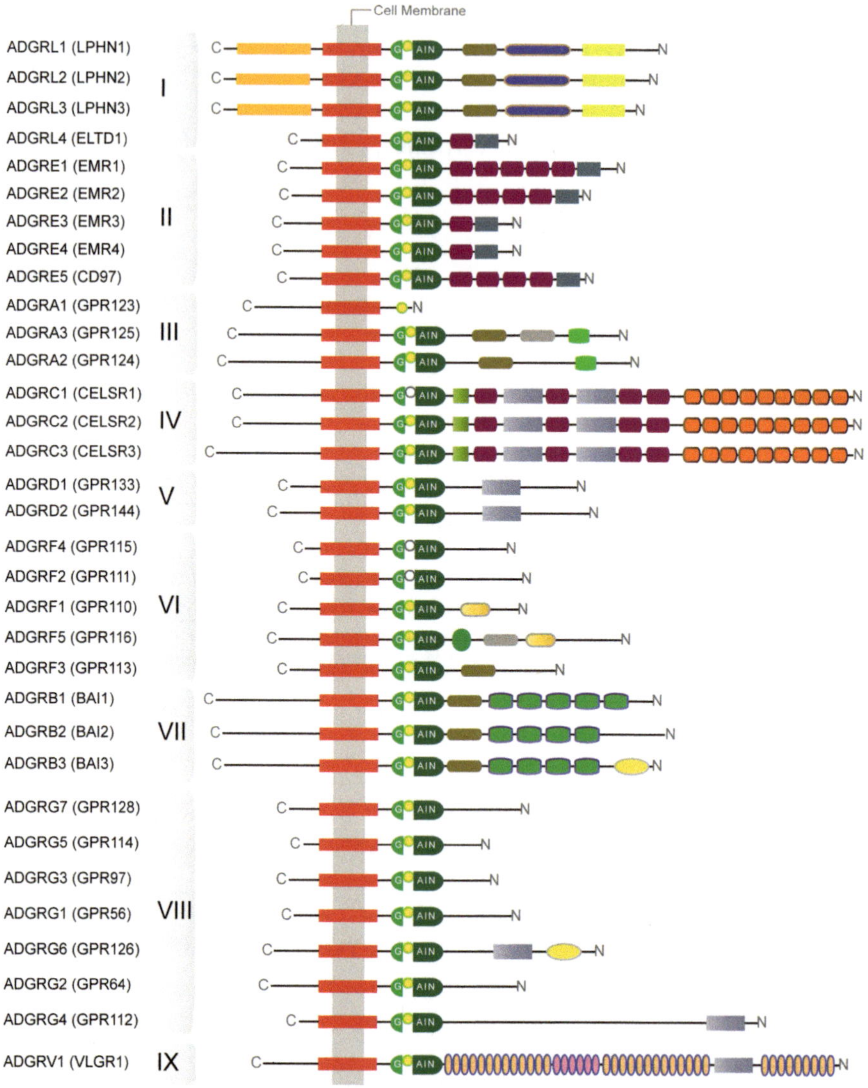

Representation of the nine distinct aGPCR subfamilies and their unique N-terminal domain architecture. The illustration also shows the extracellular structural feature shared by all aGPCRs (except ADGRA1), known as the GPCR autoproteolysis-inducing (GAIN) domain, that mediates autoproteolysis and subsequent attachment of the cleaved NTF and CTF fragments

Abstract

The adhesion family of G protein-coupled receptors (aGPCRs) is unique among all GPCR families with long N-termini and multiple domains that are implicated in cell–cell and cell–matrix interactions. Initially, aGPCRs in the human genome were phylogenetically classified into nine distinct subfamilies based on their 7TM sequence similarity. This phylogenetic grouping of genes into subfamilies was found to be in congruence in closely related mammals and other vertebrates as well. Over the years, aGPCR repertoires have been mapped in many species including model organisms, and, currently, there is a growing interest in exploring the pharmacological aspects of aGPCRs. Nonetheless, the aGPCR nomenclature has been highly diverse because experts in the field have used different names for different family members based on their characteristics (e.g., epidermal growth factor-seven-span transmembrane (EGF-TM7)), but without harmonization with regard to nomenclature efforts. In order to facilitate naming of orthologs and other genetic variants in different species in the future, the Adhesion-GPCR Consortium, together with the International Union of Basic and Clinical Pharmacology Committee on Receptor Nomenclature and Drug Classification, proposed a unified nomenclature for aGPCRs. Here, we review the classification and the most recent/current nomenclature of aGPCRs and as well discuss the structural topology of the extracellular domain (ECD)/N-terminal fragment (NTF) that is comparable with this 7TM subfamily classification. Of note, we systematically describe the structural domains in the ECD of aGPCR subfamilies and highlight their role in aGPCR-protein interactions.

Keywords

Adhesion GPCRs • Nomenclature • Classification • Pharmacology • Drug targets • Homologs • Mammals • Vertebrates • Model organisms • GAIN domain

1 Introduction

The G protein-coupled receptor (GPCR) superfamily is the largest family of cell surface receptors and is grouped into five major families: glutamate, rhodopsin, adhesion, frizzled, and secretin [1, 2]. Among all classes, the adhesion GPCRs (aGPCRs) comprise the second largest family with 33 members in the human genome [1, 2]. Prior to the release of the human genome, aGPCRs were not considered as a separate family of GPCRs. Indeed, at that time, only a few genes were identified that constituted a long extracellular region and a seven transmembrane segment (7TM) characteristic to GPCRs. One of the first aGPCRs cloned was the epidermal growth factor (EGF)-like molecule containing mucin-like hormone receptor 1 and later similar molecules were identified [3–6]. This paved the way for the recognition of these molecules as EGF-TM7-like receptors because of the presence of EGF-like domains in their extracellular region [6]. Similarly, other names that were initially termed for aGPCRs include LN-7TM [7] (for the presence of long N-terminal regions), LNB-7TM [8] (for their similarity to family B secretin-like GPCRs), or family B2 receptors [9]. Shortly after the release of the human genome, several novel genes were identified, and subsequent gene mining showed that at least 30 GPCR-like sequences exist with a long extracellular region and GPCR proteolysis site (GPS) motif that induce autocatalytic processing [10]. Some of these receptors were often initially denoted or thought as secretin-like GPCRs and were placed in proximity to family B receptors [9]. Nonetheless, the large-scale effort to comprehensively classify the GPCRs in the human genome showed convincing phylogenetic evidence that aGPCRs constitute a separate family of GPCRs [1]. This made clearer that adhesion and secretin families are indeed distinct from each other, although these molecules share vague similarities and are often placed together as family B GPCRs [9]. This view is strengthened as the largest of differences were observed in their extracellular region and in particular aGPCRs are also distinct from secretin GPCRs in molecular function. For example, most aGPCRs contain the GPS motif, which is found in close proximity to the 7TM region (for review see [11]). Moreover, it is currently understood that the GPS motif is a part of a much larger GPCR autoproteolysis-inducing (GAIN) domain, which induces the autocatalytic processing of aGPCRs into an N-terminal fragment (NTF) and a C-terminal fragment (CTF) [12]. In addition, the NTF of aGPCRs contains numerous protein domains implicated in cell and matrix interactions [11, 13], and, thus, the established name of "adhesion" family GPCRs was initially coined to refer this feature. Conversely, the secretin GPCRs do not undergo such autocatalytic processing in their N-termini, however, contains a hormone-binding domain (HBD) to mediate hormonal responses [13, 14]. In this chapter, we review the classification of aGPCRs into nine families in the human genome and briefly discuss the classification and potential homologs of these families in other vertebrate and invertebrate genomes. Of note, we address the recently recommended nomenclature of aGPCRs [15] that aim to provide a coherent and systematic naming system independent of the species and subfamily names. Also, we discuss the similarities between aGPCR subclasses with respect to the organization of their structural topology (e.g., olfactomedin, cadherin, EGF-like and thrombospondin type 1 domain). The analysis nevertheless indicates remarkable differences that demonstrate the structural diversity of aGPCRs, but could also hint at potential interaction partners for orphan aGPCRs or provide information on NTF–CTF interactions. We therefore systematically describe the unique (sub) family-specific structural features and their protein interactions.

2 Classification of aGPCRs

2.1 Human aGPCR Subfamilies

Based on phylogenetic criteria, the human aGPCR repertoire is categorized into nine distinct subfamilies (considering ADGRV1 (VLGR1) as subfamily IX) according to the molecular signature of their 7TM region [1]. The number of genes belonging to the subfamilies I to VIII vary from two genes in "subfamily

V" to seven genes in "subfamily VIII" [1, 16]. However, the subfamily IX contains only a single gene ADGRV1, also known as the very large GPCR1, referring to its unusually large extracellular region of about 6000 amino acids comprising multiple Calx-β repeats [1, 16]. Intriguingly, except for subfamily VI and VIII, each subfamily has a characteristic unique pattern of extracellular domain architecture (see Sect. 4.2) and this molecular signature is used as a marker to distinguish between the subfamilies and as well to categorize novel homologs in other species. Nonetheless, before the classification of human aGPCRs was established in 2003 [1, 10], several important factors were considered to determine the aGPCR classification. One of the major factors strongly taken into account was the strength of the phylogenetic nodes forming distinct groups, as this is central for the classification of subfamilies [1, 10]. This is vital because the tree topology supporting the classification largely depends on the choice of the alignment methods and phylogenetic reconstruction software [17]. Nonetheless, the grouping of aGPCRs into nine distinct families was found reliable when aGPCR repertoires from other mammalian genomes were curated and compared with the human aGPCR repertoire using different phylogenetic algorithms that included maximum parsimony, maximum likelihood, and neighbor-joining analyses [16, 18, 19]. Also, a consensus view of tree topologies obtained from human–mouse, human–rat, and human–dog phylogenetic comparisons showed identical grouping pattern, although the hierarchy within the subfamilies displayed considerable degrees of variation for subfamilies that comprised divergent members (e.g., within subfamily VIII). Overall, these comparative analyses further strengthened the classification, and the human aGPCR classification is now currently utilized in several studies as a reference to further classify aGPCRs in vertebrate and invertebrate genomes.

2.2 aGPCRs in Mammals and Other Vertebrates

Mining of aGPCRs in mammalian and other vertebrate genomes resolved one-to-one orthologous relationships with the human aGPCR members and as well provided insights into the evolution and diversity of aGPCRs. These gene mining studies showed that homologs of most human subfamilies are found in several vertebrate genomes. Nonetheless, the actual gene count of aGPCRs varied from species to species. For example, the mouse genome encodes one-to-one orthologs for 31 of the 33 aGPCRs found in human; however, orthologs of ADGRE2 (EMR2) and ADGRE3 (EMR3) are absent [16, 20, 21]. The same subset of aGPCRs was also observed in the rat genome with no orthologs found for human ADGRE2 and ADGRE3 [18]. Conversely, mining of aGPCRs in other mammalian genomes such as the dog genome showed that all 33 human aGPCRs are found and as well contain additional full-length genes that are homologs of ADGRE2 and ADGRE3 [19]. Additional gene mining studies discovered that the chicken genome contains 21 of 33 aGPCRs, but clearly lack orthologs of other 12 human aGPCRs [22]. Similarly, the fugu genome was surveyed and contains at least 29 aGPCR-like sequences [23]. These robust gene mining studies also suggested that aGPCRs are by no means a coherent group of equally divergent clusters and they show high degrees of variability in both, between the members of the subfamilies and one-to-one orthologous gene pairs. For example, the overall amino acid similarity within the TM regions of the human aGPCRs belonging to subfamily VII is relatively high similarly (60–70 %), whereas members of subfamily VIII share less percentage sequence identity (20–30 %). Moreover, the percentage of amino acid identity between the orthologous pairs in human–mouse, human–rat, and rat–mouse genomes also shows high degrees of variability from one subfamily to the other. For instance, the human–mouse orthologous pairs of members belonging to subfamily VII (brain-specific angiogenesis inhibitor (BAIs) or ADGRBs) share 98 % identity within their 7TM, while the orthologous pairs of subfamily VI and VIII members share a relatively low percentage identity (50–60 %). Taken together, these differences occasionally make it complicated to classify aGPCRs in distant species and in organisms that constitute local expansions. This was also observed in a recent attempt to classify aGPCRs in the zebra fish genome that constituted local expansions in subfamilies II, VI, and VIII. A few sequences had unstable positioning between the topologies observed using maximum likelihood (ML) and the Bayesian approach [24]. In such cases, aGPCR sequences were classified using a more acute approach of performing separate phylogenetic analyses on each cluster and considering the multiple alignments and extracellular domain architecture [24]. In addition, the exon–intron boundaries with intron phase/positions and the overall conservation of neighboring genes in the genomic scaffolds can serve as additional genetic markers to classify aGPCR orthologs in invertebrates.

2.3 aGPCRs in Invertebrates

In the last decade, robust efforts to sequence the genomes of several invertebrates and some non-bilaterian organisms provided large datasets to further mine and classify aGPCRs across most metazoans (animals). Several earlier studies that sought to determine aGPCRs in these distant species found homologs of a few human aGPCR subfamilies and catalogued many species-specific expansions as well. Genomic survey in closest relatives of vertebrates such as *Ciona* intestinalis and Branchiostoma floridae (amphioxus) exposed the diversity of aGPCRs before the emergence of vertebrates [25, 26]. Amphioxus contained a total of 37 aGPCRs, which is an estimate much similar to that found in the vertebrate genomes. However, amphioxus lacks homologs of human aGPCR subfamilies I, II, IV, VI, and VII and only contains a few aGPCRs that shared some similarities with subfamilies III, V, and VIII. The remaining aGPCRs in amphioxus are most likely species-specific and contained a diverse array of extracellular domains including a few domains that are not commonly found in vertebrate aGPCRs [26]. Similarly, *Ciona* contains 30 aGPCRs, of which only six genes constituted orthologous relationships with the human aGPCR subfamilies (subfamilies I, III, IV, and VIII) [25]. The remaining 24 genes are diverse from the human aGPCR dataset and formed five paralogous groups that are most likely specific to *Ciona* [25]. Mapping of aGPCRs in *Saccoglossus kowalevskii*, a hemichordate (a deuterostome animal, sister group to echinoderms, and closely related to chordates) found 18 aGPCR-like sequences, which comprise homologs of subfamilies I, III, IV, VII, and VIII, but lack representatives of other aGPCR subfamilies [27]. In addition, previous studies have also estimated the number of aGPCRs in other metazoans including, echinoderms (sea urchin) [28, 29], nematodes (*Caenorhabditis elegans*) [30], arthropods (*Drosophila melanogaster*) [31], cnidarians (*Nematostella vectensis*) [32], placozoans (*Trichoplax adhaerens*) [33], and sponges (*Amphimedon queenslandica*) [34, 35]. These ancient and primitive animals constituted several species-specific aGPCRs, but also contain a few homologs of some human aGPCR subfamilies. For instance, *N. vectensis* contains homologs of subfamilies III, IV, and IX, while *T. adhaerens* has a homolog of subfamily IV [32, 35]. Nonetheless, it must be mentioned here that these phylogenetically basal species contained highly diverse aGPCR datasets and earlier attempts to phylogenetically resolve their relationships with human and other vertebrate aGPCRs) have resulted in either unresolved topologies or lack of sufficient support to categorize them into the known subfamilies. This scenario is also observed in our recent attempt where we classified sponge aGPCRs and resolved their relationships with aGPCR datasets from other non-bilaterian species [34]. Taken together, these studies elucidate the difficulties in classifying aGPCRs in distant species. Moreover, the complexity is also aggravated in naming of these homologs and other genetic variants in different species, due to the lack of harmonized nomenclature for aGPCRs. Nonetheless, the current nomenclature (see below) will provide a better platform to classify these ancient GPCRs and to further examine the aGPCR diversity.

3 Recommended Nomenclature of aGPCRs

As described in previous sections, aGPCRs are highly diverse and are found in a wide range of species, where they perform important and diverse biological functions. However, aGPCRs lack a harmonized nomenclature and over the years diverse names have been used for the genes belongings to different subfamilies. Names such as cadherin EGF LAG seven-pass G-type receptor (CELSR), EGF-TM7 (epidermal growth factor-seven-span transmembrane), brain-specific angiogenesis inhibitor (BAI), and very large GPCR (VLGR) were initially created by pioneers of this research field much before the time point where aGPCRs were not considered as a separate family of GPCRs. Later after the release of the human genome, many new aGPCRs genes were identified and GPR# names were assigned in collaboration with the Human Genome Organization (HUGO) Gene Nomenclature Committee (HGNC). Nonetheless, these GPR# names were also considered as temporary identifiers until additional information about the protein function was elucidated. Currently, the aGPCRs are considered to assemble a separate family of GPCRs. Over the years the research field has expanded widely and more genome-wide studies including genomics/genetics, expression, and epigenetics are consistently been carried out. This permanent growth of the aGPCR research field raised the requirement for a harmonized naming system that would clearly illustrate the relationship between these proteins/genes for diverse research groups. The highly diverse aGPCR naming system was discussed in IUPHAR meetings and this has been taken into consideration by the IUPHAR Committee on Receptor Nomenclature and Drug Classification (NC-IUPHAR). The Adhesion-GPCR Consortium with assistance from HGNC worked on a coherent naming system, sought to provide aGPCRs a prefix that identifies any adhesion GPCR homolog, independent of species or subfamily [15]. Providing a unique and most appropriate prefix for all aGPCR families was considered highly important because any such naming system will help to name orthologs and other genetic variants in different species in the future [15].

The most appropriate and unique prefix found was ADGR, which stands for *ad*hesion *G* protein-coupled *r*eceptor. Moreover for easy recognition, each subfamily was then assigned a letter to relate to their previous names. For example, L for the latrophilins, E for the EGF-TM7 receptors, C for the CELSRs, B for the BAIs, and V for VLGR, while the subfamilies with GPR# names have been given a letter in alphabetic order (A, D, F, G). As stated above the members within each subfamily (I to IX) were assigned a sequential number based on the phylogenetic clustering pattern according to the molecular signature of the 7TM regions. For example, the members of the subfamily I, the latrophilins, LPHN1, LPHN2, LPHN3, and ELTD1, were named as ADGR*L* for latrophilins, and the members were assigned numbers as such where LPHN1 is now ADGRL1, LPHN2 is ADGRL2, LPHN3 is ADGRL3, and ELTD1 is ADGRL4 [15]. Likewise members of the other subfamilies contain the prefix ADGR, the associated letter to relate to previous names, followed by numbers for the paralogs. This nomenclature (Table 1) has been accepted by both HGNC and NC-IUPHAR, and both organizations encourage the use of this nomenclature in all literature and databases [15]. Upon recommendation, the new nomenclature is currently being used together with the old names, for example, "ADGRE5 (CD97)" or "ADGRL1 (LPHN1)," until the new names are fully established.

4 Structural Aspects of aGPCRs

The 33 human aGPCRs can be divided into nine subfamilies based on their 7TM sequence similarity. Interestingly, the 7TM-based classification reflects similarities within aGPCR subfamilies with respect to the organization of their extracellular domains. Class-specific features, such as the GAIN domain and often-occurring sequences like the HBD and EGF-like domains are present in several aGPCR subfamilies. In contrast, cadherin and thrombospondin type 1 repeats are subfamily specific. A subfamily ordered description of all domain structures and their identified interaction partners gives insight in the structural diversity of aGPCR and also highlights the dual function of both the NTF and the CTF.

Table 1 Currently used nomenclature of adhesion GPCRs

Subfamilies	Previous/old gene names	Proposed new names for subfamilies	Currently accepted nomenclature/ used new gene names
I	LPHN1	L (Latrophilin)	ADGRL1
	LPHN2		ADGRL2
	LPHN3		ADGRL3
	ELTD1		ADGRL4
II	EMR1	E (EGF-TM7)	ADGRE1
	EMR2		ADGRE2
	EMR3		ADGRE3
	EMR4		ADGRE4
	CD97		ADGRE5
III	GPR123	A	ADGRA1
	GPR124		ADGRA2
	GPR125		ADGRA3
IV	CELSR1	C (CELSR)	ADGRC1
	CELSR2		ADGRC2
	CELSR3		ADGRC3
V	GPR133	D	ADGRD1
	GPR144		ADGRD2
VI	GPR110	F	ADGRF1
	GPR111		ADGRF2
	GPR113		ADGRF3
	GPR115		ADGRF4
	GPR116		ADGRF5
VII	BAI1	B (BAI)	ADGRB1
	BAI2		ADGRB2
	BAI3		ADGRB3
VIII	GPR56	G	ADGRG1
	GPR64		ADGRG2
	GPR97		ADGRG3
	GPR112		ADGRG4
	GPR114		ADGRG5
	GPR126		ADGRG6
	GPR128		ADGRG7
IX	GPR98	V	ADGRV1

BAI brain-specific angiogenesis inhibitor, *CD* cluster of differentiation, *CELSR* cadherin EGF LAG seven-pass G-type receptor, *EGF-TM7* epidermal growth factor-seven-span transmembrane, *ELTD* EGF, latrophilin, and seven-transmembrane domain-containing protein, *EMR* EGF-like molecule-containing mucin-like hormone receptor

4.1 General Structural Features of aGPCRs

4.1.1 7TM

All aGPCRs possess the classical 7TM architecture. Based on sequence conservation and gene splicing sites localization [32], aGPCRs are more closely related to family B secretin receptors than to family A rhodopsin GPCRs (see Chapter IV for a complete aGPCR/secretin family B structural comparison). Seven TM sequence homology between aGPCR subfamilies classifies them in nine distinct subfamilies [16, 36]. Intriguingly, this classification can be extended to the extracellular domain (ECD), suggesting an evolutionary conserved relationship between the extracellular domains and 7TM functioning. Indeed, for some aGPCRs it has been described that protein binding to the extracellular domain results in 7TM-dependent signaling [11, 15]. Recently, an activation mechanism has been proposed in which the 7TM N-terminus (Stachel; see below) is exposed upon removal of the non-covalently attached N-terminal fragment (NTF). This tethered sequence can subsequently interact with the 7TM structure and induce receptor activation [37–40]. This activation mechanism seems applicable for the whole aGPCR family, since the majority of aGPCRs possess this conserved sequence motif (see [41]).

4.1.2 ECD

Each aGPCR has a typically large N-terminal extracellular domain (ECD) that encompasses both aGPCR subfamily-dependent structural features, e.g., adhesion domains (see Sect. 4.2 and [42]), and aGPCR-specific features like the GAIN domain. Here, we will briefly address all features and, moreover, give a tabular overview of the different aGPCR subfamilies with their characteristic domains (Table 2).

4.1.3 GAIN

The GAIN domain, which is conserved in 32 out of 33 human aGPCR (absent in ADGRA1/GPR123), can be found directly N-terminally of the first transmembrane helix (TM1). In 2012 the crystal structures of the GAIN domain in ADGRL1 (latrophilin-1) and ADGRB3 (BAI3) were published [12, 44] (see also [42]). The GAIN domain consists of domain A that contains 6 alpha-helices and domain B that has a twisted beta-sandwich with 13 beta-strands and 2 alpha-helices [12, 44]. Within in the GAIN domain lays a conserved GPCR proteolytic site (GPS) where aGPCRs are autoproteolytically cleaved into an N-terminal fragment (NTF) and C-terminal fragment (CTF) during protein maturation (Table 2). GPS cleavage takes place between the last two (12–13) beta-strands. Importantly, it was shown that the GPS of ADGRB3 (BAI3) is situated in a sharply kinked loop and is therefore non-cleavable [12]. Indeed, not all aGPCRs are actually cleaved at their GPS: ADGRF2 (GPR111), ADGRG5 (GPR114) [40], ADGRF4 (GPR115) [45], and ADGRC1 (CELSR1) [46] have been shown to be cleavage deficient, whereas ADGRE1 (EMR1), ADGRA1, ADGRA2 (GPR124), ADGRA3 (GPR125), ADGRC3 (CELSR3), and ADGRB3 (BAI3) [12] are expected to be non-cleavable due to a non-consensus sequence (i.e., the general base H is absent

Table 2 Overview of aGPCR structural NTF features

Structural Feature -> / aGPCR	GPS (aa position)	GAIN	Hormone binding domain	Olfactomedin	Lectin	STP rich region	EGF Ca2+	EGF	RGD motif	Immunoglobulin	Leucine rich repeats	Laminin-EGF	Laminin G	Cadherin	Pentraxin	SEA	Thrompopsondin type 1 repeat	CUB	Calx-beta	EPTP	Additional cleavage sites	Binding partners identified	Stachel-induced activation
ADGRL1 (LPHN1)	838-839	1	1	1	1	1																	
ADGRL2 (LPHN2)	824-825	1	1	1	1	1																	
ADGRL3 (LPHN3)	841-842	1	1	1	1	1																	
ADGRL4 (ELTD1)	406-407	1					1	1															
ADGRE1 (EMR1)	584-585*	1					5	1															
ADGRE2 (EMR2)	517-518	1					4	1															
ADGRE3 (EMR3)	338-339	1					1	1															
ADGRE4 (EMR4)	173-174	1					1	1															
ADGRE5 (CD97)	530-531	1					4	1	1														
ADGRA1 (GPR123)																							
ADGRA2 (GPR124)	746-747*	1	1						1		1												
ADGRA3 (GPR125)	737-738	1	1							1	1												
ADGRC1 (CELSR1)	2448-2449*	1	1				4					1	2	9									
ADGRC2 (CELSR2)	2356-2357*	1	1				4					1	2	9									
ADGRC3 (CELSR3)	2517-2518*	1	1				4					2	2	9									
ADGRD1 (GPR133)	544-545	1																					
ADGRD2 (GPR144)	636-637	1													1								
ADGRF1 (GPR110)	566-567	1														1							
ADGRF2 (GPR111)	426-427*	1																					
ADGRF3 (GPR113)	752-753	1	1					1															
ADGRF4 (GPR115)	381-382*	1																					
ADGRF5 (GPR116)	990-991	1								1						1							
ADGRB1 (BAI1)	926-927	1	1						1								5						
ADGRB2 (BAI2)	911-912	1	1														4						
ADGRB3 (BAI3)	856-857	1	1														4	1					
ADGRG1 (GPR56)	282-283	1																					
ADGRG2 (GPR64)	606-607	1																					
ADGRG3 (GPR97)	249-250	1																					
ADGRG4 (GPR112)	3078-3080	1							1						1								
ADGRG5 (GPR114)	226-227	1																					
ADGRG6 (GPR126)	840-841	1	1										**		1			1					
ADGRG7 (GPR128)	415-416	1																					
ADGRV1 (VLGR1)	5890-5891	1							1						1				35	6			

Structural features present in extracellular domains of aGPCR family members. Numbers indicate the amount of repeats. In *blue*, receptors that have additional cleavage sites identified in NTF, identified binding partners, and/or Stachel-induced receptor activation are reported. *Red box*: Non-cleavage (experimental evidence). *Orange box*: theoretically non-cleavable. *According to GPS sequence alignment Lin et al. [43], **Novel laminin domain in ADGRG6

or replaced by an interfering residue) at their GPS [15]. Except for the ADGRAs, this cleavage deficiency seems random and subfamily independent (Table 2 and [47]).

4.1.4 HBD

Directly N-terminal of the GAIN domain is a so-called hormone-binding domain (HBD; ~70 aa) in 12 of the 33 aGPCRs [12] (Table 2 and [42]). This domain consists of two long and two short antiparallel beta-strands that are connected with random loop structures. Interestingly, such an HBD is also found in family B secretin receptors, the most closely related GPCRs [32]. This domain has been considered as potential ligand-binding domain [12, 48]. However, the crystal structure shows a conformation of the HBD that does not allow hormone binding due to steric hindrance of the GAIN domain [12]. It should be noted however that this conformation is just a snapshot and may not represent a dynamic real-life situation. In contrast to family B secretin receptors, the aGPCR HBDs lack an N-terminal alpha-helix, which has proven important, but not essential for peptide and hormone interactions in family B secretin receptors (e.g., GLP-1 receptor [49]).

4.1.5 EGF-Like Domain

The EGF-like domain (with or without calcium binding) is another domain that is present in several aGPCRs (i.e., ADGRL4, ADGREs, and ADGRCs). EGF-like domains consist of a large N- and short C-terminal-located two-stranded beta-sheet, which are connected by a loop structure [50]. Multiple EGF-like domains fold together to form a functional solenoid architecture. The structural organization of these domains is facilitated by domain–domain contacts that can be stabilized by calcium ion binding (see also [42]).

4.2 aGPCR Subfamily-Specific Structural Features and Protein Interactions of NTF

4.2.1 ADGRL

The NTF of the ADGRL (I) subfamily has a common structure for ADGRL1–3, namely, a cysteine-rich region homologous to a galactose-binding lectin (GBL or lectin) domain (108 aa) [51, 52]; olfactomedin domain (260 aa) (a glycoprotein of the extracellular matrix of the olfactory neuroepithelium) [53]; serine, threonine, and proline (STP)-rich region (79 aa); and an HBD [32, 54].

The structure of the lectin domain was resolved by NMR, showing a beta-sandwich including two antiparallel sheets, surrounded by a 10-residue alpha-helix and two extended loops [55]. L-Rhamnose can bind with low affinity to one of the loops [55, 56]; however, no data on ADGRL function has been tested. The lectin domain is conserved in most ADGRL1–3 orthologs, except coelenterates which could hint at an important role in receptor functioning [57]. Crystallographic data for the lectin and olf domains of the mouse ADGRL3 showed that olf-b has a propeller fold and a conserved calcium-binding site [58]. Note that the olfactomedin domain is only present in vertebrate orthologs aGPCRs, which could indicate that this domain is acquired during early vertebrate evolution [59]. The STP domain is present in vertebrate and insect orthologs, however, is absent in *C. elegans* [59]. In contrast, the fourth member of the ADGRL class, ADRGL4 (ELTD1), has a different structural composition of its NTF (i.e., EGF-like domain and Ca^{2+}-binding EGF-like domain) [60, 61] (see Sect. 4.1). The ADGRL receptors were discovered due to binding of an exogenous binding partner to ADGRL1, α-latrotoxin—a black widow spider venom [62]. This venom binds to the HBD and GAIN domains in the NTF of ADGRL1 [63]. In later years three endogenous binding partners have been found to interact with ADGRL1–3. All three belong to single-span transmembrane protein families: teneurins, fibronectin leucine-rich transmembranes (FLRTs), and neurexins.

The large postsynaptic glycoprotein teneurin-2 [also named Lasso (latrophilin-associated synaptic surface organizer)] binds to the lectin domain of presynaptic

ADGRL1 "in trans" ($K_d < 2$ nM) [59, 64]. Interestingly, while ADGRL1-3 have similar domains in their NTF, teneurin-2 only binds ADGRL1 and does only bind weakly (ADGRL2) or does not bind at all (ADGRL3) to the other subfamily members [64]. In addition, ADGRL3 binds to teneurin-3 [65] and teneurin-4 binds to ADGRL1 [66]. This indicates that, although the structural similarity is present, there is room for receptor-specific targeting.

Very recently, it was shown that a short amino acid sequence on the distal extracellular tip of each teneurin named teneurin C-terminal-associated peptide (TCAP) exhibits an effect on ADGLRs, independent of the teneurin protein. Interestingly, TCAP has remarkable structural homology with peptide ligands that bind to the secretin GPCR subfamily (i.e., CRF peptide family). TCAP binds to the lectin domain of ADGRL1–3 [67].

Cell adhesion proteins fibronectin leucine-rich transmembranes (FLRTs) were initially thought to bind "in trans" to the lectin domain of ADGRL1–3 [65]. However, recent elucidation of the propeller-like olf domain showed a major role for this domain in FLRT interaction [58, 68, 69]. ADGRL1 and ADGRL3 bind both FLRT1 and FLRT3, but ADGRL2 only interacts with FLRT3. Their extracellular regions interact with ADGRLs and with uncoordinated-5 (UNC5/netrin) receptors [68].

ADGRL1 also interacts "in trans" with a family of highly variable presynaptically localized neuronal cell surface receptors (i.e., neurexins) [70, 71]. Neurexin-1a, neurexin-1b, neurexin-2a, neurexin-2b, and neurexin-3b proteins bind ADGRLs via the receptor its olf domain. This results in a transsynaptic adhesion complex [66].

Of interest, the ADGRL family has been extensively studied in *C. elegans*. The ADGRL1 homolog LAT-1 has been shown to exhibit an NTF- and CTF-independent function. The NTF is anchored to the membrane and mediates fertility, independently from the CTF [45]. Likewise, the role of the CTF part in tissue polarity is independent of the NTF. Note that these effects are GPS cleavage independent, since a non-cleavable mutant did still possess the two independent activities [45]. ADGRL1–3 have the ability to reassociate upon protein binding to their NTF. Interestingly, this reassociation does not require a CTF from the original ADGRL subfamily member, but the protein-bound NTF can bind to a CTF from another ADGRL subfamily member [57]. Since the GAIN domain has high structural homology within the aGPCR class, even functional interactions with different subfamily members have been observed (named crisscross associations) [57].

No interaction data is known to date regarding the NTF of ADGRL4. The question arises whether this aGPCR truly belongs in this aGPCR subfamily. Structure-wise ADGRL4 is able to form homodimers at the cell surface consisting of two full-length ADGRL4 proteins [61].

4.2.2 ADGRE

Members of the ADGRE subfamily have a very similar NTF consisting of 2–5 tandemly arranged EGF-like domains (see Sect. 4.1). The 1–4 EGF-like domains that are located C-terminal of the first EGF domain have a calcium-binding site and belong to the fibrillin-like family I type [6]. The NTF similarity in ADGREs is mainly caused by their evolutionary history of gene duplication and exon shuffling [6]. ADGRE4 is most likely a pseudo-gene evolved after separation from the great apes [72]. ADGRE2 has been present from early placental mammals. Both ADGRE2 and ADGRE3 have no homologs in Murinae [73]. The 7TM of ADGRE2 have high amino acid similarity to ADGRE3, but the NTF part is most similar to ADGRE5 (6 aa difference) [74]. Of interest is the fact that ADGREs (i.e., 1, 2, 5, but also EGF-like domain-containing ADGRL4) have multiple isoforms. These isoforms vary in their EGF-like domain numbers and thus the structural formation of their NTF.

Based on the NTF structural features, ADGRL4 was previously believed to be part of the ADGRE subfamily [6]. However, based on chromosome localization (all genes except of ADGRL4 are located on chromosome 19), ADGRL4 does not fit in this family [6]. In addition, an ADGRL4-like gene in zebra fish (where ADGRE members are restricted to mammals) and the expression pattern of ADGRL4 also exclude ADGRL4 from the ADGRE subfamily [6].

CD55 interacts with the first two EGF-like domains in ADGRE5, whereas the third EGF-like domain is needed for structural integrity of the binding interface [75, 76]. Strikingly, ADGRE2 which differs only three amino acids in its EGF-like domains does not bind CD55 [74], demonstrating the high specificity in CD55 binding interaction to ADGRE5 (CD97). Different isoforms (i.e., different numbers of EGF-like domains) have different affinity for CD55 [76, 77]. Also, glycosaminoglycan side chain chondroitin sulfate B [78], $\alpha5\beta1$ and $\alpha v\beta3$ integrins [79], as well as CD90 [80] interact with ADGRE5. Chondroitin sulfate binds to the fourth EGF-like domain of ADGRE5 and ADGRE2, which is only present in the largest isoform of this receptor. Integrins bind to the RGD motif and CD90 binds to the GAIN domain. It was also shown that NTF of ADGRE5 can simultaneously bind CD55 and chondroitin sulfate B [81], although it has been suggested that this dual interaction is not likely to occur in vivo due to affinity for different isoforms [76, 77]. ADGRE5 also interacts "in *cis*" with lysophosphatidic acid receptor 1 (LPA) [82, 83], although structural requirements for this interaction are not yet known. Exogenous ligands have been developed for all members of this aGPCR subfamily by means of NTF-targeting antibodies [29]. ADGRE5-directed antibodies bind to different EGF domains [84].

4.2.3 ADGRA

The ADGRA subfamily has three members and all three have quite different NTF compositions. ADGRA1 is the only aGPCR that does not include the characteristic GAIN domain. There are also no known functional domains in the short NTF region of ADGRA1 [85]. ADGRA2 and ADGRA3 both have very similar leucine-rich repeats (LRRs), an immunoglobulin (Ig) domain, and an HBD [85]. In addition, ADGRA2 also contains an Arg-Gly-Asp (RGD) motif [86]. The LRRs (~25 aa) form a characteristic alpha/beta horseshoe fold with a beta-sheet inside and helices pointing outside [87]. Ig domains in ADGRA2 and ADGRA3 have some similarity to this domain in ADGRF5 (GPR116) [85]. The HBD of ADGRA2 and ADGRA3 is most closely related to the HBD of ADGRB3 (BAI3) receptor [85].

ADGRA2 has a GPS site, but is predicted to be non-cleavable. Interestingly, shed NTF of this receptor could be detected in cell culture medium. This is caused by a cleavage site in the NTF that is an interaction point for thrombin-induced shedding promoted by cell surface protein disulfide isomerase [88]. The RGD motif in ADGRA2 can bind $\alpha v\beta 3$ integrin [86].

4.2.4 ADGRC

ADGRC (CELSR) subfamily members have several protein domains in their NTF. Their former name (i.e., CELSR) is actually an abbreviation of these different domains: cadherin, EGF LAG, and seven-pass G-type receptor [89]. ADGRCs are among the oldest aGPCRs. ADGRC consists of nine cadherin repeats (110 aa), followed by six EGF-like domains (see Sect. 4.1), two LAG domains, and an HBD (see Sect. 4.1). The cadherin repeats play a role in extracellular calcium binding [90]. The 3D structure of several cadherin domains in other proteins has been solved lately. These cadherin domains are formed by seven beta-strands and show similarity to immunoglobulin constant domains [91, 92]. Comparable with EGF-like repeats, the quaternary structural composition of several repeats is mediated by calcium binding [93]. Five calcium-binding EGF-like domains (40 aa) interrupted by two laminin G-like (LAG) domains follow the cadherin repeats. The LAG domains have a jelly roll fold structure and could also bind calcium [94]. EGF-Lam domains (60 aa) are situated between the EGF-like and LAG domain structures and the HBD (see Sect. 4.1). ADGRC1 and ADGRC2 contain one EGF-Lam domain and ADGRC3 has two EGF-Lam domains. EGF-Lam domains contain eight conserved cysteine residues [89].

Although the large NTF of ADGRC subfamily members has multiple protein domains and therefore potential interaction interfaces, no endogenous protein-binding partners have been found to date. Activation is however modulated by calcium-induced *trans*-homodimer formation by cadherin–cadherin interactions [95, 96]. Also *cis*-homodimers have been reported, and these can also be formed in absence of calcium. We could therefore speculate that one ADGRC is the "ligand" for another ADGRC and vice versa. Unfortunately, no information is yet known about the exact interaction points and/or mechanisms by which ADGRC form *trans*- and/or *cis*-dimers.

4.2.5 ADGRD

ADGRD have quite short NTFs compared to other aGPCR subfamilies. ADGRD1 does not have any known additional structural features in its NTF, but ADGRD2 has a pentraxin (PTX) domain [16]. This PTX domain has similarities (33 %) with the PTX domain in ADGRG4 (GPR112) [16]. Pentraxins consist of up to five non-covalently bound identical subunits that form a flat pentameric disk [97]. An interesting observation is that the ADGRD subgroup is the most closely related to the family B secretin receptors [32].

No binding partners have been identified for ADGRDs. However, deletion of the NTF activates ADGRD1 by exposing the tethered agonist to the 7TM [37].

4.2.6 ADGRF

ADGRF1 (GPR110) has an additional cleavage site in its NTF, which is named sperm protein/enterokinase/agrin (SEA) domain (~80 aa) [98]. This motif might have a successive beta-strand structure that is interrupted by one alpha-helix [99]. ADGRF2 and ADGRF4 show high sequence similarity and both have no functional domains in their NTF [85]. ADGRF3 (GPR113) [100] has one EGF domain and an HBD (see Sect. 4.1). There is some structural similarity between ADGRF3 and the ADGRE subfamily, as well as with ADGRL4 [85]. The HBD is mostly related to ADGRL subfamily, especially ADGRL3 [85]. ADGRF5 has a SEA box and contains two immunoglobulin-like repeats [85, 101]. ADGRF5 also has an additional cleavage site in its NTF that is sensitive for furin [102]. Moreover, ADGRF5 can form homodimers on the cell surface in *cis* [103] and has been suggested to interact with the surfactant component: surfactant protein D [104].

4.2.7 ADGRB

The NTF of the ADGRB subfamily consists of four or five thrombospondin type 1 repeats (TSRs) and an HBD (see Sect. 4.1). In addition, ADGRB1 (BAI1) contains an RGD motif (231–233 aa) (see ADGRA2) in its NTF [105]. ADGRB3 has an additional Cs1 and Csr/Uegf/BMP-1 (CUB) domain (110 aa) at the N-terminus [106].

CUB domains are made up of a beta-sandwich with a jelly roll fold. They furthermore contain four conserved cysteines that are thought to make disulfide bonds [107].

Thrombospondin type 1 repeats (60 aa) are conserved domains that are comprised of an elongated, three stranded beta-sheet [108]. One site contains a helical groove that contains a high amount of positively charged residues. Variations in the positive charge (distribution or density) is often observed within different TSRs and also within the ADGRBs and is believed to contribute to protein-binding specificity [109].

An interesting feature of the NTF of ADGRB members is their ability to be cleaved, in addition to GPS autoproteolysis [109]. Cleavage results in two fragments of 120 kDa (vasculostatin) [105, 110] and 40 kDa (vasculostatin-40). Vasculostatin-40 contains one TSR [111]. Phosphatidylserine is an endogenous ligand for ADGRB1 (BAI1) and binds to thrombospondin type 1 repeats [112, 113]. ADGRB1 can also bind the exogenous lipopolysaccharide (LPS) of Gram-negative bacteria in macrophages [109]. ADGRB1 and ADGRB2 both showed increased constitutive activity upon deletion of their NTF domains, compared to full-length receptors. This indicates that the NTF has an inhibiting effect on CTF functioning [114]. The RGD motif from ADGRB1 (BAI1) can interact with integrins [105, 115]. The shed NTF fragments vasculostatin and vasculostatin-40 interact with $\alpha v\beta 5$ integrins [105, 110, 111]. Vasculostatin binds to CD36 or histidine-rich glycoprotein (HRGP) [116]. ADGRB3 (BAI3) binds secreted C1q-like proteins with its thrombospondin type 1 repeats and CUB domain [106, 117].

4.2.8 ADGRG

The majority of ADGRG subfamily members are characterized by their lack of known functional domains [i.e., ADGRG1 (GPR56), ADGRG3 (GPR97), ADGRG5, and ADGRG7 (GPR128)] [10]. Only ADGRG6 (GPR126), which contains a CUB domain, PTX domain, and HBD, and ADGRG4 that possesses a PTX domain and RGD motif have additional features besides the GAIN domain. Interestingly, ADGRG6 NTF is additionally cleaved by furin during intracellular processing steps, which leads to the release of sub-NTF fragments [118]. Recently, ADGRG1, ADGRG2 (GPR64), ADGRG5, and ADGRG6 were shown to be activated by a tethered peptide agonist upon NTF removal [37–39, 119]. Curiously, ADGRG2 full-length proteins activate different signaling pathways than truncated ADGRG2 in which the NTF is removed [120].

Collagen III and major cross-linking enzyme transglutaminase 2 (TG2) have been identified as binding partners for ADGRG1 [121, 122]. Collagen III binds in the NTF region range 27–160 aa, and, noteworthy, glycosylation is not a prerequisite for interaction to take place [123]. TG2 binds between 108 and 177 aa in the NTF of ADGRG1 [121, 124]. It is not clear whether collagen III and TG2 can bind simultaneously to the NTF.

ADGRG3 is the only aGPCR for which an interaction with a drug (i.e., corticosteroid beclomethasone dipropionate) has been reported [125].

ADGRG6 interacts with collagen IV and laminin-211. Collagen IV binds to the CUB and PTX domains [126]. Laminin-211 has been shown to interact with a novel laminin-binding domain (446–807 aa) in the ADGRG6 NTF and thereby prevents Stachel-induced receptor activation [127].

4.2.9 ADGRV

ADGRV is the only receptor in this subfamily and has the largest NTF of all aGPCRs. The NTF consists of 35 Calx-beta motifs (45 aa) that are interrupted by a pentraxin (PTX) domain (200 aa) (between the ninth and tenth Calx-beta motif) and an epilepsy-associated repeat (EAR) domain between the 22nd and 23rd Calx-beta motifs [128]. The Calx-beta motif has most likely an alpha, beta, beta, alpha, beta topology. The EAR domain has a predicted structure that consists of seven-bladed beta propellers [129].

The Calx-beta motifs 32–35 of ADGRV can bind calcium [130]. Other ligands or binding partners have not yet been identified. The beta propeller structures of the EAR, PTX, and Calx-beta domains and motif are potential binding interfaces for peptides and proteins. In addition, this has been suggested that ADGRV NTF interacts with ADGRV NTF via its Calx-repeats on other cells via NTF homodimerization [131].

5 Summary

aGPCRs are an object of a currently growing research field, and exploration of the pharmacological potential of these important signaling molecules is developing. This family of GPCRs was phylogenetically shown to form a separate cluster from other GPCR families and was categorized into nine distinct subfamilies. Repertoires of aGPCRs have been curated in several species including some model organisms. However, naming these homologs in a harmonized manner was found to be difficult, as a coherent nomenclature system of aGPCRs is not available. Moreover, the complexity of aGPCRs such as sharing low sequence similarity within some subfamily members and their diverse long N-termini with multiple functional domains adds on to the difficulty in naming and classification of homologs in distant species. This is certainly important as proper classification and curation of homologs in important established model organisms and other emerging model organisms (distant metazoan species) is essential in exploring the aGPCR biology. Recently, the Adhesion-GPCR Consortium with assistance from HGNC worked on a coherent naming system and provided aGPCRs a prefix that identifies any adhesion GPCR homolog, independent of species or subfamily [15]. In such context, ADGR was found to be the most appropriate and unique prefix, standing for adhesion G protein-coupled receptor. And each subfamily was assigned a single-letter acronym based on their previous subfamily names (e.g., E for the EGF-TM7 receptors, C for the CELSRs). This nomenclature is currently used and will be essential for easy naming of orthologs and other genetic variants in different species, including those that encode large expansions of aGPCRs. With the current nomenclature system, future comparisons of aGPCRs repertoires will help to better classify and define novel aGPCRs in important emerging model organisms, and this would provide a key foundation for elucidating aGPCR functions and pursuing aGPCRs as therapeutic targets. The current aGPCR classification into nine distinct subfamilies is based on their 7TM sequence similarity. Moreover, aGPCRs have family-specific features, like the GAIN domain, but also possess subfamily-specific domains in their NTF. Interestingly, overlap is observed with respect to the organization of aGPCR structural NTF features and the 7TM-based classification. A systematic description of the structural topology in aGPCR NTF regions provides information on protein-binding domains and NTF–CTF interactions and could ultimately help to get insight in (novel) aGPCR interaction partners for orphan aGPCRs.

Acknowledgments The studies were supported by the Swedish Research Council and the Novo Nordisk Foundation (to A. K. and H. B. S.), the Netherlands Organization for Scientific Research (NWO VENI grant 722.014.011 to S. N.), and the Netherlands eScience Center (NLeSC)/NWO (Enabling Technologies project: 3D-e-Chem, grant 027.014.201 to C. d. G.). S. N. and C. d. G. participate in the European Cooperation in Science and Technology Action CM1207 [GPCR-Ligand Interactions, Structures, and Transmembrane Signalling: A European Research Network (GLISTEN)].

References

1. Fredriksson R, Lagerstrom MC, Lundin LG, Schioth HB (2003) The G-protein-coupled receptors in the human genome form five main families. Phylogenetic analysis, paralogon groups, and fingerprints. Mol Pharmacol 63(6):1256–1272. doi:10.1124/mol.63.6.1256
2. Schioth HB, Fredriksson R (2005) The GRAFS classification system of G-protein coupled receptors in comparative perspective. Gen Comp Endocrinol 142(1–2):94–101. doi:10.1016/j.ygcen.2004.12.018
3. Baud V, Chissoe SL, Viegas-Pequignot E, Diriong S, N'Guyen VC, Roe BA et al (1995) EMR1, an unusual member in the family of hormone receptors with seven transmembrane segments. Genomics 26(2):334–344
4. Gray JX, Haino M, Roth MJ, Maguire JE, Jensen PN, Yarme A et al (1996) CD97 is a processed, seven-transmembrane, heterodimeric receptor associated with inflammation. J Immunol 157(12):5438–5447
5. Hamann J, Eichler W, Hamann D, Kerstens HM, Poddighe PJ, Hoovers JM et al (1995) Expression cloning and chromosomal mapping of the leukocyte activation antigen CD97, a new seven-span transmembrane molecule of the secretion receptor superfamily with an unusual extracellular domain. J Immunol 155(4):1942–1950
6. Kwakkenbos MJ, Kop EN, Stacey M, Matmati M, Gordon S, Lin HH et al (2004) The EGF-TM7 family: a postgenomic view. Immunogenetics 55(10):655–666. doi:10.1007/s00251-003-0625-2
7. Zendman AJ, Cornelissen IM, Weidle UH, Ruiter DJ, van Muijen GN (1999) TM7XN1, a novel human EGF-TM7-like cDNA, detected with mRNA differential display using human melanoma cell lines with different metastatic potential. FEBS Lett 446(2–3):292–298
8. Stacey M, Lin HH, Gordon S, McKnight AJ (2000) LNB-TM7, a group of seven-transmembrane proteins related to family-B G-protein-coupled receptors. Trends Biochem Sci 25(6):284–289
9. Harmar AJ (2001) Family-B G-protein-coupled receptors. Genome Biol 2(12): REVIEWS3013. PubMed PMID: 11790261; PubMed Central PMCID: PMCPMC138994
10. Fredriksson R, Gloriam DE, Hoglund PJ, Lagerstrom MC, Schioth HB (2003) There exist at least 30 human G-protein-coupled receptors with long Ser/Thr-rich N-termini. Biochem Biophys Res Commun 301(3):725–734
11. Langenhan T, Aust G, Hamann J (2013) Sticky signaling–adhesion class G protein-coupled receptors take the stage. Sci Signal 6(276):re3. doi:10.1126/scisignal.2003825
12. Arac D, Boucard AA, Bolliger MF, Nguyen J, Soltis SM, Sudhof TC et al (2012) A novel evolutionarily conserved domain of cell-adhesion GPCRs mediates autoproteolysis. EMBO J 31(6):1364–1378. doi:10.1038/emboj.2012.26, PubMed PMID: 22333914; PubMed Central PMCID: PMCPMC3321182
13. Lagerstrom MC, Schioth HB (2008) Structural diversity of G protein-coupled receptors and significance for drug discovery. Nat Rev Drug Discov 7(4):339–357. doi:10.1038/nrd2518
14. Poyner DR, Hay DL (2012) Secretin family (Class B) G protein-coupled receptors – from molecular to clinical perspectives. Br J Pharmacol 166(1):1–3. doi:10.1111/j.1476-5381.2011.01810.x, PubMed PMID: 22489621; PubMed Central PMCID: PMCPMC3415632
15. Hamann J, Aust G, Arac D, Engel FB, Formstone C, Fredriksson R et al (2015) International Union of Basic and Clinical Pharmacology. XCIV. Adhesion G protein-coupled receptors. Pharmacol Rev 67(2):338–367. doi:10.1124/pr.114.009647, PubMed PMID: 25713288, PubMed Central PMCID: PMC4394687
16. Bjarnadottir TK, Fredriksson R, Hoglund PJ, Gloriam DE, Lagerstrom MC, Schioth HB (2004) The human and mouse repertoire of the adhesion family of G-protein-coupled receptors. Genomics 84(1):23–33. doi:10.1016/j.ygeno.2003.12.004
17. Yang Z, Rannala B (2012) Molecular phylogenetics: principles and practice. Nat Rev Genet 13(5):303–314. doi:10.1038/nrg3186

18. Gloriam DE, Fredriksson R, Schioth HB (2007) The G protein-coupled receptor subset of the rat genome. BMC Genomics 8:338. doi:10.1186/1471-2164-8-338, PubMed PMID: 17892602, PubMed Central PMCID: PMC2117022
19. Haitina T, Fredriksson R, Foord SM, Schioth HB, Gloriam DE (2009) The G protein-coupled receptor subset of the dog genome is more similar to that in humans than rodents. BMC Genomics 10:24. doi:10.1186/1471-2164-10-24, PubMed PMID: 19146662, PubMed Central PMCID: PMC2651185
20. Bjarnadottir TK, Gloriam DE, Hellstrand SH, Kristiansson H, Fredriksson R, Schioth HB (2006) Comprehensive repertoire and phylogenetic analysis of the G protein-coupled receptors in human and mouse. Genomics 88(3):263–273. doi:10.1016/j.ygeno.2006.04.001
21. Haitina T, Olsson F, Stephansson O, Alsio J, Roman E, Ebendal T et al (2008) Expression profile of the entire family of Adhesion G protein-coupled receptors in mouse and rat. BMC Neurosci 9:43. doi:10.1186/1471-2202-9-43, PubMed PMID: 18445277, PubMed Central PMCID: PMC2386866
22. Lagerstrom MC, Hellstrom AR, Gloriam DE, Larsson TP, Schioth HB, Fredriksson R (2006) The G protein-coupled receptor subset of the chicken genome. PLoS Comput Biol 2(6), e54. doi:10.1371/journal.pcbi.0020054, PubMed PMID: 16741557, PubMed Central PMCID: PMC1472694
23. Metpally RP, Sowdhamini R (2005) Genome wide survey of G protein-coupled receptors in Tetraodon nigroviridis. BMC Evol Biol 5:41. doi:10.1186/1471-2148-5-41, PubMed PMID: 16022726, PubMed Central PMCID: PMC1187884
24. Harty BL, Krishnan A, Sanchez NE, Schioth HB, Monk KR (2015) Defining the gene repertoire and spatiotemporal expression profiles of adhesion G protein-coupled receptors in zebrafish. BMC Genomics 16:62. doi:10.1186/s12864-015-1296-8, PubMed PMID: 25715737, PubMed Central PMCID: PMC4335454
25. Kamesh N, Aradhyam GK, Manoj N (2008) The repertoire of G protein-coupled receptors in the sea squirt Ciona intestinalis. BMC Evol Biol 8:129. doi:10.1186/1471-2148-8-129, PubMed PMID: 18452600, PubMed Central PMCID: PMC2396169
26. Nordstrom KJ, Fredriksson R, Schioth HB (2008) The amphioxus (Branchiostoma floridae) genome contains a highly diversified set of G protein-coupled receptors. BMC Evol Biol 8:9. doi:10.1186/1471-2148-8-9, PubMed PMID: 18199322, PubMed Central PMCID: PMC2246102
27. Krishnan A, Almen MS, Fredriksson R, Schioth HB (2013) Remarkable similarities between the hemichordate (Saccoglossus kowalevskii) and vertebrate GPCR repertoire. Gene 526 (2):122–133. doi:10.1016/j.gene.2013.05.005
28. Raible F, Tessmar-Raible K, Arboleda E, Kaller T, Bork P, Arendt D et al (2006) Opsins and clusters of sensory G-protein-coupled receptors in the sea urchin genome. Dev Biol 300 (1):461–475. doi:10.1016/j.ydbio.2006.08.070
29. Yona S, Lin HH, Siu WO, Gordon S, Stacey M (2008) Adhesion-GPCRs: emerging roles for novel receptors. Trends Biochem Sci 33(10):491–500. doi:10.1016/j.tibs.2008.07.005
30. Nagarathnam B, Kalaimathy S, Balakrishnan V, Sowdhamini R (2012) Cross-genome clustering of human and C elegans G-protein coupled receptors. Evol Bioinform Online 8:229–259. doi:10.4137/EBO.S9405, PubMed PMID: 22807621, PubMed Central PMCID: PMC3396462
31. Metpally RP, Sowdhamini R (2005) Cross genome phylogenetic analysis of human and Drosophila G protein-coupled receptors: application to functional annotation of orphan receptors. BMC Genomics 6:106. doi:10.1186/1471-2164-6-106, PubMed PMID: 16091152, PubMed Central PMCID: PMC1192796
32. Nordstrom KJ, Lagerstrom MC, Waller LM, Fredriksson R, Schioth HB (2009) The Secretin GPCRs descended from the family of Adhesion GPCRs. Mol Biol Evol 26(1):71–84. doi:10.1093/molbev/msn228
33. Nordstrom KJ, Sallman Almen M, Edstam MM, Fredriksson R, Schioth HB (2011) Independent HHsearch, Needleman–Wunsch-based, and motif analyses reveal the overall hierarchy

for most of the G protein-coupled receptor families. Mol Biol Evol 28(9):2471–2480. doi:10.1093/molbev/msr061
34. Krishnan A, Dnyansagar R, Almen MS, Williams MJ, Fredriksson R, Manoj N et al (2014) The GPCR repertoire in the demosponge Amphimedon queenslandica: insights into the GPCR system at the early divergence of animals. BMC Evol Biol 14:270. doi:10.1186/s12862-014-0270-4, PubMed PMID: 25528161, PubMed Central PMCID: PMC4302439
35. Krishnan A, Schioth HB (2015) The role of G protein-coupled receptors in the early evolution of neurotransmission and the nervous system. J Exp Biol 218(Pt 4):562–571. doi:10.1242/jeb.110312
36. Bjarnadottir TK, Fredriksson R, Schioth HB (2007) The adhesion GPCRs: a unique family of G protein-coupled receptors with important roles in both central and peripheral tissues. Cell Mol Life Sci 64(16):2104–2119. doi:10.1007/s00018-007-7067-1
37. Liebscher I, Schon J, Petersen SC, Fischer L, Auerbach N, Demberg LM et al (2014) A tethered agonist within the ectodomain activates the adhesion G protein-coupled receptors GPR126 and GPR133. Cell Rep 9(6):2018–2026. doi:10.1016/j.celrep.2014.11.036, PubMed PMID: 25533341, PubMed Central PMCID: PMC4277498
38. Stoveken HM, Hajduczok AG, Xu L, Tall GG (2015) Adhesion G protein-coupled receptors are activated by exposure of a cryptic tethered agonist. Proc Natl Acad Sci U S A 112 (19):6194–6199. doi:10.1073/pnas.1421785112, PubMed PMID: 25918380, PubMed Central PMCID: PMC4434738
39. Demberg LM, Rothemund S, Schoneberg T, Liebscher I (2015) Identification of the tethered peptide agonist of the adhesion G protein-coupled receptor GPR64/ADGRG2. Biochem Biophys Res Commun 464(3):743–747. doi:10.1016/j.bbrc.2015.07.020
40. Wilde C, Fischer L, Lede V, Kirchberger J, Rothemund S, Schoneberg T et al (2016) The constitutive activity of the adhesion GPCR GPR114/ADGRG5 is mediated by its tethered agonist. FASEB J 30(2):666–673. doi:10.1096/fj.15-276220
41. Liebscher I, Schöneberg T (2016) Tethered agonism: a common activation mechanism of adhesion GPCRs. In: Langenhan T, Schöneberg T (eds) Adhesion G protein-coupled receptors: molecular, physiological and pharmacological principles in health and disease. Springer, Heidelberg
42. Araç D, Sträter N, Seiradake E (2016) Understanding the structural basis of adhesion GPCR functions. In: Langenhan T, Schöneberg T (eds) Adhesion G protein-coupled receptors: molecular, physiological and pharmacological principles in health and disease. Springer, Heidelberg
43. Lin HH, Stacey M, Yona S, Chang GW (2010) GPS proteolytic cleavage of adhesion-GPCRs. Adv Exp Med Biol 706:49–58, PMID: 21618825
44. Promel S, Langenhan T, Arac D (2013) Matching structure with function: the GAIN domain of adhesion-GPCR and PKD1-like proteins. Trends Pharmacol Sci 34(8):470–478. doi:10.1016/j.tips.2013.06.002
45. Promel S, Frickenhaus M, Hughes S, Mestek L, Staunton D, Woollard A et al (2012) The GPS motif is a molecular switch for bimodal activities of adhesion class G protein-coupled receptors. Cell Rep 2(2):321–331. doi:10.1016/j.celrep.2012.06.015, PubMed PMID: 22938866, PubMed Central PMCID: PMC3776922
46. Formstone CJ, Moxon C, Murdoch J, Little P, Mason I (2010) Basal enrichment within neuroepithelia suggests novel function(s) for Celsr1 protein. Mol Cell Neurosci 44 (3):210–222. doi:10.1016/j.mcn.2010.03.008
47. Nieberler M, Kittel RJ, Petrenko AG, Lin H-H, Langenhan T (2016) Control of adhesion GPCR function through proteolytic processing. In: Langenhan T, Schöneberg T (eds) Adhesion G protein-coupled receptors: molecular, physiological and pharmacological principles in health and disease. Springer, Heidelberg
48. Paavola KJ, Hall RA (2012) Adhesion G protein-coupled receptors: signaling, pharmacology, and mechanisms of activation. Mol Pharmacol 82(5):777–783. doi:10.1124/mol.112.080309, PubMed PMID: 22821233, PubMed Central PMCID: PMC3477231

49. Underwood CR, Garibay P, Knudsen LB, Hastrup S, Peters GH, Rudolph R et al (2010) Crystal structure of glucagon-like peptide-1 in complex with the extracellular domain of the glucagon-like peptide-1 receptor. J Biol Chem 285(1):723–730. doi:10.1074/jbc.M109.033829, PubMed PMID: 19861722, PubMed Central PMCID: PMC2804221
50. Wouters MA, Rigoutsos I, Chu CK, Feng LL, Sparrow DB, Dunwoodie SL (2005) Evolution of distinct EGF domains with specific functions. Protein Sci 14(4):1091–1103. doi:10.1110/ps.041207005, PubMed PMID: 15772310, PubMed Central PMCID: PMC2253431
51. Krasnoperov VG, Bittner MA, Beavis R, Kuang Y, Salnikow KV, Chepurny OG et al (1997) alpha-Latrotoxin stimulates exocytosis by the interaction with a neuronal G-protein-coupled receptor. Neuron 18(6):925–937
52. Lelianova VG, Davletov BA, Sterling A, Rahman MA, Grishin EV, Totty NF et al (1997) Alpha-latrotoxin receptor, latrophilin, is a novel member of the secretin family of G protein-coupled receptors. J Biol Chem 272(34):21504–21508
53. Tomarev SI, Nakaya N (2009) Olfactomedin domain-containing proteins: possible mechanisms of action and functions in normal development and pathology. Mol Neurobiol 40(2):122–138. doi:10.1007/s12035-009-8076-x, PubMed PMID: 19554483, PubMed Central PMCID: PMC2936706
54. Matsushita H, Lelianova VG, Ushkaryov YA (1999) The latrophilin family: multiply spliced G protein-coupled receptors with differential tissue distribution. FEBS Lett 443(3):348–352
55. Vakonakis I, Langenhan T, Promel S, Russ A, Campbell ID (2008) Solution structure and sugar-binding mechanism of mouse latrophilin-1 RBL: a 7TM receptor-attached lectin-like domain. Structure 16(6):944–953. doi:10.1016/j.str.2008.02.020, PubMed PMID: 18547526, PubMed Central PMCID: PMC2430599
56. Terada T, Watanabe Y, Tateno H, Naganuma T, Ogawa T, Muramoto K et al (2007) Structural characterization of a rhamnose-binding glycoprotein (lectin) from Spanish mackerel (Scomberomorous niphonius) eggs. Biochim Biophys Acta 1770(4):617–629. doi:10.1016/j.bbagen.2006.11.003
57. Silva JP, Lelianova V, Hopkins C, Volynski KE, Ushkaryov Y (2009) Functional cross-interaction of the fragments produced by the cleavage of distinct adhesion G-protein-coupled receptors. J Biol Chem 284(10):6495–6506. doi:10.1074/jbc.M806979200, PubMed PMID: 19124473, PubMed Central PMCID: PMC2649109
58. Jackson VA, del Toro D, Carrasquero M, Roversi P, Harlos K, Klein R et al (2015) Structural basis of latrophilin-FLRT interaction. Structure 23(4):774–781. doi:10.1016/j.str.2015.01.013, PubMed PMID: 25728924, PubMed Central PMCID: PMC4396693
59. Silva JP, Ushkaryov YA (2010) The latrophilins, "split-personality" receptors. Adv Exp Med Biol 706:59–75, PubMed PMID: 21618826, PubMed Central PMCID: PMC3145135
60. Favara DM, Banham AH, Harris AL (2014) A review of ELTD1, a pro-angiogenic adhesion GPCR. Biochem Soc Trans 42(6):1658–1664. doi:10.1042/BST20140216
61. Nechiporuk T, Urness LD, Keating MT (2001) ETL, a novel seven-transmembrane receptor that is developmentally regulated in the heart. ETL is a member of the secretin family and belongs to the epidermal growth factor-seven-transmembrane subfamily. J Biol Chem 276 (6):4150–4157. doi:10.1074/jbc.M004814200
62. Rohou A, Nield J, Ushkaryov YA (2007) Insecticidal toxins from black widow spider venom. Toxicon 49(4):531–549. doi:10.1016/j.toxicon.2006.11.021, PubMed PMID: 17210168, PubMed Central PMCID: PMC2517654
63. Krasnoperov V, Bittner MA, Holz RW, Chepurny O, Petrenko AG (1999) Structural requirements for alpha-latrotoxin binding and alpha-latrotoxin-stimulated secretion. A study with calcium-independent receptor of alpha-latrotoxin (CIRL) deletion mutants. J Biol Chem 274(6):3590–3596
64. Silva JP, Lelianova VG, Ermolyuk YS, Vysokov N, Hitchen PG, Berninghausen O et al (2011) Latrophilin 1 and its endogenous ligand Lasso/teneurin-2 form a high-affinity transsynaptic receptor pair with signaling capabilities. Proc Natl Acad Sci U S A 108

(29):12113–12118. doi:10.1073/pnas.1019434108, PubMed PMID: 21724987, PubMed Central PMCID: PMC3141932
65. O'Sullivan ML, de Wit J, Savas JN, Comoletti D, Otto-Hitt S, Yates JR 3rd et al (2012) FLRT proteins are endogenous latrophilin ligands and regulate excitatory synapse development. Neuron 73(5):903–910. doi:10.1016/j.neuron.2012.01.018, PubMed PMID: 22405201, PubMed Central PMCID: PMC3326387
66. Boucard AA, Maxeiner S, Sudhof TC (2014) Latrophilins function as heterophilic cell-adhesion molecules by binding to teneurins: regulation by alternative splicing. J Biol Chem 289(1):387–402. doi:10.1074/jbc.M113.504779, PubMed PMID: 24273166, PubMed Central PMCID: PMC3879561
67. Woelfle R, D'Aquila AL, Pavlovic T, Husic M, Lovejoy DA (2015) Ancient interaction between the teneurin C-terminal associated peptides (TCAP) and latrophilin ligand-receptor coupling: a role in behavior. Front Neurosci 9:146. doi:10.3389/fnins.2015.00146, PubMed PMID: 25964737, PubMed Central PMCID: PMC4408839
68. Lu YC, Nazarko OV, Sando R III, Salzman GS, Sudhof TC, Arac D (2015) Structural basis of latrophilin-FLRT-UNC5 interaction in cell adhesion. Structure 23(9):1678–1691. doi:10.1016/j.str.2015.06.024
69. Ranaivoson FM, Liu Q, Martini F, Bergami F, von Daake S, Li S et al (2015) Structural and mechanistic insights into the latrophilin3-FLRT3 complex that mediates glutamatergic synapse development. Structure 23(9):1665–1677. doi:10.1016/j.str.2015.06.022
70. Petrenko AG, Surkova IN, Shamotienko OG, Kovalenko VA, Krasnoperov VN, Tret'iakov LA et al (1990) Study of the receptor for black widow spider neurotoxin. II. Isolation and characteristics of the receptor from bovine brain membranes. Bioorg Khim 16(2):158–165
71. Geppert M, Khvotchev M, Krasnoperov V, Goda Y, Missler M, Hammer RE et al (1998) Neurexin I alpha is a major alpha-latrotoxin receptor that cooperates in alpha-latrotoxin action. J Biol Chem 273(3):1705–1710
72. Hamann J, Kwakkenbos MJ, de Jong EC, Heus H, Olsen AS, van Lier RA (2003) Inactivation of the EGF-TM7 receptor EMR4 after the Pan-Homo divergence. Eur J Immunol 33 (5):1365–1371. doi:10.1002/eji.200323881
73. Kwakkenbos MJ, Matmati M, Madsen O, Pouwels W, Wang Y, Bontrop RE et al (2006) An unusual mode of concerted evolution of the EGF-TM7 receptor chimera EMR2. FASEB J 20 (14):2582–2584. doi:10.1096/fj.06-6500fje
74. Lin HH, Stacey M, Hamann J, Gordon S, McKnight AJ (2000) Human EMR2, a novel EGF-TM7 molecule on chromosome 19p13.1, is closely related to CD97. Genomics 67 (2):188–200. doi:10.1006/geno.2000.6238
75. Hamann J, Vogel B, van Schijndel GM, van Lier RA (1996) The seven-span transmembrane receptor CD97 has a cellular ligand (CD55, DAF). J Exp Med 184(3):1185–1189, PubMed PMID: 9064337, PubMed Central PMCID: PMC2192782
76. Hamann J, Stortelers C, Kiss-Toth E, Vogel B, Eichler W, van Lier RA (1998) Characterization of the CD55 (DAF)-binding site on the seven-span transmembrane receptor CD97. Eur J Immunol 28(5):1701–1707. doi:10.1002/(SICI)1521-4141(199805)28:05<1701::AID-IMMU1701>3.0.CO;2-2
77. Lin HH, Stacey M, Saxby C, Knott V, Chaudhry Y, Evans D et al (2001) Molecular analysis of the epidermal growth factor-like short consensus repeat domain-mediated protein-protein interactions: dissection of the CD97-CD55 complex. J Biol Chem 276(26):24160–24169. doi:10.1074/jbc.M101770200
78. Stacey M, Chang GW, Davies JQ, Kwakkenbos MJ, Sanderson RD, Hamann J et al (2003) The epidermal growth factor-like domains of the human EMR2 receptor mediate cell attachment through chondroitin sulfate glycosaminoglycans. Blood 102(8):2916–2924. doi:10.1182/blood-2002-11-3540
79. Wang T, Ward Y, Tian L, Lake R, Guedez L, Stetler-Stevenson WG et al (2005) CD97, an adhesion receptor on inflammatory cells, stimulates angiogenesis through binding integrin

counter receptors on endothelial cells. Blood 105(7):2836–2844. doi:10.1182/blood-2004-07-2878
80. Wandel E, Saalbach A, Sittig D, Gebhardt C, Aust G (2012) Thy-1 (CD90) is an interacting partner for CD97 on activated endothelial cells. J Immunol 188(3):1442–1450. doi:10.4049/jimmunol.1003944
81. Kwakkenbos MJ, Pouwels W, Matmati M, Stacey M, Lin HH, Gordon S et al (2005) Expression of the largest CD97 and EMR2 isoforms on leukocytes facilitates a specific interaction with chondroitin sulfate on B cells. J Leukoc Biol 77(1):112–119. doi:10.1189/jlb.0704402
82. Ward Y, Lake R, Yin JJ, Heger CD, Raffeld M, Goldsmith PK et al (2011) LPA receptor heterodimerizes with CD97 to amplify LPA-initiated RHO-dependent signaling and invasion in prostate cancer cells. Cancer Res 71(23):7301–7311. doi:10.1158/0008-5472.CAN-11-2381
83. Ward Y, Lake R, Martin PL, Killian K, Salerno P, Wang T et al (2013) CD97 amplifies LPA receptor signaling and promotes thyroid cancer progression in a mouse model. Oncogene 32 (22):2726–2738. doi:10.1038/onc.2012.301
84. Veninga H, de Groot DM, McCloskey N, Owens BM, Dessing MC, Verbeek JS et al (2011) CD97 antibody depletes granulocytes in mice under conditions of acute inflammation via a Fc receptor-dependent mechanism. J Leukoc Biol 89(3):413–421. doi:10.1189/jlb.0510280
85. Fredriksson R, Lagerstrom MC, Hoglund PJ, Schioth HB (2002) Novel human G protein-coupled receptors with long N-terminals containing GPS domains and Ser/Thr-rich regions. FEBS Lett 531(3):407–414
86. Vallon M, Essler M (2006) Proteolytically processed soluble tumor endothelial marker (TEM) 5 mediates endothelial cell survival during angiogenesis by linking integrin alpha (v)beta3 to glycosaminoglycans. J Biol Chem 281(45):34179–34188. doi:10.1074/jbc.M605291200
87. Enkhbayar P, Kamiya M, Osaki M, Matsumoto T, Matsushima N (2004) Structural principles of leucine-rich repeat (LRR) proteins. Proteins 54(3):394–403. doi:10.1002/prot.10605
88. Vallon M, Aubele P, Janssen KP, Essler M (2012) Thrombin-induced shedding of tumour endothelial marker 5 and exposure of its RGD motif are regulated by cell-surface protein disulfide-isomerase. Biochem J 441(3):937–944. doi:10.1042/BJ20111682
89. Wang XJ, Zhang DL, Xu ZG, Ma ML, Wang WB, Li LL et al (2014) Understanding cadherin EGF LAG seven-pass G-type receptors. J Neurochem 131(6):699–711. doi:10.1111/jnc.12955, PubMed PMID: 25280249, PubMed Central PMCID: PMC4261025
90. Shapiro L, Weis WI (2009) Structure and biochemistry of cadherins and catenins. Cold Spring Harb Perspect Biol 1(3):a003053. doi:10.1101/cshperspect.a003053, PubMed PMID: 20066110, PubMed Central PMCID: PMC2773639
91. Overduin M, Harvey TS, Bagby S, Tong KI, Yau P, Takeichi M et al (1995) Solution structure of the epithelial cadherin domain responsible for selective cell adhesion. Science 267(5196):386–389
92. Overduin M, Tong KI, Kay CM, Ikura M (1996) 1H, 15N and 13C resonance assignments and monomeric structure of the amino-terminal extracellular domain of epithelial cadherin. J Biomol NMR 7(3):173–189
93. Takeichi M (1995) Morphogenetic roles of classic cadherins. Curr Opin Cell Biol 7 (5):619–627
94. Hohenester E, Tisi D, Talts JF, Timpl R (1999) The crystal structure of a laminin G-like module reveals the molecular basis of alpha-dystroglycan binding to laminins, perlecan, and agrin. Mol Cell 4(5):783–792
95. Shima Y, Kawaguchi SY, Kosaka K, Nakayama M, Hoshino M, Nabeshima Y et al (2007) Opposing roles in neurite growth control by two seven-pass transmembrane cadherins. Nat Neurosci 10(8):963–969. doi:10.1038/nn1933
96. Zhang Y, Sivasankar S, Nelson WJ, Chu S (2009) Resolving cadherin interactions and binding cooperativity at the single-molecule level. Proc Natl Acad Sci U S A 106

(1):109–114. doi:10.1073/pnas.0811350106, PubMed PMID: 19114658, PubMed Central PMCID: PMC2629205
97. Goodman AR, Cardozo T, Abagyan R, Altmeyer A, Wisniewski HG, Vilcek J (1996) Long pentraxins: an emerging group of proteins with diverse functions. Cytokine Growth Factor Rev 7(2):191–202
98. Lum AM, Wang BB, Beck-Engeser GB, Li L, Channa N, Wabl M (2010) Orphan receptor GPR110, an oncogene overexpressed in lung and prostate cancer. BMC Cancer 10:40. doi:10.1186/1471-2407-10-40, PubMed PMID: 20149256, PubMed Central PMCID: PMC2830182
99. Bork P, Patthy L (1995) The SEA module: a new extracellular domain associated with O-glycosylation. Protein Sci 4(7):1421–1425. doi:10.1002/pro.5560040716, PubMed PMID: 7670383, PubMed Central PMCID: PMC2143162
100. LopezJimenez ND, Sainz E, Cavenagh MM, Cruz-Ithier MA, Blackwood CA, Battey JF et al (2005) Two novel genes, Gpr113, which encodes a family 2 G-protein-coupled receptor, and Trcg1, are selectively expressed in taste receptor cells. Genomics 85(4):472–482. doi:10.1016/j.ygeno.2004.12.005
101. Abe J, Suzuki H, Notoya M, Yamamoto T, Hirose S (1999) Ig-hepta, a novel member of the G protein-coupled hepta-helical receptor (GPCR) family that has immunoglobulin-like repeats in a long N-terminal extracellular domain and defines a new subfamily of GPCRs. J Biol Chem 274(28):19957–19964
102. Abe J, Fukuzawa T, Hirose S (2002) Cleavage of Ig-Hepta at a "SEA" module and at a conserved G protein-coupled receptor proteolytic site. J Biol Chem 277(26):23391–23398. doi:10.1074/jbc.M110877200
103. Davies JQ, Chang GW, Yona S, Gordon S, Stacey M, Lin HH (2007) The role of receptor oligomerization in modulating the expression and function of leukocyte adhesion-G protein-coupled receptors. J Biol Chem 282(37):27343–27353. doi:10.1074/jbc.M704096200
104. Fukuzawa T, Ishida J, Kato A, Ichinose T, Ariestanti DM, Takahashi T et al (2013) Lung surfactant levels are regulated by Ig-Hepta/GPR116 by monitoring surfactant protein D. PLoS One 8(7), e69451. doi:10.1371/journal.pone.0069451, PubMed PMID: 23922714, PubMed Central PMCID: PMC3726689
105. Koh JT, Kook H, Kee HJ, Seo YW, Jeong BC, Lee JH et al (2004) Extracellular fragment of brain-specific angiogenesis inhibitor 1 suppresses endothelial cell proliferation by blocking alphavbeta5 integrin. Exp Cell Res 294(1):172–184. doi:10.1016/j.yexcr.2003.11.008
106. Duman JG, Tu Y-K, Tolias KF (2016) Emerging roles of BAI adhesion-GPCRs in synapse development and plasticity. Neural Plast 2016:8301737. doi:10.1155/2016/8301737
107. Blanc G, Font B, Eichenberger D, Moreau C, Ricard-Blum S, Hulmes DJ et al (2007) Insights into how CUB domains can exert specific functions while sharing a common fold: conserved and specific features of the CUB1 domain contribute to the molecular basis of procollagen C-proteinase enhancer-1 activity. J Biol Chem 282(23):16924–16933. doi:10.1074/jbc.M701610200
108. Tan K, Duquette M, Liu JH, Dong Y, Zhang R, Joachimiak A et al (2002) Crystal structure of the TSP-1 type 1 repeats: a novel layered fold and its biological implication. J Cell Biol 159(2):373–382. doi:10.1083/jcb.200206062, PubMed PMID: 12391027, PubMed Central PMCID: PMC2173040
109. Das S, Owen KA, Ly KT, Park D, Black SG, Wilson JM et al (2011) Brain angiogenesis inhibitor 1 (BAI1) is a pattern recognition receptor that mediates macrophage binding and engulfment of Gram-negative bacteria. Proc Natl Acad Sci U S A 108(5):2136–2141. doi:10.1073/pnas.1014775108, PubMed PMID: 21245295, PubMed Central PMCID: PMC3033312
110. Kaur B, Brat DJ, Devi NS, Van Meir EG (2005) Vasculostatin, a proteolytic fragment of brain angiogenesis inhibitor 1, is an antiangiogenic and antitumorigenic factor. Oncogene 24(22):3632–3642. doi:10.1038/sj.onc.1208317
111. Cork SM, Kaur B, Devi NS, Cooper L, Saltz JH, Sandberg EM et al (2012) A proprotein convertase/MMP-14 proteolytic cascade releases a novel 40 kDa vasculostatin from tumor

suppressor BAI1. Oncogene 31(50):5144–5152. doi:10.1038/onc.2012.1, PubMed PMID: 22330140, PubMed Central PMCID: PMC3355202

112. Park D, Tosello-Trampont AC, Elliott MR, Lu M, Haney LB, Ma Z et al (2007) BAI1 is an engulfment receptor for apoptotic cells upstream of the ELMO/Dock180/Rac module. Nature 450(7168):430–434. doi:10.1038/nature06329
113. Mazaheri F, Breus O, Durdu S, Haas P, Wittbrodt J, Gilmour D et al (2014) Distinct roles for BAI1 and TIM-4 in the engulfment of dying neurons by microglia. Nat Commun 5:4046. doi:10.1038/ncomms5046
114. Okajima D, Kudo G, Yokota H (2010) Brain-specific angiogenesis inhibitor 2 (BAI2) may be activated by proteolytic processing. J Recept Signal Transduct Res 30(3):143–153. doi:10.3109/10799891003671139
115. D'Souza SE, Ginsberg MH, Plow EF (1991) Arginyl-glycyl-aspartic acid (RGD): a cell adhesion motif. Trends Biochem Sci 16(7):246–250
116. Klenotic PA, Huang P, Palomo J, Kaur B, Van Meir EG, Vogelbaum MA et al (2010) Histidine-rich glycoprotein modulates the anti-angiogenic effects of vasculostatin. Am J Pathol 176(4):2039–2050. doi:10.2353/ajpath.2010.090782, PubMed PMID: 20167858, PubMed Central PMCID: PMC2843491
117. Bolliger MF, Martinelli DC, Sudhof TC (2011) The cell-adhesion G protein-coupled receptor BAI3 is a high-affinity receptor for C1q-like proteins. Proc Natl Acad Sci U S A 108 (6):2534–2539. doi:10.1073/pnas.1019577108, PubMed PMID: 21262840, PubMed Central PMCID: PMC3038708
118. Moriguchi T, Haraguchi K, Ueda N, Okada M, Furuya T, Akiyama T (2004) DREG, a developmentally regulated G protein-coupled receptor containing two conserved proteolytic cleavage sites. Genes Cells 9(6):549–560. doi:10.1111/j.1356-9597.2004.00743.x
119. Wilde C, Fischer L, Lede V, Kirchberger J, Rothemund S, Schoneberg T et al (2015) The constitutive activity of the adhesion GPCR GPR114/ADGRG5 is mediated by its tethered agonist. FASEB J. doi:10.1096/fj.15-276220
120. Peeters MC, Fokkelman M, Boogaard B, Egerod KL, van de Water B, IJzerman AP et al (2015) The adhesion G protein-coupled receptor G2 (ADGRG2/GPR64) constitutively activates SRE and NFkappaB and is involved in cell adhesion and migration. Cell Signal 27 (12):2579–2588. doi:10.1016/j.cellsig.2015.08.015
121. Xu L, Begum S, Hearn JD, Hynes RO (2006) GPR56, an atypical G protein-coupled receptor, binds tissue transglutaminase, TG2, and inhibits melanoma tumor growth and metastasis. Proc Natl Acad Sci U S A 103(24):9023–9028. doi:10.1073/pnas.0602681103, PubMed PMID: 16757564, PubMed Central PMCID: PMC1474142
122. Luo R, Jeong SJ, Jin Z, Strokes N, Li S, Piao X (2011) G protein-coupled receptor 56 and collagen III, a receptor-ligand pair, regulates cortical development and lamination. Proc Natl Acad Sci U S A 108(31):12925–12930. doi:10.1073/pnas.1104821108, PubMed PMID: 21768377, PubMed Central PMCID: PMC3150909
123. Luo R, Jin Z, Deng Y, Strokes N, Piao X (2012) Disease-associated mutations prevent GPR56-collagen III interaction. PLoS One 7(1), e29818. doi:10.1371/journal.pone.0029818, PubMed PMID: 22238662, PubMed Central PMCID: PMC3251603
124. Yang L, Chen G, Mohanty S, Scott G, Fazal F, Rahman A et al (2011) GPR56 regulates VEGF production and angiogenesis during melanoma progression. Cancer Res 71 (16):5558–5568. doi:10.1158/0008-5472.CAN-10-4543, PubMed PMID: 21724588, PubMed Central PMCID: PMC3156271
125. Gupte J, Swaminath G, Danao J, Tian H, Li Y, Wu X (2012) Signaling property study of adhesion G-protein-coupled receptors. FEBS Lett 586(8):1214–1219. doi:10.1016/j.febslet.2012.03.014
126. Paavola KJ, Sidik H, Zuchero JB, Eckart M, Talbot WS (2014) Type IV collagen is an activating ligand for the adhesion G protein-coupled receptor GPR126. Sci Signal 7(338): ra76. doi:10.1126/scisignal.2005347, PubMed PMID: 25118328; PubMed Central PMCID: PMC4159047

127. Petersen SC, Luo R, Liebscher I, Giera S, Jeong SJ, Mogha A et al (2015) The adhesion GPCR GPR126 has distinct, domain-dependent functions in Schwann cell development mediated by interaction with laminin-211. Neuron 85(4):755–769. doi:10.1016/j.neuron.2014.12.057, PubMed PMID: 25695270, PubMed Central PMCID: PMC4335265
128. McMillan DR, White PC (2010) Studies on the very large G protein-coupled receptor: from initial discovery to determining its role in sensorineural deafness in higher animals. Adv Exp Med Biol 706:76–86
129. Sun JP, Li R, Ren HZ, Xu AT, Yu X, Xu ZG (2013) The very large G protein coupled receptor (Vlgr1) in hair cells. J Mol Neurosci 50(1):204–214. doi:10.1007/s12031-012-9911-5
130. Nikkila H, McMillan DR, Nunez BS, Pascoe L, Curnow KM, White PC (2000) Sequence similarities between a novel putative G protein-coupled receptor and Na+/Ca2+ exchangers define a cation binding domain. Mol Endocrinol 14(9):1351–1364. doi:10.1210/mend.14.9.0511
131. McGee J, Goodyear RJ, McMillan DR, Stauffer EA, Holt JR, Locke KG et al (2006) The very large G-protein-coupled receptor VLGR1: a component of the ankle link complex required for the normal development of auditory hair bundles. J Neurosci 26(24):6543–6553. doi:10.1523/JNEUROSCI.0693-06.2006, PubMed PMID: 16775142, PubMed Central PMCID: PMC2682555

7TM Domain Structure of Adhesion GPCRs

Saskia Nijmeijer, Steffen Wolf, Oliver P. Ernst, and Chris de Graaf

Graphical Abstract

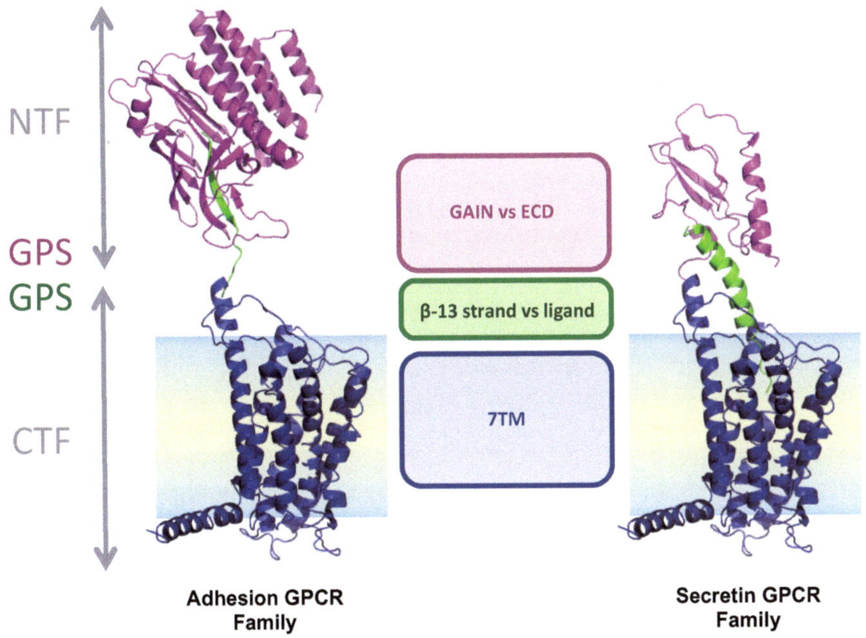

Schematic presentation of the overall adhesion G Protein-Coupled Receptor (aGPCR) structure and functional domains, covering an extracellular N-terminal fragment (NTF), a membrane-spanning C-terminal fragment (CTF) and a GPCR proteolysis site (GPS). (*Left side*) aGPCR

The original version of this chapter was revised. An erratum to this chapter can be found at DOI 10.1007/978-3-319-41523-9_18.

S. Nijmeijer • C. de Graaf (✉)
Department of Medicinal Chemistry, Amsterdam Institute for Molecules, Medicines and Systems, Vrije Universiteit Amsterdam, Amsterdam 1081HV, The Netherlands
e-mail: s.nijmeijer@vu.nl; c.de.graaf@vu.nl

T. Langenhan, T. Schöneberg (eds.), *Adhesion G Protein-coupled Receptors*, Handbook of Experimental Pharmacology 234, DOI 10.1007/978-3-319-41523-9_3

model constructed based on the seven-transmembrane (7TM) structure (*blue*) of secretin family glucagon receptor (GCGR) (PDB, 4L6R) [11] and the GPCR autoproteolysis inducing (GAIN) domain (*magenta*) structure of latrophilin 1 (PDB, 4DLQ) [9]. The β-13 strand residues are depicted in *green*. (*Right side*) The experimentally validated full-length secretin family GCGR structure combining structural and experimental information from the GCGR 7TM crystal structure (PDB, 4L6R) (*blue*), the GCGR extracellular domain (ECD) structure (PDB, 4ERS) (*magenta*) and the ECD structure of glucagon-like peptide-1 (GLP-1)-bound glucagon-like peptide-1 receptor (GLP-1R) (PDB, 3IOL) (*green*), complemented by site-directed mutagenesis, electron microscopy (EM), hydrogen–deuterium exchange (HDX) and cross-linking studies [11–13])

Contents

Abstract

Despite the recent breakthroughs in the elucidation of the three-dimensional structures of the seven transmembrane (7TM) domain of the G protein-coupled receptor (GPCR) superfamily, a corresponding structure of a member of the adhesion GPCR (aGPCR) family has not yet been solved. In this chapter, we give an overview of the current knowledge of the 7TM domain of aGPCRs by

S. Wolf
Department of Biophysics, CAS-MPG Partner Institute for Computational Biology, Key Laboratory of Computational Biology, Shanghai Institutes for Biological Sciences, Chinese Academy of Sciences, Shanghai, P.R. China

Institute of Physics, Albert Ludwigs University, Freiburg, Germany
e-mail: swolf@picb.ac.cn

O.P. Ernst
Department of Biochemistry, University of Toronto, 1 King's College Circle, Toronto, ON, Canada, M5S 1A8

Department of Molecular Genetics, University of Toronto, 1 King's College Circle, Toronto, ON, Canada, M5S 1A8
e-mail: oliver.ernst@utoronto.ca

comparative structure-based sequence similarity analyses between aGPCRs and GPCRs with known crystal structure. Of the GPCR superfamily, only the secretin family shares some sequence similarity with aGPCRs. This chapter will therefore emphasize on the comparison of these two GPCR families. Two 7TM domain structures of secretin family GPCRs are known that provide insight into the structure-function relationships of conserved sequence motifs that play important roles and are also present in most aGPCRs. This suggests that the 7TM domains of aGPCRs and secretin family GPCRs share a similar structural fold and that the conserved residues in both families may be involved in similar intermolecular interaction networks and facilitate similar conformational changes. Comparison of the residues that line the large peptide hormone binding pocket in the 7TM domain of secretin family GPCRs with corresponding residues in aGPCRs indicates that in the latter, the corresponding pocket in the 7TM domain is relatively hydrophobic and may be even larger. Improved knowledge on these conserved sequence motifs will help to understand the interactions of the aGPCR 7TM domain with ligands and gain insight into the activation mechanism of aGPCRs.

Keywords
GPCR sequence alignment • Adhesion GPCR residue nomenclature • Adhesion GPCR sequence-structure relationship • Druggability adhesion GPCRs

1 Introduction

1.1 Structural Topology of Adhesion GPCRs

aGPCRs are cell-surface proteins that belong to the large GPCR superfamily comprising the five main families, named according to the GRAFS classification system glutamate, rhodopsin, adhesion, frizzled/taste2, and secretin [1, 2]. Alternatively, GPCRs can be classified into families A, B, C and F (Table 1) [3]. A phylogenetic analysis of the seven-transmembrane (7TM) α helices of the human GPCR superfamily classified aGPCRs as a separate family [4] (the 'A' in the GRAFS or B2 in A–F classification). The GRAFS classification system clearly distinguishes aGPCRs from secretin-like GPCRs, and therefore this nomenclature

Table 1 Overview of GPCR family classifications, which allows easy comparison between the GRAFS and A–F classifications and unique receptors (unique orthologues) for which a crystal structure is available in the Protein Data Bank

Classification of GPCR superfamily			
General family name	GRAFS	A–F families	Unique receptors for which a crystal structure is available[a]
Glutamate-like GPCRs	G	C	2
Rhodopsin-like GPCRs	R	A	27
Adhesion GPCRs	A	B2	0
Frizzled/taste2 GPCRs	F	F	1
Secretin-like GPCRs	S	B1	2

[a]accessed 17 Jan 2016

will be used throughout this chapter. Note that at least one GPCR crystal structure is available for all subfamilies, except for the aGPCR and Taste2 receptor family [5].

The aGPCR family can be further divided into nine distinct subfamilies that all share typical aGPCR features [6]. aGPCRs have a twofold protein structure consisting of a large extracellular domain (ECD, ~320–5878 residues) and a membrane-spanning 7TM domain followed by an intracellular domain (ICD) (~300–600 residues) [6]. A unique feature of the aGPCR family is their autoproteolytical cleavage at the GPCR proteolysis site (GPS) [7, 8], which is embedded in the conserved GPCR autoproteolysis inducing (GAIN) domain [9, 10] that forms an important part of the ECD. This cleavage results in a split aGPCR in which an extracellular N-terminal fragment (NTF) and a membrane-spanning C-terminal fragment (CTF) that includes the ICD are non-covalently associated (Fig. 1).

It is still under debate whether NTF and CTF complement each others' function by working as a single entity and/or that both fragments possess separate functions

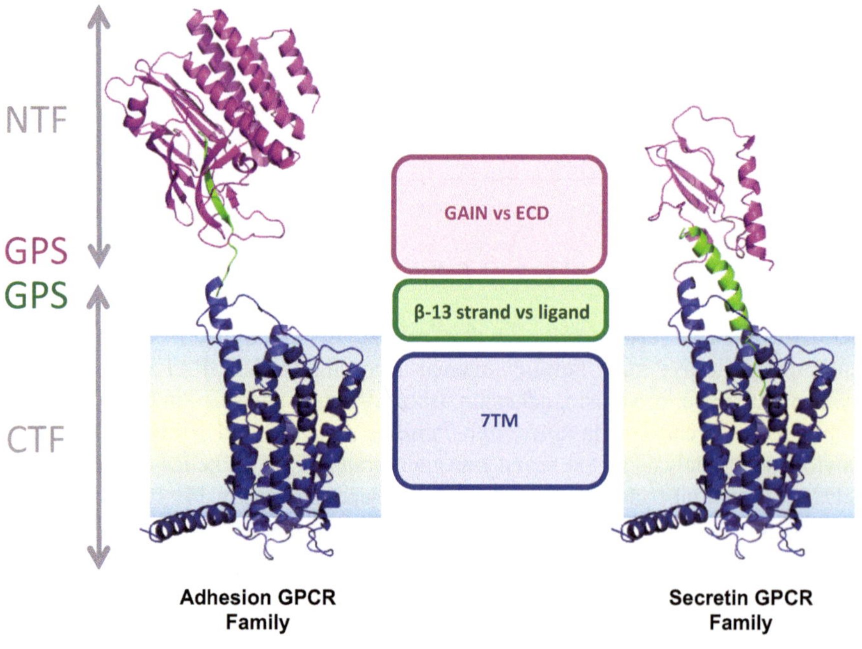

Fig. 1 Schematic presentation of the overall aGPCR structure and functional domains, covering an extracellular N-terminal fragment (NTF), a membrane-spanning C-terminal fragment (CTF) and a GPCR proteolysis site (GPS). (*Left side*) aGPCR model constructed based on the seven-transmembrane (7TM) structure (*blue*) of secretin family glucagon receptor (GCGR) (PDB, 4L6R) [11] and GPCR autoproteolysis inducing (GAIN) domain (*magenta*) structure of latrophilin 1 (PDB, 4DLQ) [9]. The β-13 strand residues are depicted in *green*. (*Right side*) The experimentally validated full-length secretin family GCGR structure combining structural and experimental information from the GCGR 7TM crystal structure (PDB, 4L6R) (*blue*), the GCGR extracellular domain (ECD) structure (PDB, 4ERS) (*magenta*) and the ECD structure of glucagon-like peptide-1 (GLP-1)-bound glucagon-like peptide-1 receptor (GLP-1R) (PDB, 3IOL) (*green*), complemented by site-directed mutagenesis, electron microscopy (EM), hydrogen–deuterium exchange (HDX) and cross-linking studies [11–13])

[6]. The NTF contains functional domains that often mediate cell-cell or cell-matrix adhesion [2] by binding to endogenous extracellular interaction partners. These interacting proteins can be part of the extracellular matrix or situated on the cell outer membrane of opposing cells (trans) as well as the same cell (cis) [6, 14]. Although binding interfaces have been investigated with purified NTF proteins, a detailed view of the interactions between NTF and its binding partners is often absent (see [15]). Only recently the binding site for a cell-adhesion protein, fibronectin leucine-rich repeat transmembrane protein 3 (FLRT3), to the latrophilin 3 aGPCR (LPHN3, ADGRL3) was described in detail where the interaction hotspot was located at the second and third blades of the olfactomedin domain propeller in the LPHN3 NTF [16]. Importantly, not all binding events at the NTF result in aGPCR signalling, and it remains to be determined whether ligand binding to the NTF is a prerequisite for aGPCR activation.

The only extracellular domain that is shared by almost all aGPCRs [excl. ADGRA1 (GPR123)] is the ~320 residues large GAIN domain, which is located immediately N-terminal of TM1 (see [15]). The importance of the GAIN domain structure in the autoproteolysis of aGPCRs has recently been highlighted for the latrophilin 1 receptor (ADGRL1) and ADGRB3 (BAI3) [9, 10]. Noteworthy, cleavage takes place at the L↓T/S/C site [17] within the GPS motif that encompasses the last five beta strands of the GAIN domain, leaving the last beta strand (β-13) of the GAIN domain still attached to the CTF [9, 18–20].

The CTF of aGPCRs resembles the classical 7TM structure of the rhodopsin- or secretin-like GPCR families [2, 17]. Increasing evidence shows that the CTF possesses signalling properties such as G-protein coupling [21–24], β-arrestin interaction [24–26] and receptor internalization [27] that clearly hints at conformational changes general to the GPCR superfamily. A more detailed look at the CTF shows that it can be divided into four parts: (1) the β-13 strand (also called the Stachel sequence), (2) the 7TM helices that cover the predominant part of the CTF, (3) the extracellular loops (ECLs) and (4) the intracellular loops (ICLs) and C-tail.

The protruding and hydrophobic β-13 strand/Stachel, which is the remaining part of the GAIN domain after cleavage, has been recently highlighted as tethered agonist sequence capable to activate aGPCRs (i.e. GPR56, GPR110, GPR126, GPR133, GPR114, GPR64) [18, 20, 28, 29] (see [30]). While this very recent work on aGPCR activation has been rapidly progressing, it remains to be revealed in what way the β-13 strand/Stachel interacts with amino acids within the 7TM helices and/or ECLs and how this leads to an active aGPCR conformation.

To date, no evidence is available on the role of ECLs and ICLs on aGPCR functioning. This is understandable, considering the limited information on aGPCR signal transduction for the majority of aGPCRs and hence very rare site-directed mutagenesis data. Potential protein-protein interaction sites/motifs have been predicted in aGPCR C-tail sequences, which led to the identification of non-canonical signalling partners (i.e. ELMO/Dock180 or dishevelled) for ADGRB1 (BAI1) [31, 32] and ADGRA3 (GPR125), respectively [33]. The 7TM region has been investigated in comparative sequence alignment studies, but experimental evidence further supporting the importance of specific residues and/or motifs is absent.

1.2 Sequence Similarity Between the 7TM Domains of aGPCR and Other GPCR Subfamilies

In earlier amino acid sequence alignment analyses of aGPCRs, a high sequence similarity in the TM helices (25–70 % amino acid identity) between members of the family was observed [17]. Previously reported aGPCR sequence alignments with secretin-like GPCRs showed a 25–30 % amino acid identity with aGPCRs [17], and sharing of several conserved residues [2, 17]. This is in line with earlier studies that not only showed sequence similarities but also a similar splice site setup [34]. Indeed, it was even suggested that secretin-like GPCRs are descendants from aGPCRs [34]. In particular aGPCRs ADGRG5 (GPR114) and ADGRG7 (GPR128) share the closest sequence similarity with the secretin family GPCRs glucagon receptor (GCGR or GLR) and corticotropin-releasing factor receptor 1 (CRFR1).

Figure 2 shows a NeighborNet analysis [35] of a MUSCLE [36] alignment of aGPCR sequences and 7TM domain sequences from crystallized GPCRs following the method of Wolf and Grünewald [37]. The NeighborNet-based network, a superposition of multiple possible phylogenetic trees, shows that aGPCRs indeed are most related to the GCGR and CRFR1 receptors. The recently solved crystal structures of the 7TM domains of GCGR [11, 40] and CRFR1 [41] therefore offer useful structural templates for the 7TM regions of aGPCRs as will be further demonstrated in this chapter. In contrast, the sequence similarities between aGPCRs and rhodopsin, glutamate and frizzled families, for which 7TM crystal structures have been solved [42–44], are relatively low (Fig. 2), although few structural features may also be conserved between aGPCRs and these GPCR families (Fig. 3).

To allow comparison between the residues at different positions in the TM helices of different GPCRs within and between different families, residues are numbered according to the *Ballesteros-Weinstein* numbering scheme [50] where the single most conserved residue in each TM helix is designated X.50 (Table 2). X is the TM helix number, and all other residues in that helix are numbered relative to this conserved position [45]. The *Ballesteros-Weinstein* numbering scheme has been initially defined for family A/rhodopsin family members (X.50a [50]), and similar residue numbering schemes are used to allow comparison between B1/secretin family (*Wootten*, defining reference position X.50b [51]), family C/glutamate family (*Pin*, X.50c [50]) or family F/frizzled family (*Wang*, X.50f [44]) GPCRs. Residues located in extracellular loop 2 are numbered based on their relative position to a conserved cysteine ($C^{45.50}$) that forms a disulphide bond with a cysteine in TM3 ($C^{3.25}/C^{3.29}$) in all GPCR families [51]. The *structural* alignment of the TM helices of the crystal structures of rhodopsin, secretin, glutamate and frizzled receptors allows the definition of a *structure-based* sequence alignment of these GPCR families (Table 2) that can be extended to aGPCRs based on the sequence alignment between the aGPCRs and homologous secretin-like receptors (Fig. 3).

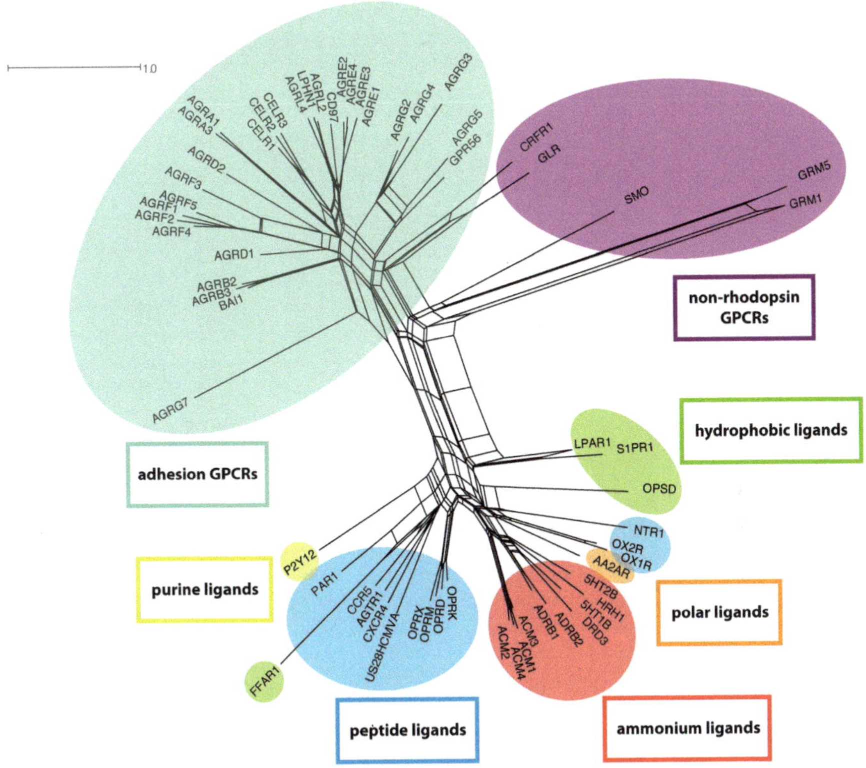

Fig. 2 NeighborNet analysis [35] of a MUSCLE [36] alignment of adhesion GPCR sequences and 7TM domain sequences from crystallized GPCRs following the method of Wolf and Grünewald [37]. The displayed network was calculated in SplitsTree [38] using maximum-likelihood distances [39]. Receptor abbrevations are given according to their Uniprot entry names. Full GPCR sequences were used to obtain the alignment, but only the 7TM domain sequences according to crystal structures were used for the analysis

Table 2 shows that the reference residues of the Wootten numbering scheme for family B1/secretin-like GPCRs are also the most conserved for five of the family B2/aGPCR TM helices ($S^{1.50b}$, $H^{2.50b}$, $N^{5.50b}$, $G^{6.50b}$, $G^{7.50b}$), and the remaining two residues in TM3–4 still display a high conservation within the aGPCR family ($E^{3.50b}$ 67 % and $W^{4.50b}$ 48 % within human aGPCRs). For example, for the glucagon receptor with its known crystal structure, the residues that are most conserved in secretin family GPCRs ($S152^{1.50b}$, $H177^{2.50b}$, $E244^{3.50b}$, $W271^{4.50b}$, $N317^{5.50b}$, $G359^{6.50b}$, $G391^{7.50b}$) are also present in most aGPCRs, for example, CD97 (ADGRE5: $S558^{1.50b}$, $H583^{2.50b}$, $E633^{3.50b}$, $Y666^{4.50b}$, $N707^{5.50b}$, $G750^{6.50b}$, $G779^{7.50b}$) or GPR56 ($S414^{1.50b}$, $H446^{2.50b}$, $E496^{3.50b}$, $W524^{4.50b}$, $N585^{5.50b}$, $G620^{6.50b}$, $G651^{7.50b}$). The highly conserved tryptophan residue in TM3 of aGPCRs (e.g. $W629^{3.46b}$ in CD97) and secretin-like receptors ($W^{3.46b}$) is the most conserved residue in TM3 of frizzled receptors ($W^{3.50f}$), while the conserved tryptophan residue in TM4 of aGPCRs (e.g. W623 in GPR56) and secretin-like

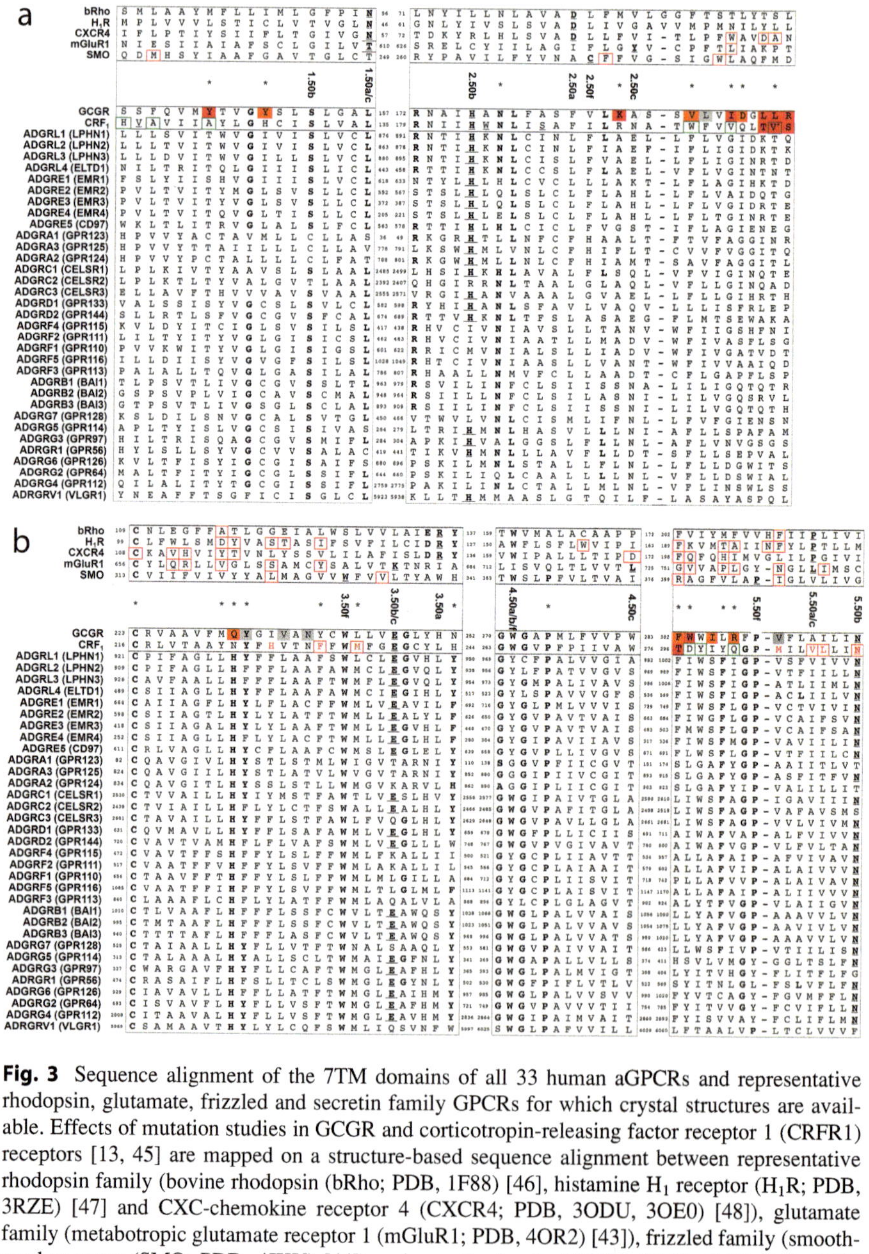

Fig. 3 Sequence alignment of the 7TM domains of all 33 human aGPCRs and representative rhodopsin, glutamate, frizzled and secretin family GPCRs for which crystal structures are available. Effects of mutation studies in GCGR and corticotropin-releasing factor receptor 1 (CRFR1) receptors [13, 45] are mapped on a structure-based sequence alignment between representative rhodopsin family (bovine rhodopsin (bRho; PDB, 1F88) [46], histamine H_1 receptor (H_1R; PDB, 3RZE) [47] and CXC-chemokine receptor 4 (CXCR4; PDB, 3ODU, 3OE0) [48]), glutamate family (metabotropic glutamate receptor 1 (mGluR1; PDB, 4OR2) [43]), frizzled family (smoothened receptor (SMO; PDB, 4JKV) [44]) and secretin family [GCGR (PDB, 4L6R [11], 5EE7 [40])] CRFR1 (PDB, 4K5Y) [41] GPCRs. Only specific parts of TM1–2 (panel a), TM3-5 (panel b), and TM6–7 (panel c) are shown, separated by *grey dashed lines*, and residue numbers of defining missing regions in the alignment are indicated. The alignment considers helix bulges and constrictions as described in [5]. Mutated residues that show four- to tenfold (*orange*) and >tenfold (*red*) changes of K_i/IC_{50} values for ligand binding (or ligand potency/EC_{50} value if no K_i/IC_{50} value has been reported) are marked as *full coloured box* (peptide ligands) and *open*

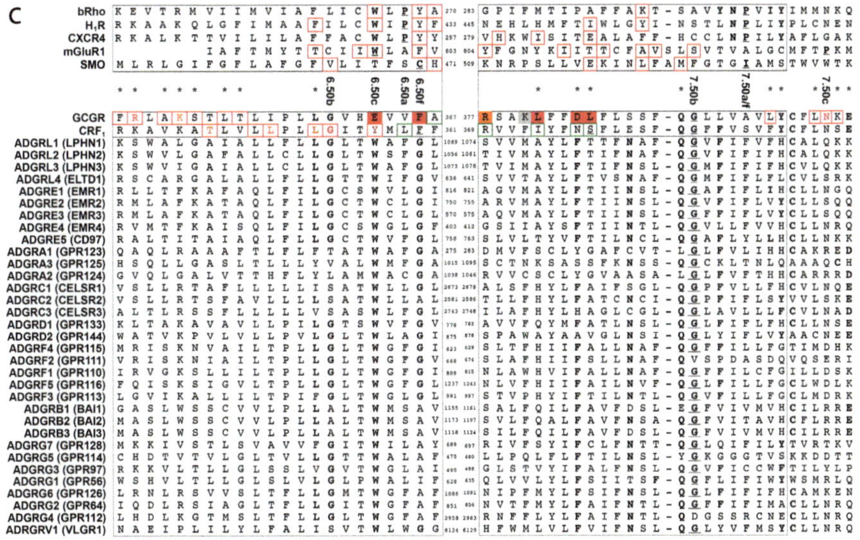

Fig. 3 (continued) coloured box (non-peptide ligands). Mutants that show receptor expression <30 % of wild type are marked *grey*. Receptor residues that covalently bind peptide ligands in photo-cross-linking or cysteine-trapping studies [13, 45, 49] are boxed *green*. Ligand contact residues in GCGR, CRFR1, bRho, H_1R, CXCR4, mGluR1 and SMO crystal structures are boxed *red*. The most conserved residues in TM1–7 of family A (X.50), family B (X.50b), family C (X.50c) and family F (X.50f) GPCRs are *underlined* and shown in *bold*. The positions of other residues described in the text are indicated with an *asterisk*

receptors ($W^{4.50b}$) is also defined as most conserved reference residue in family A/rhodopsin ($W^{4.50a}$) and family F/frizzled ($W^{4.50f}$) GPCRs. The most conserved proline residue in TM5 of frizzled ($P^{5.50f}$) is also present in most aGPCRs (e.g. $P699^{6.42b}$ in CD97) and secretin receptors ($P^{5.42b}$), while the most conserved tryptophan residue in TM6 of glutamate family GPCRs ($W^{6.50b}$) is also conserved in rhodopsin family GPCRs ($W^{6.48b}$) and aGPCRs (e.g. $W753^{6.53b}$ in CD97). The number of shared conserved residues between aGPCRs and other GPCR families reflects the network analysis in Fig. 2 and indicates that aGPCRs are most homologous to secretin-like receptors, but that also frizzled, rhodopsin and glutamate GPCRs share some local sequence similarity in specific TM helices.

Based on (1) the fact that aGPCRs share the highest overall sequence similarity to secretin-like receptors (Fig. 2) (2) the fact that aGPCRs and secretin-like receptors share almost all highly conserved residues defined as reference positions in secretin-like receptors (or other GPCR families, Table 2) and (3) the fact that aGPCRs and secretin-like receptors have historically been defined as the same GPCR family B, we propose to consistently use the family B/secretin-like numbering scheme [51, 53] for aGPCRs. Therefore, to allow systematic comparison of the residues at different positions in the TM helices of different GPCRs within and between different families, receptor residue numbers are annotated throughout this

Table 2 Comparison of Ballesteros-Weinstein reference positions X.50 [50] for representative family A/rhodopsin-like [bovine rhodopsin (bRho)], family B1/secretin-like [glucagon receptor (GCGR)], family C/glutamate [metabotropic glutamate receptor 1 (mGluR1)] and family F/frizzled [smoothened (SMO)] to representative family B2/aGPCRs CD97 and GPR56

Helix	Family B2 adhesion (CD97)	Family B2 adhesion (GPR56)	Family B1 secretin (GCGR)	Family A Rhodopsin (bRho)	Family C Glutamate (mGluR1)	Family F Frizzled (SMO)
TM1	S558$^{1.50b}$	S414$^{1.50b}$	S152$^{1.50b}$	G51$^{1.46a}$	G603$^{1.50c}$	T241$^{1.43f}$
	L562$^{1.54b}$	C418$^{1.54b}$	L156$^{1.54b}$	N55$^{1.50a}$	T607$^{1.54c}$	T245$^{1.47f}$
	C565$^{1.57b}$	T421$^{1.57b}$	A159$^{1.57b}$	T58$^{1.53a}$	V610$^{1.57b}$	T248$^{1.50f}$
TM2	H583$^{2.50b}$	H446$^{2.50b}$	H177$^{2.50b}$	L76$^{2.43a}$	C631$^{1.39c}$	L267$^{2.42f}$
	L590$^{2.57b}$	V453$^{2.57b}$	F184$^{2.57b}$	D83$^{2.50a}$	I638$^{1.46c}$	A272$^{2.49f}$
	F591$^{2.58b}$	F454$^{2.58b}$	V185$^{2.58b}$	L84$^{2.51a}$	F639$^{1.47c}$	C273$^{2.50f}$
	T595$^{2.62b}$	F454$^{2.58b}$	S189$^{2.62b}$	F88$^{2.55a}$	Y642$^{2.50c}$	V276$^{2.53f}$
TM3	C612$^{3.29b}$	C475$^{3.29b}$	C224$^{3.29b}$	C110$^{3.25a}$	C657$^{3.29c}$	C314$^{3.25f}$
	W629$^{3.46b}$	W492$^{3.46b}$	W241$^{3.46b}$	S127$^{3.42a}$	A674$^{3.46c}$	W323$^{3.50f}$
	E633$^{3.50b}$	E496$^{3.50b}$	E245$^{3.50b}$	L131$^{3.46a}$	K678$^{3.50c}$	A327$^{3.54f}$
	L637$^{3.54b}$	L500$^{3.54b}$	L249$^{3.54b}$	R135$^{3.50a}$	I354$^{3.54c}$	W331$^{3.58f}$
TM4	Y666$^{4.50b}$	W524$^{4.50b}$	W272$^{4.50b}$	W161$^{4.50a}$	I714$^{4.40c}$	W365$^{4.50f}$
	S670$^{4.60b}$	V534$^{4.60b}$	W282$^{4.60b}$	P171$^{4.60a}$	L724$^{4.50c}$	I375$^{4.60f}$
EL2	C684$^{45.50}$	C562$^{45.50}$	C294$^{45.50}$	C187$^{45.50}$	C746$^{45.50}$	C390$^{45.50}$
TM5	P699$^{5.42b}$	L577$^{5.42b}$	P310$^{5.42b}$	V211$^{5.45x46a}$	Y759$^{5.46c}$	P407$^{5.50f}$
	I703$^{5.46b}$	V581$^{5.46b}$	A314$^{5.46b}$	P215$^{5.50a}$	L763$^{5.50c}$	V411$^{5.54f}$
	N707$^{5.50b}$	N585$^{5.50b}$	N318$^{5.50b}$	I219$^{5.54a}$	C767$^{5.54c}$	G415$^{5.58f}$
TM6	G750$^{6.50b}$	G620$^{6.50b}$	G359$^{6.50b}$	L262$^{6.45a}$	C795$^{6.47c}$	V463$^{6.47f}$
	W753$^{6.53b}$	W623$^{6.53b}$	E362$^{6.53b}$	W265$^{6.48a}$	W798$^{6.50c}$	T466$^{6.47f}$
	F755$^{6.55b}$	L625$^{6.55b}$	V364$^{6.55b}$	P267$^{6.50a}$	A800$^{6.52c}$	S468$^{6.49f}$
	G756$^{6.56b}$	I626$^{6.56b}$	F365$^{6.56b}$	Y268$^{6.51a}$	F801$^{6.53c}$	C469$^{6.50f}$
TM7	G779$^{7.50b}$	G651$^{7.50b}$	G393$^{7.50b}$	A299$^{7.46a}$	V823$^{7.40c}$	F526$^{7.46f}$
	Y783$^{7.54b}$	F655$^{7.54b}$	A397$^{7.54b}$	P303$^{7.50a}$	L827$^{7.44c}$	I530$^{7.50f}$
	H786$^{7.57b}$	Y658$^{7.57b}$	Y400$^{7.57b}$	Y306$^{7.53a}$	C829$^{7.46x47c}$	S533$^{7.53f}$
	L789$^{7.60b}$	M661$^{7.60b}$	L403$^{7.60b}$	M309$^{7.56a}$	P833$^{7.50c}$	V536$^{7.56f}$

Receptor-specific Uniprot numbers are coloured *grey*, and their Ballesteros-Weinstein residue numbers appended by the corresponding GPCR family as small letter are indicated as superscript (e.g. S558$^{1.50b}$, serine residue 558 of CD97 located at Ballesteros-Weinstein position 1.50 in family B GPCRs; N55$^{1.50a}$, asparagine residue 55 in bovine rhodopsin at position 1.50 in family A GPCRs). In addition to Ballesteros-Weinstein reference positions, conserved residues C$^{3.25a/f}$/C$^{3.29b/c}$, C$^{45.50}$ and Y$^{7.53a/7.57b}$ are indicated. Conserved residues (>45 %) between GPCR families are marked *orange*, while conserved reference residues for a specific family that is not conserved in other families are marked *cyan*

chapter by their Uniprot numbers (for specific receptors) as well as their Ballesteros-Weinstein/Wootten residue number and secondary structure motif (as superscript), according to IUPHAR guidelines [55] and secretin family GPCR residue numbering [51, 54] guidelines, respectively.

1.3 Aims of This Chapter

Assuming that the tethered agonist sequence directly influences the 7TM structure, no other endogenous ligands have been described that bind to the 7TM domain of aGPCRs. This absence of suitable 7TM ligands affects the progress of aGPCR crystallization efforts, and hence we are still waiting for the first crystal structure of aGPCR proteins. Until then, the aGPCR field relies on structure-based homology models based on different GPCR superfamily members to give insight in the presence and accessibility of potential aGPCR 7TM binding cavities.

As indicated in Sect. 1.2 and described in more detail in the following paragraph, aGPCRs share sequence similarity with the secretin GPCR family, which justifies the analysis and extrapolation of sequence-structure relationships from the crystal structures for this subfamily [45], CRFR1 [41] and GCGR [11, 40], as well as the wealth of mutation data available for the 7TM domain of secretin family receptors, in particular the regions that share sequence similarity with aGPCRs.

We therefore aim to make aGPCR structure-based sequence alignments, based on our expertise in both rhodopsin and secretin family GPCR structures. Moreover, based on the similarity with secretin family GPCRs, we will identify important/conserved amino acids (motifs) in the 7TM domain of aGPCRs. In addition, the well-established Ballesteros-Weinstein numbering scheme [50] for amino acid residues within GPCR TM helices will be introduced for aGPCR family members.

2 Conserved Structure-Sequence Relationships Between Adhesion, Secretin and Rhodopsin Family GPCRs

The following section provides a detailed description of conserved structural motifs in the 7TM domains of aGPCRs and secretin family GPCRs and comparison to functionally and structurally important sequence motifs in similar regions in the 7TM domains of rhodopsin, glutamate and frizzled family GPCRs, based on structure-based sequence alignments of the crystal structures of secretin (GCGR and CRFR1), rhodopsin, glutamate and frizzled family GPCRs [5] (Fig. 3). This GPCR crystal structure-based aGPCR to secretin family alignment is in line with the previous sequence alignment studies between a selection of aGPCRs (CTFs) and the secretin-like human calcitonin receptor sequence, leading to the identification of potentially conserved residues/motifs: TM3 (WMLxE$^{3.50b}$G), TM4 (GW$^{4.50b}$GxP), ECL2 (C$^{45.50b}$WL), TM5 (GPVxxN$^{5.50b}$), TM6 (LLG$^{6.50b}$) and TM7 (QG$^{7.50b}$xF, C$^{7.54b}$, V$^{7.64b}$) [17]. In a more extensive amino acid sequence alignment of aGPCR 7TM regions, involving all 33 human aGPCRs, three residues were conserved in all aGPCRs (i.e. H$^{3.37b}$ and W$^{3.46b}$) in TM3 and P$^{4.42b}$ in TM4 (Table 2, Fig. 3) [2].

2.1 Transmission Switch: TM3, TM5, TM6 and TM7

$W^{6.53b}$ in TM6 is conserved in aGPCRs and structurally aligned with the highly conserved $W^{6.48b}$ in the rhodopsin family and $W^{6.50c}$ in glutamate family GPCRs (Table 2, Fig. 3). In rhodopsin family GPCRs, $W^{6.48b}$ acts as a rotameric toggle switch [56] that controls the structural transduction of ligand-induced receptor signalling between the extracellular 7TM ligand binding site and the intracellular G-protein recognition site. The GCGR crystal structure shows that the TM3–TM6 interface in secretin family GPCRs (one helical turn below $W^{6.48b}$) contains large hydrophobic residues $Y/F^{3.44b}$, $M/L^{3.47b}$ and $L/F^{6.49b}$ that make similar hydrophobic interactions as structurally aligned $I/V/L^{3.40a}$, $L/I^{3.43a}$ and $F^{6.44a}$ residues present in most rhodopsin family GPCRs (Fig. 4b). aGPCRs contain similar large apolar residues ($F/L^{3.44b}$, $M/L^{3.47b}$, $L/F^{6.49b}$) at these positions. The GCGR crystal structures show that this interface (defined as receptor signal transmission switch in rhodopsin family GPCRs) [42, 57] is further stabilized in secretin family GPCRs by close contact between the conserved $Y^{3.44b}$ and $G^{6.50b}$, as well as a secretin family-specific inter-helical hydrogen bond between the conserved $N^{5.50b}$ and the backbone of $L^{3.47b}$ at the TM3–TM5 interface (Fig. 4b). The same or homologous

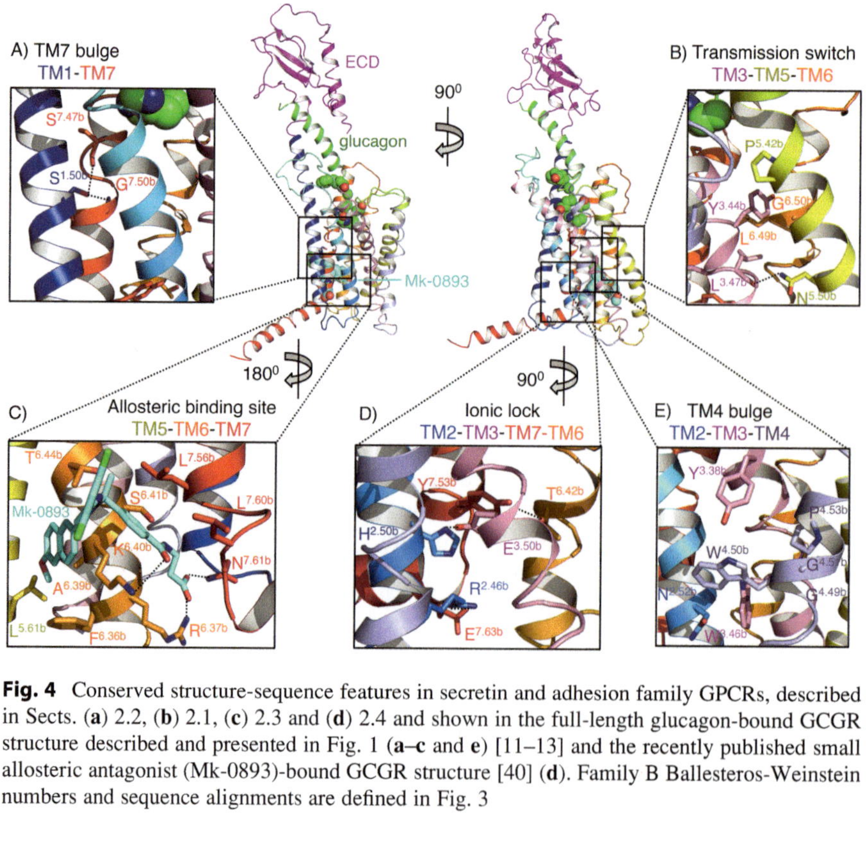

Fig. 4 Conserved structure-sequence features in secretin and adhesion family GPCRs, described in Sects. (**a**) 2.2, (**b**) 2.1, (**c**) 2.3 and (**d**) 2.4 and shown in the full-length glucagon-bound GCGR structure described and presented in Fig. 1 (**a–c** and **e**) [11–13] and the recently published small allosteric antagonist (Mk-0893)-bound GCGR structure [40] (**d**). Family B Ballesteros-Weinstein numbers and sequence alignments are defined in Fig. 3

residues involved in these contacts are also present in aGPCRs (e.g. $F625^{3.44b}$, $M628^{3.47b}$, $N707^{5.50b}$ and $G750^{6.50b}$ in CD97), suggesting that similar interaction networks are present in the 7TM of aGPCRs (Fig. 3). Moreover, the conserved $P^{5.42b}$ in secretin and adhesion family members (position 5.46a in rhodopsin family) and $P^{5.50a}$ in rhodopsin-like GPCRs (position 5.46b in secretin family) stabilize a similar bulge in TM5 that can mediate this conformational switch.

2.2 TM7 Bulge: TM1 and TM7

A serine residue ($S^{1.50b}$) in TM helix 1 (TM1) is conserved in most aGPCRs, which is also conserved in secretin family GPCRs (e.g. $S558^{1.50b}$ in CD97, Table 2, Fig. 3). In the crystal structures of GCGR and CRFR1, $S^{1.50b}$ interacts with the backbone of TM7 at $S^{7.47b}$, $G^{7.50b}$ and $F/L^{7.51b}$, thereby stabilizing the bulge in the TM7 helix of secretin family GPCRs [11] (Fig. 4a). In rhodopsin family GPCRs, the kink in TM7 is located one helical turn lower by a conserved $P^{7.50a}$ residue (aligned with position 7.54b in aGPCRs). The importance of $S^{1.50b}$ and $S^{7.47b}$ is consistent with mutation studies of secretin-like GPCRs showing that mutation of these residues alters receptor signalling [51]. $S^{1.50b}$ likely plays a similar structural role in aGPCRs by interacting in a similar fashion with the backbone of the small polar residue at position 7.47b (A/S/T) and the conserved $G^{7.50b}$ (e.g. $G779^{7.50b}$ in CD97). Most aGPCRs furthermore contain a medium-sized polar residue (N/E/D/H/Q) at position 2.61b (aligned with position 2.54 in rhodopsin family GPCRs) that may participate in this polar interaction network between TM1 and TM7 (Fig. 3). The only four aGPCRs that do not contain a serine residue at position 1.50b contain alternative small polar residues that can mediate similar H-bond networks, namely, $C^{1.50b}$ [GPR123 (ADGRA1), GPR124 (ADGRA2), GPR125 (ADGRA3)] and $T^{1.50b}$ [CELSR2 (ADGRC2)]. Position 1.50b is structurally aligned to position 1.46, which is a Gly residue in most family A GPCRs [58]. $G^{1.46a}$ is located one helical turn before the conserved residue $N^{1.50a}$ that controls rhodopsin family GPCR signalling by forming an H-bond network with $N^{7.49a}$ in TM7. $N^{7.49a}$ is structurally aligned with position 7.53b in aGPCRs (Fig. 4c) located one helical turn lower than the conserved $G^{7.50b}$ residue.

2.3 Ionic Lock: TM2, TM3, TM6, TM7 and Helix 8

At the intracellular side of the 7TM domain, $H^{2.50b}$ and $E^{3.50b}$ are conserved in secretin family GPCRs and present in most aGPCRs (e.g. $H583^{2.50b}$ and $E633^{3.50b}$ and CD97, Table 2, Fig. 3). These residues form an H-bond network within GCGR and CRFR1 crystal structures [11, 41] (Fig. 4c). Similar polar interaction networks are present in distinct, but closely located regions in rhodopsin family GPCRs, including (1) a water-mediated, sodium-stabilized H-bond interaction between $D^{2.50a}$ and $N^{7.49a}$ [59], structurally aligned with hydrophobic residues $L/F^{2.57b}$ and $I/V^{7.53b}$ that form a hydrophobic interface between TM2 and TM7 in secretin family GPCRs and (2) an ionic H-bond interaction network between $R^{3.50a}$ and $D/E^{6.30a}$, aligned with variable residues at positions 3.54b and 6.35b that do not form interactions in GCGR and CRFR1 crystal structures. $Y^{3.55b}$ in TM3, a part of a conserved $E^{3.50b}XXXXY^{3.55b}$ motif in aGPCRs, which is present in 21 out of 33 human aGPCRs [2], is however aligned with $Y^{3.51a}$ which is part of a conserved $D/ER^{3.50a}Y^{3.51a}$ motif in rhodopsin family GPCRs [60] (Fig. 3). Glutamate-like GPCRs contain a conserved charged lysine residue ($K^{3.50c}$) in TM3 that is structurally aligned with the charged $E^{3.50b}$ residue in secretin family GPCRs and most aGPCRs and forms an ionic/H-bond interaction network with a conserved glutamate residue in TM6 ($E^{6.35c}$) in mGluR1 [43] and mGluR5 crystal structures [61], which resembles the intracellular ionic locks observed in rhodopsin ($R^{3.50a}$ and $D/E^{6.30a}$) and secretin ($E^{3.50b}$ and $H^{2.50b}$, present in most aGPCRs) family crystal structures. In the secretin family GCGR crystal structure, $Y400^{7.57b}$ forms an H-bond with $T351^{6.42b}$ and $E245^{3.50b}$ (Fig. 4c) in a conformation that rhodopsin family GPCRs is linked to activation and interaction with the G protein via the structurally aligned $Y^{7.53a}$ residue at the intracellular end of TM7 [62–64]. Seven aGPCRs contain the same Tyr residue at position 7.57b, while 16 other aGPCRs contain a homologous Phe (2) or His (14) residue at this position (Fig. 3). For rhodopsin family GPCRs, the role of $Y^{7.53a}$ became evident from mutagenesis studies [65, 66] and the crystal structure of the active state rhodopsin in complex with a peptide derived from the C-terminus of the Galpha subunit [64]. In the GCGR crystal structure [11] $E406^{7.63b}$ located at the start of helix 8 forms an H-bond network with $R173^{2.46b}$ (Fig. 4c), a residue that is conserved in secretin family GPCRs and present in about half of the aGPCRs. $R173^{2.46b}A$ is one of the eleven thermostabilizing mutations in another GCGR crystal structure [40]. Seven human aGPCRs contain a positively charged residue at position 2.46b and a negatively charged residue at the position aligned with $E406^{7.63b}$ in GCGR (i.e. five positions from the conserved $C^{7.58b}$ residue), namely, ADGRB1–3 (BAI1–3), ADGRD1–2 (GPR133, GPR144) and ADGRA1–2 (GPR123, GPR124), suggesting that a similar H-bond network may be present in these aGPCRs.

2.4 TM4 Bulge: TM2, TM3 and TM4

TM4 of aGPCRs and secretin family GPCRs contains a conserved $GW/Y^{4.50b}GxP$ motif, at the same position as the highly conserved $W^{4.50a}$ in rhodopsin family GPCRs (Fig. 3) that is proposed to form a cholesterol binding site [67]. Despite structural differences in the $GW^{4.50b}GxP$ region in the GCGR and CRFR1 crystal structures, all structures show that $W^{4.50b}$ stabilizes the TM4 bulge by interacting with $Y^{3.38b}$, $W^{3.46b}$ and $N^{2.52b}$, and identical/homologous residues are conserved in most aGPCRs (e.g. $F484^{3.38b}$, $W492^{3.46b}$, $N446^{2.52b}$ in GPR56, Table 2, Fig. 4d). The interaction sites of $W^{4.50b}$ with TM2 and TM3 are located close to residues that are part of the transmission switch ($M/L^{3.47b}$, Fig. 4b) and ionic lock ($H^{2.50b}$, Fig. 4c) in secretin and adhesion family GPCRs.

2.5 Extracellular Loops

The extracellular loops 1 (ECL1) and 3 (ECL3) of aGPCRs are of similar short length as the ECL1 and ECL3 of glutamate family GPCRs [43], but significantly shorter than the ECL1 of rhodopsin family GPCRs [58], secretin-like receptors (including CRFR1 [41] and GCGR [11]) and frizzled [44] family GPCRs (Fig. 3). All GPCR families contain a disulphide bond between a cysteine residue in extracellular loop 2 ($C^{45.50}$) and a cysteine residue in TM3 ($C^{3.29b}$ in aGPCRs and secretin family GPCRs (Table 2), $C^{3.25a}$ in rhodopsin, $C^{3.29c}$ in glutamate and $C^{3.25f}$ in frizzled family GPCRs) [53]. It should be noted however that several GPCRs do not contain such disulphide bond [53], including the crystallized S1P1 and LPA1 receptors which form an internal disulphide bridge in ECL2 [68, 69]. The length of the ECL2 region upstream from $C^{45.50}$ (i.e. from TM4 to the disulphide bridge) is at least two residues longer in aGPCRs, and also the downstream ECL2 length (i.e. from the disulphide bridge to TM5) is variable between secretin family and adhesion family members. Especially ADRGR1–7 has significantly longer (upstream in most cases, also downstream) ECL2 length than the GCGR and CRFR1 crystal structure templates (Fig. 4). Nevertheless $W^{45.51}$, located one position downstream from $C^{45.50}$, is conserved between secretin family and adhesion family GPCRs, and this residue plays a role in peptide ligand binding (and/or structural integrity of ECL2) in several secretin family GPCRs [11, 45]. As Fig. 3 shows, we can indeed observe that the ECL2 sequence length of aGPCRs is quite comparable to the one of secretin receptors and contains a cysteine residue at a position comparable to CRFR1 and GCGR. As this cysteine is forming a disulphide bond to $C^{3.50a}$ and $C^{3.50b}$, it introduces an important structural constraint for the folding of ECL2. However, several members of the family exhibit extended lengths of ECL2, while still containing only a single cysteine.

3 Druggability of aGPCRs

Rational design of aGPCR ligands that interact with a 7TM binding cavity relies on good receptor models. As proven in the last decades, rhodopsin family members have been successfully targeted by a variety of molecules binding to either orthosteric or allosteric binding sites in the 7TM bundle. Likewise, glutamate GPCRs that possess a large extracellular Venus flytrap domain can be allosterically targeted in their 7TM regions. In previous years, aGPCRs have been successfully targeted (at their extracellular domains) with aGPCR-specific antibodies. Interestingly, some of these antibodies modulate aGPCR functioning, resulting in therapeutic relevant outcomes: monoclonal antibodies against the NTF of ADGRE1 (EMR1) [70] and ADGRE5 (CD97) [71] are affective in eosinophilic disorders and inflammatory disease models, respectively. Thus far, only ADGRG3 (GPR97) was shown to be activated by a drug molecule (i.e. the corticosteroid beclomethasone dipropionate) [23]. Improved insights in aGPCR 7TM organization, structure and potential ligand interaction sites will form the basis to target aGPCRs 7TM regions in a pharmacological relevant manner. As presented in Sects. 1 and 2, secretin family GPCRs represent a template for the structural fold of aGPCRs and hence are considered as a useful structural template to characterize the 7TM binding pocket of aGPCRs as described below.

3.1 Comparison of 7TM Pocket of Secretin Family GPCRs and aGPCRs

Mutations of residues at positions $Y^{1.43b}$, $Y/H^{1.47b}$, $R/K^{2.60b}$, $V^{2.46b}$, $I/K^{2.67b}$, $D^{2.68b}$, $L/V^{2.71b}$, $Q^{3.37b}$, $W^{45.51}$, $F/Y/I/T^{5.35b}$, $W^{5.36b}$, $I^{5.38b}$, $R^{5.40b}$, $E^{6.53b}$, $F^{6.56b}$, $R^{7.35b}$, $L^{7.39b}$, $D^{7.42b}$ and $L^{7.43b}$ have been shown to affect peptide ligand binding and/or potency in multiple secretin family GPCRs, including GCGR [13, 72–74], CRFR1 [41, 75] GLP-1 [13, 76–80], GIP [81, 82], secretin [83, 84], $VPAC_1$ [85, 86] and PTH1 [87, 88] receptors (Fig. 3). These residues line the proposed peptide ligand binding site in the 7TM of GCGR and CRFR1 crystal structures [11, 41] and are expected to line the 7TM binding site of aGPCRs as well based on conservation of residues that determine the TM fold (e.g. $G^{1.46b}$, $G^{7.49b}$XGXP, $P^{5.46b}$, $G^{6.50b}$, $G^{7.50b}$). In most aGPCRs, large-/medium-sized residues at position 1.44b (Y/W/H/N/Q) may also be accessible from this binding site, as this residue is located next to a small residue (T/S) at position 1.43b. In the same way $E/N/D/Q^{2.61b}$ may be accessible from the 7TM pocket of aGPCRs as this residue is located next to the relatively small $A/S^{2.62b}$, while $L/V/I^{6.57b}$ located next to $G/A^{6.56b}$ may be accessible from the aGPCR 7TM pocket as well. Several of the polar residues that line the 7TM binding site of secretin family GPCRs are hydrophobic residues in most aGPCRs ($A/L^{2.60b}$, $I^{2.68b}$, $I/V/L/A^{5.40b}$, $W^{6.53b}$, $F^{7.42b}$), while there are no conserved hydrophobic binding pocket residues in secretin family GPCRs that are polar residues in most aGPCRs (Fig. 3). Assuming that aGPCRs and secretin-like receptors share a similar large and open 7TM binding site (Figs. 1–4), the putative 7TM pockets of aGPCRs contain less buried polar groups than in related secretin-like receptors. The implications of this with respect to druggability of the 7TM pocket of aGPCRs are discussed in Sect. 3.2.

3.2 Druggability 7TM Pocket of aGPCRs

Despite the lack of sequence conservation, comparison of secretin family members CRFR1 and GCGR structures with those of rhodopsin, glutamate and frizzled family GPCR shows that the orientations and positions of TM helices are conserved [11, 40, 41]. This common fold is stabilized by similar contacts between TM helices in both families, but involves distinct patterns of conserved residues in rhodopsin family [42] and secretin family [11] GPCRs. The distances between the extracellular ends of TM2 and TM7 and TM3 and TM7 of CRFR1 (PDB, 4K5Y) and GCGR (PDB, 4L6R, 5EE7) are however (among) the largest observed in GPCR structures. Consequently, the orthosteric 7TM pockets of these two receptors are wider and deeper than those of any rhodopsin family member [11]. In addition, several residues located deep in the orthosteric pocket between TM1 (1.47b) and TM2 (2.60b) play an important role in peptide ligand binding in secretin family GPCRs, while the corresponding structurally aligned residues (positions 1.43 and 2.53) are *not* involved in ligand binding in rhodopsin family receptors. Figure 5 shows a comparison of the 7TM pocket of the available secretin family GPCR crystal structures and two representative rhodopsin family GPCRs, CXCR4 and H_1R. The CXCR4 structure is the only GPCR crystal structure in which a (non-peptide) ligand (IT1t) binds only in the subpocket between TM1–3 and TM7 (PDB, 3ODU) [48] that is also primarily targeted by the chemokine ligand vMIP-II (PDB, 4RWS) [89]. The peptide ligand CVX15 targets an overlapping but distinct binding site region in CXCR4 located between TM3 and TM7 (PDB, 3OE0) [48], which is occupied by non-peptide ligands in all other rhodopsin family GPCR crystal structures [42, 90] and is also targeted in the ligand-bound 7TM crystal structures of glutamate family GPCRs (mGluR1, mGLuR5) [43, 61, 91] and frizzled family GPCRs (SMO) (Fig. 3) [44, 92]. The H_1R ligand doxepin binds in a pocket that is located closest to the cytoplasm of all the ligand binding sites observed in rhodopsin family GPCR crystal structures (PDB, 3RZE) [47]. The crystal structure of CRFR1 revealed an unexpected small-molecule binding pocket located in the cytoplasmic half of the receptor, more than 7 Å further down than the doxepin binding site in the rhodopsin family GPCR H_1R (Fig. 5). CP-376395 binds in this druggable site defined by residues of TM3, TM5 and TM6 showing a combination of hydrophobic and hydrophilic features compatible with drug-like small organic molecules. In this region, the sequence identity in secretin family GPCRs is remarkably high (Fig. 3). Of the 14 residues directly interacting with CP-376395, 7 are identical in CRFR1 and GCGR, and 8 are conserved/similar among aGPCRs. Among them is $N^{5.50b}$, which forms an essential hydrogen bond with the ligand, while CRFR1 and most aGPCRs share homologous residues at positions $Y/F^{3.44b}$, $M^{3.47b}$, $V/I^{5.46b}$, $L/V/I^{5.47b}$, $L^{6.45b}$, $L^{6.46b}$, $L^{6.49b}$ and $G^{6.50b}$ which provide hydrophobic interactions with the antagonist. A second recently solved crystal structure of GCGR revealed another unexpected allosteric small-molecule binding pocket in the secretin family of GPCRs at an interface between the cytoplasmic end of TM5, TM6, TM7 and the membrane bilayer (Figs. 4d and 5) [40]. In this GCGR crystal structure (PDB, 5EE7), the small-molecule allosteric antagonist Mk-0893 forms H-bonds with the side chains of residues $R346^{6.37b}$, $K349^{6.40b}$, $S350^{6.41b}$ and $N404^{7.61b}$ and makes apolar interactions with $L329^{5.61b}$, $F345^{6.36b}$, $A348^{6.39b}$, $K349^{6.40b}$, $L352^{6.43b}$, $T353^{6.44b}$, $L399^{7.56b}$, $L403^{7.60b}$ and four co-crystallized oleic acid molecules. Mk-0893 and chemically similar GCGR antagonists (large, hydrophobic, negatively charged) are proposed to adopt similar binding modes and prevent glucagon-induced activation of GCGR by restriction of the outward helical movement of TM6 required for G-protein coupling [40]. Most of the residues comprising this shallow pocket at the membrane interface at the outside of GCGR are not conserved among aGPCRs (Fig. 3). Several aGPCRs share $N^{7.61b}$ that forms an H-bond with the carboxylic acid moiety of Mk-0893 in GCGR, and few aGPCRs contain $K^{6.40b}$ that forms an H-bond with the amide carbonyl and

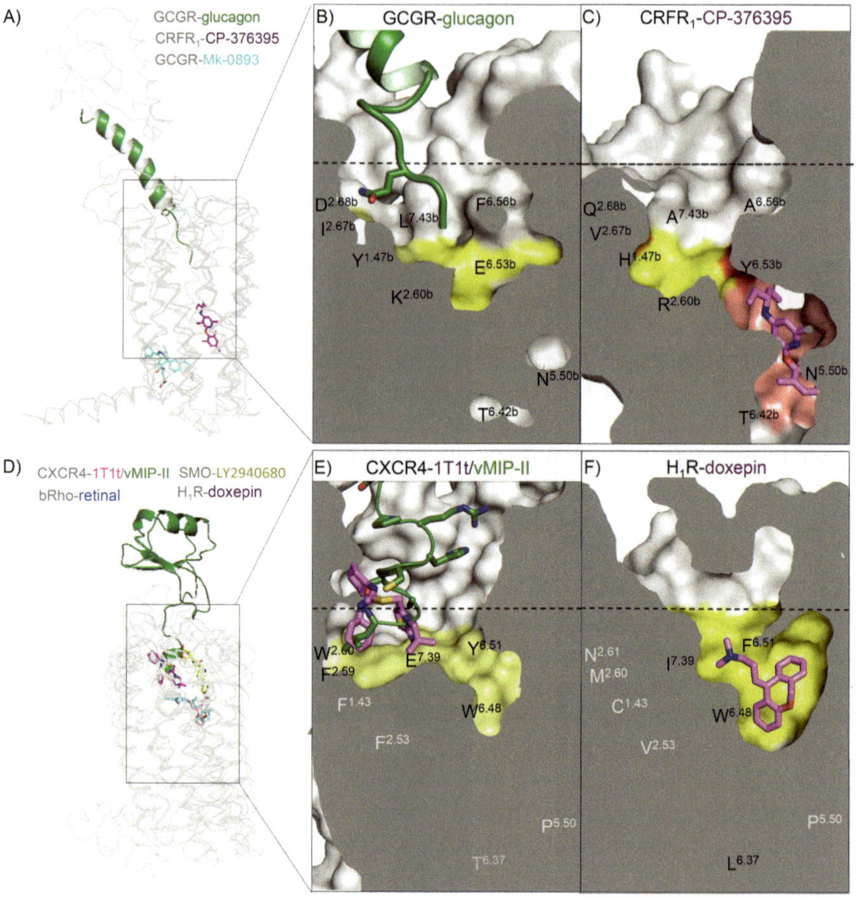

Fig. 5 Structural alignments of (**a**) secretin family (GCGR [11, 40] and CRFR1 [41]) and (**d**) rhodopsin (bRho [46], H_1R [47], CXCR4 [48, 88]), frizzled (SMO [44]) and glutamate family (mGluR1 [43]) crystal structures, and comparison of druggable binding sites of (**a**) glucagon receptor (GCGR; PDB, 4L6R), (**b**) corticotropin-releasing factor receptor 1 (CRFR1; PDB, 4K5Y), (**e**) CXC-chemokine receptor 4 (CXCR4; PDB, 3ODU, 4RWS) and (**f**) histamine H_1 receptor (H_1R; PDB, 3RZE). GCGR and CRFR1 are secretin family GPCRs that share sequence similarity and structural fold with aGPCRs (see Fig. 3 for corresponding sequence alignment). The surfaces of binding sites that are considered druggable are coloured *yellow* (orthosteric pocket) and salmon (CP-376395 binding pocket). Of nine structurally aligned residues, the residues that line the binding pocket shown in the figures are labelled *black*; the residues that are not part of the depicted pockets are labelled *light grey*. The approximate position of the extracellular membrane boundary is shown as a *dashed black line*. The binding mode of glucagon is derived from an experimentally validated full-length secretin family GCGR structure combining structural and experimental information from the GCGR 7TMD crystal structure (PDB, 4L6R), the GCGR ECD structure (PDB, 4ERS) and the ECD structure of GLP-1-bound GLP-1R (PDB, 3IOL), complemented by site-directed mutagenesis, electron microscopy, HDX and cross-linking studies [11–13]. Carbon atoms of peptide ligands (glucagon in the full-length GCGR model and vMIP-II in the CXCR4 crystal structure) and non-peptide ligands (IT1t in the CXCR4 crystal structure and doxepin in the H_1R crystal structure) are coloured *green* and *magenta*, respectively

makes apolar contacts with the naphthalene and phenylacetamide moieties of Mk-0893. None of the aGPCRs however combines these residues with a positively charged (R/K) and a small polar residue (S/T/N) at positions 6.37b and 6.41b (R346$^{6.37b}$ and S350$^{6.41b}$ in GCGR) that is required to allow the negatively charged beta-alanine group of Mk-0893 to target the intracellular pocket of GCGR. GPR110, GPR111 and GPR115, for example, share R$^{6.37b}$ and K$^{6.40b}$ in combination with S/N$^{6.41b}$ but contain a negatively charged D/E residue aligned with N404$^{7.61b}$ of GCGR. Although it is unlikely that aGPCRs can be targeted via the same ligand binding mode observed in the Mk-0893-bound GCGR crystal structure, similar hydrophobic, charged allosteric binding sites at extrahelical interface with the membrane may be accessible in aGPCRs as well. The buried orthosteric site of H_1R represents an excellent example of a druggable pocket, whereas the open binding region of CXCR4 is more challenging from a drug design perspective (Fig. 5) [93]. The orthosteric sites of both GCGR and CRFR1 are similar to CXCR4 and open and mostly occupied by bulk-like solvent, with only a single druggable site at the bottom of the pocket. This hotspot in the 7TM pocket of secretin family GPCRs is lined with residues that play an important role in peptide ligand binding, representing a mix of polar (e.g. Y$^{1.47b}$, R/K$^{2.60b}$, E/Y$^{6.53b}$, D/N$^{7.42b}$) and apolar (I/V$^{2.67b}$, F$^{6.56b}$, L$^{7.43b}$) residues (Fig. 4). In contrast, aGPCRs contain only hydrophobic (I/L/V$^{1.47b}$, I/L$^{2.60b}$, W$^{6.53b}$, F$^{7.42b}$) or smaller (G/A$^{2.68b}$, T/A$^{6.56b}$, T/A$^{7.43b}$) residues at the positions surrounding these hotspots and lack other conserved polar interaction sites in the 7TM pocket, suggesting that this binding pocket is even less druggable than in secretin family GPCRs.

Acknowledgements This work was supported by the Netherlands Organization for Scientific Research (NWO VENI grant 722.014.011 to SN), the Netherlands eScience Center (NLeSC)/NWO (Enabling Technologies project, 3D-e-Chem, grant 027.014.201 to CdG), the National Natural Science Foundation of China (NSFC, Research Fund for International Young Scientists, grant No. 31250110070 to SW) and the Canada Excellence Research Chair program (OPE). OPE holds the Anne and Max Tanenbaum Chair in Neuroscience at the University of Toronto. SN and CdG participate in the European Cooperation in Science and Technology Action CM1207 [GPCR-Ligand Interactions, Structures, and Transmembrane Signalling: A European Research Network (GLISTEN)] and the GPCR Consortium (gpcrconsortium.org).

References

1. Fredriksson R, Lagerström MC, Höglund PJ, Schiöth HB (2002) Novel human G protein-coupled receptors with long N-terminals containing GPS domains and Ser/Thr-rich regions. FEBS Lett 531:407–414
2. Bjarnadóttir TK, Fredriksson R, Schiöth HB (2007) The adhesion GPCRs: a unique family of G protein-coupled receptors with important roles in both central and peripheral tissues. Cell Mol Life Sci 64:2104–2119. doi:10.1007/s00018-007-7067-1
3. Kolakowski LF (1994) GCRDb: a G-protein-coupled receptor database. Recept Channels 2:1–7
4. Fredriksson R, Lagerström MC, Lundin L-G, Schiöth HB (2003) The G-protein-coupled receptors in the human genome form five main families. Phylogenetic analysis, paralogon groups, and fingerprints. Mol Pharmacol 63:1256–1272. doi:10.1124/mol.63.6.1256

5. Isberg V, de Graaf C, Bortolato A, Cherezov V, Katritch V, Marshall FH et al (2015) Generic GPCR residue numbers – aligning topology maps while minding the gaps. Trends Pharmacol Sci 36:22–31. doi:10.1016/j.tips.2014.11.001
6. Hamann J, Aust G, Araç D, Engel FB, Formstone C, Fredriksson R et al (2015) International Union of Basic and Clinical Pharmacology. XCIV. Adhesion G protein-coupled receptors. Pharmacol Rev 67:338–367. doi:10.1124/pr.114.009647
7. Krasnoperov VG, Bittner MA, Beavis R, Kuang Y, Salnikow KV, Chepurny OG et al (1997) α-Latrotoxin stimulates exocytosis by the interaction with a neuronal G-protein-coupled receptor. Neuron 18:925–937. doi:10.1016/S0896-6273(00)80332-3
8. Lin H-H, Chang G-W, Davies JQ, Stacey M, Harris J, Gordon S (2004) Autocatalytic cleavage of the EMR2 receptor occurs at a conserved G protein-coupled receptor proteolytic site motif. J Biol Chem 279:31823–31832. doi:10.1074/jbc.M402974200
9. Araç D, Boucard AA, Bolliger MF, Nguyen J, Soltis SM, Südhof TC et al (2012) A novel evolutionarily conserved domain of cell-adhesion GPCRs mediates autoproteolysis. EMBO J 31:1364–1378. doi:10.1038/emboj.2012.26
10. Prömel S, Langenhan T, Araç D (2013) Matching structure with function: the GAIN domain of Adhesion-GPCR and PKD1-like proteins. Trends Pharmacol Sci 34:470–478. doi:10.1016/j.tips.2013.06.002
11. Siu FY, He M, de Graaf C, Han GW, Yang D, Zhang Z et al (2013) Structure of the human glucagon class B G-protein-coupled receptor. Nature 499:444–449. doi:10.1038/nature12393
12. Yang L, Yang D, de Graaf C, Moeller A, West GM, Dharmarajan V et al (2015) Conformational states of the full-length glucagon receptor. Nat Commun 6:7859. doi:10.1038/ncomms8859
13. Yang D, de Graaf C, Yang L, Song G, Dai A, Cai X, et al (2016) Structural determinants of binding the seven-transmembrane domain of the glucagon-like peptide-1 receptor. J Biol Chem 291:12991–13004. doi:10.1074/jbc.M116.721977
14. Langenhan T, Aust G, Hamann J (2013) Sticky signaling–adhesion class G protein-coupled receptors take the stage. Sci Signal 6:re3. doi: 10.1126/scisignal.2003825
15. Araç D, Sträter N, Seiradake E (2016) Understanding the structural basis of adhesion GPCR functions. In: Langenhan T, Schöneberg T (eds) Adhesion G protein-coupled receptors: molecular, physiological and pharmacological principles in health and disease. Springer, Heidelberg
16. Lu YC, Nazarko OV, Sando R, Salzman GS, Südhof TC, Araç D (2015) Structural basis of latrophilin-FLRT-UNC5 interaction in cell adhesion. Structure. doi:10.1016/j.str.2015.06.024
17. Stacey M, Lin H-H, Gordon S, McKnight AJ (2000) LNB-TM7, a group of seven-transmembrane proteins related to family-B G-protein-coupled receptors. Trends Biochem Sci 25:284–289. doi:10.1016/S0968-0004(00)01583-8
18. Stoveken HM, Hajduczok AG, Xu L, Tall GG (2015) Adhesion G protein-coupled receptors are activated by exposure of a cryptic tethered agonist. Proc Natl Acad Sci U S A. doi:10.1073/pnas.1421785112
19. Liebscher I, Monk KR, Schöneberg T (2015) How to wake a giant. Oncotarget 6:23038–23039. doi: 10.18632/oncotarget.5112
20. Liebscher I, Schön J, Petersen SC, Fischer L, Auerbach N, Demberg LM et al (2014) A tethered agonist within the ectodomain activates the adhesion G protein-coupled receptors GPR126 and GPR133. Cell Rep 9:2018–2026. doi:10.1016/j.celrep.2014.11.036
21. Paavola KJ, Sidik H, Zuchero JB, Eckart M, Talbot WS (2014) Type IV collagen is an activating ligand for the adhesion G protein-coupled receptor GPR126. Sci Signal 7:ra76. doi:10.1126/scisignal.2005347
22. Bohnekamp J, Schöneberg T (2011) Cell adhesion receptor GPR133 couples to Gs protein. J Biol Chem 286:41912–41916. doi:10.1074/jbc.C111.265934
23. Gupte J, Swaminath G, Danao J, Tian H, Li Y, Wu X (2012) Signaling property study of adhesion G-protein-coupled receptors. FEBS Lett 586:1214–1219. doi:10.1016/j.febslet.2012.03.014

24. Stephenson JR, Paavola KJ, Schaefer SA, Kaur B, Van Meir EG, Hall RA (2013) Brain-specific angiogenesis inhibitor-1 signaling, regulation, and enrichment in the postsynaptic density. J Biol Chem 288:22248–22256. doi:10.1074/jbc.M113.489757
25. Paavola KJ, Stephenson JR, Ritter SL, Alter SP, Hall RA (2011) The N terminus of the adhesion G protein-coupled receptor GPR56 controls receptor signaling activity. J Biol Chem 286:28914–28921. doi:10.1074/jbc.M111.247973
26. Southern C, Cook JM, Neetoo-Isseljee Z, Taylor DL, Kettleborough CA, Merritt A et al (2013) Screening β-arrestin recruitment for the identification of natural ligands for orphan G-protein-coupled receptors. J Biomol Screen 18:599–609. doi:10.1177/1087057113475480
27. Yang L, Yang L, Friedland S, Friedland S, Corson N, Corson N et al (2014) GPR56 inhibits melanoma growth by internalizing and degrading its ligand TG2. Cancer Res 74:1022–1031. doi:10.1158/0008-5472.CAN-13-1268
28. Wilde C, Fischer L, Lede V, Kirchberger J, Rothemund S, Schöneberg T et al (2015) The constitutive activity of the adhesion GPCR GPR114/ADGRG5 is mediated by its tethered agonist. FASEB J. doi:10.1096/fj.15-276220
29. Demberg LM, Rothemund S, Schöneberg T, Liebscher I (2015) Identification of the tethered peptide agonist of the adhesion G protein-coupled receptor GPR64/ADGRG2. Biochem Biophys Res Commun 464:743–747. doi:10.1016/j.bbrc.2015.07.020
30. Liebscher I, Schöneberg T (2016) Tethered agonism: a common activation mechanism of adhesion GPCRs. In: Langenhan T, Schöneberg T (eds) Adhesion G protein-coupled receptors: molecular, physiological and pharmacological principles in health and disease. Springer, Heidelberg
31. Hamoud N, Tran V, Croteau L-P, Kania A, Côté J-F (2014) G-protein coupled receptor BAI3 promotes myoblast fusion in vertebrates. Proc Natl Acad Sci U S A 111:3745–3750. doi:10.1073/pnas.1313886111
32. Duman JG, Tzeng CP, Tu Y-K, Munjal T, Schwechter B, Ho TS-Y et al (2013) The adhesion-GPCR BAI1 regulates synaptogenesis by controlling the recruitment of the Par3/Tiam1 polarity complex to synaptic sites. J Neurosci 33:6964–6978. doi:10.1523/JNEUROSCI.3978-12.2013
33. Li X, Roszko I, Sepich DS, Ni M, Hamm HE, Marlow FL et al (2013) Gpr125 modulates Dishevelled distribution and planar cell polarity signaling. Development 140:3028–3039. doi:10.1242/dev.094839
34. Nordström KJV, Lagerström MC, Waller LMJ, Fredriksson R, Schiöth HB (2009) The secretin GPCRs descended from the family of adhesion GPCRs. Mol Biol Evol 26:71–84. doi:10.1093/molbev/msn228
35. Bryant D, Moulton V (2004) Neighbor-net: an agglomerative method for the construction of phylogenetic networks. Mol Biol Evol 21:255–265. doi:10.1093/molbev/msh018
36. Edgar RC (2004) MUSCLE: multiple sequence alignment with high accuracy and high throughput. Nucleic Acids Res 32:1792–1797. doi:10.1093/nar/gkh340
37. Wolf S, Grünewald S (2015) Sequence, structure and ligand binding evolution of rhodopsin-like G protein-coupled receptors: a crystal structure-based phylogenetic analysis. PLoS One 10, e0123533. doi:10.1371/journal.pone.0123533
38. Huson DH, Bryant D (2006) Application of phylogenetic networks in evolutionary studies. Mol Biol Evol 23:254–267. doi:10.1093/molbev/msj030
39. Kuhner MK, Felsenstein J (1994) A simulation comparison of phylogeny algorithms under equal and unequal evolutionary rates. Mol Biol Evol 11:459–468
40. Jazayeri A, Doré AS, Lamb D, Krishnamurthy H, Southall SM, Baig AH et al (2016) Extra-helical binding site of a glucagon receptor antagonist. Nature. doi:10.1038/nature17414
41. Hollenstein K, Kean J, Bortolato A, Cheng RKY, Doré AS, Jazayeri A et al (2013) Structure of class B GPCR corticotropin-releasing factor receptor 1. Nature 499:438–443. doi:10.1038/nature12357
42. Venkatakrishnan AJ, Deupi X, Lebon G, Tate CG, Schertler GF, Babu MM (2013) Molecular signatures of G-protein-coupled receptors. Nature 494:185–194. doi:10.1038/nature11896

43. Wu H, Wang C, Gregory KJ, Han GW, Cho HP, Xia Y et al (2014) Structure of a class C GPCR metabotropic glutamate receptor 1 bound to an allosteric modulator. Science 344:58–64. doi:10.1126/science.1249489
44. Wang C, Wu H, Katritch V, Han GW, Huang X-P, Liu W et al (2013) Structure of the human smoothened receptor bound to an antitumour agent. Nature 497:338–343. doi:10.1038/nature12167
45. Hollenstein K, de Graaf C, Bortolato A, Wang M-W, Marshall FH, Stevens RC (2014) Insights into the structure of class B GPCRs. Trends Pharmacol Sci 35:12–22. doi:10.1016/j.tips.2013.11.001
46. Palczewski K, Kumasaka T, Hori T, Behnke CA, Motoshima H, Fox BA et al (2000) Crystal structure of rhodopsin: a G protein-coupled receptor. Science 289:739–745. doi: 10.1126/science.289.5480.739
47. Shimamura T, Shiroishi M, Weyand S, Tsujimoto H, Winter G, Katritch V et al (2011) Structure of the human histamine H1 receptor complex with doxepin. Nature 475:65–70. doi:10.1038/nature10236
48. Wu B, Chien EYT, Mol CD, Fenalti G, Liu W, Katritch V et al (2010) Structures of the CXCR4 chemokine GPCR with small-molecule and cyclic peptide antagonists. Science. doi:10.1126/science.1194396
49. Coin I, Katritch V, Sun T, Xiang Z, Siu FY, Beyermann M, Stevens RC et al (2013) Genetically encoded chemical probes in cells reveal the binding path of urocortin-I to CRF Class B GPCR. Cell 155:1258–1269. doi:10.1016/j.cell.2013.11.008
50. Ballesteros JA, Weinstein H (1995) Integrated methods for the construction of three-dimensional models and computational probing of structure-function relations in G protein-coupled receptors. Methods Neurosci 25:366–428. doi:10.1016/S1043-9471(05)80049-7
51. Wootten D, Simms J, Miller LJ, Christopoulos A, Sexton PM (2013) Polar transmembrane interactions drive formation of ligand-specific and signal pathway-biased family B G protein-coupled receptor conformations. Proc Natl Acad Sci 110:5211–5216. doi:10.1073/pnas.1221585110
53. Pin J-P, Galvez T, Prezeau L (2003) Evolution, structure, and activation mechanism of family 3/C G-protein-coupled receptors. Pharmacol Ther 98:325–354
52. de Graaf C, Foata N, Engkvist O, Rognan D (2008) Molecular modeling of the second extracellular loop of G-protein coupled receptors and its implication on structure-based virtual screening. Proteins 71:599–620. doi:10.1002/prot.21724
54. Isberg V, Mordalski S, Munk C, Rataj K, Harpsøe K, Hauser AS et al (2015) GPCRdb: an information system for G protein-coupled receptors. Nucleic Acids Res. doi:10.1093/nar/gkv1178
55. Pawson AJ, Sharman JL, Benson HE, Faccenda E, Alexander SPH, Buneman OP et al (2014) The IUPHAR/BPS Guide to PHARMACOLOGY: an expert-driven knowledgebase of drug targets and their ligands. Nucleic Acids Res 42:D1098–D1106. doi:10.1093/nar/gkt1143
56. Shi L, Liapakis G, Xu R, Guarnieri F, Ballesteros JA, Javitch JA (2002) Beta2 adrenergic receptor activation. Modulation of the proline kink in transmembrane 6 by a rotamer toggle switch. J Biol Chem 277:40989–40996. doi:10.1074/jbc.M206801200
57. Tehan BG, Bortolato A, Blaney FE, Weir MP, Mason JS (2014) Unifying family A GPCR theories of activation. Pharmacol Ther 143:51–60. doi:10.1016/j.pharmthera.2014.02.004
58. Mirzadegan T, Benkö G, Filipek S, Palczewski K (2003) Sequence analyses of G-protein-coupled receptors: similarities to rhodopsin. Biochemistry 42:2759–2767. doi:10.1021/bi027224
59. Katritch V, Fenalti G, Abola EE, Roth BL, Cherezov V, Stevens RC (2014) Allosteric sodium in class A GPCR signaling. Trends Biochem Sci 39:233–244. doi:10.1016/j.tibs.2014.03.002
60. Hofmann KP, Scheerer P, Hildebrand PW, Choe H-W, Park JH, Heck M et al (2009) A G protein-coupled receptor at work: the rhodopsin model. Trends Biochem Sci 34:540–552. doi:10.1016/j.tibs.2009.07.005

61. Doré AS, Okrasa K, Patel JC, Serrano-Vega M, Bennett K, Cooke RM et al (2014) Structure of class C GPCR metabotropic glutamate receptor 5 transmembrane domain. Nature 511:557–562. doi:10.1038/nature13396
62. Rasmussen SGF, Devree BT, Zou Y, Kruse AC, Chung KY, Kobilka TS et al (2011) Crystal structure of the β2 adrenergic receptor-Gs protein complex. Nature 477:549–555. doi:10.1038/nature10361
63. Scheerer P, Park JH, Hildebrand PW, Kim YJ, Krauss N, Choe H-W et al (2008) Crystal structure of opsin in its G-protein-interacting conformation. Nature 455:497–502. doi:10.1038/nature07330
64. Choe H-W, Kim YJ, Park JH, Morizumi T, Pai EF, Krauss N et al (2011) Crystal structure of metarhodopsin II. Nature 471:651–655. doi:10.1038/nature09789
65. Fritze O, Filipek S, Kuksa V, Palczewski K, Hofmann KP, Ernst OP (2003) Role of the conserved NPxxY(x)5,6F motif in the rhodopsin ground state and during activation. Proc Natl Acad Sci U S A 100:2290–2295. doi:10.1073/pnas.0435715100
66. Prioleau C, Visiers I, Ebersole BJ, Weinstein H, Sealfon SC (2002) Conserved helix 7 tyrosine acts as a multistate conformational switch in the 5HT2C receptor. identification of a novel "locked-on" phenotype and double revertant mutations. J Biol Chem 277:36577–36584. doi:10.1074/jbc.M206223200
67. Hanson MA, Cherezov V, Griffith MT, Roth CB, Jaakola V-P, Chien EYT et al (2008) A specific cholesterol binding site is established by the 2.8 A structure of the human beta2-adrenergic receptor. Structure 16:897–905. doi:10.1016/j.str.2008.05.001
68. Chrencik JE, Roth CB, Terakado M, Kurata H, Omi R, Kihara Y et al (2015) Crystal structure of antagonist bound human lysophosphatidic acid receptor 1. Cell 161:1633–1643. doi:10.1016/j.cell.2015.06.002
69. Hanson MA, Roth CB, Jo E, Griffith MT, Scott FL, Reinhart G et al (2012) Crystal structure of a lipid G protein-coupled receptor. Science 335:851–855. doi:10.1126/science.1215904
70. Legrand F, Tomasevic N, Simakova O, Lee C-CR, Wang Z, Raffeld M, et al (2014) The eosinophil surface receptor epidermal growth factor-like module containing mucin-like hormone receptor 1 (EMR1): a novel therapeutic target for eosinophilic disorders. J Allergy Clin Immunol 133:1439–1447, 1447.e1–e8. doi:10.1016/j.jaci.2013.11.041
71. Veninga H, de Groot DM, McCloskey N, Owens BM, Dessing MC, Verbeek JS et al (2011) CD97 antibody depletes granulocytes in mice under conditions of acute inflammation via a Fc receptor-dependent mechanism. J Leukoc Biol 89:413–421. doi:10.1189/jlb.0510280
72. Perret J, Craenenbroeck M, Langer I, Vertongen P (2002) Mutational analysis of the glucagon receptor: similarities with the vasoactive intestinal peptide (VIP)/pituitary adenylate cyclase-activating peptide (PACAP). Biochem J 362:389–394
73. Koth CM, Murray JM, Mukund S, Madjidi A, Minn A, Clarke HJ et al (2012) Molecular basis for negative regulation of the glucagon receptor. Proc Natl Acad Sci 109:14393–14398. doi:10.1073/pnas.1206734109
74. Unson CG, Wu C-R, Jiang Y, Yoo B, Cheung C, Sakmar TP et al (2002) Roles of specific extracellular domains of the glucagon receptor in ligand binding and signaling. Biochemistry 41:11795–11803. doi:10.1021/bi025711j
75. Hoare SRJ, Brown BT, Santos MA, Malany S, Betz SF, Grigoriadis DE (2006) Single amino acid residue determinants of non-peptide antagonist binding to the corticotropin-releasing factor1 (CRF1) receptor. Biochem Pharmacol 72:244–255. doi: 10.1016/j.bcp.2006.04.007
76. Donnelly D (2012) The structure and function of the glucagon-like peptide-1 receptor and its ligands. Br J Pharmacol 166:27–41. doi:10.1111/j.1476-5381.2011.01687.x
77. Wootten D, Savage EE, Willard FS, Bueno AB, Sloop KW, Christopoulos A et al (2013) Differential activation and modulation of the glucagon-like peptide-1 receptor by small molecule ligands. Mol Pharmacol 83:822–834. doi:10.1124/mol.112.084525
78. Coopman K, Wallis R, Robb G, Brown AJH, Wilkinson GF, Timms D et al (2011) Residues within the transmembrane domain of the glucagon-like peptide-1 receptor involved in ligand

binding and receptor activation: modelling the ligand-bound receptor. Mol Endocrinol 25:1804–1818. doi:10.1210/me.2011-1160
79. Underwood CR, Garibay P, Knudsen LB, Hastrup S, Peters GH, Rudolph R et al (2010) Crystal structure of glucagon-like peptide-1 in complex with the extracellular domain of the glucagon-like peptide-1 receptor. J Biol Chem 285:723–730. doi:10.1074/jbc.M109.033829
80. Wootten D, Reynolds CA, Koole C, Smith KJ, Mobarec JC, Simms J et al (2016) A hydrogen-bonded polar network in the core of the glucagon-like peptide-1 receptor is a fulcrum for biased agonism: lessons from class B crystal structures. Mol Pharmacol 89:335–347. doi:10.1124/mol.115.101246
81. Yaqub T, Tikhonova IG, Lättig J, Magnan R, Laval M, Escrieut C et al (2010) Identification of determinants of glucose-dependent insulinotropic polypeptide receptor that interact with N-terminal biologically active region of the natural ligand. Mol Pharmacol 77:547–558. doi:10.1124/mol.109.060111
82. Tseng C-C, Lin L (1997) A point mutation in the glucose-dependent insulinotropic peptide receptor confers constitutive activity. Biochem Biophys Res Commun 232:96–100. doi:10.1006/bbrc.1997.6231
83. Di Paolo E, Petry H, Moguilevsky N, Bollen A, De Neef P, Waelbroeck M et al (1999) Mutations of aromatic residues in the first transmembrane helix impair signalling by the secretin receptor. Recept Channels 6:309–315
84. Di Paolo E, De Neef P, Moguilevsky N, Petry H, Bollen A, Waelbroeck M et al (1998) Contribution of the second transmembrane helix of the secretin receptor to the positioning of secretin. FEBS Lett 424:207–210. doi: 10.1016/S0014-5793(98)00175-6
85. Solano RM, Langer I, Perret J, Vertongen P, Juarranz MG, Robberecht P et al (2001) Two basic residues of the h-VPAC1 receptor second transmembrane helix are essential for ligand binding and signal transduction. J Biol Chem 276:1084–1088. doi:10.1074/jbc.M007696200
86. Ceraudo E, Hierso R, Tan Y-V, Murail S, Rouyer-Fessard C, Nicole P et al (2012) Spatial proximity between the VPAC1 receptor and the amino terminus of agonist and antagonist peptides reveals distinct sites of interaction. FASEB J 26:2060–2071. doi:10.1096/fj.11-196444
87. Gensure RC, Shimizu N, Tsang J, Gardella TJ (2013) Identification of a contact site for residue 19 of parathyroid hormone (PTH) and PTH-related protein analogs in transmembrane domain two of the type 1 PTH receptor. Mol Endocrinol 17:2647–2658. doi:10.1210/me.2003-0275
88. Gardella TJ, Jüppner H (2001) Molecular properties of the PTH/PTHrP receptor. Trends Endocrinol Metab 12:210–217. doi:10.1016/S1043-2760(01)00409-X
89. Qin L, Kufareva I, Holden LG, Wang C, Zheng Y, Zhao C et al (2015) Structural biology. Crystal structure of the chemokine receptor CXCR4 in complex with a viral chemokine. Science 347:1117–1122. doi:10.1126/science.1261064
90. Katritch V, Cherezov V, Stevens RC (2013) Structure-function of the G protein-coupled receptor superfamily. Annu Rev Pharmacol Toxicol 53:531–556. doi:10.1146/annurev-pharmtox-032112-135923
91. Christopher JA, Aves SJ, Bennett KA, Doré AS, Errey JC, Jazayeri A et al (2015) Fragment and structure-based drug discovery for a C GPCR: discovery of the mGlu5 negative allosteric modulator HTL14242 (3-chloro-5-[6-(5-fluoropyridin-2-yl)pyrimidin-4-yl]benzonitrile). J Med Chem 58:6653–6664. doi:10.1021/acs.jmedchem.5b00892
92. Wang C, Wu H, Evron T, Vardy E, Han GW, Huang X-P et al (2014) Structural basis for Smoothened receptor modulation and chemoresistance to anticancer drugs. Nat Commun 5:4355. doi:10.1038/ncomms5355
93. Mason JS, Bortolato A, Congreve M, Marshall FH (2012) New insights from structural biology into the druggability of G protein-coupled receptors. Trends Pharmacol Sci 33:249–260. doi:10.1016/j.tips.2012.02.005

Understanding the Structural Basis of Adhesion GPCR Functions

Demet Araç, Norbert Sträter, and Elena Seiradake

Graphical Abstract

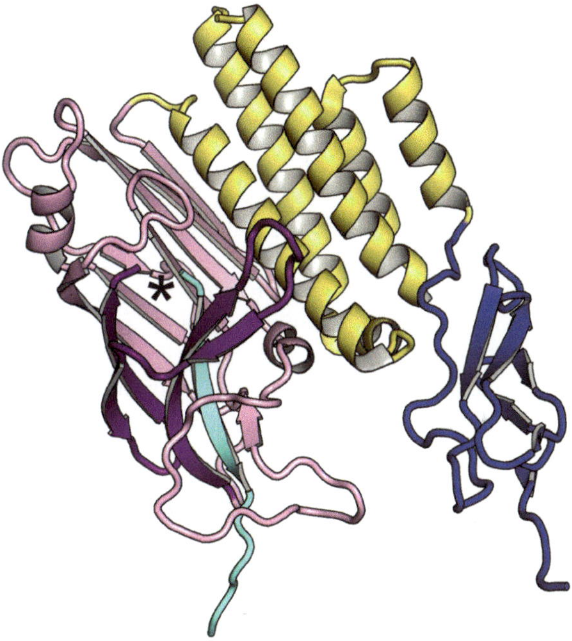

D. Araç (✉)
Department of Biochemistry and Molecular Biology, University of Chicago, 929 E. 57th Street, GCIS W210, Chicago, IL 60637, USA
e-mail: arac@uchicago.edu

N. Sträter
Institute for Bioanalytical Chemistry, Leipzig University, Deutscher Platz 5, Leipzig 04103, Germany
e-mail: strater@bbz.uni-leipzig.de

E. Seiradake
Department of Biochemistry, University of Oxford, South Parks Road, Oxford OX1 3QU, UK
e-mail: elena.seiradake@bioch.ox.ac.uk

T. Langenhan, T. Schöneberg (eds.), *Adhesion G Protein-coupled Receptors*,
Handbook of Experimental Pharmacology 234, DOI 10.1007/978-3-319-41523-9_4

Contents

Abstract

Unlike conventional G-protein-coupled receptors (GPCRs), adhesion GPCRs (aGPCRs) have large extracellular regions that are autoproteolytically cleaved from their membrane-embedded seven-pass transmembrane helices. Autoproteolysis occurs within the conserved GPCR-Autoproteolysis INducing (GAIN) domain that is juxtaposed to the transmembrane domain and cleaves the last beta strand of the GAIN domain. The other domains of the extracellular region are variable and specific to each aGPCR and are likely involved in adhering to various ligands. Emerging evidence suggest that extracellular regions may modulate receptor function and that ligand binding to the extracellular regions may induce receptor activation via multiple mechanisms. Here, we summarize current knowledge about the structural understanding for the extracellular regions of aGPCRs and discuss their possible functional roles that emerge from the available structural information.

Keywords

GAIN • Horm • Olfactomedin • Lectin • Stachel • Tethered agonist • Unc5 • FLRT • Super-complex • Latrophilin

1 Introduction

The aGPCR ectodomain consists of an N-terminal region with a family-specific set of domains, which probably mediate interactions with binding partners, followed by a conserved region that in almost all aGPCRs contains a GPCR-Autoproteolysis INducing (GAIN) domain just upstream of the seven-transmembrane helix (7TM) domain (Fig. 1) [1]. Direct experimental structural information on the architecture and mode of action of aGPCR ectodomain is still scarce. However, crystal structures are available for the hormone receptor (HormR) and GAIN domains of B3 and L1 [1]. Conformational changes induced by ligand binding must be transmitted to the 7TM domain. This is most likely the crucial function of the GAIN

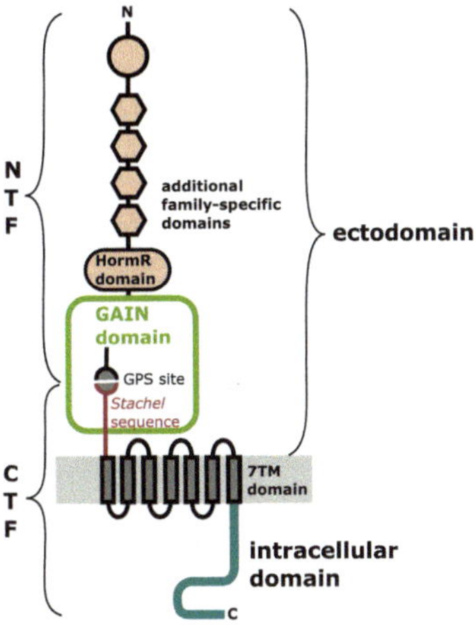

Fig. 1 Scheme of the architecture of a prototypical aGPCR. The GAIN domain contains an autoproteolytic cleavage site (GPS site), resulting in an N-terminal fragment (NTF) and a C-terminal fragment (CTF). As the N-terminal part of the *Stachel* sequence (sequence between the GPS site and the start of the 7TM domain) is tightly bound to the GAIN domain, NTF and CTF do not dissociate. The HormR domain is present in many, but not all, aGPCRs

domain. Liebscher et al. [2] demonstrated that the sequence stretch after the GPS autoproteolytic cleavage site of the GAIN domain activates the GPR126 (ADGRG6) and GPR133 (ADGRD1) receptors in a sequence-specific manner if the rest of the GAIN domain is missing. This sequence stretch has been designated as the *Stachel* sequence. In addition, peptides corresponding to the *Stachel* sequence could also activate variants of these receptors lacking the GAIN domain and *Stachel* sequence with sequence specificity but relatively low affinity. A similar mechanism of activation by the region following the GPS cleavage site was also demonstrated for ADGRG1 (GPR56), ADGRF1 (GPR110), ADGRG5 (GPR114), and ADGRG2 (GPR64) [3–5]. These studies demonstrated that aGPCRs are activated by a tethered agonist which resides in the GAIN domain.

Latrophilin is the only aGPCR whose N-terminal ligand-binding domains have been determined structurally. Nuclear magnetic resonance (NMR) models of the N-terminal lectin domain [6], a recent crystal structure of lectin and olfactomedin domains [7] and first ligand-complex structures [8–10], have provided exciting first insights into the structural repertoire of the N-terminal aGPCR domains. These suggest that the N-terminal domains of latrophilin have at least two distinct functions: They allow tethering of the ectodomain to specific extracellular ligands through high-affinity protein-protein interaction. In addition, they encode a structural mechanism for ligand-induced receptor clustering, thereby controlling the multimerization state of the aGPCR latrophilin. In the following, we summarize the key features of the available crystal structures of aGPCR ectodomains as well as some insights obtained from sequence comparisons to related structurally known domain folds.

2 The GAIN Domain

2.1 A Conserved Domain in All aGPCR Extracellular Regions

Unlike other GPCRs, aGPCRs have large extracellular regions that are autoproteolytically cleaved from their seven-pass transmembrane regions at a conserved GPCR proteolysis site (GPS) [11, 12] (see also [13]). The GPS is well described and recognized as a vital unit for receptor function [14–16]. However, a recent study showed that—as opposed to the previous belief—the GPS does not constitute a functional folded domain by itself, but rather is an integral part of a much larger novel domain that is termed the GPCR-Autoproteolysis INducing (GAIN) domain [1]. Of great importance, the tethered agonist *Stachel* peptide resides within the GAIN domain (see also [17, 18]).

Strikingly, the GAIN domain is shared by all human aGPCRs (except GPR123/ADGRA1 which has only a short sequence of ~20 residues before the first transmembrane helix). GAIN domain is unique in that it is the only domain that exists in all members of the aGPCRs in humans indicating an essential role in aGPCR function. Moreover, database searches revealed that all five members of another human protein family, the polycystic kidney disease proteins, an unrelated family of membrane proteins that are also autoproteolyzed, contain GAIN domains. In addition, primitive organisms, such as *Dictyostelium discoideum* that arose early in evolution before animals emerged, encode GAIN domains although they lack most other autoproteolytic domains, important adhesion and signaling domains, and critical signaling pathways. Among all domains that are found in aGPCRs, LRR (leucine-rich repeats), EGF (epidermal growth factor), and TSP (thrombospondin) are the only other domains that exist in such primitive organisms. These results showed that the GAIN domain is widespread and conserved in higher eukaryotes as well as in ancient organisms. Intriguingly, analysis of numerous sequences also revealed that the GAIN domain always immediately precedes the first transmembrane helix by a short linker of about seven residues raising the question whether the GAIN domain regulates receptor signaling via intramolecular interactions with the nearby transmembrane helices.

2.2 Structure of the GAIN Domain

The crystal structures of GAIN domains from two distantly related aGPCRs, latrophilin 1 (LPHN1/ADGRL1) and brain angiogenesis inhibitor 3 (BAI3/ADGRB3), revealed a novel fold that was previously unidentified. The GAIN domain (of LPHN1) contains an N-terminal subdomain A that is composed of six alpha helices and a C-terminal subdomain B that is composed of a twisted beta sandwich including 13 beta strands and two small helices (Fig. 2). The last five beta strands of subdomain B constitute the GPS motif. The GPS motif is the most conserved region of the GAIN domain (and thus it was chronologically noticed first and was mistakenly called a domain). The conservation of primary sequence of the

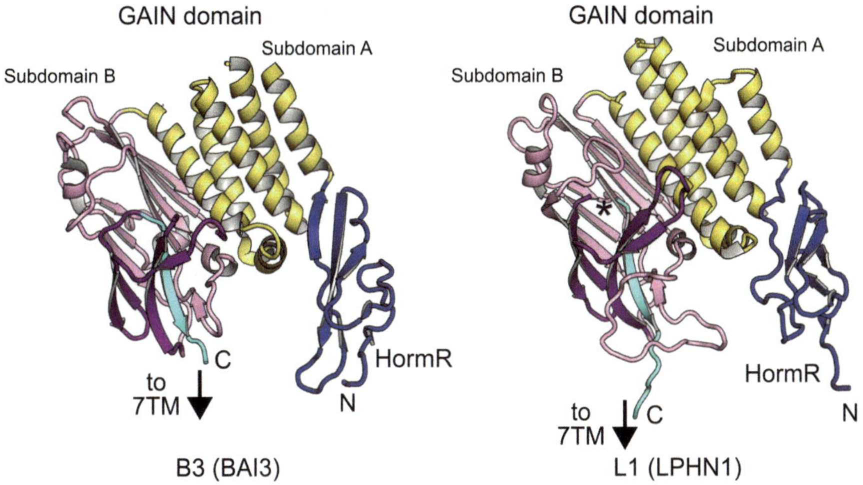

Fig. 2 Atomic structures of the GAIN and HormR domains. Cartoon representations of the uncleaved ADGRB3 GAIN and HormR domains (*left*, PDB ID 4DLO) and the cleaved ADGRL1 GAIN and HormR domains (*right*, PDB ID 4DLQ) revealed the novel GAIN domain fold. GAIN subdomain A and subdomain B are colored *yellow* and *light-pink*, respectively. The GPS motif (*magenta*) is an integral part of subdomain B. *Stachel* peptide (*cyan*) corresponds to the last beta strand of the GAIN domain and is deeply buried within the GAIN domain. Autoproteolysis (indicated by *asterisk*) occurs between the last two beta strands of the GAIN domain cleaving, but not releasing the *Stachel* peptide. The HormR domain (*blue*) has a rigid orientation relative to the GAIN domain in both ADGRL1 and ADGRB3

GAIN domains increases from the N-terminus to the C-terminus. In spite of the low sequence identity between LPHN1 and BAI3 GAIN domains (24 %), the very high similarity of the GAIN domain structures indicate that the three-dimensional structure is conserved more strictly than primary sequence throughout evolution.

2.3 Autoproteolysis Cleaves Off the *Stachel* Peptide But Does Not Release It from the GAIN Domain

GAIN domain is an autoproteolytic fold that is both required and sufficient for autoproteolysis, whereas the GPS motif without the rest of the GAIN domain is not functional. Autoproteolysis occurs in the short loop between the last two beta strands of the GAIN domain and cleaves the C-terminal beta strand from the rest of the domain. Indeed, the C-terminal beta strand of the GAIN domain corresponds to the *Stachel* peptide and acts as the tethered agonist that activates the aGPCR (Fig. 2). However, autoproteolysis does not cause the dissociation of the *Stachel* peptide from the rest of the GAIN domain, and it is unclear how it can be exposed to the transmembrane region to activate the receptor. The *Stachel* peptide is highly conserved and has a hydrophobic nature (TNFAVLM in LPHN1) and is involved in approximately 15 strong hydrogen bonds and numerous hydrophobic interactions that tightly keep it in the GAIN domain. It is suggested that mechanical force applied on the extracellular domains may lead to shedding of the extracellular regions and release of the *Stachel* peptide from the GAIN domain enabling the *Stachel* peptide to activate the receptor (see also [17–20]).

2.4 Possible Roles of the GAIN Domain in aGPCR Activation

It is possible that, in addition to modulating *Stachel* peptide exposure to the transmembrane region, the GAIN domain is involved in more complicated mechanisms for regulating aGPCR function. Analysis of disease-causing mutations in different GAIN domain-containing proteins revealed that GAIN domains are hot spots for disease mutations. The GAIN domains of CL/BAI, polycystic kidney disease, and GPR56 and CL3 proteins are mutated in cancer [21], polycystic kidney disease, bilateral frontoparietal polymicrogyria, and attention deficit hyperactivity disorder [22], respectively. Monitoring the effect of polycystic kidney disease and cancer mutations on autoproteolysis and plasma membrane localization (thus presumably proper protein folding) of the receptors suggested that the mechanisms by which GAIN domain mutations lead to different diseases might be various. Thus, the GAIN domain might be involved in many functions in addition to autoproteolysis such as binding to other proteins, regulating receptor signaling via intramolecular interactions with the transmembrane helices, etc. (see also [19]).

The extracellular regions of several aGPCRs were suggested to regulate aGPCR activation. For example, the deletion of the N-terminal fragment (NTF) of GPR56 exhibits enhanced signaling ability when compared to the wild-type GPR56, and thus, the extracellular region is proposed to have an inhibitory role on receptor activation [23, 24]. The observation that the GPR126 GAIN domain binds to laminin suggests that GAIN domain functions as a protein-binding domain in addition to an autoproteolytic domain pointing to a multipurpose role in aGPCR function. However, other mechanisms by which aGPCRs are regulated and which domains of their very large extracellular regions are involved in their downstream signaling are unclear.

3 The HormR Domain

The hormone receptor (HormR) domain is the second most frequently observed domain in aGPCRs (found in 12 out of 33 human aGPCRs, specifically in LPHN1–3, GPR124/ADGRA2, GPR125/ADGRA3, CELSR1–3/ADGRC1–3, GPR113/ADGRF3, BAI1–3/ADGRB1–3). HormR domain-containing aGPCRs are homologous to family B GPCRs that always include an HormR domain. The crystal structures show that the HormR domains of LPHN1 and BAI3 are ~70-residue domains formed by two antiparallel β-sheets with conserved disulfide bonds and tryptophan residues (Fig. 2) and yielded a low RMSD (1.1 Å) in spite of the low sequence identity (24 %). Since no hormone ligand has yet been found for aGPCRs, it has been believed that the HormR domain in aGPCRs may not be a true hormone-binding region. Remarkably, a DALI search revealed that N-terminal hormone-binding extracellular domain of the corticotrophin-releasing factor receptor (CRFR) (PDB ID: 3EHU) is strikingly similar to the HormR domains of LPHN1 and BAI3 yielding RMSDs of 0.7 Å and 1.1 Å, respectively, in spite of the low sequence identity (29 %) [25]. The unusually high structural similarity of the HormR domains of LPHN1 and BAI3 to that of the CRF receptor raised the possibility that aGPCRs can be real hormone receptors.

Superposition of the LPHN1 structure with the structure of the CRFR HormR domain bound to corticotrophin-releasing factor (CRF), a 41-amino acid peptide hormone, showed that a similar hormone could not bind to LPHN1 HormR because the GAIN domain is blocking the homologous hormone-binding site on the HormR domain of LPHN1 [1]. Clearly, hormone binding would require a conformational change in LPHN1 to expose the putative hormone-binding site (such as the reduction of one of the conserved disulfide bonds to switch to an open conformation). In family B GPCRs, the HormR domain precedes the transmembrane helices and is juxtaposed to the membrane. According to the two-domain model of family B GPCR activation, the interaction of the extracellular hormone-binding domain with the hormone promotes the interaction of the hormone with the transmembrane helices, leading to the activation of the receptor [26]. However, the mechanism of aGPCR activation upon hormone binding may be different, because the GAIN domain lies between the HormR domain and the transmembrane helices.

4 Olfactomedin and Lectin Domains of Latrophilins

Based on phylogenetic information of the 7TM parts, aGPCRs have been categorized into nine subfamilies [27] (see also [13]). First structural information on the N-terminal ligand-binding domains has become available for the subfamily I aGPCR member latrophilin. The three members of the latrophilin (LPHN, ADGRL) aGPCR subfamily have conserved domain architecture including N-terminal lectin domain, olfactomedin domain, serine/threonine-rich region, HormR domain, juxtamembrane GAIN domain, 7TM, and intracellular domains (Fig. 3a). The LPHN ectodomains have been reported to interact with various ligands including alpha-latrotoxin [11, 28], a potent toxin produced by the black widow spider, teneurin/ODZ family proteins [29], neurexins [30], and the single-pass transmembrane fibronectin-leucine-rich repeat protein FLRT3 [31]. Importantly, these interactions are mediated chiefly via the N-terminal domains of the latrophilin ectodomain. The first extracellular domain of an aGPCR to be determined at high resolution was the murine LPHN1 lectin domain: Nuclear magnetic resonance (NMR) revealed a compact structure, decorated by largely ordered loops, which harbor a weak glycan-binding site of unknown biological significance

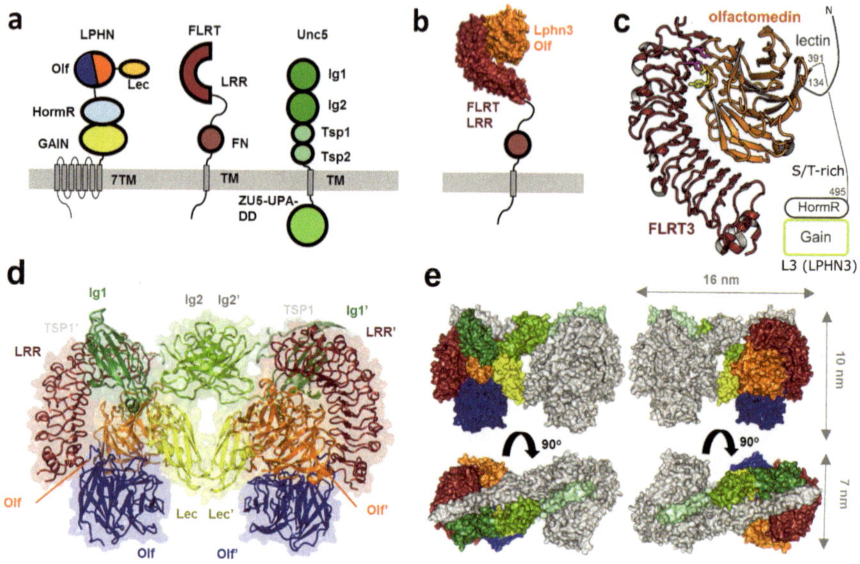

Fig. 3 Structures of latrophilin ectodomain interactions. (**a**) Domain overview of LPHN, FLRT, and Unc5. *Lec* lectin domain, *Olf* olfactomedin domain, *Horm* hormone domain, *GAIN* GPCR-Autoproteolysis INducing domain, *7TM* GPCR domain, *LRR* leucine-rich repeat domain, *FN* fibronectin domain, *Ig1* immunoglobulin domain 1, *Ig2* immunoglobulin domain 2, *TSP1* thrombospondin domain 1, *TSP2* thrombospondin domain 2, *DD* death domain. (**b**) FLRT3 LRR: LPHN3 Olf complex [9]. (**c**) Zoomed view of the FLRT3 LRR: LPHN3 Olf complex [9]. (**d**) The super-complex formed by extracellular domains of FLRT2 (LRR), Unc5D (Ig1, Ig2, TSP1), and LPHN3 (Lec, Olf) reveals a large assembly with an unusual stoichiometry (2:2:4) [10]. (**e**) The super-complex shown in panel (**d**) consists of two pseudosymmetric halves, joined together by the antiparallel arrangement of Unc5D

[6]. More recently, X-ray crystallography revealed the structure of LPHN3/ADGRL3 lectin plus olfactomedin domains. The olfactomedin domain has a five-bladed beta-propeller structure with tightly bound, presumably structural metal ions at its center [7–9]. The olfactomedin domain of LPHN3 has been shown to promote glutamatergic synapse formation by interacting with FLRT3 [31]. Surface mutagenesis and X-ray crystallography approaches demonstrated an extended binding surface across the "top" of the domain, which interacts with the concave surface of the horseshoe-shaped human FLRT2/3 leucine-rich repeat domains [7–9] (Fig. 3b, c). The crystal structures of FLRT3 leucine-rich repeat domain in complex with isolated LPHN3 olfactomedin domain revealed a simple 1:1 mode of interaction [8, 9]. The formation of a ternary complex between LPHN3, FLRT3, and another receptor involved in axon guidance, Uncoordinated-5B/D (Unc5B/D), which does not interact with LPHN3 in the absence of FLRT, was demonstrated suggesting LPHN3 may function via interacting with multiple ligands simultaneously [9, 10]. Surprisingly, crystal structures showed that the addition of Unc5D to the FLRT2 leucine-rich repeat domain and LPHN3 lectin-olfactomedin

domains resulted in an unexpected large assembly with a 2:2:4 stoichiometry [10] (Fig. 3d). A defining characteristic of the "super-complex" is that it demonstrates a molecular mechanism for high-affinity receptor oligomerization via the extracellular regions of the three proteins: FLRT2 acts as an initial scaffold by providing the previously described distinct high-affinity binding sites for LPHN3 and Unc5D [8, 9, 32]. This minimal ternary complex triggers the docking of a second copy of LPHN3 onto a newly formed binding site provided by FLRT2 and the first copy of LPHN3 (olfactomedin and lectin domains). The resulting tetrameric arrangement further dimerizes via the Unc5D Ig2 and TSP1 domains, thus forming an octameric super-complex. Unc5D adopts an antiparallel arrangement that holds together the sandwich formed by FLRT2 and four copies of LPHN3 (Fig. 3e). Sequence analysis revealed that other homologues of the Unc5 and FLRT family (Unc5A,B,C; FLRT1,3) are likely to produce smaller ternary complexes, including only one copy of each FLRT and Unc5 and two copies of LPHN [10].

What may the functions of different LPHN-FLRT and LPHN-FLRT-Unc5 assemblies be in the neural system, where these receptors are expressed in overlapping regions? Cell aggregation experiments have suggested that the interaction between full-length LPHN3 and FLRT3 occurs in trans [9], which is in agreement with previous cell stripe assay data [7] and the suggested adhesive role at synapses [31]. Given the recent speculation that mechanical force on the ectodomain may play a key role in activation of aGPCRs, it is interesting to consider that the FLRT3-expressing cell may exert force on the LPHN3-expressing cell and thereby activate LPHN3 in glutamatergic synapse formation. In contrast to the adhesive LPHN:FLRT interaction, the 1:1 Unc5:FLRT interaction triggers strong repulsion of radially migrating cells in the developing cortex and also induces neurite repulsion/collapse in vitro [32, 33]. These apparently opposing functions (FLRT:LPHN-adhesion, FLRT:Unc5-repulsion) make the ternary super-complex functionally intriguing. Pull-down experiments using murine cortical lysates are in agreement with the ternary complex forming in vivo, and recent stripe assay data suggested that the co-expression of Unc5D in FLRT2-expressing cells reduces their adhesive response to externally presented LPHN3 protein [10]. Thus, the ternary complex is likely to play distinct roles in vivo, possibly involving functions that require higher-order receptor multimerization.

5 Ectodomain Structures of Subfamily II–IX aGPCR

In contrast to subfamily I aGPCR, no crystal structures are available for the ectodomains except for the HormR and GAIN domains, as discussed above. In this section, we will focus on the domains which are located at the N-terminal side of the HormR and GAIN domains.

Subfamily II (E) aGPCRs contain one EGF-like domain at the N-terminus followed by one, four, or five calcium-binding EGF-like domains (Fig. 4a). EGF-like domains are found in many extracellular domains of membrane-bound proteins, and they often form repeats that fold to a linear solenoid architecture. Each

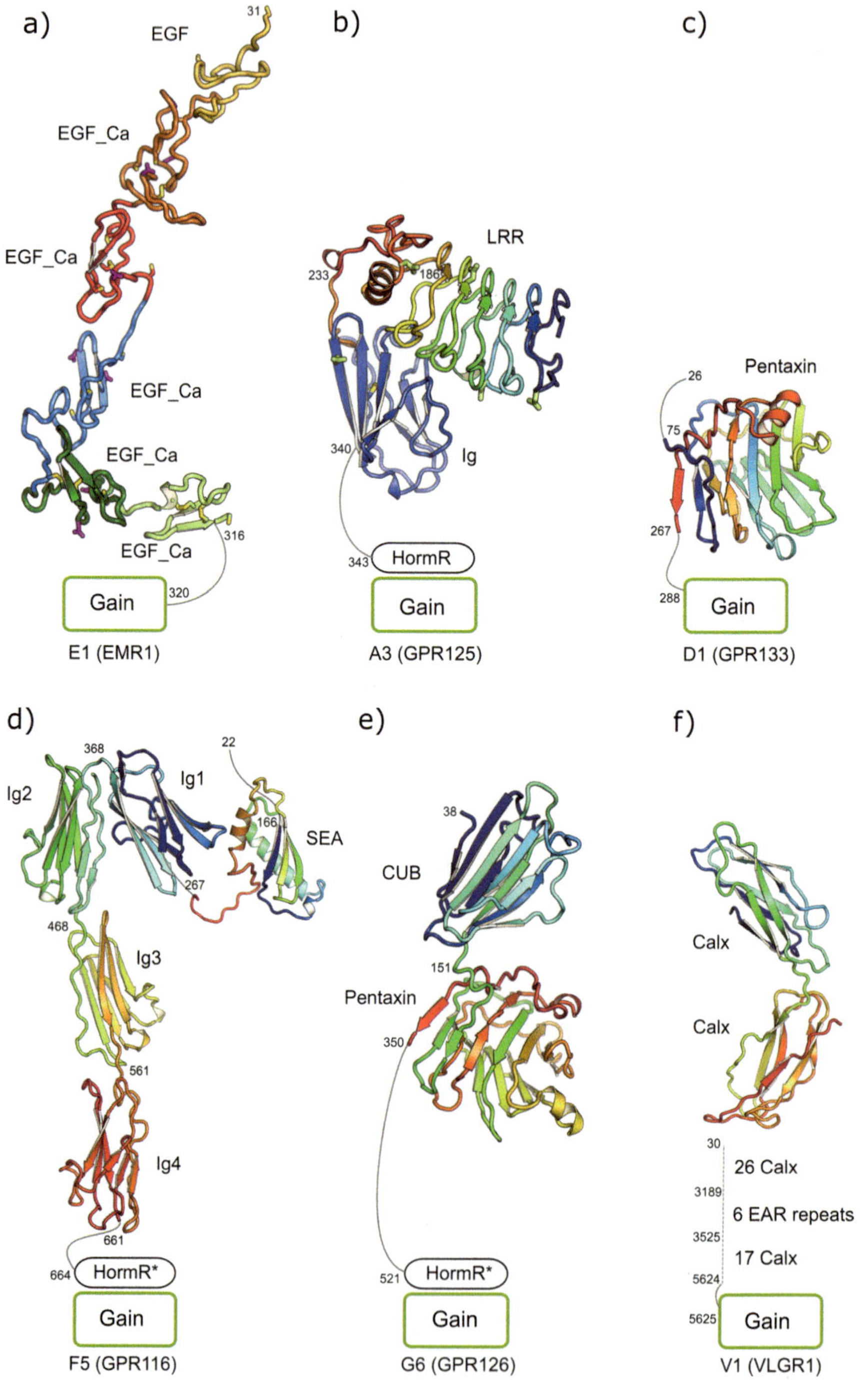

Fig. 4 Models of selected aGPCR ectodomains. (**a**) ADGRE1 (EMR1) of subfamily II aGPCR, (**b**) ADGRA3 of subfamily III aGPCR, (**c**) ADGRD1 of subfamily V aGPCR, (**d**) ADGRF5 (GPR116) of subfamily VI, (**e**) ADGRG6 (GPR126) of subfamily VIII, aGPCR, and (**f**)

EGF domain is typically cross-linked by three intradomain cystine bridges [35]. The calcium-binding EGF-like domains have a size of ~45 residues and typically contain a single calcium-binding site of the consensus sequence D-X-D-Q/E-C [36]. The ectodomain of L4 (ELTD1/ADGRL4) of receptor subfamily I also contains an EGF and an EGF_Ca domain, and it thus resembles more of a subfamily II aGPCR concerning the ectodomain structure.

Subfamily III consists of three receptors, of which A1 (GPR123) is unique in that it does not contain a GAIN domain but only a ~20-residue-long peptide sequence before the start of the first transmembrane helix. Interestingly, this length corresponds to the typical length of the *Stachel* sequence. The other two receptors of subfamily III, A2 (GPR124) and A3 (GPR125), share 42 % sequence identity displaying homology over the whole sequence range. They consist of a leucine-rich repeat (LRR) domain at the N-terminus, followed by an immunoglobulin (Ig) domain and HormR domain before the GAIN domain (Fig. 4b).

The three subfamily IV receptors share about 35 % sequence identity and have a similar domain structure. At the N-terminus, they contain a region of 237 (C1), 150 (C2), and 292 (C3) residues with low sequence similarity to each other and no predictable domain structures. This region is followed by nine cadherin domains and in total eight calcium-binding EGF domains, two laminin G domains and one laminin EGF domain (Fig. 5a). Each of the domains usually has a compact structure, and the termini are positioned at opposite ends of the domain (except for the laminin G domains) to form an extended chain of domains. This chain may be bent, but it is usually not folded itself into a compact ectodomain structure. For the C1–C3 ectodomains, an extended conformation of the domains would have a length of around 70 nm (in addition to the ~240 residues at the N-terminus and the GAIN/HormR domains). However, the arrangement of the individual domains cannot be predicted with confidence, and the structure in Fig. 5a is meant to illustrate the approximate architecture of the domains forming these large ectodomains.

The two receptors D1 (GPR133) and D2 (GPR144/ADGRD2) of subfamily V aGPCR share only 20 % sequence identity for the complete ectodomain, and both contain a pentaxin domain (Fig. 4c). A HormR domain appears not to be present in the subfamily V receptors.

Subfamily VI comprises five receptors with quite diverse ectodomain structures compared to the subfamilies already described. F3 is the only receptor in this

Fig. 4 (continued) ADGRV1 (VLGR1) of subfamily IX. Domain structures have been modeled with phyre2 [34], and groups of domains have either been arranged in an arbitrary orientation to form an extended chain or are based on a multidomain template structure, if available. The overall sequence identity for many of these models to the template is less than 20 %. Therefore, this figure illustrates the predicted domain architecture of the aGPCR ectodomains shown here, but accurate models for most of the ectodomains cannot be obtained based on current structural information. HormR* indicates that, while the domain is HormR-like in sequence, it is only distantly related to the structurally characterized HormR domains and probably contains only one of the two conserved disulfide bridges

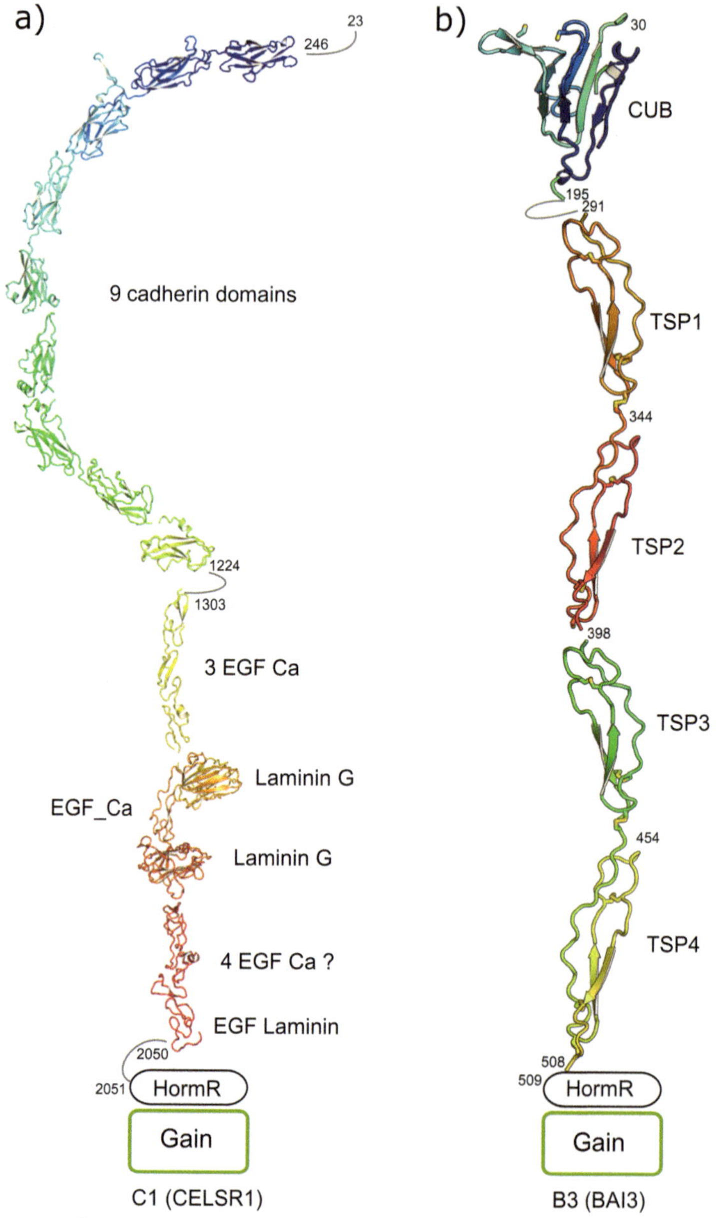

Fig. 5 Models of the ectodomain architecture of C1 and B3. (**a**) ADGRC1 of subfamily IV aGPCR and (**b**) ADGRB3 of subfamily 7. See Fig. 4 legend for further information

subfamily which contains a clearly identifiable HormR domain containing both conserved disulfide bridges. However, also for the other four receptors, a HormR-like fold is probably present, with at least one disulfide bridge conserved. F2 and F4 contain rather short ectodomains comprising only the HormR/GAIN domains. In addition, F2 contains an N-terminal region of less than 100 residues with no predicted homologous domain structures. F5 contains the largest ectodomain in this receptor subfamily (Fig. 4d). It comprises ~140 residues at the N-terminus, with no identifiable homologous folds, followed by a SEA domain (named after its occurrence in the three proteins: sperm protein, enterokinase, and agrin), four Ig-like domains, and the HormR/GAIN domains. F1 also contains an N-terminal fold which is related to that of F5, followed by a SEA domain and the HormR/GAIN domains. It thus lacks the Ig domains. F3 contains the N-terminal fold followed by two Ig domains and the HormR/GAIN domains.

In subfamily VII, the three receptors ADGRB1–3 all contain a HormR domain and upstream four or five thrombospondin (TSP) domains (Fig. 5b). The N-terminal regions of ~250–270 probably contain a CUB domain (for complement C1r/C1s, Uegf, Bmp1) in B1 and B3. As detailed above, the structure of the HormR/GAIN domain of B3 has been elucidated (Fig. 2a). The very N-terminal sequence RCSEQRCP of the crystallized construct probably belongs to the preceding TSP domain, as indicated by sequence comparisons.

Subfamily VIII comprises seven receptors. Some of these do not contain complete GAIN domains. This is most obvious for G5 (GPR114) and G3 (GPR97), which have ectodomains of only 260 and 280 residues, respectively, and a sequence comparison indicates that the first three helices of the helical GAIN subdomain A are missing. Likewise, for G1 (GPR56) and G2 (GPR64), only the C-terminal part of subdomain A appears to be conserved. These receptors contain longer N-terminal regions, but these probably do not correspond to the first three helices of GAIN subdomain A. Domain folds cannot be clearly identified for these sequence regions. In contrast, G4 (GPR112), G6 (GPR126), and G7 (GPR128) contain complete GAIN domains and probably also HormR-like domains with two (G6 and G7) or one (G4) disulfide bridge conserved. G6 contains a CUB and a pentaxin domain at the N-terminus (Fig. 4e). G4 also contains a pentaxin domain at the N-terminus, followed by a long stretch of more than 2000 residues with no significant homology to known domain folds.

Subfamily IX comprises only one receptor, but it has the largest ectodomain with 5878 residues. In addition to the GAIN domain, it probably contains in total 43 Calx-beta motifs, which have been characterized as a tandem repeat in the cytoplasmic domains of Calx sodium-calcium exchangers (Fig. 4f). After the 16th Calx domain, six EARs (epilepsy-associated repeats) are inserted. Each motif contains four β-strands, and it has been speculated that these form a seven-bladed β-propeller, as other proteins usually contain seven EARs.

6 Concluding Remarks

A flurry of recent structural data has recently been published on the conserved C-terminal regions of aGPCRs and the family-specific N-terminal domains of LPHN3. The conserved GAIN domain, harboring the autoproteolytic GPS motif, is of key interest with regard to the recently established activation mechanism of aGPCRs via mechanical force [2, 3, 37, 38]. The ligand-binding domains are thought to act as anchors, enabling mechanical tethering to other cell surface receptors/ligands or matrix proteins through high-affinity interaction. Many aGPCRs are orphan receptors still, but insights into the structures of LPHN3 N-terminal domains have revealed first examples for such a tethering mechanism, using the binding partner FLRT. High-affinity interaction with FLRT is ensured through an extended and highly specific binding interface. Recent structural information [10] also revealed a surprising novel mechanism for higher-order LPHN multimerization via the N-terminal domains, when in complex with two receptors, FLRT and Unc5. This exciting finding suggests that the N-terminal ligand-binding domains have functions beyond mediating simple adhesive interactions and that, at least in the case of LPHNs, they likely determine the receptor's multimerization state. GPCR multimerization is an emerging theme in the field and has so far been suggested for a series of non-aGPCRs [39]. The functional consequences of aGPCR multimerization and how this compares mechanistically with the multimerization seen for other GPCRs via their transmembrane domains remain to be explored.

Taken together, this chapter has aimed to summarize the significant recent progress made with understanding the structural/functional properties of aGPCR extracellular domains, a rapidly progressing field that is receiving increasing attention.

Acknowledgments We would like to thank Gabriel Salzman (University of Chicago) for his help with some of the figures and critical reading of the manuscript. Supported by grants from the Brain Research Foundation (to D.A.) and Big Ideas Generator (to D.A.). N.S. thanks the Deutsche Forschungsgemeinschaft for support in the Research Unit 2149 (Project 5, STR 477/15-1). E.S. was supported by the UK Medical Research Council (MR/L018039/1).

References

1. Araç D, Boucard AA, Bolliger MF, Nguyen J, Soltis SM, Südhof TC et al (2012) A novel evolutionarily conserved domain of cell-adhesion GPCRs mediates autoproteolysis. EMBO J 31:1364–1378. doi:10.1038/emboj.2012.26
2. Liebscher I, Schön J, Petersen SC, Fischer L, Auerbach N, Demberg LM et al (2014) A tethered agonist within the ectodomain activates the adhesion G protein-coupled receptors GPR126 and GPR133. Cell Rep 9:2018–2026. doi:10.1016/j.celrep.2014.11.036
3. Stoveken HM, Hajduczok AG, Xu L, Tall GG (2015) Adhesion G protein-coupled receptors are activated by exposure of a cryptic tethered agonist. Proc Natl Acad Sci U S A 112:6194–6199. doi:10.1073/pnas.1421785112
4. Demberg LM, Rothemund S, Schöneberg T, Liebscher I (2015) Identification of the tethered peptide agonist of the adhesion G protein-coupled receptor GPR64/ADGRG2. Biochem Biophys Res Commun 464:743–747. doi:10.1016/j.bbrc.2015.07.020
5. Wilde C, Fischer L, Lede V, Kirchberger J, Rothemund S, Schöneberg T et al (2016) The constitutive activity of the adhesion GPCR GPR114/ADGRG5 is mediated by its tethered agonist. FASEB J 30:666–673. doi:10.1096/fj.15-276220

6. Vakonakis I, Langenhan T, Prömel S, Russ A, Campbell ID (2008) Solution structure and sugar-binding mechanism of mouse latrophilin-1 RBL: a 7TM receptor-attached lectin-like domain. Structure 16:944–953. doi:10.1016/j.str.2008.02.020
7. Jackson VA, del Toro D, Carrasquero M, Roversi P, Harlos K, Klein R et al (2015) Structural basis of latrophilin-FLRT interaction. Structure 23:774–781. doi:10.1016/j.str.2015.01.013
8. Ranaivoson FM, Liu Q, Martini F, Bergami F, von Daake S, Li S et al (2015) Structural and mechanistic insights into the latrophilin3-FLRT3 complex that mediates glutamatergic synapse development. Structure 23:1665–1677. doi:10.1016/j.str.2015.06.022
9. Lu YC, Nazarko OV, Sando R, Salzman GS, Südhof TC, Araç D (2015) Structural basis of latrophilin-FLRT-UNC5 interaction in cell adhesion. Structure 23:1678–1691. doi:10.1016/j.str.2015.06.024
10. Jackson VA, Mehmood S, Chavent M, Roversi P, Carrasquero M, del Toro D et al (2016) Super-complexes of adhesion GPCRs and neural guidance receptors. Nat Commun 7:11184
11. Krasnoperov VG, Bittner MA, Beavis R, Kuang Y, Salnikow KV, Chepurny OG et al (1997) alpha-Latrotoxin stimulates exocytosis by the interaction with a neuronal G-protein-coupled receptor. Neuron 18:925–937
12. Lin H-H, Chang G-W, Davies JQ, Stacey M, Harris J, Gordon S (2004) Autocatalytic cleavage of the EMR2 receptor occurs at a conserved G protein-coupled receptor proteolytic site motif. J Biol Chem 279:31823–31832. doi:10.1074/jbc.M402974200
13. Krishnan A, Nijmeijer S, de Graaf C, Schiöth HB (2016) Classification, nomenclature and structural aspects of adhesion GPCRs. In: Langenhan T, Schöneberg T (eds) Adhesion G protein-coupled receptors: molecular, physiological and pharmacological principles in health and disease. Springer, Heidelberg
14. Gray JX, Haino M, Roth MJ, Maguire JE, Jensen PN, Yarme A et al (1996) CD97 is a processed, seven-transmembrane, heterodimeric receptor associated with inflammation. J Immunol 157:5438–5447
15. Shashidhar S, Lorente G, Nagavarapu U, Nelson A, Kuo J, Cummins J et al (2005) GPR56 is a GPCR that is overexpressed in gliomas and functions in tumor cell adhesion. Oncogene 24:1673–1682. doi:10.1038/sj.onc.1208395
16. Stacey M, Chang G-W, Sanos SL, Chittenden LR, Stubbs L, Gordon S et al (2002) EMR4, a novel epidermal growth factor (EGF)-TM7 molecule up-regulated in activated mouse macrophages, binds to a putative cellular ligand on B lymphoma cell line A20. J Biol Chem 277:29283–29293. doi:10.1074/jbc.M204306200
17. Liebscher I, Schöneberg T (2016) Tethered agonism: a common activation mechanism of adhesion GPCRs. In: Langenhan T, Schöneberg T (eds) Adhesion G protein-coupled receptors: molecular, physiological and pharmacological principles in health and disease. Springer, Heidelberg
18. Kishore A, Hall RA (2016) Versatile signaling activity of adhesion GPCRs. In: Langenhan T, Schöneberg T (eds) Adhesion G protein-coupled receptors: molecular, physiological and pharmacological principles in health and disease. Springer, Heidelberg
19. Nieberler M, Kittel RJ, Petrenko AG, Lin H-H, Langenhan T (2016) Control of adhesion GPCR function through proteolytic processing. In: Langenhan T, Schöneberg T (eds) Adhesion G protein-coupled receptors: molecular, physiological and pharmacological principles in health and disease. Springer, Heidelberg
20. Sigoillot SM, Monk KR, Piao X, Selimi F, Harty BL (2016) Adhesion G protein-coupled receptors in the nervous system: from synapse and dendrite morphogenesis to myelination. In: Langenhan T, Schöneberg T (eds) Adhesion G protein-coupled receptors: molecular, physiological and pharmacological principles in health and disease. Springer, Heidelberg
21. Kan Z, Jaiswal BS, Stinson J, Janakiraman V, Bhatt D, Stern HM et al (2010) Diverse somatic mutation patterns and pathway alterations in human cancers. Nature 466:869–873. doi:10.1038/nature09208

22. Arcos-Burgos M, Jain M, Acosta MT, Shively S, Stanescu H, Wallis D et al (2010) A common variant of the latrophilin 3 gene, LPHN3, confers susceptibility to ADHD and predicts effectiveness of stimulant medication. Mol Psychiatry 15:1053–1066. doi:10.1038/mp.2010.6
23. Paavola KJ, Stephenson JR, Ritter SL, Alter SP, Hall RA (2011) The N terminus of the adhesion G protein-coupled receptor GPR56 controls receptor signaling activity. J Biol Chem 286:28914–28921. doi:10.1074/jbc.M111.247973
24. Yang L, Chen G, Mohanty S, Scott G, Fazal F, Rahman A et al (2011) GPR56 regulates VEGF production and angiogenesis during melanoma progression. Cancer Res 71:5558–5568. doi:10.1158/0008-5472.CAN-10-4543
25. Grace CRR, Perrin MH, Gulyas J, Digruccio MR, Cantle JP, Rivier JE et al (2007) Structure of the N-terminal domain of a type B1 G protein-coupled receptor in complex with a peptide ligand. Proc Natl Acad Sci U S A 104:4858–4863. doi:10.1073/pnas.0700682104
26. Hoare SRJ (2005) Mechanisms of peptide and nonpeptide ligand binding to Class B G-protein-coupled receptors. Drug Discov Today 10:417–427. doi:10.1016/S1359-6446(05)03370-2
27. Bjarnadóttir TK, Fredriksson R, Höglund PJ, Gloriam DE, Lagerström MC, Schiöth HB (2004) The human and mouse repertoire of the adhesion family of G-protein-coupled receptors. Genomics 84:23–33. doi:10.1016/j.ygeno.2003.12.004
28. Davletov BA, Shamotienko OG, Lelianova VG, Grishin EV, Ushkaryov YA (1996) Isolation and biochemical characterization of a Ca2+-independent alpha-latrotoxin-binding protein. J Biol Chem 271:23239–23245
29. Silva J-P, Lelianova VG, Ermolyuk YS, Vysokov N, Hitchen PG, Berninghausen O et al (2011) Latrophilin 1 and its endogenous ligand Lasso/teneurin-2 form a high-affinity transsynaptic receptor pair with signaling capabilities. Proc Natl Acad Sci U S A 108:12113–12118. doi:10.1073/pnas.1019434108
30. Boucard AA, Ko J, Südhof TC (2012) High affinity neurexin binding to cell adhesion G-protein-coupled receptor CIRL1/latrophilin-1 produces an intercellular adhesion complex. J Biol Chem 287:9399–9413. doi:10.1074/jbc.M111.318659
31. O'Sullivan ML, de Wit J, Savas JN, Comoletti D, Otto-Hitt S, Yates JR et al (2012) FLRT proteins are endogenous latrophilin ligands and regulate excitatory synapse development. Neuron 73:903–910. doi:10.1016/j.neuron.2012.01.018
32. Seiradake E, del Toro D, Nagel D, Cop F, Härtl R, Ruff T et al (2014) FLRT structure: balancing repulsion and cell adhesion in cortical and vascular development. Neuron 84:370–385. doi:10.1016/j.neuron.2014.10.008
33. Yamagishi S, Hampel F, Hata K, Del Toro D, Schwark M, Kvachnina E et al (2011) FLRT2 and FLRT3 act as repulsive guidance cues for Unc5-positive neurons. EMBO J 30:2920–2933. doi:10.1038/emboj.2011.189
34. Kelley LA, Mezulis S, Yates CM, Wass MN, Sternberg MJE (2015) The Phyre2 web portal for protein modeling, prediction and analysis. Nat Protoc 10:845–858. doi:10.1038/nprot.2015.053
35. Wouters MA, Rigoutsos I, Chu CK, Feng LL, Sparrow DB, Dunwoodie SL (2005) Evolution of distinct EGF domains with specific functions. Protein Sci 14:1091–1103. doi:10.1110/ps.041207005
36. Handford P, Downing AK, Rao Z, Hewett DR, Sykes BC, Kielty CM (1995) The calcium binding properties and molecular organization of epidermal growth factor-like domains in human fibrillin-1. J Biol Chem 270:6751–6756
37. Scholz N, Gehring J, Guan C, Ljaschenko D, Fischer R, Lakshmanan V et al (2015) The adhesion GPCR latrophilin/CIRL shapes mechanosensation. Cell Rep 11:866–874. doi:10.1016/j.celrep.2015.04.008
38. Hamann J, Aust G, Araç D, Engel FB, Formstone C, Fredriksson R et al (2015) International Union of Basic and Clinical Pharmacology. XCIV. Adhesion G protein-coupled receptors. Pharmacol Rev 67:338–367. doi:10.1124/pr.114.009647
39. Zhang D, Zhao Q, Wu B (2015) Structural studies of G protein-coupled receptors. Mol Cells 38:836–842, doi:10.14348/molcells.2015.0263

Control of Adhesion GPCR Function Through Proteolytic Processing

Matthias Nieberler, Robert J. Kittel, Alexander G. Petrenko, Hsi-Hsien Lin, and Tobias Langenhan

Graphical Abstract

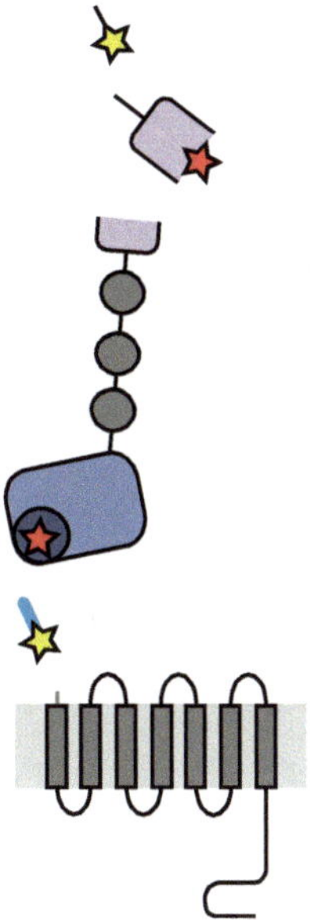

Proteolytic processing events in adhesion GPCRs. aGPCRs can undergo multiple autoproteolytic (*red asterisks*) and proteolytic processing events by exogenous proteases (*yellow asterisks*) that may be involved in signaling events of the receptors.

Contents

M. Nieberler • R.J. Kittel • T. Langenhan (✉)
Department of Neurophysiology, Institute of Physiology, University of Würzburg, Röntgenring 9, Würzburg 97070, Germany
e-mail: tobias.langenhan@gmail.com

T. Langenhan, T. Schöneberg (eds.), *Adhesion G Protein-coupled Receptors*,
Handbook of Experimental Pharmacology 234, DOI 10.1007/978-3-319-41523-9_5

Abstract

Proteolytic processing is an unusual property of adhesion family G protein-coupled receptors (aGPCRs) that was observed upon their cloning and biochemical characterization. Ever since, much effort has been dedicated to delineate the mechanisms and requirements for cleavage events in the control of aGPCR function. Most notably, all aGPCRs possess a juxtamembrane protein fold, the GPCR autoproteolysis-inducing (GAIN) domain, which operates as an autoprotease for many aGPCR homologs investigated thus far. Analysis of its autoproteolytic reaction, the consequences for receptor fate and function, and the allocation of physiological effects to this peculiar feature of aGPCRs has occupied the experimental agenda of the aGPCR field and shaped our current understanding of the signaling properties and cell biological effects of aGPCRs. Interestingly, individual aGPCRs may undergo additional proteolytic steps, one of them resulting in shedding of the entire ectodomain that is secreted and can function independently. Here, we summarize the current state of knowledge on GAIN domain-mediated and GAIN domain-independent aGPCR cleavage events and their significance for the pharmacological and cellular actions of aGPCRs. Further, we compare and contrast the proteolytic profile of aGPCRs with known signaling routes that are governed through proteolysis of surface molecules such as the Notch and ephrin pathways.

A.G. Petrenko
Laboratory of Receptor Cell Biology, Shemyakin-Ovchinnikov Institute of Bioorganic Chemistry, Moscow, Russia

H. Lin (✉)
Department of Microbiology and Immunology, College of Medicine, Chang Gung University, 259 Wen-Hwa 1st Road, Kwei-San, Tao-Yuan, Taiwan

Chang Gung Immunology Consortium and Department of Anatomic Pathology, Chang Gung Memorial Hospital-Linkou, Tao-Yuan 333, Taiwan
e-mail: hhlin@mail.cgu.edu.tw

Keywords
Adhesion GPCR • GAIN domain • GPS • Proteolysis • Autoproteolysis

1 Forms of Proteolytic Processing Events in aGPCRs

1.1 GAIN-Mediated GPS Cleavage

One of the structural and functional hallmarks of aGPCRs is the juxtamembrane localization of a highly conserved GPCR proteolysis site (GPS) motif (Fig. 1) [1–3]. All aGPCRs, except GPR123/ADGRA1, contain the GPS motif [1]. Proteolytic modification of aGPCRs was first reported for CD97/ADGRE5 in 1996 by Kelly and colleagues [4]. They revealed a novel two-subunit structure of CD97, consisting of an extracellular fragment and a seven-transmembrane (7TM) fragment derived from a proprotein precursor. Petrenko et al. later identified the cleavage site and coined the term GPS to describe the proteolytic processing of CIRL/latrophilin/ADGRL1 [5, 6].

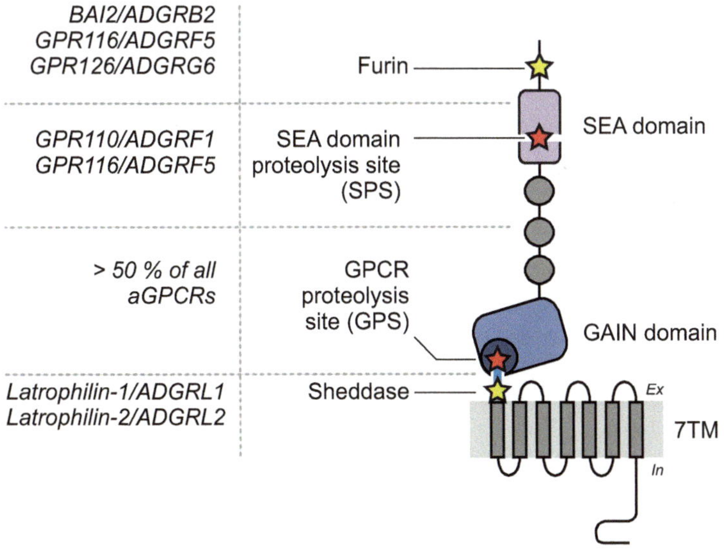

Fig. 1 Proteolytic processing events of adhesion GPCRs. aGPCRs can undergo multiple autoproteolytic (*red asterisks*) and proteolytic processing events by exogenous proteases (*yellow asterisks*) that may be involved in signaling events of the receptors. The most prevalent cleavage of aGPCR family occurs at the GPCR proteolytic site (GPS; *dark blue circle*) and is catalyzed by the GPCR autoproteolysis-inducing (GAIN) domain (*light blue*). Another type of autoproteolysis is governed by the SEA domain (*pink box*) and shares similarities with GAIN domain cleavages. aGPCRs can also be substrates for proteases and release parts of the ECD. Exemplary aGPCR homologs and associated proteolytic processing events are indicated on the *left*

The GPS motif of ~50 amino acids contains a highly conserved tripeptide cleavage sequence and several canonical cysteine and tryptophan residues [2]. Moreover, the 6–8 residues C-terminal to the cleavage site are usually small and hydrophobic [7]. The cleavage tripeptide almost always starts with His, followed by Leu/Ile and Ser/Thr, with proteolysis occurring between Leu/Ile and Ser/Thr (HL/I↓S/T) [2, 3]. Most interestingly, GPS proteolysis is not executed by any proteinases, but is brought about by an autocatalytic mechanism analogous to that of *hedgehog* morphogens [8, 9] and Ntn-hydrolases [10–14]. It is concluded that the GPS proteolytic reaction is most likely initiated by the deprotonation of the hydroxyl group of the P^{+1} residue (Ser/Thr) by the P^{-2} His residue. This is followed by a *cis*-nucleophilic attack on the α-carbonyl carbon of the P^{-1} Leu/Ile residue, producing a tetrahedral intermediate. An ester intermediate is subsequently generated via an N→O acyl shift. Finally, the attack by H_2O cleaves the ester bond splitting the receptor into two protein fragments (Fig. 2) [11]. The two fragments usually do not separate after proteolysis, but instead associate non-covalently to form a mature heterodimeric receptor complex on the cell surface [6, 15]. Interestingly, the GPS motif is absolutely necessary for proteolysis, but is insufficient to mediate the autoproteolytic reaction on its own [7].

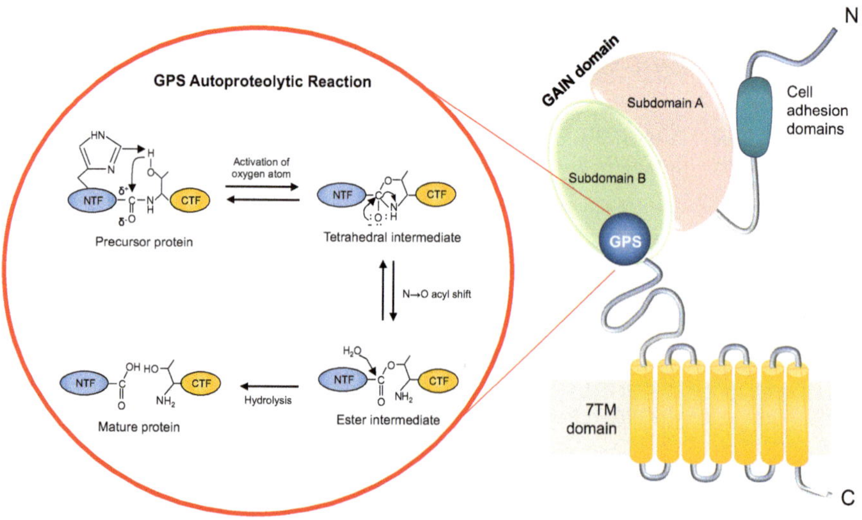

Fig. 2 The GAIN domain and GPS autoproteolysis of adhesion GPCRs. A schematic diagram of an aGPCR is shown. Located at the C-terminal half of the NTF, the GAIN domain is divided into the subdomain A (*pink*) and subdomain B (*light green*). The GPS motif (*blue*) is a part of subdomain B. The proposed mechanism of the GPS autoproteolytic reaction is shown inside the *red circle*. A His or another general base withdraws a proton from the hydroxyl group of a Ser or Thr at position +1. The resulting negatively charged oxygen makes a nucleophilic attack on the carbonyl group of the residue at position −1 (e.g., a Leu), yielding a tetrahedral intermediate and subsequently an ester intermediate. The resulting ester is then hydrolyzed to produce the NTF and CTF that form the mature protein

Interestingly, while it is well accepted now that the GPS proteolysis is a self-catalytic intramolecular reaction, the proteolytic efficiency is not always complete. Indeed, both processed and unprocessed GPR56/ADGRG1 and polycystin-1 receptors have been detected in vivo [16–18]. Moreover, crystals of two structurally similar GPS-containing fragments of CIRL/latrophilin and BAI3/ADGRB3 were described, the former in a completely cleaved and the latter in a non-cleaved conformation [19]. Hence, it is suggested that the GPS domain-containing receptors might adopt receptor folding conditions that either promote or demote GPS cleavage depending on cell types and cellular environments [20, 21].

One critical factor regulating the GPS proteolysis is found to be the first step of N-glycosylation during receptor biosynthesis in the ER [20, 21]. Other potential factors include the specific conditions of aGPCR expression, including the cell type and expression levels. While in the brain tissue only the cleaved form of CIRL/latrophilin is detected, its heterologous expression in transfected cells yields only a minor portion of the processed form that depends on the cell line used [5, 15]. Interestingly, the receptor cleavage yield in transfected cells can be regulated by pharmacological agents such as PMA and ionomycin, regulators of the protein kinase pathway, suggesting the existence of intracellular signaling mechanisms to fine-tune the autoproteolysis [22]. As with any chemical reaction, one can also anticipate a contribution of local pH and ionic changes, as well as a direct involvement of available nucleophilic molecules that serve as cofactors.

Pulse-chase experiments and use of various recombinant receptors and protein trafficking inhibitors have identified the ER lumen as the major subcellular localization of GPS proteolysis [2, 4, 11, 15]. However, due to the highly regulated nature of the GPS proteolytic reaction mentioned above, it is likely that GPS proteolysis could also occur at a later time point during protein maturation or perhaps even at the cell surface. Indeed, cell type-specific location of the GPS cleavage was reported for polycystin-1, whose proteolysis may occur within the ER or post-ER [21].

Recent structural analyses finally delineate the peculiar requirement and characteristics of GPS autoproteolysis. Crystallization of two aGPCRs, Latrophilin-1/ADGRL1 and BAI3, identifies a much larger extracellular GPCR autoproteolysis-inducing (GAIN) domain (~320 residues) that is sufficient and minimally required for GPS autoproteolytic reaction [19] (see also [23]). In fact, the GPS motif is an integral part of the GAIN domain. The crystal structure of the GAIN domain shows a subdomain A of 6 α-helices and a subdomain B consisting of a twisted β-sandwich of 13 β-strands and 2 small α-helices [19, 24]. The GPS motif is enclosed in the last five β-strands of subdomain B, and the cleavage takes place in a short kinked loop between the last two β-strands (Fig. 2) [19, 24]. The proper folding of the GAIN domain, hence the arrangement of the scissile bond in a unique configuration, provides an essential environment for the GPS autoproteolytic reaction. Due to the lack of the conformational constraint and chemical environment required for the proteolytic reaction, the GPS motif alone cannot mediate autoproteolysis. In addition, the cleaved last β-strand is tightly embedded within the rest of the GAIN domain, hence it is energetically unfavorable for the two fragments to dissociate [19, 24].

Interestingly, phylogenetic analysis shows that the GAIN domain is evolutionarily conserved from tetrahymena to mammals. In fact, it is believed that the GAIN domain is one of the most evolutionarily ancient and functional autoproteolytic protein folds identified to date [19, 24]. The close proximity to the TM region and the unique structural requirement for the GPS autoproteolytic reaction all suggest an important role for the GAIN domain in aGPCR function [25]. Furthermore, the GAIN domain is also present in all members of human polycystic kidney disease 1 (PKD1) protein family, suggesting a much wider usage of this novel domain in receptor biology [19, 24]. Indeed, sequence mutations in the GAIN domain have been linked to various human diseases, a clear indication of its role in regulating receptor activities [17, 24, 26, 27]. How the GAIN domain-mediated autoproteolysis may regulate receptor signaling and function will be discussed in the later sections.

1.2 Other Autoproteolytic Cleavages of aGPCRs

Apart from the GPS autoproteolysis, additional autoproteolytic reactions were noted for certain aGPCRs. Abe et al. showed that Ig-Hepta (GPR116/ADGRF5) undergoes two specific proteolytic events in the extracellular region [28]. One of the proteolytic sites identified is at the GPS motif, while the other is at the SEA module located at the N-terminus of the receptor (Fig. 1). Identified first in three different proteins (sea urchin sperm protein, enterokinase, and agrin) [29], the SEA module is a conserved extracellular protein motif of ~80–110 residues usually found in O-glycosylated mucin-like membrane proteins such as MUC1, MUC3, MUC12, MUC13, and MUC17 [30–32]. A highly conserved G↓S[V/I]VV sequence is identified as the SEA domain proteolysis site (SPS) [32].

Interestingly, the SEA module-mediated proteolysis shares many similar characteristics with GPS autoproteolysis. First, although some SEA module-containing molecules are soluble proteins, the SEA module is mostly found in cell-surface proteins and is located at the extracellular region of the molecule, near or close to the TM region [32]. Second, the proteolytic modification takes place within the ER during early protein biosynthesis. Third, proteolysis only proceeds when the P^{+1} cleavage site is a residue containing a terminal hydroxyl group (Ser and Thr) or thiol group (Cys) [33]. Fourth, proteolysis at the SEA domain is an autocatalytic intramolecular reaction likely mediated by a series of nucleophilic attacks and the formation and hydrolysis of an ester intermediate via an N→O acyl shift and H_2O, respectively [33–35]. Fifth, the autoproteolytic reaction is achieved by conformational strain and requires strict and proper protein folding [34, 36, 37]. Finally, the resulting cleaved fragments remain associated non-covalently following proteolysis [33, 35].

Two aGPCRs, GPR110/ADGRF1 and GPR116, are known to contain both the SEA module and the GAIN domain [1]. Indeed, multiple proteolytic modifications of GPR116 have been identified, leading to the formation of a mature receptor with many non-covalently associated fragments [28, 38]. GPR116 has been linked to a number of physiological and pathological processes such as pulmonary surfactant homeostasis, insulin insensitivity, and breast cancer metastasis [39–44]. However, the role of autoproteolysis in the SEA module and GAIN domain in GPR116 function has yet to be investigated. Little is known regarding the proteolytic modification of GPR110.

1.3 Cleavage of aGPCRs by Other Proteases

With aGPCR research on the rise, more and more homologs are identified as targets of classical proteases such as furin or matrix metalloproteinase (MMP; Fig. 1). These include BAI1/ADGRB1, BAI2/ADGRB2, GPR116, GPR126/ADGRG6, and Latrophilin-1/ADGRL1 [38, 45–48]. Furin, a subtilisin-like proprotein convertase, is a calcium-dependent serine endoprotease enriched in Golgi and is involved predominantly in intracellular protein processing within the secretory pathway [49]. Consistent with previous reports, the proteolytic site of BAI2, GPR116, and GPR126 by furin was identified right after an Arg residue of a consensus furin-cleavage sequence [38, 47, 48]. Interestingly, these furin-cleavage sites are all located at the extracellular region N-terminal to the GPS and SEA domain. One exception is the furin processing of Latrophilin-1, which occurs before an Arg residue located C-terminal of the GPS motif within the CTF [46]. The furin-cleaved aGPCR fragment was shown to either remain associated with the rest of the molecule (GPR116) or released to the extracellular milieu (BAI2, GPR126, Latrophilin-1). The functional significance of the furin-mediated proteolysis of aGPCRs is currently unknown, but additional functions exerted by the shed receptor ectodomain remain a possibility. Modulation of aGPCR activity by furin-mediated shedding is also an alternative.

BAI1, initially identified as a brain-specific p53-regulated gene, is highly expressed in normal but not tumor brain cells [50, 51]. GPS proteolysis of BAI1 released a 120 kDa thrombospondin type-1 repeat (TSR)-containing "vasculostatin" fragment with anti-angiogenic and anti-tumorigenic function [50]. Later studies revealed another extracellular cleavage mediated by MMP-14 at a more N-terminal region, producing a 40 kDa (vasculostatin-40) fragment also with very potent anti-angiogenic activity [45]. In fact, the second cleavage of BAI1 is processed by a two-step protease activation cascade in which the latent MMP-14 is activated by furin [45]. Interestingly, the generation of vasculostatin-120 by GPS autoproteolysis is not a prerequisite for vasculostatin-40 production by MMP-14. Hence, intra- and extracellular proteolytic processing of BAI1 to distinct ectodomain fragments by GPS autoproteolysis and MMP-14, respectively, represents important activation and regulatory mechanisms for the BAI1 receptor function [45].

Another interesting example of aGPCR cleavage involving a sheddase is the dissociation of a CIRL/Latrophilin-1 two-subunit complex at the cell surface that results in the secretion of its ectodomain that contains the intact GAIN domain (Fig. 1). About 5 % of the endogenous brain-expressed CIRL/latrophilin undergoes this processing. The soluble receptor form is comprised of NTF linked to a small peptide fragment of CTF. This peptide was identified by mass spectrometry indicating the location of the second cleavage site at the border between the GAIN domain and the 7TM core. Similar processing was also shown for CIRL-2/ADGRL2 [46].

2 Biological Effects Controlled Through aGPCR Proteolysis

The consequences and roles of the autoproteolytic processing of aGPCRs have been under intense scrutiny since its discovery. Multiple experimental approaches have been implemented to grasp this biochemical peculiarity of aGPCRs, and several conclusions have been drawn from the results. We will discuss the most popular ones below.

2.1 Trafficking

Several cell physiological consequences have been ascribed to the autoproteolytic processing of aGPCRs at the GPS. Insights into these features derived from studies of aGPCR and polycystin-1 homologs, in which the consensus site was mutated at different positions in order to disable the autocatalytic reaction. The Latrophilin-1 homolog with a GPS disrupting mutation was the first receptor that was scrutinized this way. It was noted that the GPS-deficient Latrophilin-1 variant did not traffic to the cell surface lending support to a model, in which the posttranslational cleavage event may function as a maturation signal during the biosynthesis of the receptor molecule in the ER [15]. Later on, this hypothesis was further explored in several other aGPCRs and polycystins returning mixed results: while impeded surface expression was found for proteolysis-deficient versions of Latrophilin-1 [15] and GPR126 [52], no such effect was noted for polycystin-1 [17], GPR133/ADGRG1 [53], and the nematode latrophilin homolog LAT-1 [54]. Also Latrophilin-1 was reprobed and several GPS cleavage mutations did not affect cell-surface transport of the receptor [19].

Also the GAIN-mediated cleavage of polycystin-1 has drawn interest to its physiological requirement, and its investigation contributed insights into the role of the proteolysis event. An allele of *PKD1*, which encodes for a cleavage-deficient polycystin-1 product, leads to strong hypomorphic phenotypes that manifested through defects in the development of kidney tubules [18] (see below). Follow-up work on this effect suggests that the CTF of polycystin-1 may act as a cofactor that is required for membrane trafficking of the NTF. The NTF subsequently detaches from the CTF, but remains associated to the membrane, probably through other surface receptors [55]. Similar findings were obtained for the NTF of the aGPCR Latrophilin-1, whose CTF may also exist as a separate protomer at the cell surface [56].

Hence, it remains controversial whether GPS autoproteolysis is functioning as a gatekeeping step in the biosynthesis and maturation of aGPCRs. One solution to this puzzle may be offered by the observation that several potentially GPS-disabling mutations rather lead to reduced stability and unfolding of the GAIN domain and consequently do not traffic properly to the cell membrane. Further, GPS cleavage appears to be dependent on cell context and other posttranslational modifiers such as glycosylation [17, 20]. Therefore, recombinant expression of aGPCRs in heterologous cell lines—the classical test system utilized for cleavage assays—may not provide the necessary cofactors or conditions that are required for efficient GAIN proteolysis.

2.2 Terminating Adhesion

An obvious role for the autoproteolytic cleavage of aGPCRs is one that has remained unexplored thus far. Movements during proliferation, migration, polarity establishment, but also postmitotic motion of cells or their context impose considerable forces on cells, which are counteracted by adhesion molecules such as cadherins, laminins, or integrins [57].

In this vein, aGPCRs possess an extensive repertoire of adhesion domains that are located in the ectodomain of most of the receptor homologs (see also [23]). aGPCRs are exposed to and likely engage in binding events with adhesive partner molecules that are affixed either within the extracellular matrix lattice or anchored on opposite cell surfaces [25]. Thus, autoproteolytic cleavage of aGPCRs may determine a threshold for forces transmitted onto the receptor expressing cells, above which the NTF and CTF are separated and relieved of their adhesive interaction. Such a role was suggested for other surface-mounted molecules such as mucins (see above), which line the surface of mucous epithelia. By means of an autocatalytically active SEA domain, potentially damaging shear forces that endanger the epithelial barrier are limited to the energy that is necessary to split the two non-covalently bound cleavage fragments of mucins [35].

2.3 Triggering Metabotropic Signaling

With the advent of molecular models on the activation mechanism of aGPCRs, and their suspected role as mechanoreceptors, receptor autoproteolysis receives increasing attention as a potentially crucial component in these processes.

As discussed in detail in [58, 59], several aGPCRs possess a tethered agonist that is an integral part of the receptor molecule. Structure-function studies of GPR56 implied that the NTF of an aGPCR exerts an inhibitory role on the metabotropic and biological activity of its CTF. This conclusion was based on receptor variants that either contained a shortened or no NTF at all, which displayed increased activation of cellular behaviors [44] and downstream effectors [60], respectively. These observations were explained by two models: either the NTF directly suppresses metabotropic activity of the CTF consistent with the function of a tethered inverse agonist, or alternatively, the NTF counteracts the activity of a tethered agonist of the CTF [61]. Both models account for the disinhibiting effects of NTF removal. Studies on LAT-1 (see also [62]) provided evidence for the latter model. A panel of LAT-1 receptor variants was scored for their capacity to rescue the penetrant developmental lethality caused through removal of the *lat-1* gene in *C. elegans*. In the course of this study, it was noted that neither a receptor that lacks the 7TM domain nor a full-length chimeric version containing a foreign GPS motif of the GAIN domain was able to remedy the lethal effects of *lat-1* deletion. However, when both receptor variants were co-expressed, they complemented each other intermolecularly to reestablish the full biological functionality of the wild-type receptor. The conclusion drawn from this set of experiments suggested that the GPS motif interacts with the 7TM domain in an agonistic fashion [54].

Further investigations unveiled the molecular underpinnings of this effect and supplied further evidence for the model that aGPCR signaling can be activated through a tethered agonist. The stalk region that links the GPS with the first TM helix, a peptide of approximately 15–25 amino acids in length depending on individual receptor homologs, comprises an agonistic activity that stimulates metabotropic signaling of aGPCRs. When truncated receptor versions that lack the NTF are expressed, the agonist (termed *Stachel*; German: sting, or alternatively *stalk*) is exposed and conceivably interacts with the 7TM continuously leading to high signaling activity as observed before. Receptor layouts that lack the entire ECD (i.e., including the *Stachel*/*stalk*), however, are muted, but can be reactivated by high amounts of soluble *Stachel*/*stalk* peptide indicating that the tethered agonist is necessary and sufficient for receptor activation. This was shown first for GPR126 and GPR133 [52] and subsequently for additional receptors including GPR56 [63], GPR64/ADGRG2 [64], GPR114/ADGRG5 [65], Latrophilin-1, and LAT-1 [66].

Interestingly, the agonistic property of the peptide appears to reside in its N-terminal half [65], which also represents the last beta-sheet of the GAIN domain that is severed through the autocatalytic event from the much larger rest of the fold. In cleavage-competent receptor homologs, the *Stachel*/*stalk* therefore constitutes the very beginning of the CTF, which also mediates the non-covalent lock between NTF and CTF that results in the heterodimeric configuration in which aGPCRs are found on the cell membrane [19].

How is exposure of the *Stachel*/*stalk* enacted under physiological conditions? As the agonist is buried inside the GAIN domain, the simplest mode would see the NTF removed through a combination of firm ligand engagement with the extracellular adhesion domains through which mechanical force is transmitted onto the NTF that pulls it off the CTF. This way, the *Stachel*/*stalk* sequence would become instantly exposed. Corroboration of the interplay between mechanical challenge and transmembrane signal transduction has recently been found in EMR2/ADGRE2: Boyden et al. identified two kindreds that displayed symptoms of severe vibratory urticaria, a condition associated with degranulation of mast cells upon dermal challenge with physical force. In this study, an autosomal-dominant missense mutation in EMR2/ADGRE2 was shown to underlie these effects. In vitro experiments with mast cells transfected with the mutated receptor variant indicated that removal of the NTF through vibratory shear stress was increased [67]. This is consistent with model in which elevated exposure of the tethered agonist (*Stachel*/stalk) triggers subsequent downstream signaling events and is further discussed in [68]. Also for other protease-activated membrane receptor systems, e.g., the Notch-DSL pathway (see below), similar mechanisms, executed through proteolysis by an exogenous protease, were proposed [69].

In this context, GAIN autoproteolysis would be an essential precondition for the liberation of the tethered agonist upon mechanical stimulus encounter and a satisfactory explanation for its evolutionary conservation. However, also non-cleavable aGPCRs appear to possess agonistic activity in the *Stachel*/*stalk* peptide and are sensitive to mechanical stimulation, at least in vitro, as recently shown for GPR114 [65]. To complicate matters, recent studies indicate that aGPCR engage in *Stachel*-independent metabotropic (CTF-dependent) signaling [70], which is discussed in detail in [59].

Currently, there is no obvious explanation for how the encounter between the agonist and the 7TM may be facilitated assuming that the available GAIN domain structures are representing the physiological conformation of the fold (see also [58, 71]). Alternatively, there exist steric layouts of the GAIN domain that allow access of the *Stachel*/*stalk* to its cognate 7TM interface even if the agonist is an integral part of a contiguous polypeptide chain rather than released through the autoproteolytic cleavage. Such conformations are subject to future investigations and will help answering the question for the role of aGPCR autoproteolysis.

2.4 Liberation of NTF for Cell-Non-autonomous Effects

An interesting addition to the cell-autonomous information fed into the Notch-expressing cell, the Notch-DSL interactions also appear to drive cell-non-autonomous events in the ligand-expressing cells. Also this feature of the Notch pathway may compare to properties of several aGPCR homologs and their capacity to not only act as signal sensors but also senders of information. A well-studied example is the effect of the N-terminal fragment (NTF) of Gpr126/ADGRG6 during mouse and zebra fish heart development [72, 73]. Gpr126 is expressed by endocardial cells but not cardiomyocytes and is essential for cardiac mitochondrial function and trabeculation of the heart. Genetic structure-function studies have indicated that the C-terminal fragment of Gpr126, which contains the metabotropic signaling unit of the aGPCR [52], is dispensable for these effects while they critically depend on the NTF of Gpr126. Interestingly, this requirement is shared by endocardial cells and cardiomyocytes, of which the latter do not express the receptor molecule. Immunolocalization studies further detailed that Gpr126 may work in a paracrine mode to exert its function on cardiomyocytes, possibly by shedding its NTF and thereby governing the development of these cells in a cell-non-autonomous fashion [72].

2.4.1 Split Personality Hypothesis

While the NTF and CTF of most aGPCRs are associated non-covalently, it was found that the two fragments could also be expressed separately on the cell surface as independent entities in some aGPCRs such as Latrophilin-1 and EMR2/ADGRE2 [56, 74, 75]. The so-called split personality hypothesis was coined to reflect the fact that the NTF remains free even though the CTF is pulled down exhaustively by immunoprecipitation [56, 76]. Furthermore, expression of the CTF-truncated recombinant Latrophilin-1 was found to remain tethered on the cell membrane. Most interestingly, the NTF could be efficiently removed from the membrane without solubilizing any CTF when cells were treated with perfluorooctanoic acid, a weak detergent that does not disrupt the lipid bilayer. These results strongly suggest that some NTF is self-anchored on the membrane independently of the CTF.

Indeed, subsequent studies showed that membrane localization of the two fragments does not completely overlap and the fragments could even be internalized independently [56]. Further, it is possible to detect ligand-induced interaction of individual NTF and CTF from the same receptor molecule fused with different tags (so-called homogeneric heterodimers), or even from two distinct aGPCRs (heterogeneric heterodimers; e.g., $NTF^{Latrophilin\text{-}1}$::$CTF^{GPR56}$ or NTF^{EMR4}::CTF^{EMR2}) [56, 74, 75]. For EMR2, it was shown that the NTF and CTF were differentially distributed in lipid raft microdomains and ligation of the NTF by EMR2-specific monoclonal antibodies induced the translocation and interaction of NTF with CTF to the lipid rafts for receptor activation and signaling [74]. Consistent with these findings, GPS proteolysis of aGPCRs could possibly create diverse functional receptor complexes by cross association of independent NTFs and CTFs of different aGPCRs.

Several possibilities exist as to how such molecular cross-chimerization may come about. Receptor fragments may either recombine after GAIN cleavage at the GPS. For this scenario GPS cleavage is absolutely necessary. Alternatively, aGPCRs may form heterodimers at the level of the 7TM domain, ECD, or ICD that may lead to crosswise pulldown results interpreted as heterogeneric heterodimer formation. Only in one study thus far, these possibilities were tested by the use of GPS cleavage-incompetent receptor forms, which still showed co-immunoprecipitation [77]. The authors thus concluded that homo- and heterogeneric cross talk of aGPCRs is likely the result of receptor oligomerization that does not involve NTF-CTF re-pairing at the GPS, but rather the lateral interaction of several aGPCR molecules.

Future investigations will need to further define the properties of GPS proteolysis for separate fates and biological activities of aGPCR fragments.

3 Similarities and Differences to Other Proteolysis-Dependent Signaling Pathways

aGPCRs are by far not the only group of biomolecules whose actions are controlled through proteolytic cleavage (Fig. 3, Table 1). Here, we only concentrate on those that are governed by the proteolytic processing of surface receptors. However, we note that also a wealth of other biological signals depend on the proteolytic activation of precursor states of intracellular or secreted substrates. This includes the shedding of N-terminal signal peptides through signal peptide peptidases or the functionalization of prohormones and proenzymes into active molecule species, such as proinsulin in pancreatic β cells or serine proteases in the gastrointestinal system, respectively.

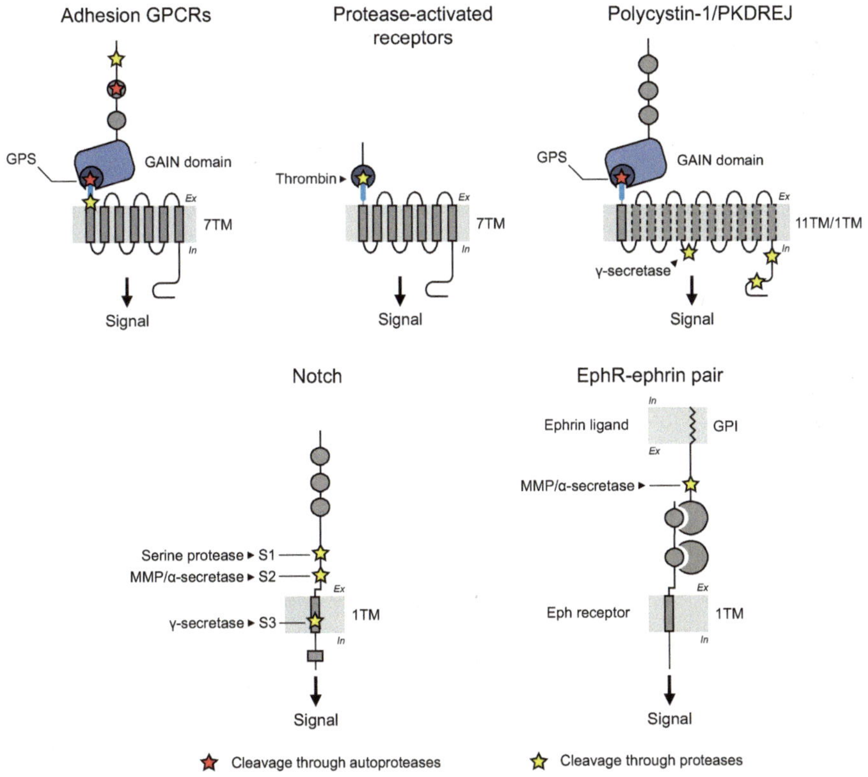

Fig. 3 Molecular pathways controlled though proteolytic processing. aGPCR processing through self-cleavage and cleavage by proteases is implicated in several biological properties of these receptors including critical steps in their signaling cascade. In this respect, aGPCRs may share signaling principles with other receptor systems that rely on proteolysis to trigger and/or transduce extracellular events into intracellular information. These encompass, among others, the protease-activated receptor group of GPCRs, polycystin-1/PKDREJ, Notch, and ephrin receptor families of cell-surface receptors

Table 1 Signaling pathways executed and/or modulated through non-GAIN-domain-mediated proteolysis

Substrate	Serine protease	MMP/ α-secretase	β-secretase	γ-secretase	Selected physiological effects of proteolysis
EGF, TGF-α, neuregulins, etc.		●			Shedding of substrate for EGFR: cell growth, migration, proliferation, differentiation [92, 93]
TNF-α		●			Shedding of substrate for cytokine receptor: inflammation, apoptosis, necrosis [94, 95]
APP		●		●	Physiological tasks: learning, memory, synaptic plasticity (LTP) [96–98]
			●	●	Pathological effects: accumulation of Amyloid-β protein in Alzheimer's Disease [98, 99]
Notch	●	●		●	Binary developmental commands involved in cell fate decisions; NCID: transcriptional regulation [100] of inner ear development [101], pancreatic β cell production [102], specification of goblet cells in the intestine [103], various functions in B and T cell development [104, 105], etc.
N-cadherin		●		●	Cell-cell adhesion [106]; β-catenin transcriptional control: neuronal differentiation, synaptogenesis, axonal remodeling (synaptic plasticity) [107]; cell proliferation and cell survival (cancer) [106]
Ephrin-A2		●			Contact-mediated axon repulsion [108, 109]
Neuroligin-1		●		●	Negative regulation of spine remodeling at excitatory synapses [110]
Neurexin-3β		●		●	Glutamatergic synapse assembly and possibly also function [111, 112]
PAR1	●				Fibrin conversion (platelet aggregation), endothelial cell activation, vasoconstriction/vasodilation, immune regulation [81, 82, 113]
BAI1/ADGRB1		●			Release of vasculostatin-40: inhibition of angiogenesis [45]
BAI2/ADGRB2	●				Unknown [48]
GPR116/ADGRF5	●				α-Fragment: unknown role in cellular signaling [38]
LPHN1/ADGRL1	●				Regulation of surface expression, generation of soluble receptor forms [46, 77]
GPR126/ADGRG6	●				Release of secreted ligand which regulates myocardial energy metabolism, contractility, and trabeculation [47, 72]

3.1 Protease-Activated Receptors (PARs)

Apart from aGPCRs, there are also other members of the GPCR superfamily that require cleavage for their biological activity, e.g., for the initiation of their signaling cascade. Thus far, four protease-activated receptors have been identified, PAR1–4 [78], each following a canonical activation principle. PAR1 is considered the prototype receptor of the family. It is activated by the serine protease thrombin, a key regulator of platelet aggregation, endothelial cell activation, and further vascular effects [79–81].

For activation of human PAR1, thrombin cleaves the ectodomain of the receptor at a specific recognition site (LDPR↓S) that is located at position 41 of the receptor molecule [82]. The resulting new N-terminus contains a tethered agonist that becomes unmasked upon the proteolytic event (Fig. 3, Table 1). The agonist physically interacts with the 7TM domain of PAR1 and activates its signaling cascade. Furthermore, synthetic peptides comprising the first six residues of the unmasked tethered agonist are capable of activating PAR1, even without prior receptor cleavage [82]. Additionally, genetic exchange of the cleavage site, e.g., to a trypsin cleavage site, resulted in receptor activation through trypsin under heterologous expression conditions [83, 84], confirming the hypothesis that the role of thrombin comprises the exposure of the receptor's tethered agonist [85, 86].

An activation mechanism which shares similarities with the proteolytic activation of PAR1 has recently been unraveled for GPR126, GPR133 [52], and further aGPCRs [63–65] (see above). However, several differences to PAR activation have to be considered: while PAR1 processing through the exogenous thrombin protease directly leads to the exposure of its tethered ligand, GAIN domain-mediated GPS cleavage of aGPCRs alone may not be sufficient to unmask the *Stachel*/*stalk* agonist of selected aGPCRs as the cleavage fragments remain attached to each other. Further structural changes in their extracellular domain, e.g., through ligand binding to the adhesion domains within the receptor ectodomain similar to the situation of the Notch receptor, and/or mechanical removal of the NTF may be required for *Stachel*/*stalk* exposure and aGPCR activation [87–89].

However, studying properties of PAR receptors may reveal additional parallels to the signaling paradigm(s) utilized by aGPCRs. For example, rapid phosphorylation-dependent internalization of activated PAR molecules and subsequent lysosomal degradation terminate PAR1 signaling [90, 91]. At least one aGPCR study suggests that ligand contact and mechanical challenge of CD97 trigger removal and degradation of the receptor's CTF, thereby providing means to quench signaling through aGPCRs [87].

3.2 Notch

Developmental signals governed through the activation of the Notch receptor are arguably the best researched and understood functions that result from proteolytic processing of a receptor molecule. The Notch receptor consists of the single-pass transmembrane protein and contains a species-specific array of up to 36 epidermal growth factor-like (EGF) repeats strung along the length of its extensive ectodomain. Through the ectodomain, the receptor interacts with DSL ligands (Delta, Serrate/Jagged, LAG-2), which themselves are large type I transmembrane molecules mounted on neighboring cell surfaces to the Notch-bearing cell [114, 115], thereby resembling the interaction scenario of several aGPCRs and their cellular ligands, e.g., CD97 with CD55 [116] or Latrophilins with FLRTs, teneurins, and neurexins [117–119].

An important consequence imposed by the positional Notch receptor-ligand configuration is the restriction of signaling events to cellular neighbors [120], which may also figure in the physiological roles of aGPCRs. This restriction is critical for the developmental switches governed by the Notch pathway, as it regulates binary cell fate decisions during embryogenesis, organogenesis, and cell differentiation, and many examples across the tree of metazoan life bear witness to the generality of this concept [104, 121]. In the classical paradigm of the Notch receptor-DSL ligand interplay, two daughter cells deriving from a precursor blastomere inherit equal amounts of both Notch and DSL. Engagement of Notch and ligand at the cell contact faces initiates an iteratively looping feedback cycle, which culminates in downregulation of the receptor in only one of the two cells. In the 'winner cell', the intracellular actions of Notch repress a proneural gene battery and drive it into the epidermal cell lineage. Conversely, the cell that has lost Notch on its surface becomes a neuronal precursor cell [122].

Intriguingly, the activation of the Notch receptor molecule is the consequence of a cascade of at least four cleavage events that sequentially process the receptor molecule along its N→C axis (Fig. 3, Table 1). First, after biosynthesis and en route to the cell surface, the receptor is cleaved by a furin-like convertase at the S1 cleavage site severing most of the receptor's ectodomain including the ligand-binding EGF repeats from a fragment holding the juxtamembrane, transmembrane, and intracellular receptor portion. S1 cleavage, however, does not result in physical separation of the cleavage fragments as they form a heterodimer held together through non-covalent interactions [123, 124], resembling the situation of aGPCRs that have undergone GAIN autoproteolysis but appear as heterodimers at the plasma membrane [4, 125].

The second proteolytic step occurs at the S2 site, which is positioned C-terminal to the S1 site just above the transmembrane helix. Before activation, the S2 site is protected by an arrangement of three LIN-12/Notch (LNR) repeats that are grouped around the cleavage site blocking access for the cognate S2 metalloproteases Kuzbanian and TACE/ADAM10 [126, 127]. When Notch engages with its DSL ligand, DSL endocytosis is thought to generate mechanical forces pulling at the receptor molecule, which eventually leads to conformational unwinding of the LNR repeats and exposure of the S2 site and its cleavage [128, 129]. While S1 cleavage is dispensable for Notch function [130], S2 cleavage appears as the gatekeeping step in Notch activation rendering the pathway a developmental command control system that may be triggered by mechanical input and may thus share similarities with the role of aGPCRs in development.

After S2 proteolysis, the remaining transmembrane-intracellular fragment of the Notch receptor [Notch extracellular truncated (NEXT)] undergoes further regulated intramembrane cleavage (RIP) catalyzed by the γ-secretase complex. This large multi-protein enzyme cleaves the NEXT intermediate at two further sites inside the membrane (S3 and S4 sites) [126, 127, 131, 132]. Ultimately, S3/S4 proteolysis results in the release of the Notch intracellular domain (NICD), which heteromerizes with DNA-binding and transcriptional activation partners and travels into the nucleus, where the complex controls the expression of target genes [133–136].

3.3 Ephrins

Apart from Notch, proteolysis through ADAM10 assumes a central position in the processing of a number of other neuronal proteins like APP, N-cadherin, neuroligins, or ephrins [96, 106, 108, 110]. Among those, the proteolytic activation and physiological relevance of the signaling mode of ephrin-A2 are exemplary (Fig. 3, Table 1).

Ephrin-A2 is a GPI-anchored molecule that is cleaved by ADAM10 upon binding its endogenous receptor EphA3. The binding and cleavage event consequently disrupts the cell-cell contact mediated through the Eph/ephrin interaction [108]. Upon the formation of the ligand-receptor complex, the molecular recognition motif in ephrin-A2 is rendered accessible for ADAM10, which then associates with this complex and cleaves ephrin-A2 in a *trans* mode, as protease and substrate are expressed in different cells [137, 138]. It appears that this mechanism ensures the exclusive cleavage of receptor-bound ligands [137]. Following the proteolytic rupture of the intercellular connection, the Eph/ephrin complex is rapidly internalized into the receptor expressing cell [138], which has been shown for ephrin-A5, a related member of the ephrin family [139]. Blocking the Eph/ephrin complex binding site of ADAM10 using specific monoclonal antibodies resulted in impaired internalization and EphA3-mediated cell function, suggesting a physiological role for the cleavage [140].

Ephrins and their Eph receptors are generally involved in the guidance of cell migration and neural development, tissue separation, and synaptic plasticity [139, 141], but also in extraneuronal processes including vascular development, epithelial cell response, and inflammation [142–144]. One particular physiological function of ephrin-A2 is the control of axon guidance [108] and involves proteolysis through ADAM10. Migrating EphA3-presenting axons come in contact with cells expressing ephrin-A2. Upon this encounter, the EphA3-positive neurites are actively repelled by the proteolytic disruption of the Eph/ephrin connection and thereby lead to axon withdrawal and precise spatio-mechanical control of neurogenesis [108, 145]. Inside the cell, regulation of this signal is mainly communicated through the intrinsic tyrosine kinase activity of ephrin receptors. Phosphorylation-dependent activation of EphA3 triggers a conformational change shifting the kinase domain away from the plasma membrane into its active form

[146, 147], where it no longer obstructs the alignment with ADAM10 and therefore allows ephrin shedding [146, 148]. Accordingly, EphA3 mutants carrying a constitutively released kinase domain showed increased ephrin cleavage by ADAM10, even when kinase function was disabled [146]. Thus, tyrosine kinase activity of ephrin receptors is an intracellularly regulated means to switch between cell-cell repulsion (high activity) and cell-cell adhesion (low activity) [146, 148, 149].

This binary signaling of Eph receptors may bear functional and cell biological similarities to aGPCRs. Their variety of extracellular adhesion motifs are predestined for intercellular cell-cell interactions like those observed for Eph/ephrin, although the majority of aGPCRs are still orphaned without known ligands or intracellular interactors [1]. aGPCR-ligand complexes could conceivably be shed involving an exogenous protease (see discussion about furin-mediated cleavages of individual aGPCR homologs above). Equally possible, mechanical force may solely govern receptor fragment (NTF-CTF) separation at the breakpoint originating from receptor autoproteolysis at the GPS. This way, aGPCR-expressing cells may be able to switch from an adhesion to signaling state.

3.4 Polycystins

PKD1 and *PKD2* are genes encoding polycystins, which are multitransmembrane proteins with a large amino-terminal extracellular domain [150]. *PKD* mutations have been demonstrated to cause one of the most common genetic diseases worldwide, the autosomal-dominant polycystic kidney disease (ADPKD) [151], which is characterized by the formation of multiple fluid-filled cysts that lead to renal failure in patients [152]. Loss-of-function mutations in polycystin-1 (*PKD1*) are responsible for a vast majority of ADPKD cases with physiological functions of *PKD1* found in cell adhesion and cell junction formation [153, 154]. Consistent with these findings, polycystin-1 appears involved in mechanical coupling between cells and in the regulation of tubular lumen diameter along the nephron [155].

Interestingly, polycystin-1 shares several structural features with aGPCRs (Fig. 3). They possess a multi-pass transmembrane domain (with 11 instead of 7 helices), an extended ectodomain with arrays of PKD motifs, and most notably a GAIN domain [19]. Similar to aGPCRs, polycystin-1 undergoes autoproteolysis resulting in the generation of an NTF and CTF [17], which remain non-covalently attached after cleavage. The physiological role of the polycystin-1 GPS cleavage is unknown. However, cleavage-deficient mutants exhibit impaired function in vitro [17] and in vivo [18]. A *PKD1* GPS proteolysis-deficient mouse mutant shows abnormal renal development after the first days of postnatal life apparent in reduced size and weight, as well as grossly enlarged cystic kidneys. Furthermore, the mutation is lethal within 6 weeks after birth presumably due to renal insufficiency [18]. This is only partly compatible with defects displayed by *PKD1*$^{-/-}$ mice, which show severe embryonic phenotypes and die already a few hours after birth [156]. Therefore, it was concluded that GPS cleavage of polycystin-1 is required for postnatal renal maturation, while it is not essential for embryonic nephrogenesis [18]. This is supported by the fact that cleaved and uncleaved polycystin-1 can coexist under physiological conditions [155]. Interestingly, polycystin-1 is also substrate to proteolytic events in addition to GAIN domain autoproteolysis, as it, too, is cleaved by the γ-secretase complex, resembling S3 and S4 proteolyses of the Notch receptor [157, 158].

4 Conclusions

aGPCRs are by far not the only group of biomolecules whose actions are controlled through proteolytic cleavage. Here, we only concentrated on those that are governed by the classical proteases and autoproteolytic events. As shown here, autoproteolysis and proteolytic cleavage seem neither mutually exclusive nor are their functional implications in aGPCRs sufficiently understood. Considering the tremendous number of surface receptors controlled through proteolysis, and the requirements for their function, there is no doubt that elucidation of the physiology of aGPCRs requires further investigation of their proteolytic properties. This should include a better understanding how GAIN domain-mediated cleavage is involved in receptor signaling and resolve the question if and how it may be modulated, e.g., through allosteric mechanisms. Further interest should be directed toward the study of other aGPCR domains that entertain non-GAIN domain autoproteolytic steps and the role of other proteases in the processing of the receptor molecule and pin down their physiological roles in receptor trafficking, cell adhesion, metabotropic, and non-cell-autonomous signaling. The structural and physiological properties of other surface molecule systems including the PAR, Notch, ephrin, or polycystin pathways should be considered as examples of how proteolytic processing can shape the function of receptor modules. The extensive body of work accumulated on their biological significance can instruct new experimental avenues and working models that should be explored in the quest to elucidate the interplay between the proteolytic processing of aGPCRs and their diverse signaling profiles.

Acknowledgments The writing of this manuscript was supported by grants from the Deutsche Forschungsgemeinschaft to R.J.K. and T.L. (Research Unit FOR 2149, Projects P1 [LA 2861/4-1] and P3 [LA 2861/5-1, KI 1460/2-1]; SFB 1047, Project A5; SFB-TR 166 Projects B4 and C3; LA 2861/7-1. A.G.P. acknowledges support from the Russian Science Foundation (14-14-01195), H.-H.L. acknowledges support from the Ministry of Science and Technology, Taiwan (MOST-104-2320-B-182-035-MY3), and the Chang Gung Memorial Hospital (CMRPD1C0633, CMRPD1D0072-3, CMRPD1D0392).

References

1. Hamann J, Aust G, Arac D, Engel FB, Formstone C et al (2015) International Union of Basic and Clinical Pharmacology. XCIV. Adhesion G protein-coupled receptors. Pharmacol Rev 67:338–367

2. Lin HH, Stacey M, Yona S, Chang GW (2010) GPS proteolytic cleavage of Adhesion-GPCRs. Adv Exp Med Biol 706:49–58
3. Stacey M, Lin HH, Gordon S, McKnight AJ (2000) LNB-TM7, a group of seven-transmembrane proteins related to family-B G-protein-coupled receptors. Trends Biochem Sci 25:284–289
4. Gray JX, Haino M, Roth MJ, Maguire JE, Jensen PN et al (1996) CD97 is a processed, seven-transmembrane, heterodimeric receptor associated with inflammation. J Immunol 157:5438–5447
5. Krasnoperov V, Bittner MA, Beavis R, Kuang Y, Salnikow KV et al (1997) alpha-Latrotoxin stimulates exocytosis by the interaction with a neuronal G-protein-coupled receptor. Neuron 18:925–937
6. Krasnoperov V, Bittner MA, Holz RW, Chepurny O, Petrenko AG (1999) Structural requirements for alpha-latrotoxin binding and alpha-latrotoxin-stimulated secretion. A study with calcium-independent receptor of alpha-latrotoxin (CIRL) deletion mutants. J Biol Chem 274:3590–3596
7. Chang GW, Stacey M, Kwakkenbos MJ, Hamann J, Gordon S et al (2003) Proteolytic cleavage of the EMR2 receptor requires both the extracellular stalk and the GPS motif. FEBS Lett 547:145–150
8. Lee JJ, Ekker SC, von Kessler DP, Porter JA, Sun BI et al (1994) Autoproteolysis in hedgehog protein biogenesis. Science 266:1528–1537
9. Porter JA, von Kessler DP, Ekker SC, Young KE, Lee JJ et al (1995) The product of hedgehog autoproteolytic cleavage active in local and long-range signalling. Nature 374:363–366
10. Guan C, Cui T, Rao V, Liao W, Benner J et al (1996) Activation of glycosylasparaginase. Formation of active N-terminal threonine by intramolecular autoproteolysis. J Biol Chem 271:1732–1737
11. Lin HH, Chang GW, Davies JQ, Stacey M, Harris J et al (2004) Autocatalytic cleavage of the EMR2 receptor occurs at a conserved G protein-coupled receptor proteolytic site motif. J Biol Chem 279:31823–31832
12. Oinonen C, Tikkanen R, Rouvinen J, Peltonen L (1995) Three-dimensional structure of human lysosomal aspartylglucosaminidase. Nat Struct Biol 2:1102–1108
13. Tikkanen R, Riikonen A, Oinonen C, Rouvinen R, Peltonen L (1996) Functional analyses of active site residues of human lysosomal aspartylglucosaminidase: implications for catalytic mechanism and autocatalytic activation. EMBO J 15:2954–2960
14. Xu Q, Buckley D, Guan C, Guo HC (1999) Structural insights into the mechanism of intramolecular proteolysis. Cell 98:651–661
15. Krasnoperov V, Lu Y, Buryanovsky L, Neubert TA, Ichtchenko K et al (2002) Post-translational proteolytic processing of the calcium-independent receptor of alpha-latrotoxin (CIRL), a natural chimera of the cell adhesion protein and the G protein-coupled receptor. Role of the G protein-coupled receptor proteolysis site (GPS) motif. J Biol Chem 277:46518–46526
16. Iguchi T, Sakata K, Yoshizaki K, Tago K, Mizuno N et al (2008) Orphan G protein-coupled receptor GPR56 regulates neural progenitor cell migration via a G alpha 12/13 and Rho pathway. J Biol Chem 283:14469–14478
17. Qian F, Boletta A, Bhunia AK, Xu H, Liu L et al (2002) Cleavage of polycystin-1 requires the receptor for egg jelly domain and is disrupted by human autosomal-dominant polycystic kidney disease 1-associated mutations. Proc Natl Acad Sci U S A 99:16981–16986
18. Yu S, Hackmann K, Gao J, He X, Piontek K et al (2007) Essential role of cleavage of Polycystin-1 at G protein-coupled receptor proteolytic site for kidney tubular structure. Proc Natl Acad Sci U S A 104:18688–18693
19. Arac D, Boucard AA, Bolliger MF, Nguyen J, Soltis SM et al (2012) A novel evolutionarily conserved domain of cell-adhesion GPCRs mediates autoproteolysis. EMBO J 31:1364–1378

20. Hsiao CC, Cheng KF, Chen HY, Chou YH, Stacey M et al (2009) Site-specific N-glycosylation regulates the GPS auto-proteolysis of CD97. FEBS Lett 583:3285–3290
21. Wei W, Hackmann K, Xu H, Germino G, Qian F (2007) Characterization of cis-autoproteolysis of polycystin-1, the product of human polycystic kidney disease 1 gene. J Biol Chem 282:21729–21737
22. Deyev IE, Petrenko AG (2010) Regulation of CIRL-1 proteolysis and trafficking. Biochimie 92:418–422
23. Araç D, Sträter N, Seiradake E (2016) Understanding the structural basis of adhesion GPCR functions. In: Langenhan T, Schöneberg T (eds) Adhesion G protein-coupled receptors: molecular, physiological and pharmacological principles in health and disease. Springer, Heidelberg
24. Prömel S, Langenhan T, Arac D (2013) Matching structure with function: the GAIN domain of adhesion-GPCR and PKD1-like proteins. Trends Pharmacol Sci 34:470–478
25. Langenhan T, Aust G, Hamann J (2013) Sticky signaling–adhesion class G protein-coupled receptors take the stage. Sci Signal 6:re3
26. Kan Z, Jaiswal BS, Stinson J, Janakiraman V, Bhatt D et al (2010) Diverse somatic mutation patterns and pathway alterations in human cancers. Nature 466:869–873
27. Piao X, Hill RS, Bodell A, Chang BS, Basel-Vanagaite L et al (2004) G protein-coupled receptor-dependent development of human frontal cortex. Science 303:2033–2036
28. Abe J, Fukuzawa T, Hirose S (2002) Cleavage of Ig-Hepta at a "SEA" module and at a conserved G protein-coupled receptor proteolytic site. J Biol Chem 277:23391–23398
29. Bork P, Patthy L (1995) The SEA module: a new extracellular domain associated with O-glycosylation. Protein Sci 4:1421–1425
30. Khatri IA, Wang R, Forstner JF (2003) SEA (sea-urchin sperm protein, enterokinase and agrin)-module cleavage, association of fragments and membrane targeting of rat intestinal mucin Muc3. Biochem J 372:263–270
31. Palmai-Pallag T, Khodabukus N, Kinarsky L, Leir SH, Sherman S et al (2005) The role of the SEA (sea urchin sperm protein, enterokinase and agrin) module in cleavage of membrane-tethered mucins. FEBS J 272:2901–2911
32. Wreschner DH, McGuckin MA, Williams SJ, Baruch A, Yoeli M et al (2002) Generation of ligand-receptor alliances by "SEA" module-mediated cleavage of membrane-associated mucin proteins. Protein Sci 11:698–706
33. Levitin F, Stern O, Weiss M, Gil-Henn C, Ziv R et al (2005) The MUC1 SEA module is a self-cleaving domain. J Biol Chem 280:33374–33386
34. Johansson DG, Wallin G, Sandberg A, Macao B, Aqvist J et al (2009) Protein autoproteolysis: conformational strain linked to the rate of peptide cleavage by the pH dependence of the N → O acyl shift reaction. J Am Chem Soc 131:9475–9477
35. Macao B, Johansson DG, Hansson GC, Hard T (2006) Autoproteolysis coupled to protein folding in the SEA domain of the membrane-bound MUC1 mucin. Nat Struct Mol Biol 13:71–76
36. Johansson DG, Macao B, Sandberg A, Hard T (2008) SEA domain autoproteolysis accelerated by conformational strain: mechanistic aspects. J Mol Biol 377:1130–1143
37. Sandberg A, Johansson DG, Macao B, Hard T (2008) SEA domain autoproteolysis accelerated by conformational strain: energetic aspects. J Mol Biol 377:1117–1129
38. Fukuzawa T, Hirose S (2006) Multiple processing of Ig-Hepta/GPR116, a G protein-coupled receptor with immunoglobulin (Ig)-like repeats, and generation of EGF2-like fragment. J Biochem 140:445–452
39. Ariestanti DM, Ando H, Hirose S, Nakamura N (2015) Targeted disruption of Ig-Hepta/Gpr116 causes emphysema-like symptoms that are associated with alveolar macrophage activation. J Biol Chem 290:11032–11040
40. Bridges JP, Ludwig MG, Mueller M, Kinzel B, Sato A et al (2013) Orphan G protein-coupled receptor GPR116 regulates pulmonary surfactant pool size. Am J Respir Cell Mol Biol 49:348–357

41. Fukuzawa T, Ishida J, Kato A, Ichinose T, Ariestanti DM et al (2013) Lung surfactant levels are regulated by Ig-Hepta/GPR116 by monitoring surfactant protein D. PLoS One 8, e69451
42. Nie T, Hui X, Gao X, Li K, Lin W et al (2012) Adipose tissue deletion of Gpr116 impairs insulin sensitivity through modulation of adipose function. FEBS Lett 586:3618–3625
43. Tang X, Jin R, Qu G, Wang X, Li Z et al (2013) GPR116, an adhesion G-protein-coupled receptor, promotes breast cancer metastasis via the Galphaq-p63RhoGEF-Rho GTPase pathway. Cancer Res 73:6206–6218
44. Yang MY, Hilton MB, Seaman S, Haines DC, Nagashima K et al (2013) Essential regulation of lung surfactant homeostasis by the orphan G protein-coupled receptor GPR116. Cell Rep 3:1457–1464
45. Cork SM, Kaur B, Devi NS, Cooper L, Saltz JH et al (2012) A proprotein convertase/MMP-14 proteolytic cascade releases a novel 40 kDa vasculostatin from tumor suppressor BAI1. Oncogene 31:5144–5152
46. Krasnoperov V, Deyev IE, Serova OV, Xu C, Lu Y et al (2009) Dissociation of the subunits of the calcium-independent receptor of alpha-latrotoxin as a result of two-step proteolysis. Biochemistry 48:3230–3238
47. Moriguchi T, Haraguchi K, Ueda N, Okada M, Furuya T et al (2004) DREG, a developmentally regulated G protein-coupled receptor containing two conserved proteolytic cleavage sites. Genes Cells 9:549–560
48. Okajima D, Kudo G, Yokota H (2010) Brain-specific angiogenesis inhibitor 2 (BAI2) may be activated by proteolytic processing. J Recept Signal Transduct Res 30:143–153
49. Thomas G (2002) Furin at the cutting edge: from protein traffic to embryogenesis and disease. Nat Rev Mol Cell Biol 3:753–766
50. Kaur B, Brat DJ, Devi NS, Van Meir EG (2005) Vasculostatin, a proteolytic fragment of brain angiogenesis inhibitor 1, is an antiangiogenic and antitumorigenic factor. Oncogene 24:3632–3642
51. Nishimori H, Shiratsuchi T, Urano T, Kimura Y, Kiyono K et al (1997) A novel brain-specific, p53-target gene, BAI1, containing thrombospondin type 1 repeats inhibits experimental angiogenesis. Oncogene 15:2145–2150
52. Liebscher I, Schön J, Petersen SC, Fischer L, Auerbach N et al (2014) A tethered agonist within the ectodomain activates the adhesion G protein-coupled receptors GPR126 and GPR133. Cell Rep 9:2018–2026
53. Bohnekamp J, Schöneberg T (2011) Cell adhesion receptor GPR133 couples to Gs protein. J Biol Chem 286:41912–41916
54. Prömel S, Frickenhaus M, Hughes S, Mestek L, Staunton D et al (2012) The GPS motif is a molecular switch for bimodal activities of adhesion class G protein-coupled receptors. Cell Rep 2:321–331
55. Kurbegovic A, Kim H, Xu H, Yu S, Cruanès J et al (2014) Novel functional complexity of polycystin-1 by GPS cleavage in vivo: role in polycystic kidney disease. Mol Cell Biol 34:3341–3353
56. Volynski KE, Silva JP, Lelianova VG, Atiqur Rahman M, Hopkins C et al (2004) Latrophilin fragments behave as independent proteins that associate and signal on binding of LTX(N4C). EMBO J 23:4423–4433
57. Heisenberg CP, Bellaiche Y (2013) Forces in tissue morphogenesis and patterning. Cell 153:948–962
58. Liebscher I, Schöneberg T (2016) Tethered agonism: a common activation mechanism of adhesion GPCRs. In: Langenhan T, Schöneberg T (eds) Adhesion G protein-coupled receptors: molecular, physiological and pharmacological principles in health and disease. Springer, Heidelberg
59. Kishore A, Hall RA (2016) Versatile signaling activity of adhesion GPCRs. In: Langenhan T, Schöneberg T (eds) Adhesion G protein-coupled receptors: molecular, physiological and pharmacological principles in health and disease. Springer, Heidelberg

60. Paavola KJ, Stephenson JR, Ritter SL, Alter SP, Hall RA (2011) The N terminus of the adhesion G protein-coupled receptor GPR56 controls receptor signaling activity. J Biol Chem 286:28914–28921
61. Arac D, Aust G, Calebiro D, Engel FB, Formstone C et al (2012) Dissecting signaling and functions of adhesion G protein-coupled receptors. Ann N Y Acad Sci 1276:1–25
62. Strutt D, Schnabel R, Fiedler F, Prömel S (2016) Adhesion GPCRs govern polarity of epithelia and cell migration. In: Langenhan T, Schöneberg T (eds) Adhesion G protein-coupled receptors: molecular, physiological and pharmacological principles in health and disease. Springer, Heidelberg
63. Stoveken HM, Hajduczok AG, Xu L, Tall GG (2015) Adhesion G protein-coupled receptors are activated by exposure of a cryptic tethered agonist. Proc Natl Acad Sci U S A 112:6194–6199
64. Demberg LM, Rothemund S, Schöneberg T, Liebscher I (2015) Identification of the tethered peptide agonist of the adhesion G protein-coupled receptor GPR64/ADGRG2. Biochem Biophys Res Commun 464:743–747
65. Wilde C, Fischer L, Lede V, Kirchberger J, Rothemund S et al (2016) The constitutive activity of the adhesion GPCR GPR114/ADGRG5 is mediated by its tethered agonist. FASEB J 30(2):666–673
66. Müller A, Winkler J, Fiedler F, Sastradihardja T, Binder C et al (2015) Oriented cell division in the C. elegans embryo is coordinated by G-protein signaling dependent on the adhesion GPCR LAT-1. PLoS Genet 11:e1005624
67. Boyden SE, Desai A, Cruse G, Young ML, Bolan HC et al (2016) Vibratory urticaria associated with a missense variant in ADGRE2. N Engl J Med 374:656–663
68. Hamann J, Hsiao C-C, Lee CS, Ravichandran KS, Lin H-H (2016) Adhesion GPCRs as modulators of immune cell function. In: Langenhan T, Schöneberg T (eds) Adhesion G protein-coupled receptors: molecular, physiological and pharmacological principles in health and disease. Springer, Heidelberg
69. Gordon WR, Vardar-Ulu D, Histen G, Sanchez-Irizarry C, Aster JC et al (2007) Structural basis for autoinhibition of Notch. Nat Struct Mol Biol 14:295–300
70. Kishore A, Purcell RH, Nassiri-Toosi Z, Hall RA (2016) Stalk-dependent and stalk-independent signaling by the adhesion G protein-coupled receptors GPR56 (ADGRG1) and BAI1 (ADGRB1). J Biol Chem 291:3385–3394
71. Nijmeijer S, Wolf S, Ernst OP, de Graaf C (2016) 7TM domain structure of adhesion GPCRs. In: Langenhan T, Schöneberg T (eds) Adhesion G protein-coupled receptors: molecular, physiological and pharmacological principles in health and disease. Springer, Heidelberg
72. Patra C, van Amerongen MJ, Ghosh S, Ricciardi F, Sajjad A et al (2013) Organ-specific function of adhesion G protein-coupled receptor GPR126 is domain-dependent. Proc Natl Acad Sci U S A 110:16898–16903
73. Waller-Evans H, Prömel S, Langenhan T, Dixon J, Zahn D et al (2010) The orphan adhesion-GPCR GPR126 is required for embryonic development in the mouse. PLoS One 5, e14047
74. Huang YS, Chiang NY, Hu CH, Hsiao CC, Cheng KF et al (2012) Activation of myeloid cell-specific adhesion class G protein-coupled receptor EMR2 via ligation-induced translocation and interaction of receptor subunits in lipid raft microdomains. Mol Cell Biol 32:1408–1420
75. Silva JP, Lelianova V, Hopkins C, Volynski KE, Ushkaryov Y (2009) Functional cross-interaction of the fragments produced by the cleavage of distinct adhesion G-protein-coupled receptors. J Biol Chem 284:6495–6506
76. Silva JP, Ushkaryov Y (2010) The Latrophilins, "split-personality" receptors. Adv Exp Med Biol 706:59–75
77. Serova OV, Popova NV, Petrenko AG, Deyev IE (2010) Association of the subunits of the calcium-independent receptor of alpha-latrotoxin. Biochem Biophys Res Commun 402:658–662
78. Macfarlane SR, Seatter MJ, Kanke T, Hunter GD, Plevin R (2001) Proteinase-activated receptors. Pharmacol Rev 53:245–282

79. Davey MG, Luscher EF (1967) Actions of thrombin and other coagulant and proteolytic enzymes on blood platelets. Nature 216:857–858
80. Hattori R, Hamilton KK, Fugate RD, McEver RP, Sims PJ (1989) Stimulated secretion of endothelial von Willebrand factor is accompanied by rapid redistribution to the cell surface of the intracellular granule membrane protein GMP-140. J Biol Chem 264:7768–7771
81. Sambrano GR, Weiss EJ, Zheng YW, Huang W, Coughlin SR (2001) Role of thrombin signalling in platelets in haemostasis and thrombosis. Nature 413:74–78
82. Vu TK, Hung DT, Wheaton VI, Coughlin SR (1991) Molecular cloning of a functional thrombin receptor reveals a novel proteolytic mechanism of receptor activation. Cell 64:1057–1068
83. Hammes SR, Coughlin SR (1999) Protease-activated receptor-1 can mediate responses to SFLLRN in thrombin-desensitized cells: evidence for a novel mechanism for preventing or terminating signaling by PAR1's tethered ligand. Biochemistry 38:2486–2493
84. Vu TK, Wheaton VI, Hung DT, Charo I, Coughlin SR (1991) Domains specifying thrombin-receptor interaction. Nature 353:674–677
85. Coughlin SR (1999) How the protease thrombin talks to cells. Proc Natl Acad Sci U S A 96:11023–11027
86. Coughlin SR (1998) Sol Sherry lecture in thrombosis: how thrombin 'talks' to cells: molecular mechanisms and roles in vivo. Arterioscler Thromb Vasc Biol 18:514–518
87. Karpus ON, Veninga H, Hoek RM, Flierman D, van Buul JD et al (2013) Shear stress-dependent downregulation of the adhesion-G protein-coupled receptor CD97 on circulating leukocytes upon contact with its ligand CD55. J Immunol 190:3740–3748
88. Langenhan T, Barr MM, Bruchas MR, Ewer J, Griffith LC et al (2015) Model organisms in G protein-coupled receptor research. Mol Pharmacol 88:596–603
89. Monk KR, Hamann J, Langenhan T, Nijmeijer S, Schöneberg T et al (2015) Adhesion G protein-coupled receptors: from in vitro pharmacology to in vivo mechanisms. Mol Pharmacol 88:617–623
90. Hoxie JA, Ahuja M, Belmonte E, Pizarro S, Parton R et al (1993) Internalization and recycling of activated thrombin receptors. J Biol Chem 268:13756–13763
91. Trejo J, Hammes SR, Coughlin SR (1998) Termination of signaling by protease-activated receptor-1 is linked to lysosomal sorting. Proc Natl Acad Sci U S A 95:13698–13702
92. Sahin U, Weskamp G, Kelly K, Zhou HM, Higashiyama S et al (2004) Distinct roles for ADAM10 and ADAM17 in ectodomain shedding of six EGFR ligands. J Cell Biol 164:769–779
93. Yarden Y, Sliwkowski MX (2001) Untangling the ErbB signalling network. Nat Rev Mol Cell Biol 2:127–137
94. Black RA, Rauch CT, Kozlosky CJ, Peschon JJ, Slack JL et al (1997) A metalloproteinase disintegrin that releases tumour-necrosis factor-alpha from cells. Nature 385:729–733
95. Kriegler M, Perez C, Defay K, Albert I, Lu SD (1988) A novel form of Tnf/cachectin is a cell-surface cyto-toxic transmembrane protein – ramifications for the complex physiology of Tnf. Cell 53:45–53
96. Asai M, Hattori C, Szabó B, Sasagawa N, Maruyama K et al (2003) Putative function of ADAM9, ADAM10, and ADAM17 as APP-secretase. Biochem Biophys Res Commun 301:231–235
97. Postina R, Schroeder A, Dewachter I, Bohl J, Schmitt U et al (2004) A disintegrin-metalloproteinase prevents amyloid plaque formation and hippocampal defects in an Alzheimer disease mouse model. J Clin Invest 113:1456–1464
98. Selkoe DJ (1991) The molecular pathology of Alzheimer's disease. Neuron 6:487–498
99. Vassar R, Bennett BD, Babu-Khan S, Kahn S, Mendiaz EA et al (1999) Beta-secretase cleavage of Alzheimer's amyloid precursor protein by the transmembrane aspartic protease BACE. Science 286:735–741
100. Jarriault S, Brou C, Logeat F, Schroeter EH, Kopan R et al (1995) Signalling downstream of activated mammalian Notch. Nature 377:355–358

101. Lanford PJ, Lan Y, Jiang RL, Lindsell C, Weinmaster G et al (1999) Notch signalling pathway mediates hair cell development in mammalian cochlea. Nat Genet 21:289–292
102. Apelqvist A, Li H, Sommer L, Beatus P, Anderson DJ et al (1999) Notch signalling controls pancreatic cell differentiation. Nature 400:877–881
103. van Es JH, van Gijn ME, Riccio O, van den Born M, Vooijs M et al (2005) Notch/gamma-secretase inhibition turns proliferative cells in intestinal crypts and adenomas into goblet cells. Nature 435:959–963
104. Washburn T, Schweighoffer E, Gridley T, Chang D, Fowlkes BJ et al (1997) Notch activity influences the alphabeta versus gammadelta T cell lineage decision. Cell 88:833–843
105. Robey E, Chang D, Itano A, Cado D, Alexander H et al (1996) An activated form of notch influences the choice between CD4 and CD8 T cell lineages. Cell 87:483–492
106. Reiss K, Maretzky T, Ludwig A, Tousseyn T, de Strooper B et al (2005) ADAM10 cleavage of N-cadherin and regulation of cell-cell adhesion and beta-catenin nuclear signalling. EMBO J 24:742–752
107. Uemura K, Kihara T, Kuzuya A, Okawa K, Nishimoto T et al (2006) Characterization of sequential N-cadherin cleavage by ADAM10 and PS1. Neurosci Lett 402:278–283
108. Hattori M, Osterfield M, Flanagan JG (2000) Regulated cleavage of a contact-mediated axon repellent. Science 289:1360–1365
109. Saftig P, Lichtenthaler SF (2015) The alpha secretase ADAM10: a metalloprotease with multiple functions in the brain. Prog Neurobiol 135:1–20
110. Suzuki K, Hayashi Y, Nakahara S, Kumazaki H, Prox J et al (2012) Activity-dependent proteolytic cleavage of neuroligin-1. Neuron 76:410–422
111. Bot N, Schweizer C, Ben Halima S, Fraering PC (2011) Processing of the synaptic cell adhesion molecule neurexin-3beta by Alzheimer disease alpha- and gamma-secretases. J Biol Chem 286:2762–2773
112. Saura CA, Servian-Morilla E, Scholl FG (2011) Presenilin/gamma-secretase regulates neurexin processing at synapses. PLoS One 6, e19430
113. Coughlin SR (2000) Thrombin signalling and protease-activated receptors. Nature 407:258–264
114. Fehon RG, Kooh PJ, Rebay I, Regan CL, Xu T et al (1990) Molecular interactions between the protein products of the neurogenic loci Notch and Delta, two EGF-homologous genes in Drosophila. Cell 61:523–534
115. Rebay I, Fleming RJ, Fehon RG, Cherbas L, Cherbas P et al (1991) Specific EGF repeats of Notch mediate interactions with Delta and Serrate: implications for Notch as a multifunctional receptor. Cell 67:687–699
116. Hamann J, Vogel B, van Schijndel GM, van Lier RA (1996) The seven-span transmembrane receptor CD97 has a cellular ligand (CD55, DAF). J Exp Med 184:1185–1189
117. Boucard AA, Ko J, Südhof TC (2012) High affinity neurexin binding to cell adhesion G-protein-coupled receptor CIRL1/latrophilin-1 produces an intercellular adhesion complex. J Biol Chem 287:9399–9413
118. O'Sullivan ML, de Wit J, Savas JN, Comoletti D, Otto-Hitt S et al (2012) FLRT proteins are endogenous latrophilin ligands and regulate excitatory synapse development. Neuron 73:903–910
119. Silva JP, Lelianova VG, Ermolyuk YS, Vysokov N, Hitchen PG et al (2011) Latrophilin 1 and its endogenous ligand Lasso/teneurin-2 form a high-affinity transsynaptic receptor pair with signaling capabilities. Proc Natl Acad Sci U S A 108:12113–12118
120. Bray SJ (2006) Notch signalling: a simple pathway becomes complex. Nat Rev Mol Cell Biol 7:678–689
121. Heitzler P, Simpson P (1991) The choice of cell fate in the epidermis of Drosophila. Cell 64:1083–1092
122. Cau E, Blader P (2009) Notch activity in the nervous system: to switch or not switch? Neural Dev 4:36

123. Blaumueller CM, Qi H, Zagouras P, Artavanis-Tsakonas S (1997) Intracellular cleavage of Notch leads to a heterodimeric receptor on the plasma membrane. Cell 90:281–291
124. Logeat F, Bessia C, Brou C, LeBail O, Jarriault S et al (1998) The Notch1 receptor is cleaved constitutively by a furin-like convertase. Proc Natl Acad Sci U S A 95:8108–8112
125. Obermann H, Samalecos A, Osterhoff C, Schroder B, Heller R et al (2003) HE6, a two-subunit heptahelical receptor associated with apical membranes of efferent and epididymal duct epithelia. Mol Reprod Dev 64:13–26
126. Brou C, Logeat F, Gupta N, Bessia C, LeBail O et al (2000) A novel proteolytic cleavage involved in Notch signaling: the role of the disintegrin-metalloprotease TACE. Mol Cell 5:207–216
127. Mumm JS, Schroeter EH, Saxena MT, Griesemer A, Tian X et al (2000) A ligand-induced extracellular cleavage regulates gamma-secretase-like proteolytic activation of Notch1. Mol Cell 5:197–206
128. Parks AL, Klueg KM, Stout JR, Muskavitch MA (2000) Ligand endocytosis drives receptor dissociation and activation in the Notch pathway. Development 127:1373–1385
129. Stephenson NL, Avis JM (2012) Direct observation of proteolytic cleavage at the S2 site upon forced unfolding of the Notch negative regulatory region. Proc Natl Acad Sci U S A 109: E2757–E2765
130. Kidd S, Lieber T (2002) Furin cleavage is not a requirement for Drosophila Notch function. Mech Dev 115:41–51
131. Okochi M, Steiner H, Fukumori A, Tanii H, Tomita T et al (2002) Presenilins mediate a dual intramembranous gamma-secretase cleavage of Notch-1. EMBO J 21:5408–5416
132. Struhl G, Adachi A (2000) Requirements for presenilin-dependent cleavage of notch and other transmembrane proteins. Mol Cell 6:625–636
133. Kidd S, Lieber T, Young MW (1998) Ligand-induced cleavage and regulation of nuclear entry of Notch in Drosophila melanogaster embryos. Genes Dev 12:3728–3740
134. Lecourtois M, Schweisguth F (1998) Indirect evidence for delta-dependent intracellular processing of notch in Drosophila embryos. Curr Biol 8:771–774
135. Schroeter EH, Kisslinger JA, Kopan R (1998) Notch-1 signalling requires ligand-induced proteolytic release of intracellular domain. Nature 393:382–386
136. Struhl G, Adachi A (1998) Nuclear access and action of notch in vivo. Cell 93:649–660
137. Janes PW, Saha N, Barton WA, Kolev MV, Wimmer-Kleikamp SH et al (2005) Adam meets Eph: an ADAM substrate recognition module acts as a molecular switch for ephrin cleavage in trans. Cell 123:291–304
138. Mancia F, Shapiro L (2005) ADAM and Eph: how Ephrin-signaling cells become detached. Cell 123:185–187
139. Flanagan JG, Vanderhaeghen P (1998) The ephrins and Eph receptors in neural development. Annu Rev Neurosci 21:309–345
140. Atapattu L, Saha N, Llerena C, Vail ME, Scott AM et al (2012) Antibodies binding the ADAM10 substrate recognition domain inhibit Eph function. J Cell Sci 125:6084–6093
141. Kullander K, Klein R (2002) Mechanisms and functions of Eph and ephrin signalling. Nat Rev Mol Cell Biol 3:475–486
142. Garcia-Ceca J, Alfaro D, Montero-Herradón S, Tobajas E, Munoz JJ et al (2015) Eph/ephrins-mediated thymocyte-thymic epithelial cell interactions control numerous processes of thymus biology. Front Immunol 6:333
143. Perez White BE, Getsios S (2014) Eph receptor and ephrin function in breast, gut, and skin epithelia. Cell Adh Migr 8:327–338
144. Wijeratne DT, Rodger J, Wood FM, Fear MW (2016) The role of Eph receptors and Ephrins in the skin. Int J Dermatol 55:3–10
145. Salaita K, Nair PM, Petit RS, Neve RM, Das D et al (2010) Restriction of receptor movement alters cellular response: physical force sensing by EphA2. Science 327:1380–1385

146. Janes PW, Wimmer-Kleikamp SH, Frangakis AS, Treble K, Griesshaber B et al (2009) Cytoplasmic relaxation of active Eph controls ephrin shedding by ADAM10. PLoS Biol 7, e1000215
147. Wybenga-Groot LE, Baskin B, Ong SH, Tong J, Pawson T et al (2001) Structural basis for autoinhibition of the Ephb2 receptor tyrosine kinase by the unphosphorylated juxtamembrane region. Cell 106:745–757
148. Nievergall E, Lackmann M, Janes PW (2012) Eph-dependent cell-cell adhesion and segregation in development and cancer. Cell Mol Life Sci 69:1813–1842
149. Atapattu L, Lackmann M, Janes PW (2014) The role of proteases in regulating Eph/ephrin signaling. Cell Adh Migr 8:294–307
150. Hughes J, Ward CJ, Peral B, Aspinwall R, Clark K et al (1995) The polycystic kidney disease 1 (PKD1) gene encodes a novel protein with multiple cell recognition domains. Nat Genet 10:151–160
151. Igarashi P, Somlo S (2002) Genetics and pathogenesis of polycystic kidney disease. J Am Soc Nephrol 13:2384–2398
152. Milutinovic J, Fialkow PJ, Agodoa LY, Phillips LA, Rudd TG et al (1984) Autosomal dominant polycystic kidney disease: symptoms and clinical findings. Q J Med 53:511–522
153. Streets AJ, Newby LJ, O'Hare MJ, Bukanov NO, Ibraghimov-Beskrovnaya O et al (2003) Functional analysis of PKD1 transgenic lines reveals a direct role for polycystin-1 in mediating cell-cell adhesion. J Am Soc Nephrol 14:1804–1815
154. Streets AJ, Wagner BE, Harris PC, Ward CJ, Ong AC (2009) Homophilic and heterophilic polycystin 1 interactions regulate E-cadherin recruitment and junction assembly in MDCK cells. J Cell Sci 122:1410–1417
155. Ong AC, Harris PC (2015) A polycystin-centric view of cyst formation and disease: the polycystins revisited. Kidney Int 88:699–710
156. Lu W, Peissel B, Babakhanlou H, Pavlova A, Geng L et al (1997) Perinatal lethality with kidney and pancreas defects in mice with a targetted Pkd1 mutation. Nat Genet 17:179–181
157. Low SH, Vasanth S, Larson CH, Mukherjee S, Sharma N et al (2006) Polycystin-1, STAT6, and P100 function in a pathway that transduces ciliary mechanosensation and is activated in polycystic kidney disease. Dev Cell 10:57–69
158. Merrick D, Chapin H, Baggs JE, Yu Z, Somlo S et al (2012) The gamma-secretase cleavage product of polycystin-1 regulates TCF and CHOP-mediated transcriptional activation through a p300-dependent mechanism. Dev Cell 22:197–210

Tethered Agonism: A Common Activation Mechanism of Adhesion GPCRs

Ines Liebscher and Torsten Schöneberg

Graphical Abstract

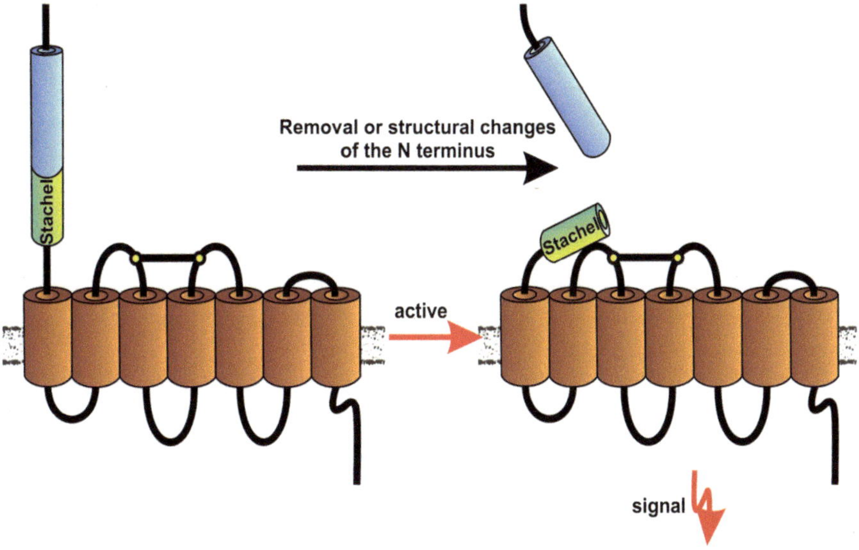

Adhesion GPCRs harbor a tethered agonist sequence (reproduced from [24])

I. Liebscher (✉) • T. Schöneberg
Medical Faculty, Institute of Biochemistry, University Leipzig, Johannisallee 30, Leipzig 04103, Germany
e-mail: ines.liebscher@medizin.uni-leipzig.de

T. Langenhan, T. Schöneberg (eds.), *Adhesion G Protein-coupled Receptors*,
Handbook of Experimental Pharmacology 234, DOI 10.1007/978-3-319-41523-9_6

Contents

Abstract

As the past years have seen a magnificent increase in knowledge on adhesion GPCR (aGPCR) signal transduction, the time had come to fill the gap on how these receptors can be activated. Based on experimental observations that deletion of the ectodomain can induce signaling, the idea arose that aGPCRs, just like other atypical GPCRs, may harbor a tethered agonist sequence. In this chapter, we describe the recent findings and characteristics of this agonist, called the *Stachel* sequence, and discuss potential mechanisms that cause liberation of this encrypted sequence. Further, we provide perspectives for application of *Stachel*-derived synthetic peptides in future studies of aGPCR function.

Keywords

Adhesion GPCR • Signal transduction • Tethered agonism • Activation mechanism • Peptide agonist

1 Adhesion GPCRs Harbor an Encrypted Agonistic Sequence, the *Stachel* Sequence

In classic pharmacology, an agonist (activating ligand) is a compound that specifically binds to its receptor, induces conformational changes (activation), and produces a biological response. This liganded and unliganded conformational stage of the receptor is formalized as:

$L + R \Leftrightarrow LR$, where the quotient of k_{on} and k_{off} is the equilibrium constant K_a. The reciprocal of K_a, the dissociation constant K_d, is a direct measure of the receptor-agonist affinity. The kinetics of most agonist-receptor pairs follows this equilibrium where the agonist reaches its receptor by diffusion and attaches to the binding pocket. However, a few exceptions exist from this classical model. Rhodopsin, a light-sensitive G-protein-coupled receptor (GPCR), is the most prominent example. Rhodopsin and other opsins contain a chromophore as a prosthetic group and are specialized for the detection of quanta of light. Here, the ligand is covalently bound to the receptor, and a single photon isomerizes the 11-*cis*- to *trans*-

retinal triggering receptor conformation changes and G-protein activation [1]. At least during rhodopsin activation, agonist diffusion and concentration are not part of the process allowing extreme fast off/on kinetics.

The concept of covalently carrying an agonist that isomerizes or is exposed by an external signal is brought to perfection in protease-activated receptors (PAR). PAR1 is a GPCR that is classically activated through cleavage of the N-terminal ectodomain by the serine protease thrombin exposing an agonistic amino acid sequence (SFLLRN) within the N terminus [2]. The agonist is covalently bound during activation, but it is not a prosthetic group as in the case of rhodopsin. Here, cleavage seems to isomerize, relax, or expose the peptidic agonist into the activating conformation. Not only specific cleavage by thrombin but also by other proteases such as trypsin and plasmin activates PARs [3]. Interestingly, matrix metalloproteases (MMP-1 and MMP-13) cleave the N-terminal ectodomain of PAR1 at noncanonical sites, which results in distinct tethered ligands that activate G-protein signaling pathways [4]. An integration of diverse signals into a defined biological response, e.g., during platelet activation and aggregation, is most probably the biological meaning behind this stimulus diversity.

All adhesion GPCRs (aGPCRs) have a large ectodomain providing manifold interaction possibilities with other molecules (see also [5]). One can speculate that the ectodomain may serve as an integrator for diverse signals. This concept, however, would require a tethered agonist as in the case of rhodopsin and PARs. This chapter features the discovery of such internal agonistic sequences—termed *Stachel* sequence—in aGPCRs and discusses the mechanisms of exposing, isomerizing, or relaxing this tethered agonist.

1.1 Identification of the *Stachel* Sequence

Taken the enormous pharmacological potential of aGPCRs, the identification of the endogenous agonists or molecules that modulate their activation levels was a major task in this field. However, the screen for ligands was limited by the unknown signal transduction of aGPCRs. This knowledge is a prerequisite to establish reliable readout systems of receptor activation.

Its structural relation of the 7TM to other GPCRs suggested G-protein coupling and a number of studies tested for the capability of aGPCRs to activate these classical signaling cascades. Overexpression of receptor variants has been used to identify the signaling pathways of several GPCRs because of increased basal activity in heterologous expression cell system [6–8]. The same approach was applied to identify the G-protein-coupling abilities of several aGPCRs: GPR56/ADGRG1 [9, 10], CD97/ADGRE5 [11], GPR133/ADGRD1 [12], GPR114/ADGRG2, GPR97/ADGRG3, GPR110/ADGRF1, GPR115/ADGRF4 [13], BAI1/ADGRB1 [14], GPR126/ADGRG6 [15], VLGR1/ADGRV1 [16], GPR64/ADGRG2 [17, 18], and LAT-1/ADGRL1 [19]. Interestingly, even with these readout systems at hand, the search for endogenous ligands did not yield any activating compounds for a long time.

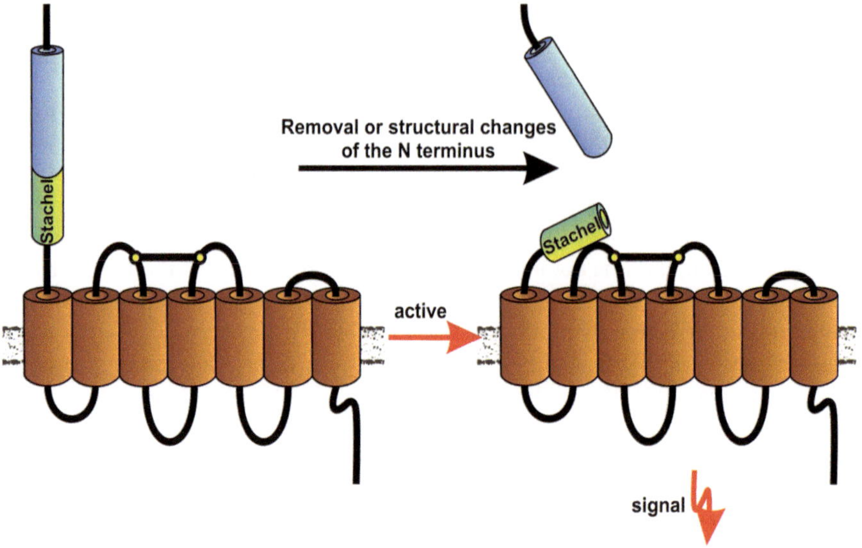

Fig. 1 Adhesion GPCRs harbor a tethered agonist sequence (reproduced from [24])

An important contribution to the identification of the aGPCR activation mechanism was the observation that receptor activation was seen upon deletion of the N-terminal fragment (NTF) of BAI2/ADGRB2 [20], GPR56 [21, 22], and CD97 [11]. Two possible scenarios were suggested to explain this deletion-induced activation [23]: (1) the N terminus acts as an inhibitor of the constitutively active C-terminal fragment (CTF) or (2) the residual part of the N terminus contains an agonistic sequence. To shed light on the underlying mechanism, the NTFs of two aGPCRs, GPR126 and GPR133, which had been shown to couple to the G_s protein/adenylyl cyclase pathway were deleted, and, as expected, subsequent increases in their signaling activity were seen [24]. However, upon deletion of the amino acids N-terminally of the transmembrane helix 1 (TM1) that remained in the CTF, receptor activity was abolished, favoring a tethered agonist model over an inhibitory NTF. These data suggested the presence of a tethered agonist between the natural cleavage site and the N terminus of TM1. Indeed, synthetic peptides derived from this sequence were capable of activating the inactive receptor mutants as well as the full-length GPR126 and GPR133 (Fig. 1). Based on the exposed position of this region after removal of the NTF and its expected piercing interaction with the potential binding pocket within the 7TM region, the tethered agonist sequence was named the "*Stachel*" sequence, which is the German word for "stinger." Mutational analysis of CTF mutants and derived peptides revealed that an exchange of the first amino acid of the tethered agonist to alanine is tolerated for sustained receptor function, while positions 2–7 are essential for CTF mutant and peptide activity. In line with these results, a mutant GPR126 zebrafish was designed, which lacked the amino acids 4 and 5 after the cleavage site of the GPS within the full-length receptor

[24]. This mutant zebrafish exhibited the same phenotype as the previously published complete GPR126 gene knockout zebrafish strain [25] presenting with a pronounced myelination defect and so-called puffy ears. All these in vitro and in vivo data support the hypothesis that the residual part of the N terminus in the CTF contains an agonistic sequence.

Later a tethered agonist sequence was identified for several other aGPCRs, namely GPR110 and GPR56 [26], GPR64 [17], GPR114 [27], and LAT-1 [19]. However, a recent study indicated that CTF mutants of GPR56 and BAI1 do not exclusively elicit constitutive activity [28] (see also [29]). While GPR56 CTF showed increased activity in a serum response factor reporter assay, all other CTF constructs examined did not show elevated activity in other activity measures. Whether these findings hold true upon stimulation with agonistic peptides needs to be determined since the repertoire of the receptors' signal transduction may significantly differ between agonist and mutationally induced activity as shown for the aGPCR GPR64 [17].

1.2 Structure and Genomic Organization of the *Stachel* Sequence

At the current state, there is no X-ray crystal structure of a full-length aGPCR available. However, partial structures of aGPCR ectodomains are solved [30, 31] (see also [5]). There are two crystal structure models available for LAT-1 and BAI3 that reach from the functional "sticky domains" to TM1, thereby including the mucin-like stalk region, the highly conserved GPCR proteolytic site (GPS), and the *Stachel* sequence [30]. Since it was shown that this crystallized domain is required and sufficient for cleavage within the GPS, it was called the GPCR autoproteolysis-inducing (GAIN) domain. Projecting our functional data onto these crystal structures, it is conceivable that the *Stachel* sequence matches the most C-terminal β-strand of the GAIN domain. Interestingly, this β-strand is encapsulated in a β-sandwich of 12 additional β-strands [30]. The crystal structures suggest that the *Stachel* sequence is buried between those β-strands prohibiting the exposure of the agonistic sequence within the full-length receptor. As the *Stachel* sequence presents with an assembly of hydrophobic amino acids (Fig. 2), exposition of this region to a hydrophilic environment is unfavorable and most probably supports interaction with its hypothetical binding pocket, whose exact position needs yet to be determined.

One important factor for *Stachel*-mediated receptor activation is the length of the β-strand. Screening of peptide libraries varying in length revealed active peptides of the *Stachel* region between 7 and 18 amino acids (see below). This is in accordance with our mutagenesis-based findings that amino acids 2–7 are essential for activity in GPR126 (see above). We therefore defined the first 8 amino acids of the *Stachel* sequence as the core region. Interestingly, the core regions of the *Stachel* sequences are always encoded by a single exon in a module-like fashion (Fig. 2). There are natural splice variants differing in the length of the *Stachel* sequence. For example, there are two isoforms of GPR114, which diverge only in the existence of a single

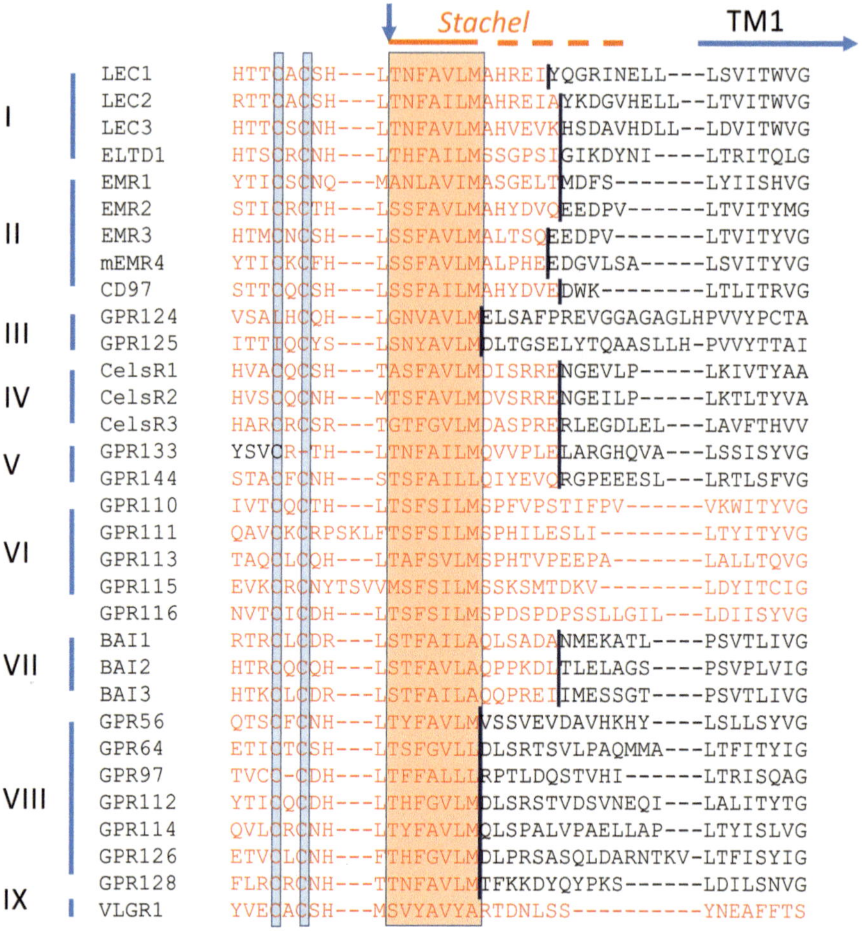

Fig. 2 *Stachel* sequence of adhesion receptors. An alignment of the amino acid sequences of all human adhesion GPCRs in close proximity to the transmembrane helix 1 (TM1) is shown (most of the N terminus is not included in the alignment). Two highly conserved cysteine residues N-terminal of the potential autocleavage site (*arrow*) are boxed in *blue*. The core sequence of the *Stachel* sequence is boxed in *orange*. Many active peptides derived from this region are longer than the core *Stachel* sequence and are marked by a *broken orange line*. Interestingly, in several families, the C terminus of the *Stachel* core sequence is defined by an exon/intron boundary (*black line*; see text). The *Stachel* core sequences are encoded by a single exon (*red characters*). Note: GPR123 was not included in the alignment since this sequence does not show any relevant sequence similarities in this region

glutamine within the *Stachel* sequence, yet display vast differences in basal full-length and mutant activity levels [27]. The isoform with the glutamine elicits very high basal activity, which can be increased through deletion of the NTF, while the isoform without the glutamine has low basal activity levels which cannot be elevated further. Deletion of the glutamine within the synthetic peptide does not affect its activation potential. As this amino acid is only important within the receptor structure, it can be assumed that it plays a role in positioning the *Stachel* sequence relative to the 7TM (see also [32]). The shortened agonistic sequence upon amino acid deletion is not able to reach its binding pocket, thereby failing to activate the receptor. This hypothesis is supported by the fact that the deletion of the adjacent leucine showed the same lack of activation, while both mutants are expressed in the cell membrane and can be activated through external peptide application [27].

Taking a closer look at the genomic structure of aGPCRs, it became evident that the GPR114 isoforms are in fact two splice variants as the glutamine is positioned at the beginning of exon 8 [27]. Here, alternate splicing can be seen as a regulatory event in order to control GPR114 function. It is therefore not surprising that the distribution of the variants shows tissue specificity. Analysis of the whole aGPCR family revealed that both families III (except GPR123, which has no GPS) and VIII have an intron/exon transition at the same location within the highly conserved *Stachel* sequence (Fig. 2), supporting the idea of an evolutionary conserved way of regulating receptor activity.

1.3 Proposed Mechanisms of *Stachel* Release

While the existence of activating peptides derived from the *Stachel* sequence is accepted, the concept of how the liberation of this sequence should happen in vivo is still a matter of debate. Current models favor an activation scenario in which the ligand binding to the NTF together with the exertion of mechanical stress will lead to its removal and subsequent exposition of the tethered agonist [33, 34], and the presumptive function of aGPCRs as mechanosensors is extensively discussed in [35]. This hypothesis is supported by the fact that GPR126 and GPR114 can be activated in vitro through mechanical forces like shaking and vibration [27, 33]. - Shaking-induced activation depends on a functional *Stachel* sequence because no mechanical activation is seen when inactivating mutations are introduced into the core *Stachel* sequence. The in vivo correlate to mechanical activation of GPR126 is the polymerization of its ligand laminin 211, which, when impaired, mimics the phenotype of the receptor knockout in zebrafish [33]. Further hints for the above suggested activation scenario come from a study that observed an enrichment of the NTF of GPR56 in supernatant after receptor stimulation with its ligand collagen III [36]. There is further in vivo evidence that aGPCRs are involved in mechanosensing. Thus, signaling of GPR56 regulates muscle hypertrophy associated with mechanical overload in mice, [37] and latrophilin/CIRL mediates perception of tactile, proprioceptive, and auditory stimuli in *Drosophila* [38].

Although binding of extracellular proteins to the NTF and the removal of the autocleaved NTF by mechanical forces are an intriguing activation model, it does not explain how non-cleavable receptors get activated through their tethered agonist or how shaking of the receptor without the addition of the ligand can already lead to an increase in signaling. It further fails to explain how it is possible that the same aGPCR that is thought to be activated through NTF removal after ligand binding can be directly activated through a different ligand. For example, the other known ligand of GPR126, collagen IV, has been shown to induce cAMP production on GPR126-expressing cells without exposure to mechanical forces [39]. Also, GPR56 has been shown to be internalized with its ligand transglutaminase 2 (TG2) meaning that there are also no cleavage or shedding processes involved since TG2 binds within the NTF of GPR56 [40].

The existence of an autoproteolytic event and its biological function has probably been one of the most controversially discussed facts in the aGPCR field in the past years and is discussed in detail in [41]. With the initial discovery of this event for latrophilin 1 (CIRL) [42] and further detailed description of the mechanism underlying this process [43], it had been assumed that the majority of aGPCRs should undergo this process and that it is a specific requirement for proper receptor function [44]. However, several studies have questioned this assumption as a general feature of aGPCRs. As for the cleavage event, there are aGPCRs that have a disrupted motif (GPR111 and GPR115; see Fig. 2) and that are therefore accepted to be non-cleavable [45]. BAI1 has a classic cleavage site, but it was shown that cleavage of this receptor does not occur in HEK293 cells but in malignant melanoma cells (discussed in [46]). Of note, BAI1 is properly expressed in the cell membrane even without cleavage. The question how an autoproteolytic event can be cell-type specific has not been answered, yet. For many other aGPCRs, it has never been tested whether cleavage actually occurs. Several studies show that cleavage is dispensable for proper receptor function [12, 47]. Interestingly, GPR114 displays high constitutive activity [13] but is not cleaved [13, 27]. Yet, GPR114 contains a tethered agonist sequence and can be activated through *Stachel* sequence-derived peptides. These findings are not compatible with a scenario in which the tethered agonist is liberated through the removal of the NTF. The high constitutive activity of GPR114, which mutually depends on the intactness of the *Stachel* sequence [27], is also not compatible with the current crystal structure of the GAIN domain [30]. Since the *Stachel* sequence is shielded by β-sheets in this structural model, it would be blocked from interacting with the 7TM to mediate the high level of basal activity. Current functional data rather favors a scenario where the *Stachel* sequence is prebound in the binding pocket of the 7TM. Upon binding of an extracellular ligand and/or exposure to mechanical stress, the inactive conformation of the *Stachel* sequence then changes into an active conformation. In the constitutively active isoform of GPR114, the *Stachel* sequence seems to be already in its active conformation. Such isomerization-induced activation of a GPCR is found in rhodopsin but needs to be proven for aGPCRs, e.g., by crystallizing full-length receptors with or without activating ligands bound to the ectodomain.

1.4 Tethered Agonism: A Common Activation Mechanism in GPCR?

As introduced above the existence of an agonistic sequence that is bound to its receptor is rare but has been shown for rhodopsin-like GPCRs. The prototypic and most intensively studied GPCR, rhodopsin, is just one example. Here, the chromophore 11-*cis*-retinal, a prosthetic group, is covalently linked to the apoprotein opsin. When exposed to light, 11-*cis*-retinal isomerizes to an all-*trans* state. The resulting metarhodopsin II confers then the signal via transducin into the cell [48]. In the case of the PARs, the receptors harbor a tethered agonist within their N termini, which is released through enzymatic cleavage of the sequence N-terminal. As in aGPCRs, this agonist sequence can be mimicked using synthetic peptides [49]. For PARs it is not clear whether the tethered ligand is prebound and only isomerizes. Very recently we have shown that glycoprotein hormone receptors (GPHRs), also members of the rhodopsin-like GPCR family, are activated by an internal agonist upon glycoprotein hormone binding [50]. Similar to findings on aGPCRs, studies have shown that removal of the large ectodomain can lead to GPHR activation. For example, proteolytic cleavage of the GPHR ectodomain by trypsin or artificially generated ectodomain deletions and truncations increased TSH receptor activity [51–53]. We have shown that deletion of the entire ectodomain did not activate the LH/hCG receptor which supported an alternative hypothesis of an "intramolecular agonistic unit" where an internal agonist within the ectodomain is exposed upon ligand binding [54, 55]. This "intramolecular agonistic unit" was later identified as p10 region, integrating the activating actions of glycoprotein hormones, mutations, and autoantibodies to trigger GPHR signal transduction [50].

Reflecting the findings with aGPCRs and other GPCRs which contain internal agonists, one can speculate that attachment of large ectodomains to the conserved 7TM structure of GPCRs has the general functions:

1. To allow for G-protein signaling for very large proteins (e.g., matrix or cell-bound proteins) which do not fit into the classical 7TM binding pocket
2. To integrate diverse extracellular signals (different proteins, ions, mechanical forces) into the same G-protein signaling pathway of a given cell/tissue

Latter point is further diversified by the fact that alternative splicing allows for modularly defining the signals that may interact with the ectodomain and, as shown for GPR114, fine-tunes the basal activity of a given receptor. It is therefore conceivable that aGPCRs might present a collection of all mechanisms known to activate the tethered agonist sequence. Current knowledge allows for the conclusion that the majority, if not all aGPCRs, carries a tethered agonist sequence, but future studies will have to focus on unraveling the molecular mechanisms exposing the tethered agonist.

2 Synthetic Peptides Derived from the *Stachel* Sequence as Tools to Modulate aGPCR Function

2.1 Pharmacologic Properties of the Peptide Agonists

Short peptide sequences mimicking the *Stachel* sequence have been shown to activate aGPCRs in vitro. The length of the most efficient peptides varies between 7 and 18 amino acids (pGPR126, 16 AA [24]; pGPR133, 13 AA [24]; pGPR110, 12 AA [26]; pGPR56, 7 AA [26]; pGPR64, 15 AA [17]; pGPR114, 18 AA [27]; pLPHN-1, 12 AA [19]). It is noteworthy that the activating peptide of GPR56 differs from all other peptide agonists in a way that it is by far the shortest peptide identified and that it was the only peptide out of the whole library tested on GPR56 that elicited some activation [26]. All other libraries featured at least one more peptide with similar activity levels. The *Stachel* sequences are highly conserved regions of aGPCRs which explain that peptides can activate more than one species. Thus, synthetic peptides have been shown to activate different receptor orthologs like human, mouse, and zebrafish. For example, a peptide derived from the human GPR126 *Stachel* sequences is able to activate also the zebrafish ortholog, and a peptide derived from the human GPR133 *Stachel* sequences can also activate the mouse GPR133 [24]. Yet, no cross-species activity was found for peptides derived from LPHN-1 of rat and *C. elegans* [19].

As the *Stachel* sequence is highly hydrophobic, the resulting peptides are similarly difficult to dilute in assay reagents making it sometimes hard to work with them. However, within the peptide library derived from the same receptor, there has been no correlation between the activity induced by and the solubility of the peptide. Another obstacle in working with these peptides is that they require large amounts in order to elicit significant activation levels in vitro with EC_{50} values between ~80 and 400 μM. The reason for this low affinity interaction can be explained by the physiologic 1:1 stoichiometry of the tethered agonist and the 7TM interface which makes high affinity dispensable. Similarly, an effective concentration of 0.4 mM of the tethered agonistic peptide was determined for PAR1 [56]. Even as such high concentrations are needed, the interaction of each peptide with its derived receptor seems to be highly specific for a given aGPCR [17, 24]. However, we cannot exclude that peptides derived from *Stachel* sequences of closely related aGPCRs can cross-activate. The *Stachel* sequences within but also between aGPCR groups are very similar (see Fig. 2). Future analyses will study possible cross activations in more detail.

2.2 Synthetic Peptides as Tools for Ex Vivo and In Vivo aGPCR Studies

Despite the difficulties that arise from the properties of the synthetic peptides, they are resourceful tools to study aGPCR function ex and in vivo. They have been shown to rescue pathological phenotypes in receptor-deficient animal models.

Thus, a GPR126 hypomorph zebrafish showed significant improvement of the myelination deficit upon treatment with 100 μM of peptide solution [24]. Further, the lethality phenotype of LAT-1 mutant nematodes was rescued with the same peptide concentration [19].

The low solubility and affinity of most *Stachel* sequence-derived peptides limit in vivo studies. Chemical modification of the peptides may help to improve their solubility. Investigations are launched which systematically test modifications of amino acids in order to increase peptide affinity to the 7TM. This may also help to define the 7TM binding pocket.

Due to the peptide nature and the high concentration needed for stimulation, in vivo experiments essentially require the development of more potent and specific small molecule ligands. As in the case of the GPHRs [57–59], such small molecule compounds may bind in the 7TM at ortho- but also allosteric positions and modulate 7TM activity. Beclomethasone which apparently activates GPR97 [13] is a promising example for such small molecule screening strategies.

3 Summary and Future Perspectives

Based on the currently available information, activation of several aGPCRs is conducted via a tethered agonist sequence, which resides within the very N-terminal part of the CTF. It is conceivable that this could be applicable for all members of the aGPCR class, yet it still needs to be proven for the majority of aGPCRs. While synthetic peptides that are mimicking the agonistic sequence can readily activate the receptor they are derived from, the question remains how this would be exercised under physiological conditions. According to the available crystal structure, the *Stachel* sequence, being equivalent to the last β-strand of the GAIN domain, is embedded in other β-strands shielding it from exposure to the 7TM interface. It is estimated that only structure-disruptive forces could liberate this sequence. This notion is supported by the fact that the combination of mechanical force and ligand addition lead to receptor activation which was not elicited under ligand addition itself. This model, however, requires a cleavable receptor. Taken that there are several non-cleavable receptors out of which at least one (GPR114) is highly constitutively active; it is eminent that there have to be other activation scenarios that need to be considered. The isomerization of a prebound agonist is one possible scenario. However, future studies are needed to answer this question. Especially the establishment of crystal structures that encompass the full-length aGPCR in its active and inactive state should provide clarity.

Even though the molecular characterization of aGPCR function has leaped forward tremendously in the past years, there are still several gaps to close in our understanding. First of all, the intracellular pathways following receptor activation need to be unraveled for the remaining orphan aGPCRs. Further, after looking at activation mechanisms, the processes involved in ending the signaling cascades need to be examined more closely. Interaction with β-arrestin and ubiquitination has only been shown for BAI1 [14], and studies dealing with internalization of aGPCRs are effectively missing. Knowing that there is most likely an interaction of aGPCRs with β-arrestin, the question arises if these receptors just like rhodopsin-like GPCRs have biased ligands. Is it possible that endogenous ligands under different mechanical conditions could activate distinct signaling cascades? In order to answer that question, it is mandatory to determine and define the specific kinds of mechanical stress that are needed to elicit receptor activation.

As peptides are not ideal activators for in vivo studies for several reasons (low affinity, solubility, and degradation issues), the development of small molecule agonists, antagonists, and inverse agonists is decisive. Another aspect for future experiments is to identify the potential of *trans* signaling for pharmacological studies. In vivo models are already distinguishing between the consequences of a signal that is mediated by the CTF within the cell (*cis* signaling) and the effects transduced by the NTF (*trans* signaling) [33, 47, 60]. In vitro studies could test for auto- or cross activation of purified functional domains of the NTF of aGPCRs. Such an approach was already undertaken for processed parts of the NTF of GPR116 [61]. Here, a purified part of the NTF, called the alpha fragment, was used to show influence on the expression levels of intracellular signaling molecules. As most aGPCRs display highly modular ectodomains, it is conceivable that similar observations can be seen in the future.

Acknowledgments This work was supported by the Deutsche Forschungsgemeinschaft (FOR2149, Projects 4 & 5) and the BMBF (IFB AdipositasDiseases Leipzig ADI-K767).

References

1. Ritter E, Zimmermann K, Heck M, Hofmann KP, Bartl FJ (2004) Transition of rhodopsin into the active metarhodopsin II state opens a new light-induced pathway linked to Schiff base isomerization. J Biol Chem 279:48102–48111
2. Coughlin SR (2005) Protease-activated receptors in hemostasis, thrombosis and vascular biology. J Thromb Haemost 3:1800–1814
3. Adams MN, Ramachandran R, Yau M-K, Suen JY, Fairlie DP, Hollenberg MD et al (2011) Structure, function and pathophysiology of protease activated receptors. Pharmacol Ther 130:248–282
4. Austin KM, Covic L, Kuliopulos A (2013) Matrix metalloproteases and PAR1 activation. Blood 121:431–439
5. Araç D, Sträter N, Seiradake E (2016) Understanding the structural basis of adhesion GPCR functions. In: Langenhan T, Schöneberg T (eds) Adhesion G protein-coupled receptors: molecular, physiological and pharmacological principles in health and disease. Springer, Heidelberg
6. Schulz A, Schoneberg T (2003) The structural evolution of a P2Y-like G-protein-coupled receptor. J Biol Chem 278:35531–35541
7. Böselt I, Römpler H, Hermsdorf T, Thor D, Busch W, Schulz A et al (2009) Involvement of the V2 vasopressin receptor in adaptation to limited water supply. PLoS One 4, e5573
8. Sanchez-Mas J, Hahmann C, Gerritsen I, Garcia-Borron JC, Jimenez-Cervantes C (2004) Agonist-independent, high constitutive activity of the human melanocortin 1 receptor. Pigment Cell Res 17:386–395

9. Iguchi T, Sakata K, Yoshizaki K, Tago K, Mizuno N, Itoh H (2008) Orphan G protein-coupled receptor GPR56 regulates neural progenitor cell migration via a G alpha 12/13 and Rho pathway. J Biol Chem 283:14469–14478
10. Luo R, Jeong S-J, Jin Z, Strokes N, Li S, Piao X (2011) G protein-coupled receptor 56 and collagen III, a receptor-ligand pair, regulates cortical development and lamination. Proc Natl Acad Sci U S A 108:12925–12930
11. Ward Y, Lake R, Yin JJ, Heger CD, Raffeld M, Goldsmith PK et al (2011) LPA receptor heterodimerizes with CD97 to amplify LPA-initiated RHO-dependent signaling and invasion in prostate cancer cells. Cancer Res 71:7301–7311
12. Bohnekamp J, Schöneberg T (2011) Cell adhesion receptor GPR133 couples to Gs protein. J Biol Chem 286:41912–41916
13. Gupte J, Swaminath G, Danao J, Tian H, Li Y, Wu X (2012) Signaling property study of adhesion G-protein-coupled receptors. FEBS Lett 586:1214–1219
14. Stephenson JR, Paavola KJ, Schaefer SA, Kaur B, Van Meir EG, Hall RA (2013) Brain-specific angiogenesis inhibitor-1 signaling, regulation, and enrichment in the postsynaptic density. J Biol Chem 288:22248–22256
15. Mogha A, Benesh AE, Patra C, Engel FB, Schoneberg T, Liebscher I et al (2013) Gpr126 functions in Schwann cells to control differentiation and myelination via G-protein activation. J Neurosci 33:17976–17985
16. Hu QX, Dong JH, Du HB, Zhang DL, Ren HZ, Ma ML et al (2014) Constitutive Galphai coupling activity of very large G protein-coupled receptor 1 (VLGR1) and its regulation by PDZD7 protein. J Biol Chem 289:24215–24225
17. Demberg LM, Rothemund S, Schoneberg T, Liebscher I (2015) Identification of the tethered peptide agonist of the adhesion G protein-coupled receptor GPR64/ADGRG2. Biochem Biophys Res Commun 464:743–747
18. Peeters MC, Fokkelman M, Boogaard B, Egerod KL, van de Water B, IJzerman AP et al (2015) The adhesion G protein-coupled receptor G2 (ADGRG2/GPR64) constitutively activates SRE and NFkB and is involved in cell adhesion and migration. Cell Signal 27 (12):2579–2588
19. Müller A, Winkler J, Fiedler F, Sastradihardja T, Binder C, Schnabel R et al (2015) Oriented cell division in the C. elegans embryo is coordinated by G-protein signaling dependent on the adhesion GPCR LAT-1. PLoS Genet 11, e1005624
20. Okajima D, Kudo G, Yokota H (2010) Brain-specific angiogenesis inhibitor 2 (BAI2) may be activated by proteolytic processing. J Recept Signal Transduct Res 30:143–153
21. Yang L, Chen G, Mohanty S, Scott G, Fazal F, Rahman A et al (2011) GPR56 regulates VEGF production and angiogenesis during melanoma progression. Cancer Res 71:5558–5568
22. Paavola KJ, Stephenson JR, Ritter SL, Alter SP, Hall RA (2011) The N terminus of the adhesion G protein-coupled receptor GPR56 controls receptor signaling activity. J Biol Chem 286:28914–28921
23. Liebscher I, Schoneberg T, Promel S (2013) Progress in demystification of adhesion G protein-coupled receptors. Biol Chem 394:937–950
24. Liebscher I, Schon J, Petersen SC, Fischer L, Auerbach N, Demberg LM et al (2014) A tethered agonist within the ectodomain activates the adhesion G protein-coupled receptors GPR126 and GPR133. Cell Rep 9:2018–2026
25. Monk KR, Naylor SG, Glenn TD, Mercurio S, Perlin JR, Dominguez C et al (2009) A G protein-coupled receptor is essential for Schwann cells to initiate myelination. Science 325:1402–1405
26. Stoveken HM, Hajduczok AG, Xu L, Tall GG (2015) Adhesion G protein-coupled receptors are activated by exposure of a cryptic tethered agonist. Proc Natl Acad Sci U S A 112:6194–6199
27. Wilde C, Fischer L, Lede V, Kirchberger J, Rothemund S, Schoneberg T et al (2016) The constitutive activity of the adhesion GPCR GPR114/ADGRG5 is mediated by its tethered agonist. FASEB J 30:666–673

28. Kishore A, Purcell RH, Nassiri-Toosi Z, Hall RA (2016) Stalk-dependent and stalk-independent signaling by the adhesion G protein-coupled receptors GPR56 (ADGRG1) and BAI1 (ADGRB1). J Biol Chem 291:3385–3394
29. Kishore A, Hall RA (2016) Versatile signaling activity of adhesion GPCRs. In: Langenhan T, Schöneberg T (eds) Adhesion G protein-coupled receptors: molecular, physiological and pharmacological principles in health and disease. Springer, Heidelberg
30. Araç D, Boucard AA, Bolliger MF, Nguyen J, Soltis SM, Südhof TC et al (2012) A novel evolutionarily conserved domain of cell-adhesion GPCRs mediates autoproteolysis. EMBO J 31:1364–1378
31. Lu YC, Nazarko OV, Sando R, Salzman GS, Südhof TC, Araç D (2015) Structural basis of latrophilin-FLRT-UNC5 interaction in cell adhesion. Structure 23:1678–1691
32. Nijmeijer S, Wolf S, Ernst OP, de Graaf C (2016) 7TM domain structure of adhesion GPCRs. In: Langenhan T, Schöneberg T (eds) Adhesion G protein-coupled receptors: molecular, physiological and pharmacological principles in health and disease. Springer, Heidelberg
33. Petersen SC, Luo R, Liebscher I, Giera S, Jeong S-J, Mogha A et al (2015) The adhesion GPCR GPR126 has distinct, domain-dependent functions in Schwann cell development mediated by interaction with laminin-211. Neuron 85:755–769
34. Monk KR, Hamann J, Langenhan T, Nijmeijer S, Schöneberg T, Liebscher I (2015) Adhesion G protein-coupled receptors: from in vitro pharmacology to in vivo mechanisms. Mol Pharmacol 88:617–623
35. Scholz N, Monk KR, Kittel RJ, Langenhan T (2016) Adhesion GPCRs as a putative class of metabotropic mechanosensors. In: Langenhan T, Schöneberg T (eds) Adhesion G protein-coupled receptors: molecular, physiological and pharmacological principles in health and disease. Springer, Heidelberg
36. Luo R, Jeong S-J, Yang A, Wen M, Saslowsky DE, Lencer WI et al (2014) Mechanism for adhesion G protein-coupled receptor GPR56-mediated RhoA activation induced by collagen III stimulation. PLoS One 9, e100043
37. White JP, Wrann CD, Rao RR, Nair SK, Jedrychowski MP, You J-S et al (2014) G protein-coupled receptor 56 regulates mechanical overload-induced muscle hypertrophy. Proc Natl Acad Sci U S A 111:15756–15761
38. Scholz N, Gehring J, Guan C, Ljaschenko D, Fischer R, Lakshmanan V et al (2015) The adhesion GPCR latrophilin/CIRL shapes mechanosensation. Cell Rep 11:866–874
39. Paavola KJ, Sidik H, Zuchero JB, Eckart M, Talbot WS (2014) Type IV collagen is an activating ligand for the adhesion G protein-coupled receptor GPR126. Sci Signal 7:ra76
40. Yang L, Friedland S, Corson N, Xu L (2014) GPR56 inhibits melanoma growth by internalizing and degrading its ligand TG2. Cancer Res 74:1022–1031
41. Nieberler M, Kittel RJ, Petrenko AG, Lin H-H, Langenhan T (2016) Control of adhesion GPCR function through proteolytic processing. In: Langenhan T, Schöneberg T (eds) Adhesion G protein-coupled receptors: molecular, physiological and pharmacological principles in health and disease. Springer, Heidelberg
42. Krasnoperov V, Lu Y, Buryanovsky L, Neubert TA, Ichtchenko K, Petrenko AG (2002) Post-translational proteolytic processing of the calcium-independent receptor of alpha-latrotoxin (CIRL), a natural chimera of the cell adhesion protein and the G protein-coupled receptor. Role of the G protein-coupled receptor proteolysis site (GPS) motif. J Biol Chem 277:46518–46526
43. Lin H-H, Chang G-W, Davies JQ, Stacey M, Harris J (2004) Autocatalytic cleavage of the EMR2 receptor occurs at a conserved G protein-coupled receptor proteolytic site motif. J Biol Chem 279:31823–31832
44. Moriguchi T, Haraguchi K, Ueda N, Okada M, Furuya T, Akiyama T (2004) DREG, a developmentally regulated G protein-coupled receptor containing two conserved proteolytic cleavage sites. Genes Cells 9:549–560
45. Prömel S, Waller-Evans H, Dixon J, Zahn D, Colledge WH, Doran J et al (2012) Characterization and functional study of a cluster of four highly conserved orphan adhesion-GPCR in mouse. Dev Dyn 241:1591–1602

46. Stephenson JR, Purcell RH, Hall RA (2014) The BAI subfamily of adhesion GPCRs: synaptic regulation and beyond. Trends Pharmacol Sci 35:208–215
47. Prömel S, Frickenhaus M, Hughes S, Mestek L, Staunton D, Woollard A et al (2012) The GPS motif is a molecular switch for bimodal activities of adhesion class G protein-coupled receptors. Cell Rep 2:321–331
48. Wald G (1968) The molecular basis of visual excitation. Nature 219:800–807
49. Vu TK, Hung DT, Wheaton VI, Coughlin SR (1991) Molecular cloning of a functional thrombin receptor reveals a novel proteolytic mechanism of receptor activation. Cell 64:1057–1068
50. Bruser A, Schulz A, Rothemund S, Ricken A, Calebiro D, Kleinau G et al (2016) The Activation Mechanism of Glycoprotein Hormone Receptors with Implications in the Cause and Therapy of Endocrine Diseases. J Biol Chem 291:508–520
51. Vlaeminck-Guillem V, Ho S-C, Rodien P, Vassart G, Costagliola S (2002) Activation of the cAMP pathway by the TSH receptor involves switching of the ectodomain from a tethered inverse agonist to an agonist. Mol Endocrinol 16:736–746
52. van Sande J, Massart C, Costagliola S, Allgeier A, Cetani F, Vassart G et al (1996) Specific activation of the thyrotropin receptor by trypsin. Mol Cell Endocrinol 119:161–168
53. Zhang M, Tong KP, Fremont V, Chen J, Narayan P, Puett D et al (2000) The extracellular domain suppresses constitutive activity of the transmembrane domain of the human TSH receptor: implications for hormone-receptor interaction and antagonist design. Endocrinology 141:3514–3517
54. Sangkuhl K, Schulz A, Schultz G, Schoneberg T (2002) Structural requirements for mutational lutropin/choriogonadotropin receptor activation. J Biol Chem 277:47748–47755
55. Krause G, Kreuchwig A, Kleinau G (2012) Extended and structurally supported insights into extracellular hormone binding, signal transduction and organization of the thyrotropin receptor. PLoS One 7, e52920
56. Zhang C, Srinivasan Y, Arlow DH, Fung JJ, Palmer D, Zheng Y et al (2012) High-resolution crystal structure of human protease-activated receptor 1. Nature 492:387–392
57. Heitman LH, Oosterom J, Bonger KM, Timmers CM, Wiegerinck PHG, Ijzerman AP (2008) 3HOrg 43553, the first low-molecular-weight agonistic and allosteric radioligand for the human luteinizing hormone receptor. Mol Pharmacol 73:518–524
58. Neumann S, Huang W, Titus S, Krause G, Kleinau G, Alberobello AT et al (2009) Small-molecule agonists for the thyrotropin receptor stimulate thyroid function in human thyrocytes and mice. Proc Natl Acad Sci U S A 106:12471–12476
59. Neumann S, Nir EA, Eliseeva E, Huang W, Marugan J, Xiao J et al (2014) A selective TSH receptor antagonist inhibits stimulation of thyroid function in female mice. Endocrinology 155:310–314
60. van Amerongen PC, Machteld J, Ghosh S, Ricciardi F, Sajjad A, Novoyatleva T et al (2013) Organ-specific function of adhesion G protein-coupled receptor GPR126 is domain-dependent. Proc Natl Acad Sci U S A 110:16898–16903
61. Fukuzawa T, Hirose S (2006) Multiple processing of Ig-Hepta/GPR116, a G protein-coupled receptor with immunoglobulin (Ig)-like repeats, and generation of EGF2-like fragment. J Biochem 140:445–452

Versatile Signaling Activity of Adhesion GPCRs

Ayush Kishore and Randy A. Hall

Graphical Abstract

A. Kishore • R.A. Hall (✉)
Department of Pharmacology, Emory University School of Medicine, 1510 Clifton Road NE, Atlanta, GA 30322, USA
e-mail: rhall3@emory.edu

T. Langenhan, T. Schöneberg (eds.), *Adhesion G Protein-coupled Receptors*, Handbook of Experimental Pharmacology 234, DOI 10.1007/978-3-319-41523-9_7

Contents

Abstract

The adhesion G protein-coupled receptors (aGPCRs) are a family of 33 receptors in humans that are widely expressed in various tissues and involved in many diverse biological processes. These receptors possess extremely large N-termini (NT) containing a variety of adhesion domains. A distinguishing feature of these receptors is the presence within the NT of a highly conserved *G*PCR *a*utoproteolysis-*in*ducing (GAIN) domain, which mediates autoproteolysis of the receptors into N-terminal and C-terminal fragments that stay non-covalently associated. The downstream signaling pathways and G protein-coupling preferences of many aGPCRs have recently been elucidated, and putative endogenous ligands for some aGPCRs have also been discovered and characterized in recent years. A pivotal observation for aGPCRs has been that deletion or removal of the NT up the point of GAIN cleavage results in constitutive receptor activation. For at least some aGPCRs, this activation is dependent on the unmasking of specific agonistic peptide sequences within the N-terminal stalk region (i.e., the region between the site of GAIN domain cleavage and the first transmembrane domain). However, the specific peptide sequences involved and the overall importance of the stalk region for activation can vary greatly from receptor to receptor. An emerging theme of work in this area is that aGPCRs are capable of versatile signaling activity that may be fine-tuned to suit the specific physiological roles played by the various members of this family.

Keywords

Adhesion • Receptor • GPCR • G protein • Signaling • Activity • Agonist • Ligand • Arrestin • Pathway

1 Adhesion G Protein-Coupled Receptors Are a Diverse Group of Self-Cleaving Cell Surface Receptors

G protein-coupled receptors (GPCRs) form the largest superfamily of cell surface signaling proteins in vertebrates [1]. Within this superfamily, the adhesion GPCRs (aGPCRs) represent the second largest family, encompassing 33 receptors in humans. These receptors are broadly expressed in different tissues and involved in many diverse processes including neural development, immunity, myelination, and angiogenesis [2]. Via their large extracellular N-termini (NT), which range from 200 to 5600 amino acids in length and harbor a variety of adhesion domains, aGPCRs are thought to survey the surrounding cellular environment and transduce signals from the extracellular milieu into intracellular signaling [3]. The tremendous diversity in the NT regions of aGPCRs has led to the further categorization of the 33 aGPCRs into 9 distinct subfamilies [4] (see [5]). A new nomenclature for aGPCRs based on these subfamilies was recently approved by IUPHAR [4], and both the new names (all starting with "ADGR") and the traditional names for each receptor will be used in this review. The most commonly shared protein-protein interaction domains found in the N-termini of each of the nine aGPCR families are olfactomedin (OLF) and rhamnose-binding lectin-like (RBL) domains (subfamily I); epidermal growth factor (EGF)-like repeats (subfamily II); leucine-rich repeats (LRRs; subfamily III); cadherin repeats (subfamily IV); pentraxin domains (PTX; subfamily V); sea urchin sperm protein, enterokinase, and agrin (SEA) domains (subfamily VI); thrombospondin type 1 repeats (TSRs; subfamily VII); pentraxin domains (subfamily VIII); and calx-β repeats (subfamily IX) [4] (see [6]).

A unique feature of the aGPCRs is their autoproteolytic activity at a membrane-proximal motif of the NT called the GPS or *G*PCR *p*roteolysis *s*ite motif [7, 8] (see also [6, 9]). This ~50-amino acid, cysteine- and tryptophan-rich motif is located within a much larger functional domain that is both necessary and sufficient for aGPCR self-cleavage called the *G*PCR *a*utoproteolysis-*in*ducing (GAIN) domain [10]. The GAIN domain is the only commonly shared domain in the NT of aGPCRs (with the exception of ADGRA1/GPR123) [11]. Moreover, the GAIN domain is also one of the most ancient domains found in aGPCRs, existing in the genomes of more primitive organisms such as *Dictyostelium discoideum* and *Tetrahymena thermophila* [10, 12]. Structural studies by Arac and colleagues showed that the GAIN domain stays intact following cleavage through an extensive network of hydrogen bonding and hydrophobic side-chain interactions [10]. These insights confirmed prior biochemical observations that autoproteolysis does not necessarily result in the dissociation of the N-terminal fragment (NTF) and C-terminal fragment (CTF) that result from GAIN domain cleavage of a given aGPCR.

2 Evidence for G Protein-Mediated Signaling by Adhesion GPCRs

Notwithstanding their N-terminal diversity, all members of the aGPCR family share a similar seven-transmembrane (7TM) domain architecture, which is the molecular signature of GPCRs. However, in the early years of aGPCR research, it was not known whether these proteins were bona fide GPCRs. In studies that were facilitated by the serendipitous discovery of a potent and high-affinity agonist, ADGRL1 (latrophilin-1) was one of the first aGPCRs characterized in terms of its signaling activity [13]. It was found that α-latrotoxin (α-LTX), a component of black widow spider venom, stimulated increases in intracellular cAMP and IP3 levels in ADGRL1-transfected COS7 cells in a receptor-dependent manner

[14]. However, in addition to binding to ADGRL1, α-LTX can also form calcium-permeable pores in the plasma membrane and trigger exocytosis [15]. Therefore, a mutant version of the toxin was generated, α-LTX^{N4C}, which does not cause exocytosis but still binds to and activates ADGRL1 [15]. Further studies showed that ADGRL1 could activate phospholipase C (PLC) and increase intracellular Ca^{2+} within minutes of α-LTX^{N4C} treatment, suggesting coupling of the receptor to $G\alpha_q$ [16]. Moreover, ADGRL1 could be co-purified with $G\alpha_o$ [14, 17] and $G\alpha_{q/11}$ [17] using α-LTX affinity chromatography.

Unlike ADGRL1, the majority of aGPCRs do not have known ligands. Thus, a common method of discerning the signaling pathways downstream of aGPCRs has been to overexpress the receptors in heterologous systems and measure their constitutive activities in assays of specific G protein signaling. For example, overexpression of ADGRG1 (GPR56), a receptor that is critically involved in the development of the cerebral cortex [18, 19], was shown to robustly stimulate the activation of RhoA via coupling to the $G\alpha_{12/13}$ signaling pathway [20, 21]. Subsequent studies have demonstrated that ADGRG1 expression can upregulate the activity of a variety of downstream transcription factors, including NFκB [22], PAI-1 [22], TCF [22], SRE [20, 23–25], SRF [26], and NFAT [23, 26]. Other outputs influenced by ADGRG1 include PKCα [27], VEGF [25], and TGFα shedding [26]. In addition to these results, other lines of evidence supporting receptor G protein coupling have been provided by several groups. For example, it was demonstrated that $G\alpha_{q/11}$ could be co-immunoprecipitated with ADGRG1 in heterologous cells [28]. This interaction, however, depended on the presence of the tetraspanin CD81, which may act as a scaffold for the ADGRG1/$G\alpha_{q/11}$ signaling complex. In agreement with these data, stimulation of ADGRG1 in U87-MG cells was found to raise intracellular Ca^{2+} levels in a manner that was blocked by YM-245890, an inhibitor of $G\alpha_{q/11}$-mediated signaling [29]. Additionally, ADGRG1 has been shown to activate $G\alpha_{13}$ in a reconstituted GTPγS-binding assay [24], and an association between ADGRG1 and $G\alpha_{13}$ has also been shown via a co-immunoprecipitation approach [26].

In addition to ADGRG1, evidence for G protein coupling has also been provided for several other members of aGPCR subfamily VIII. For example, ADGRG2 (GPR64) expression in transfected cells has been demonstrated to stimulate the SRE and NFκB pathways [30], raise intracellular cAMP, and elevate IP3 levels in the presence of the chimeric G protein $G\alpha_{qi4}$, suggesting promiscuous coupling to both $G_{\alpha s}$ and $G_{\alpha i}$ [31]. Similarly, it was shown that overexpression of ADGRG3 (GPR97) in HEK293 cells stimulated IP3 accumulation only in the presence of chimeric G protein $G\alpha_{qo3}$, which converts $G\alpha_o$ signaling into $G\alpha_q$ activity, suggesting natural coupling of the receptor to $G\alpha_o$ [32]. ADGRG5 (GPR114) overexpression was shown to potentiate cAMP levels, an effect that could be blocked via knockdown of endogenous $G\alpha_s$ or overexpression of the chimeric G protein $G\alpha_{qs4}$, which converts $G\alpha_s$ signaling into $G\alpha_q$-mediated activity [33]. Another member of the subfamily, ADGRG6 (GPR126), which plays an important role in regulating peripheral nerve myelination [34], was also found to raise intracellular cAMP [35–37] as well as stimulate IP3 accumulation in the presence of chimeric G proteins to redirect either $G_{\alpha s}$ or $G_{\alpha i}$ activity toward $G_{\alpha q}$ pathways [36]. Thus, both ADGRG2 and ADGRG6 may couple to $G\alpha_s$ to raise cAMP levels while also exhibiting coupling to other G proteins to mediate pleiotropic effects on cellular physiology.

ADGRB1 (BAI1), a receptor that regulates phagocytosis [38–41], myogenesis [42], and synaptic plasticity [43, 44], has been shown to constitutively activate RhoA [45], Rac1 [41], ERK [45], SRF [26], NFAT [26], and TGFα shedding [26] when overexpressed in heterologous cells. ADGRB1 signaling to most of these downstream readouts can be greatly attenuated by co-expression of the RGS domain of p115-RhoGEF, suggesting a predominant coupling of the receptor to $G\alpha_{12/13}$. These functional data are consistent with co-immunoprecipitation data revealing the existence of cellular complexes between ADGRB1 and $G\alpha_{12/13}$ [26]. Expression of ADGRB2 (BAI2), a close relative of ADGRB1, was found to also stimulate the NFAT pathway and additionally induce IP3 accumulation in HEK293 cells, indicating a likely coupling to $G\alpha_{q/11}$ [46].

ADGRE2 (EMR2), a receptor highly enriched in immune cells, was demonstrated to stimulate IP3 accumulation in transiently transfected HEK293 cells, indicative of $G_{\alpha q}$ coupling [32]. Expression of another receptor from the same subfamily, ADGRE5 (CD97), was found to activate the SRE pathway in transfected COS7 cells in a manner that was sensitive to the presence of RGS-p115-RhoGEF, suggesting receptor coupling to $G\alpha_{12/13}$ [47]. Receptors ADGRF1 (GPR110) and ADGRF4 (GPR115) were both shown to stimulate IP3 accumulation in transiently transfected HEK293 cells [32]. In separate studies that confirmed some of these findings, ADGRF1 was shown to activate $G\alpha_q$ in a GTPγS assay [24].

ADGRV1 (VLGR1), a receptor that has a crucial role in hearing and vision and whose dysfunction is associated with the human disease known as Usher syndrome, was shown to inhibit isoproterenol-induced cAMP levels in HEK293 cells, indicative of $G\alpha_i$ coupling [48]. Moreover, co-expression of the chimeric G protein $G\alpha_{qi5}$ was able to reroute receptor activity toward a $G\alpha_{q/11}$ readout (NFAT activation), thereby providing further evidence for $G\alpha_i$ coupling. In contrast, expression of ADGRD1 (GPR133) has been shown to raise cAMP levels in multiple studies [32, 37, 49]. Moreover, it was demonstrated that ADGRD1-mediated cAMP elevation could be blocked by knocking down $G\alpha_s$ [32].

3 Ligands for Adhesion GPCRs

Potential ligands have been identified for a number of members of the aGPCR family (Table 1). As mentioned previously, α-LTX is a high-affinity agonist of ADGRL1 that has been shown to stimulate several readouts of receptor activity. Another reported ligand for ADGRL1 is teneurin-2, a large (~2800 residue) glycoprotein with a single transmembrane region that is found predominantly in the brain [50]. Teneurin-2 was first identified as a binding partner of ADGRL1 through pull-down studies in which rat brain lysates were subjected to α-LTX affinity chromatography [50]. Treatment of cultured neurons expressing ADGRL1 with a soluble,

Table 1 Adhesion GPCR ligands and/or agonists

Receptor	Ligand	Binding region	Downstream activity
Family I			
ADGRL1	α-Latrotoxin	NT (GAIN domain)	Increased cAMP [14], IP3 [14], Ca^{2+} [16], and PLC activation [16]
ADGRL1	Teneurin-2	NT	Increased Ca^{2+} in cultured hippocampal neurons [50]
ADGRL1	Neurexin1α	NT	Regulation of α-latrotoxin-mediated glutamate release [51]
ADGRL3	FLRT3	NT	Regulation of synaptic density [52]
ADGRL3	FLRT2	NT (OLF domain)	Regulation of cell adhesion/repulsion [53]
Family II			
ADGRE2	NT antibody (2A1)	NT	Increased production of inflammatory cytokines [54]
ADGRE2/ ADGRE5	Chondroitin sulfate	? (likely NT region)	Mediates cell adhesion [55]
ADGRE5	CD55	NT (EGF domains)	Alteration in ADGRE5 NT-CTF interaction [56]
ADGRE5	α5β1/αvβ3	NT	Mediates endothelial cell migration [57]
ADGRE5	CD90	NT	Mediates cell adhesion [58]
Family V			
ADGRD1	Stalk peptide(s)	? (likely 7TM region)	Increased cAMP levels [37]
Family VI			
ADGRF1	Stalk peptide(s)	? (likely 7TM region)	Increased GTPγS binding [24]
Family VII			
ADGRB1	Phosphatidylserine	NT (TSR domains)	Enhanced Rac1-dependent uptake of apoptotic cells [39]
ADGRB3	C1ql1	NT (CUB domain)	Regulation of dendritic spine density [59]
ADGRB3	C1ql3	NT (TSR domains)	Regulation of synaptic density [60]
Family VIII			
ADGRG1	Tissue transglutaminase 2	NT (STP region)	Regulation of VEGF secretion [27]
ADGRG1	Collagen III	NT (aa 27–160)	Stimulation of RhoA activation [61]
ADGRG1	NT antibody	NT	Stimulation of SRE and RhoA activity [20]
ADGRG1	Stalk peptide(s)	? (likely 7TM region)	Stimulation of SRE luciferase [24]
ADGRG2	Stalk peptide(s)	? (likely 7TM region)	Increased cAMP and IP3 accumulation [31]
ADGRG3	Beclomethasone dipropionate	?	Increased GTPγS binding [32]

(continued)

Table 1 (continued)

Receptor	Ligand	Binding region	Downstream activity
ADGRG5	Stalk peptide(s)	? (likely 7TM region)	Increased cAMP levels [33]
ADGRG6	Collagen IV	NT (CUB and PTX domains)	Increased cAMP levels [35]
ADGRG6	Laminin-211	NT (aa 446–807)	Increased cAMP levels upon mechanical shaking [62]
ADGRG6	Stalk peptide(s)	? (likely 7TM region)	Increased cAMP levels [37] and IP3 accumulation when co-expressed with chimeric Gqi [36]

? unavailable

C-terminal fragment of teneurin-2 was found to trigger the release of intracellular Ca^{2+}, possibly through a G protein-dependent mechanism [50]. In another study, coculturing cells expressing either ADGRL1 or teneurin-2 resulted in the formation of large cell aggregates, indicating that the specific interaction between the two proteins may mediate cell adhesion [64]. In the brain, ADGRL1 and teneurin-2 are enriched in the presynaptic and postsynaptic membranes, respectively. The extracellular NT of ADGRL1, however, may be large enough to span the synaptic cleft to mediate interneuronal contact through its high-affinity interaction with teneurin-2.

ADGRL1 has also been shown to interact with neurexin, a presynaptic protein implicated in synaptogenesis and function [65]. Neurexin is a binding partner of α-LTX, as is ADGRL1 [66]. A particular neurexin isoform (1α) binds α-LTX in a Ca^{2+}-dependent fashion, while the α-LTX-ADGRL1 interaction is Ca^{2+} independent [66]. Interestingly, in the absence of Ca^{2+}, knockdown of neurexin in cultured hippocampal neurons significantly diminished the α-LTX response compared to wild-type neurons, suggesting that while ADGRL1 and neurexin can independently associate with α-LTX, their interaction may synergistically enhance α-LTX-induced signaling by ADGRL1 [51]. Moreover, coculture of cells expressing either ADGRL1 or neurexin resulted in numerous cell aggregates, providing evidence that the interaction promotes adhesion complexes [67]. More work must be done, however, to demonstrate whether neurexins directly stimulate receptor signaling activity.

The fibronectin leucine-rich repeat transmembrane (FLRT) proteins are an additional class of ligands for ADGRL1 and the related receptor ADGRL3 (latrophilin-3) [52]. Direct interactions between the NT of ADGRL3 and FLRT3 were demonstrated in a non-cell-based assay [52]. In vivo, both proteins are enriched in cell-to-cell junctions and regulate synaptic density [52]. In another study, a high-affinity interaction was demonstrated for ADGRL3 and FLRT2 [53]. This interaction was found to be mediated by the OLF domain on the ADGRL3 NT and, intriguingly, promoted either adhesion of FLRT2-expressing HeLa cells or repulsion of FLRT2-expressing cultured cortical neurons. These results potentially highlight the influence that cellular environment may have on the relationship between receptor and ligand. At present, however, there is no evidence that FLRT proteins can directly instigate signaling by the latrophilin receptors.

The association between ADGRE5 and CD55 was one of the first confirmed protein-protein interactions involving an aGPCR [68]. This interaction was found to be mediated by the EGF domains on the receptor's NT [69]. Recently, it was shown that CD55 does not modulate ADGRE5-mediated signaling to ERK or Akt [56]. It remains to be determined whether CD55 can modulate other receptor-controlled pathways, such as perhaps the RhoA signaling pathway. ADGRE2 is a close relative of ADGRE5 with highly homologous EGF domains, but nonetheless ADGRE2 has been found to have a much lower binding affinity for CD55 than ADGRE5 [70]. Both ADGRE5 and ADGRE2 have also been shown to bind to extracellular matrix (ECM) components known as chondroitin sulfates [55]. These interactions are generally low affinity and Ca^{2+} dependent and have not yet been demonstrated to instigate G protein-mediated signaling for either receptor.

A number of ligands have been identified for subfamily VII aGPCRs. ADGRB1 was found to bind externalized phosphatidylserine on apoptotic cells through the thrombospondin type 1 repeat domains on its NT [38]. This interaction promoted the engulfment of the apoptotic cells in a mechanism reliant on the adaptor protein ELMO1 and signaling by the small GTPase Rac1 [38]. Another receptor from this subfamily, ADGRB3 (BAI3), was shown to bind to C1q-like (C1ql) proteins [60, 71]. Similar to the interaction of ADGRB1 and phosphatidylserine, the interaction between ADGRB3 and C1ql3 was found to be mediated by thrombospondin repeats on the receptor's NT [60]. In cultured neurons, submicromolar C1ql3 treatment significantly reduced synaptic density, an effect readily blocked by exogenous addition of purified ADGRB3 NT [60]. In a similar study, it was shown that ADGRB3 binds C1ql1 via its N-terminal CUB domain and that both proteins were necessary for normal spine density of cerebellar neurons [59]. Furthermore, the interaction between C1ql1 and ADGRB3 was demonstrated to regulate pruning in mouse cerebellum, with knockout of either protein resulting in severe motor learning deficits [72]. Future studies in this area will likely examine whether C1ql proteins have similar binding affinities for other members of subfamily VII and whether those interactions can stimulate receptor-mediated activity.

Several ligands have been identified for ADGRG1, including tissue transglutaminase 2 (TG2), a major cross-linking enzyme of the extracellular matrix implicated in cancer progression [63, 73]. TG2 binds a ~70-residue region on the NT of ADGRG1; deletion of this TG2-binding region was found to enhance receptor-mediated VEGF production in vitro and significantly increase tumor growth and angiogenesis in vivo, whereas expression of the wild-type receptor reduced both measures [27]. In a more recent study, it was demonstrated that the antagonistic relationship between ADGRG1 and TG2 may be attributed to internalization and lysosomal degradation of extracellular TG2 in a receptor-dependent mechanism [74]. It is unclear at present whether interaction with TG2 stimulates G protein-mediated signaling by ADGRG1.

Collagen III is another ligand for ADGRG1 [61]. ADGRG1 loss-of-function mutations cause the human disease bilateral frontoparietal polymicrogyria (BFPP). Patients with BFPP have a cortical malformation due to aberrant neural stem cell migration [75]. Remarkably, knockout of collagen III in mice results in a cortical phenotype similar to that observed in mice lacking ADGRG1 as well as human BFPP patients [75]. Collagen III binds a ~130-residue region in the distal half of the receptor's NT [76]. Moreover, nanomolar concentrations of collagen III have been shown to significantly reduce migration of mouse neurospheres (masses of cells containing neural stem cells) in a receptor-dependent fashion [61]. Biochemical studies revealed that collagen III could stimulate RhoA signaling in a mechanism dependent on receptor expression and likely mediated by $G\alpha_{12/13}$ [61].

Another subfamily VIII receptor, ADGRG6, has also been shown to be stimulated by collagen interactions, albeit with a distinct type of collagen. The association between ADGRG6 and collagen IV was found to be mediated by a region of the ADGRG6 NT containing the CUB and PTX domains [35]. Furthermore, the association was shown to be specific, as other types of collagen, including collagen III, did not bind the receptor. In heterologous cells, collagen IV stimulated receptor-dependent cAMP elevation. The half-maximal effective concentration for this response was 0.7 nM, indicating that collagen IV is a potent agonist for ADGRG6.

An additional ligand for ADGRG6 is laminin-211, an extracellular matrix protein that is involved in Schwann cell development and peripheral nervous system myelination [62]. Interestingly, laminin-211 was found to antagonize receptor-mediated cAMP elevation in a dose-dependent fashion in heterologous cells. Furthermore, cAMP inhibition was due to antagonism of receptor-mediated $G\alpha_s$ activity rather than through differential activation of $G\alpha_i$. Remarkably, laminin-211 treatment under the condition of mechanical shaking had the opposite effect of boosting receptor-mediated cAMP levels. Thus, laminin-211 may serve as a unique ligand that can differentially modulate receptor activity depending upon other physical cues and mechanical forces in the extracellular environment.

Most of the putative aGPCR endogenous ligands described thus far are large, ECM-derived molecules. Nonetheless, it has been shown that small molecules can be developed as aGPCR ligands. For example, screening studies revealed beclomethasone dipropionate as a ligand for ADGRG3 [32]. Beclomethasone dipropionate is a glucocorticoid steroid that can stimulate ADGRG3 with nanomolar potency. The region of the receptor that interacts with beclomethasone is unknown, but considering the molecule's hydrophobicity, it would not be surprising if it were found in future studies to directly interact with the receptor's 7TM region to modulate receptor activity.

An intriguing observation made for several aGPCRs has been that these receptors may be activated by antibodies directed against their NT regions. Antibodies may be able to mimic the binding of endogenous ligands to aGPCRs and thus may represent powerful research tools for studying aGPCR signaling, especially for those receptors with no identified ligands. An N-terminal activating antibody of ADGRG1 was first described in 2008 by Itoh and colleagues. Studies in heterologous cells revealed that antibody treatment could dose-dependently stimulate receptor signaling in the SRE luciferase assay (a commonly used assay for $G\alpha_{12/13}$ activity) [20]. Moreover, stimulation was readily blocked by exogenous addition of the receptor's NT, which presumably competed for antibody binding. Moreover, in a later study it was shown that other newly generated N-terminal antibodies for ADGRG1 could inhibit cell migration in a manner that was sensitive to inhibition of either $G\alpha_q$ or $G\alpha_{12/13}$ signaling [29]. In another example, an antibody directed against the N-terminal region of ADGRE2 was shown to dose-dependently increase inflammatory cytokine production in receptor-mediated neutrophil activation [54].

Given the importance of aGPCR N-termini in mediating binding to extracellular ligands, it is perhaps not surprising that mutations to the aGPCR N-termini can oftentimes lead to loss of receptor function and human disease. For example, there are several reported N-terminal disease-causing mutations to ADGRG1 that result in reduced plasma membrane expression of the receptor [77, 78] and/or disruption of the receptor's ability to bind collagen III [76]. Another prominent example is of ADGRV1, where several NT mutations cause cochlear and retinal defects in humans [79]. Moreover, missense NT mutations to ADGRC1 (CELSR1) impair surface trafficking of the protein and are implicated in a severe neural tube defect in humans known as craniorachischisis [80].

4 Adhesion GPCR Models of Activation

With the idea that aGPCR ligands mainly bind to the large extracellular NT regions and that the NT regions are cleaved in the GAIN domain and may be removed at some point following ligand binding, a number of groups have generated truncated versions of aGPCRs lacking most of their NT regions up to the sites of predicted GAIN cleavage. The first studies of this type were performed independently for a trio of receptors—ADGRB2 [46], ADGRG1 [21], and ADGRE5 [47]—and in each case the truncation was found to result in a substantial increase in the receptors' constitutive signaling activity. Subsequently, this phenomenon has been reported for a number of other aGPCRs, including ADGRB1 [45], ADGRG6 [35], ADGRG2 [30, 31], ADGRD1 [37], ADGRF1 [24], and ADGRV1 [48]. In light of these findings, a general model of aGPCR activation was proposed wherein the tethered NTF behaves as an antagonist of CTF-mediated signaling, with N-terminal deletion mimicking ligand-mediated removal of the NTF to result in receptor activation [81]. This model of activation, termed the *disinhibition model*, was a general model that left open the mechanistic question of precisely how removal of aGPCR NT regions might activate receptor signaling.

Subsequently, a more mechanistically specific model of aGPCR activation, termed the *tethered agonist model*, was proposed (Fig. 1; see also [82]). In this model, GAIN domain autoproteolysis (and/or conformational change) reveals a tethered cryptic agonist sequence contained within the NT region between the site of cleavage and the first transmembrane domain (i.e., the stachel or stalk region).

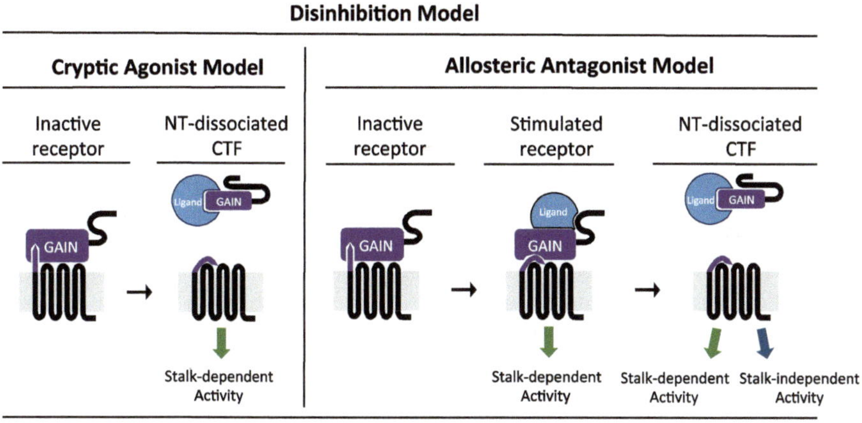

Fig. 1 Models of adhesion GPCR activation. Cryptic agonist model—*Inactive receptor*: The GAIN domain antagonizes receptor activity by concealing a cryptic agonist found in the N-terminal stalk region between the site of autoproteolysis and the first transmembrane domain. *NTF-dissociated CTF*: Following ligation of the N-terminal fragment (NTF) with an extracellular ligand and subsequent removal from the plasma membrane, the cryptic agonist sequence (the stachel) is unveiled and stimulates activity through interactions with the remaining C-terminal fragment (CTF). Allosteric antagonist model—*Inactive receptor*: In the absence of ligand engagement, the GAIN domain can inhibit receptor activity in two distinct ways: by concealing a cryptic agonist on the N-terminal stalk and also by dampening the inherent constitutive activity of the CTF. *Stimulated receptor*: Ligation of the NTF with an extracellular ligand induces a conformational change to allow for stimulation by the cryptic agonist within the stalk, even though the NTF may stay associated with the CTF for some time. *NTF-dissociated CTF*: If and when ligand binding induces NTF dissociation from the CTF, another wave of receptor activity may be unleashed, with the inherent, stalk-independent activity of the CTF being stimulated. In this stage, the receptor may achieve its maximal activity due to the summation of signals from both stalk-dependent and stalk-independent mechanisms

This mechanism of activation is conceptually similar to that of the protease-activated receptors, for which proteolysis of the N-terminal domain by an extracellular protease unveils an agonist in the remaining NT [83]. Evidence in favor of the cryptic agonist model was provided by two independent groups: Liebscher et al. and Stoveken et al. First, Liebscher et al. showed that deletion of the remaining NT (i.e., the stachel or stalk region) from constitutively active NTF-lacking versions of ADGRG6 and ADGRD1 ablated activity of both receptors in cAMP accumulation assays [37]. Moreover, synthetic peptides corresponding to the stalk regions of each receptor were able to restore activity of the stalkless mutants with varying degrees of efficacy. The most potent peptides displayed half-maximal effective concentrations in the high micromolar range. Further studies from Liebscher et al. along similar lines provided evidence for tethered agonist-mediated activation of ADGRG2 [31] and ADGRG5 [33]. Additionally, Stoveken et al. showed that stalkless versions of ADGRG1 and ADGRF1 lacked activity in reconstitution assays examining GTP binding to purified $G\alpha_{13}$ and $G\alpha_q$, respectively [24]. Synthetic peptides fashioned after the stalk of each receptor were shown to resuscitate their cognate stalkless receptors in a dose-dependent manner, with the most potent peptides displaying submicromolar half-maximal effective concentrations. Moreover, the most potent stalk peptide of ADGRG1 was shown to stimulate receptor-mediated activity in cellular SRE luciferase assays in addition to the $G\alpha_{13}$ reconstitution studies.

The finding from Stoveken et al. that stalk-deficient ADGRG1 is unable to activate SRE luciferase was confirmed in recent studies using a similar readout, SRF luciferase [26]. However, the stalkless ADGRG1 was found in these studies to be functional in other readouts of receptor signaling activity including TGFα shedding, NFAT luciferase, beta-arrestin recruitment, and receptor ubiquitination [26]. In parallel, a stalkless truncated version of ADGRB1 was examined in the same battery of assays and found to have nearly identical activity to the constitutively active truncated version of ADGRB1 that retained the stalk. A conclusion from this work was that aGPCRs are capable of both stalk-dependent and stalk-independent signaling, with the relative contribution of the stalk varying between different receptors and even between different readouts for the same receptor. These findings led to the proposal of the *allosteric antagonist model* of aGPCR activation (Fig. 1), in which aGPCR NT regions can dampen receptor activity in at least two distinct ways: (1) by masking the stalk region to prevent stalk-dependent signaling and (2) by allosterically antagonizing the inherent, stalk-independent activity of the 7TM region.

Further evidence that a proteolytically liberated agonist in the stalk region may not be required for all aspects of aGPCR signaling comes from studies on non-cleavable aGPCR mutants. The GAIN domain crystal structures from Arac et al. revealed how mutation of a key catalytic threonine in the GPS motif could block GAIN domain cleavage but allow for normal GAIN domain folding [10]. Such non-cleaving mutants of ADGRD1 [49], ADGRG1 [26], and ADGRG2 [30] have been studied and found to be capable of robust constitutive signaling, although in the case of ADGRG2 the non-cleavable mutant receptor exhibited signaling comparable to the wild-type receptor in one pathway but reduced signaling when a distinct pathway was measured. There is also evidence that certain aGPCRs may not undergo GAIN cleavage at all [84]. ADGRG5 and ADGRB1 are examples of aGPCRs that are naturally cleavage deficient (at least in some cellular contexts) and yet retain signaling ability [33, 45]. Moreover, in vivo studies on lat-1, the *C. elegans* ortholog of ADGRL1, revealed that wild-type and mutant non-cleavable versions of the receptor performed just as well in the transgenic rescue of deficits resulting from receptor knockout [85]. The requirement of the stalk region for aGPCR signaling is also uncertain due to observations that individual aGPCRs, such as ADGRL1, undergo additional proteolytic processing wherein GAIN autoproteolysis is followed by one or more additional cleavage events that remove the stalk region [86] (see also [9] for an in-depth discussion on the relationship between proteolytic processing and aGPCR activity). These findings taken together suggest that neither GAIN domain autoproteolysis nor the presence of the stalk region are absolutely required for aGPCR signaling activity but rather may be important for some receptors and certain downstream pathways.

5 Adhesion GPCR N-Termini as Sensors of Mechanical Force

There is emerging evidence that aGPCRs may be involved in sensing mechanical forces. For example, it was shown that the ADGRE5 NTF is released from the CTF after engagement with the ligand CD55, but only under mechanical shaking conditions that are meant to recapitulate the shear stress associated with circulating blood [56]. In a similar vein, laminin-211, a ligand of ADGRG6 as mentioned above, was found to only stimulate the receptor under shaking conditions and actually antagonized receptor activity under static conditions [62]. In these studies, the mechanical forces may have helped laminin-211 to disengage the NTF from its CTF, whereas without shaking, the ligand binding may have actually stabilized the inhibitory NTF-CTF interaction. These examples support the idea that, for at least some ligand-receptor pairs, mechanical force may be a key determinant of the signaling output that results from the interaction. In a key in vivo study on aGPCR-mediated mechanosensation, Scholz et al. recently demonstrated that *Drosophila* larvae lacking the ADGRL1 ortholog CIRL exhibited diminished sensitivity to mechanical stimuli [87]. The role of aGPCRs in sensing mechanical force is likely to be an active area of research in the coming years and discussed in detail in [88].

6 Associations of aGPCRs with Signaling Proteins Other Than G Proteins

In addition to the aforementioned examples of aGPCR coupling to G proteins, there have also been a number of cytoplasmic proteins other than G proteins that have been found to interact with aGPCRs (see [89] for more on this topic). In some cases, these interactions appear to modulate G protein-mediated signaling, while in other cases these associations appear to mediate G protein-independent signaling (Fig. 2). One example of the regulation of G protein signaling comes from work on ADGRV1, which was found to interact with the PDZ domain-containing protein PDZD7, a key scaffold protein in the USH2 protein complex that is known to be pivotal for stereocilial development and function [48]. Association with PDZD7 was found to antagonize ADGRV1 activity, likely by competitively disrupting receptor association with $G\alpha_i$ [48, 90]. ADGRB1 is another aGPCR that has been found to associate with PDZ scaffold proteins. One such PDZ protein, MAGI-3, was found to potentiate receptor-mediated ERK signaling, possibly by recruiting positive regulators of the pathway [45].

In terms of G protein-independent signaling by aGPCRs, ADGRB1 and ADGRB3 have both been shown to bind to the intracellular adaptor protein ELMO1 [38, 91]. For ADGRB1, this interaction has been demonstrated to result in the formation of a complex at the plasma membrane capable of activating the

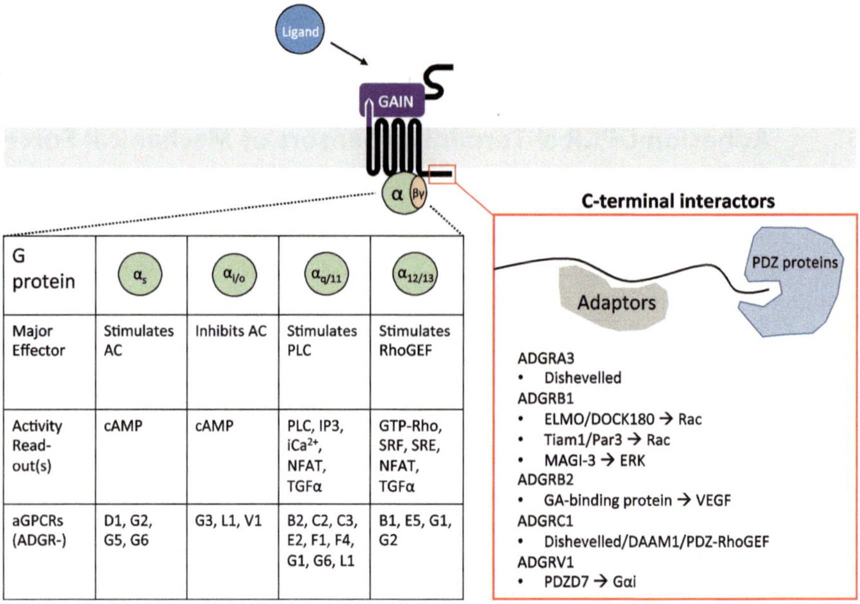

G protein	α_s	$\alpha_{i/o}$	$\alpha_{q/11}$	$\alpha_{12/13}$
Major Effector	Stimulates AC	Inhibits AC	Stimulates PLC	Stimulates RhoGEF
Activity Read-out(s)	cAMP	cAMP	PLC, IP3, iCa^{2+}, NFAT, TGFα	GTP-Rho, SRF, SRE, NFAT, TGFα
aGPCRs (ADGR-)	D1, G2, G5, G6	G3, L1, V1	B2, C2, C3, E2, F1, F4, G1, G6, L1	B1, E5, G1, G2

Fig. 2 G protein-dependent and G protein-independent signaling by adhesion GPCRs. The *left panel* shows the various G protein-dependent pathways that can be activated by aGPCRs. Also shown are the probable G protein-coupling preferences for selected members of the aGPCR family. The *right panel* displays various aGPCR C-terminal binding partners and briefly describes their influence on aGPCR signaling pathways (both G protein dependent and G protein independent)

small GTPase Rac1 in a G protein-independent manner [38]. ADGRB1-mediated activation of Rac1 has been implicated in phagocytosis and myoblast fusion [38, 42]. Intriguingly, ADGRB1 can also activate Rac in a distinct G protein-independent manner through association with the RacGEF Tiam1 [43]. Other examples of G protein-independent signaling by aGPCRs include ADGRB2 interaction with GA-binding protein (GABP) gamma to regulate VEGF expression [92]; ADGRC1 association with dishevelled, DAAM1, and PDZ-RhoGEF to regulate neural tube closure [93]; and ADGRA3 (GPR125) interaction with dishevelled to mediate the recruitment of planar cell polarity components [94].

7 Concluding Remarks

The versatility of aGPCR signaling described here highlights the need to comprehensively study the members of this family on a receptor-by-receptor basis in order to delineate the diversity of metabotropic pathways they serve. Further insights gained into the mechanisms of aGPCR activation will have important implications for drug development efforts aimed at these receptors. Given the number of human diseases linked to aGPCR mutations and the intriguing phenotypes observed upon genetic deletion of aGPCRs [4], there are compelling reasons to believe that elucidation of the activation mechanisms and downstream pathways of aGPCRs will allow for an enhanced understanding of human disease and promote the development of novel classes of therapeutics.

Acknowledgments The authors' research is supported by the National Institutes of Health.

References

1. Pierce KL, Premont RT, Lefkowitz RJ (2002) Seven-transmembrane receptors. Nat Rev Mol Cell Biol 3(9):639–650
2. Fredriksson R, Gloriam DE, Hoglund PJ, Lagerstrom MC, Schioth HB (2003) There exist at least 30 human G-protein-coupled receptors with long Ser/Thr-rich N-termini. Biochem Biophys Res Commun 301(3):725–734
3. Langenhan T, Aust G, Hamann J (2013) Sticky signaling–adhesion class G protein-coupled receptors take the stage. Sci Signal 6(276):re3
4. Hamann J, Aust G, Arac D, Engel FB, Formstone C, Fredriksson R et al (2015) International Union of Basic and Clinical Pharmacology. XCIV. Adhesion G protein-coupled receptors. Pharmacol Rev 67(2):338–367
5. Krishnan A, Nijmeijer S, de Graaf C, Schiöth HB (2016) Classification, nomenclature and structural aspects of adhesion GPCRs. In: Langenhan T, Schöneberg T (eds) Adhesion G protein-coupled receptors: molecular, physiological and pharmacological principles in health and disease. Springer, Heidelberg
6. Araç D, Sträter N, Seiradake E (2016) Understanding the structural basis of adhesion GPCR functions. In: Langenhan T, Schöneberg T (eds) Adhesion G protein-coupled receptors: molecular, physiological and pharmacological principles in health and disease. Springer, Heidelberg
7. Krasnoperov VG, Bittner MA, Beavis R, Kuang Y, Salnikow KV, Chepurny OG et al (1997) alpha-Latrotoxin stimulates exocytosis by the interaction with a neuronal G-protein-coupled receptor. Neuron 18(6):925–937
8. Lin HH, Chang GW, Davies JQ, Stacey M, Harris J, Gordon S (2004) Autocatalytic cleavage of the EMR2 receptor occurs at a conserved G protein-coupled receptor proteolytic site motif. J Biol Chem 279(30):31823–31832
9. Nieberler M, Kittel RJ, Petrenko AG, Lin H-H, Langenhan T (2016) Control of adhesion GPCR function through proteolytic processing. In: Langenhan T, Schöneberg T (eds) Adhesion G protein-coupled receptors: molecular, physiological and pharmacological principles in health and disease. Springer, Heidelberg
10. Arac D, Boucard AA, Bolliger MF, Nguyen J, Soltis SM, Sudhof TC et al (2012) A novel evolutionarily conserved domain of cell-adhesion GPCRs mediates autoproteolysis. EMBO J 31(6):1364–1378
11. Liebscher I, Ackley B, Arac D, Ariestanti DM, Aust G, Bae BI et al (2014) New functions and signaling mechanisms for the class of adhesion G protein-coupled receptors. Ann N Y Acad Sci 1333:43–64
12. Promel S, Langenhan T, Arac D (2013) Matching structure with function: the GAIN domain of adhesion-GPCR and PKD1-like proteins. Trends Pharmacol Sci 34(8):470–478
13. Davletov BA, Shamotienko OG, Lelianova VG, Grishin EV, Ushkaryov YA (1996) Isolation and biochemical characterization of a Ca2+-independent alpha-latrotoxin-binding protein. J Biol Chem 271(38):23239–23245
14. Lelianova VG, Davletov BA, Sterling A, Rahman MA, Grishin EV, Totty NF et al (1997) Alpha-latrotoxin receptor, latrophilin, is a novel member of the secretin family of G protein-coupled receptors. J Biol Chem 272(34):21504–21508

15. Ichtchenko K, Khvotchev M, Kiyatkin N, Simpson L, Sugita S, Sudhof TC (1998) alpha-Latrotoxin action probed with recombinant toxin: receptors recruit alpha-latrotoxin but do not transduce an exocytotic signal. EMBO J 17(21):6188–6199
16. Volynski KE, Silva JP, Lelianova VG, Atiqur Rahman M, Hopkins C, Ushkaryov YA (2004) Latrophilin fragments behave as independent proteins that associate and signal on binding of LTX(N4C). EMBO J 23(22):4423–4433
17. Rahman MA, Ashton AC, Meunier FA, Davletov BA, Dolly JO, Ushkaryov YA (1999) Norepinephrine exocytosis stimulated by alpha-latrotoxin requires both external and stored Ca2+ and is mediated by latrophilin, G proteins and phospholipase C. Philos Trans R Soc Lond B Biol Sci 354(1381):379–386
18. Piao X, Hill RS, Bodell A, Chang BS, Basel-Vanagaite L, Straussberg R et al (2004) G protein-coupled receptor-dependent development of human frontal cortex. Science 303 (5666):2033–2036
19. Bae BI, Tietjen I, Atabay KD, Evrony GD, Johnson MB, Asare E et al (2014) Evolutionarily dynamic alternative splicing of GPR56 regulates regional cerebral cortical patterning. Science 343(6172):764–768
20. Iguchi T, Sakata K, Yoshizaki K, Tago K, Mizuno N, Itoh H (2008) Orphan G protein-coupled receptor GPR56 regulates neural progenitor cell migration via a G alpha 12/13 and Rho pathway. J Biol Chem 283(21):14469–14478
21. Paavola KJ, Stephenson JR, Ritter SL, Alter SP, Hall RA (2011) The N terminus of the adhesion G protein-coupled receptor GPR56 controls receptor signaling activity. J Biol Chem 286(33):28914–28921
22. Shashidhar S, Lorente G, Nagavarapu U, Nelson A, Kuo J, Cummins J et al (2005) GPR56 is a GPCR that is overexpressed in gliomas and functions in tumor cell adhesion. Oncogene 24 (10):1673–1682
23. Wu MP, Doyle JR, Barry B, Beauvais A, Rozkalne A, Piao X et al (2013) G-protein coupled receptor 56 promotes myoblast fusion through serum response factor- and nuclear factor of activated T-cell-mediated signalling but is not essential for muscle development in vivo. FEBS J 280(23):6097–6113
24. Stoveken HM, Hajduczok AG, Xu L, Tall GG (2015) Adhesion G protein-coupled receptors are activated by exposure of a cryptic tethered agonist. Proc Natl Acad Sci U S A 112 (19):6194–6199
25. Kim JE, Han JM, Park CR, Shin KJ, Ahn C, Seong JY et al (2010) Splicing variants of the orphan G-protein-coupled receptor GPR56 regulate the activity of transcription factors associated with tumorigenesis. J Cancer Res Clin Oncol 136(1):47–53
26. Kishore A, Purcell RH, Nassiri-Toosi Z, Hall RA (2016) Stalk-dependent and stalk-independent signaling by the adhesion G protein-coupled receptors GPR56 (ADGRG1) and BAI1 (ADGRB1). J Biol Chem 291(7):3385–3394
27. Yang L, Chen G, Mohanty S, Scott G, Fazal F, Rahman A et al (2011) GPR56 regulates VEGF production and angiogenesis during melanoma progression. Cancer Res 71(16):5558–5568
28. Little KD, Hemler ME, Stipp CS (2004) Dynamic regulation of a GPCR-tetraspanin-G protein complex on intact cells: central role of CD81 in facilitating GPR56-Galpha q/11 association. Mol Biol Cell 15(5):2375–2387
29. Ohta S, Sakaguchi S, Kobayashi Y, Mizuno N, Tago K, Itoh H (2015) Agonistic antibodies reveal the function of GPR56 in human glioma U87-MG cells. Biol Pharm Bull 38(4):594–600
30. Peeters MC, Fokkelman M, Boogaard B, Egerod KL, van de Water B, IJzerman AP et al (2015) The adhesion G protein-coupled receptor G2 (ADGRG2/GPR64) constitutively activates SRE and NFkB and is involved in cell adhesion and migration. Cell Signal 27 (12):2579–2588
31. Demberg LM, Rothemund S, Schoneberg T, Liebscher I (2015) Identification of the tethered peptide agonist of the adhesion G protein-coupled receptor GPR64/ADGRG2. Biochem Biophys Res Commun 464(3):743–747

32. Gupte J, Swaminath G, Danao J, Tian H, Li Y, Wu X (2012) Signaling property study of adhesion G-protein-coupled receptors. FEBS Lett 586(8):1214–1219
33. Wilde C, Fischer L, Lede V, Kirchberger J, Rothemund S, Schoneberg T et al (2016) The constitutive activity of the adhesion GPCR GPR114/ADGRG5 is mediated by its tethered agonist. FASEB J 30(2):666–673
34. Monk KR, Naylor SG, Glenn TD, Mercurio S, Perlin JR, Dominguez C et al (2009) A G protein-coupled receptor is essential for Schwann cells to initiate myelination. Science 325 (5946):1402–1405
35. Paavola KJ, Sidik H, Zuchero JB, Eckart M, Talbot WS (2014) Type IV collagen is an activating ligand for the adhesion G protein-coupled receptor GPR126. Sci Signal 7(338):ra76
36. Mogha A, Benesh AE, Patra C, Engel FB, Schoneberg T, Liebscher I et al (2013) Gpr126 functions in Schwann cells to control differentiation and myelination via G-protein activation. J Neurosci 33(46):17976–17985
37. Liebscher I, Schon J, Petersen SC, Fischer L, Auerbach N, Demberg LM et al (2014) A tethered agonist within the ectodomain activates the adhesion G protein-coupled receptors GPR126 and GPR133. Cell Rep 9(6):2018–2026
38. Park D, Tosello-Trampont AC, Elliott MR, Lu M, Haney LB, Ma Z et al (2007) BAI1 is an engulfment receptor for apoptotic cells upstream of the ELMO/Dock180/Rac module. Nature 450(7168):430–434
39. Das S, Sarkar A, Ryan KA, Fox S, Berger AH, Juncadella IJ et al (2014) Brain angiogenesis inhibitor 1 is expressed by gastric phagocytes during infection with Helicobacter pylori and mediates the recognition and engulfment of human apoptotic gastric epithelial cells. FASEB J 28(5):2214–2224
40. Mazaheri F, Breus O, Durdu S, Haas P, Wittbrodt J, Gilmour D et al (2014) Distinct roles for BAI1 and TIM-4 in the engulfment of dying neurons by microglia. Nat Commun 5:4046
41. Das S, Owen KA, Ly KT, Park D, Black SG, Wilson JM et al (2011) Brain angiogenesis inhibitor 1 (BAI1) is a pattern recognition receptor that mediates macrophage binding and engulfment of Gram-negative bacteria. Proc Natl Acad Sci U S A 108(5):2136–2141
42. Hochreiter-Hufford AE, Lee CS, Kinchen JM, Sokolowski JD, Arandjelovic S, Call JA et al (2013) Phosphatidylserine receptor BAI1 and apoptotic cells as new promoters of myoblast fusion. Nature 497(7448):263–267
43. Duman JG, Tzeng CP, Tu YK, Munjal T, Schwechter B, Ho TS et al (2013) The adhesion-GPCR BAI1 regulates synaptogenesis by controlling the recruitment of the Par3/Tiam1 polarity complex to synaptic sites. J Neurosci 33(16):6964–6978
44. Zhu D, Li C, Swanson AM, Villalba RM, Guo J, Zhang Z et al (2015) BAI1 regulates spatial learning and synaptic plasticity in the hippocampus. J Clin Invest 125(4):1497–1508
45. Stephenson JR, Paavola KJ, Schaefer SA, Kaur B, Van Meir EG, Hall RA (2013) Brain-specific angiogenesis inhibitor-1 signaling, regulation, and enrichment in the postsynaptic density. J Biol Chem 288(31):22248–22256
46. Okajima D, Kudo G, Yokota H (2010) Brain-specific angiogenesis inhibitor 2 (BAI2) may be activated by proteolytic processing. J Recept Signal Transduct Res 30(3):143–153
47. Ward Y, Lake R, Yin JJ, Heger CD, Raffeld M, Goldsmith PK et al (2011) LPA receptor heterodimerizes with CD97 to amplify LPA-initiated RHO-dependent signaling and invasion in prostate cancer cells. Cancer Res 71(23):7301–7311
48. Hu QX, Dong JH, Du HB, Zhang DL, Ren HZ, Ma ML et al (2014) Constitutive Galphai coupling activity of very large G protein-coupled receptor 1 (VLGR1) and its regulation by PDZD7 protein. J Biol Chem 289(35):24215–24225
49. Bohnekamp J, Schoneberg T (2011) Cell adhesion receptor GPR133 couples to Gs protein. J Biol Chem 286(49):41912–41916
50. Silva JP, Lelianova VG, Ermolyuk YS, Vysokov N, Hitchen PG, Berninghausen O et al (2011) Latrophilin 1 and its endogenous ligand Lasso/teneurin-2 form a high-affinity transsynaptic receptor pair with signaling capabilities. Proc Natl Acad Sci U S A 108(29):12113–12118

51. Tobaben S, Sudhof TC, Stahl B (2002) Genetic analysis of alpha-latrotoxin receptors reveals functional interdependence of CIRL/latrophilin 1 and neurexin 1 alpha. J Biol Chem 277 (8):6359–6365
52. O'Sullivan ML, de Wit J, Savas JN, Comoletti D, Otto-Hitt S, Yates JR III et al (2012) FLRT proteins are endogenous latrophilin ligands and regulate excitatory synapse development. Neuron 73(5):903–910
53. Jackson VA, del Toro D, Carrasquero M, Roversi P, Harlos K, Klein R et al (2015) Structural basis of latrophilin-FLRT interaction. Structure 23(4):774–781
54. Huang YS, Chiang NY, Hu CH, Hsiao CC, Cheng KF, Tsai WP et al (2012) Activation of myeloid cell-specific adhesion class G protein-coupled receptor EMR2 via ligation-induced translocation and interaction of receptor subunits in lipid raft microdomains. Mol Cell Biol 32 (8):1408–1420
55. Stacey M, Chang GW, Davies JQ, Kwakkenbos MJ, Sanderson RD, Hamann J et al (2003) The epidermal growth factor-like domains of the human EMR2 receptor mediate cell attachment through chondroitin sulfate glycosaminoglycans. Blood 102(8):2916–2924
56. Karpus ON, Veninga H, Hoek RM, Flierman D, van Buul JD, Vandenakker CC et al (2013) Shear stress-dependent downregulation of the adhesion-G protein-coupled receptor CD97 on circulating leukocytes upon contact with its ligand CD55. J Immunol 190(7):3740–3748
57. Wang T, Ward Y, Tian L, Lake R, Guedez L, Stetler-Stevenson WG et al (2005) CD97, an adhesion receptor on inflammatory cells, stimulates angiogenesis through binding integrin counterreceptors on endothelial cells. Blood 105(7):2836–2844
58. Wandel E, Saalbach A, Sittig D, Gebhardt C, Aust G (2012) Thy-1 (CD90) is an interacting partner for CD97 on activated endothelial cells. J Immunol 188(3):1442–1450
59. Sigoillot SM, Iyer K, Binda F, Gonzalez-Calvo I, Talleur M, Vodjdani G et al (2015) The secreted protein C1QL1 and its receptor BAI3 control the synaptic connectivity of excitatory inputs converging on cerebellar Purkinje cells. Cell Rep 10(5):820–832
60. Bolliger MF, Martinelli DC, Sudhof TC (2011) The cell-adhesion G protein-coupled receptor BAI3 is a high-affinity receptor for C1q-like proteins. Proc Natl Acad Sci U S A 108 (6):2534–2539
61. Luo R, Jeong SJ, Jin Z, Strokes N, Li S, Piao X (2011) G protein-coupled receptor 56 and collagen III, a receptor-ligand pair, regulates cortical development and lamination. Proc Natl Acad Sci U S A 108(31):12925–12930
62. Petersen SC, Luo R, Liebscher I, Giera S, Jeong SJ, Mogha A et al (2015) The adhesion GPCR GPR126 has distinct, domain-dependent functions in Schwann cell development mediated by interaction with laminin-211. Neuron 85(4):755–769
63. Xu L, Hynes RO (2007) GPR56 and TG2: possible roles in suppression of tumor growth by the microenvironment. Cell Cycle 6(2):160–165
64. Boucard AA, Maxeiner S, Sudhof TC (2014) Latrophilins function as heterophilic cell-adhesion molecules by binding to teneurins: regulation by alternative splicing. J Biol Chem 289(1):387–402
65. Bang ML, Owczarek S (2013) A matter of balance: role of neurexin and neuroligin at the synapse. Neurochem Res 38(6):1174–1189
66. Geppert M, Khvotchev M, Krasnoperov V, Goda Y, Missler M, Hammer RE et al (1998) Neurexin I alpha is a major alpha-latrotoxin receptor that cooperates in alpha-latrotoxin action. J Biol Chem 273(3):1705–1710
67. Boucard AA, Ko J, Sudhof TC (2012) High affinity neurexin binding to cell adhesion G-protein-coupled receptor CIRL1/latrophilin-1 produces an intercellular adhesion complex. J Biol Chem 287(12):9399–9413
68. Hamann J, Vogel B, van Schijndel GM, van Lier RA (1996) The seven-span transmembrane receptor CD97 has a cellular ligand (CD55, DAF). J Exp Med 184(3):1185–1189
69. Hamann J, Stortelers C, Kiss-Toth E, Vogel B, Eichler W, van Lier RA (1998) Characterization of the CD55 (DAF)-binding site on the seven-span transmembrane receptor CD97. Eur J Immunol 28(5):1701–1707

70. Lin HH, Stacey M, Saxby C, Knott V, Chaudhry Y, Evans D et al (2001) Molecular analysis of the epidermal growth factor-like short consensus repeat domain-mediated protein-protein interactions: dissection of the CD97-CD55 complex. J Biol Chem 276(26):24160–24169
71. Iijima T, Miura E, Watanabe M, Yuzaki M (2010) Distinct expression of C1q-like family mRNAs in mouse brain and biochemical characterization of their encoded proteins. Eur J Neurosci 31(9):1606–1615
72. Kakegawa W, Mitakidis N, Miura E, Abe M, Matsuda K, Takeo YH et al (2015) Anterograde C1ql1 signaling is required in order to determine and maintain a single-winner climbing fiber in the mouse cerebellum. Neuron 85(2):316–329
73. Xu L, Begum S, Hearn JD, Hynes RO (2006) GPR56, an atypical G protein-coupled receptor, binds tissue transglutaminase, TG2, and inhibits melanoma tumor growth and metastasis. Proc Natl Acad Sci U S A 103(24):9023–9028
74. Yang L, Friedland S, Corson N, Xu L (2014) GPR56 inhibits melanoma growth by internalizing and degrading its ligand TG2. Cancer Res 74(4):1022–1031
75. Singer K, Luo R, Jeong SJ, Piao X (2013) GPR56 and the developing cerebral cortex: cells, matrix, and neuronal migration. Mol Neurobiol 47(1):186–196
76. Luo R, Jin Z, Deng Y, Strokes N, Piao X (2012) Disease-associated mutations prevent GPR56-collagen III interaction. PLoS One 7(1), e29818
77. Chiang NY, Hsiao CC, Huang YS, Chen HY, Hsieh IJ, Chang GW et al (2011) Disease-associated GPR56 mutations cause bilateral frontoparietal polymicrogyria via multiple mechanisms. J Biol Chem 286(16):14215–14225
78. Jin Z, Tietjen I, Bu L, Liu-Yesucevitz L, Gaur SK, Walsh CA et al (2007) Disease-associated mutations affect GPR56 protein trafficking and cell surface expression. Hum Mol Genet 16 (16):1972–1985
79. Weston MD, Luijendijk MW, Humphrey KD, Moller C, Kimberling WJ (2004) Mutations in the VLGR1 gene implicate G-protein signaling in the pathogenesis of Usher syndrome type II. Am J Hum Genet 74(2):357–366
80. Robinson A, Escuin S, Doudney K, Vekemans M, Stevenson RE, Greene ND et al (2012) Mutations in the planar cell polarity genes CELSR1 and SCRIB are associated with the severe neural tube defect craniorachischisis. Hum Mutat 33(2):440–447
81. Paavola KJ, Hall RA (2012) Adhesion G protein-coupled receptors: signaling, pharmacology, and mechanisms of activation. Mol Pharmacol 82(5):777–783
82. Liebscher I, Schöneberg T (2016) Tethered agonism: a common activation mechanism of adhesion GPCRs. In: Langenhan T, Schöneberg T (eds) Adhesion G protein-coupled receptors: molecular, physiological and pharmacological principles in health and disease. Springer, Heidelberg
83. Coughlin SR (2000) Thrombin signalling and protease-activated receptors. Nature 407 (6801):258–264
84. Promel S, Waller-Evans H, Dixon J, Zahn D, Colledge WH, Doran J et al (2012) Characterization and functional study of a cluster of four highly conserved orphan adhesion-GPCR in mouse. Dev Dyn 241(10):1591–1602
85. Promel S, Frickenhaus M, Hughes S, Mestek L, Staunton D, Woollard A et al (2012) The GPS motif is a molecular switch for bimodal activities of adhesion class G protein-coupled receptors. Cell Rep 2(2):321–331
86. Krasnoperov V, Deyev IE, Serova OV, Xu C, Lu Y, Buryanovsky L et al (2009) Dissociation of the subunits of the calcium-independent receptor of alpha-latrotoxin as a result of two-step proteolysis. Biochemistry 48(14):3230–3238
87. Scholz N, Gehring J, Guan C, Ljaschenko D, Fischer R, Lakshmanan V et al (2015) The adhesion GPCR latrophilin/CIRL shapes mechanosensation. Cell Rep 11(6):866–874
88. Scholz N, Monk KR, Kittel RJ, Langenhan T (2016) Adhesion GPCRs as a putative class of metabotropic mechanosensors. In: Langenhan T, Schöneberg T (eds) Adhesion G protein-coupled receptors: molecular, physiological and pharmacological principles in health and disease. Springer, Heidelberg

89. Knapp B, Wolfrum U (2016) Adhesion G protein-coupled receptor-related protein networks. In: Langenhan T, Schöneberg T (eds) Adhesion G protein-coupled receptors: molecular, physiological and pharmacological principles in health and disease. Springer, Heidelberg
90. Zou J, Zheng T, Ren C, Askew C, Liu XP, Pan B et al (2014) Deletion of PDZD7 disrupts the Usher syndrome type 2 protein complex in cochlear hair cells and causes hearing loss in mice. Hum Mol Genet 23(9):2374–2390
91. Lanoue V, Usardi A, Sigoillot SM, Talleur M, Iyer K, Mariani J et al (2013) The adhesion-GPCR BAI3, a gene linked to psychiatric disorders, regulates dendrite morphogenesis in neurons. Mol Psychiatry 18(8):943–950
92. Jeong BC, Kim MY, Lee JH, Kee HJ, Kho DH, Han KE et al (2006) Brain-specific angiogenesis inhibitor 2 regulates VEGF through GABP that acts as a transcriptional repressor. FEBS Lett 580(2):669–676
93. Nishimura T, Honda H, Takeichi M (2012) Planar cell polarity links axes of spatial dynamics in neural-tube closure. Cell 149(5):1084–1097
94. Li X, Roszko I, Sepich DS, Ni M, Hamm HE, Marlow FL et al (2013) Gpr125 modulates Dishevelled distribution and planar cell polarity signaling. Development 140(14):3028–3039

Adhesion GPCR-Related Protein Networks

Barbara Knapp and Uwe Wolfrum

Graphical Abstract

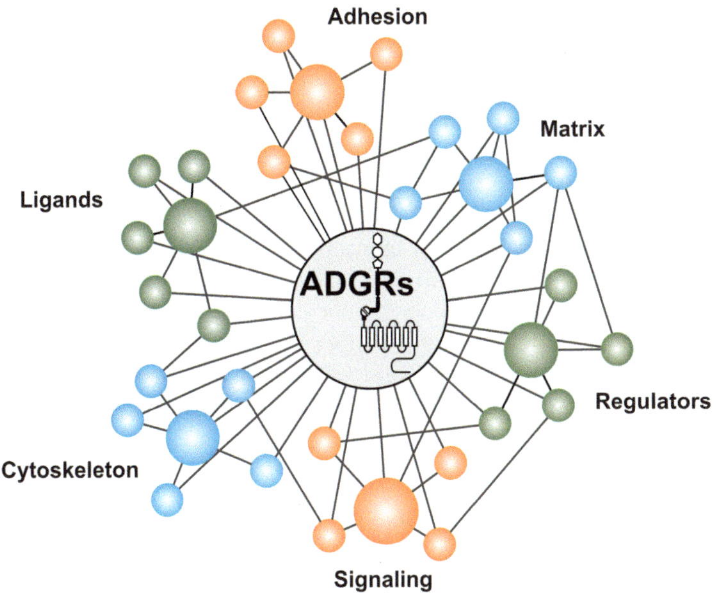

Contents

B. Knapp • U. Wolfrum (✉)
Cell and Matrix Biology, Institute of Zoology, Johannes Gutenberg University of Mainz, Johannes von Muellerweg 6, Mainz 55099, Germany
e-mail: wolfrum@uni-mainz.de

T. Langenhan, T. Schöneberg (eds.), *Adhesion G Protein-coupled Receptors*, Handbook of Experimental Pharmacology 234, DOI 10.1007/978-3-319-41523-9_8

Abstract

Adhesion G protein-coupled receptors (aGPCRs/ADGRs) are unique receptors that combine cell adhesion and signaling functions. Protein networks related to ADGRs exert diverse functions, e.g., in tissue polarity, cell migration, nerve cell function, or immune response, and are regulated via different mechanisms. The large extracellular domain of ADGRs is capable of mediating cell–cell or cell–matrix protein interactions. Their intracellular surface and domains are coupled to downstream signaling pathways and often bind to scaffold proteins, organizing membrane-associated protein complexes. The cohesive interplay between ADGR-related network components is essential to prevent severe disease-causing damage in numerous cell types. Consequently, in recent years, attention has focused on the decipherment of the precise molecular composition of ADGR protein complexes and interactomes in various cellular modules. In this chapter, we discuss the affiliation of ADGR networks to cellular modules and how they can be regulated, pinpointing common features in the networks related to the diverse ADGRs. Detailed decipherment of the composition of protein networks should provide novel targets for the development of novel therapies with the aim to cure human diseases related to ADGRs.

Keywords

Protein networks • Adhesion complexes • Adhesion GPCR • Affinity proteomics • Brain-specific angiogenesis inhibitor • Latrophillin • VLGR1 • GPR98 • ADGR • Signaling pathways

1 Introduction

Over the last years, the prominent impact of adhesion G protein-coupled receptors (aGPCRs/ADGRs) on diverse cellular functions, e.g., tissue polarity, cell migration, nerve cell function, and immune response, has been evident. The interaction of ADGRs with a myriad of other proteins and the assembly and maintenance of their cellular modules is essential for correct function and regulation of these diverse processes. The aim of the present book chapter is to outline the current knowledge of supramolecular protein complexes and protein networks related to ADGRs. Although we consequently apply the nomenclature of the ADGRs which has been recently harmonized [1], we add historic and alternative names found in the literature, where necessary.

2 The ADGR Protein Structure Reveals Multiple Protein-Binding Domains

Proteins of the ADGR family possess a unique molecular structure (Fig. 1) [1] (see also [2]). They usually display an extraordinary large N-terminal extracellular domain (ECD) featuring various types of subdomains that are generally thought to communicate with the extracellular milieu and mediate their characteristic adhesive functions. The ECD of each receptor subfamily contains a specific combination of domains. Receptors from the same subfamily mainly differ in the number of domain repeats, which results in varying sizes of their ECDs. An important common feature of ADGRs is the G protein proteolytic site (GPS). It is an integral part of the much larger GPCR autoproteolysis-inducing (GAIN) domain [3]. Here, most ADGRs undergo autocleavage into an extracellular N-terminal fragment (NTF) and a C-terminal fragment (CTF). Both fragments usually remain non-covalently associated at the cell surface (see also [4, 5]). Similar to other extracellular protein domains, NTFs are highly glycosylated and their functional roles resemble those of adhesion proteins. The CTF, which is comprised of a canonical seven transmembrane domain (7TM) and the intracellular domain (ICD), mediates the activation of intracellular signaling cascades. At the very C-terminal end of the ICD, almost half of the ADGRs exhibit a PDZ-binding

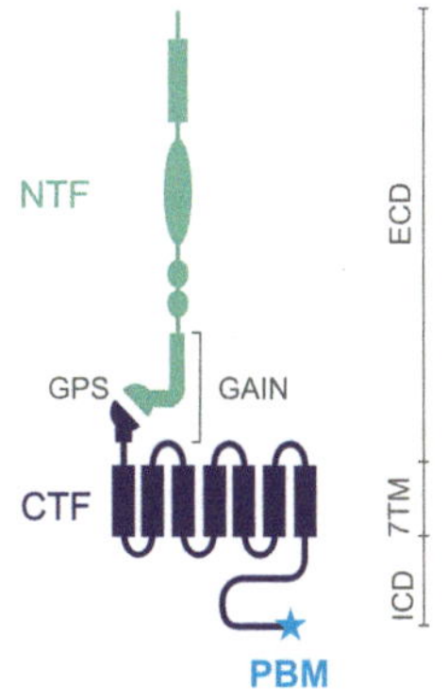

Fig. 1 Schematic representation of the ADGR protein structure. ADGRs are subdivided in an extracellular domain (ECD), a seven-transmembrane domain (7TM), and an intracellular domain (ICD), based on their topology. Autocleavage of ADGRs results in two protein fragments termed N-terminal fragment (NTF) and C-terminal fragment (CTF), respectively. The GPCR autoproteolysis-inducing (GAIN) domain includes the G protein proteolytic site (GPS); *PBM* PDZ-binding motif

motif (PBM) that allows the binding to PDZ domains. The latter domains are named after three scaffold proteins, the postsynaptic density protein PSD95/DLG4; the discs large (Dlg) protein, present in the septate junctions of arthropods; and ZO-1, a protein of the *zonula occludens* adhesion complexes. PDZ domains are characteristic for an abundant group of scaffold proteins, which organize supramolecular protein complexes and networks [6, 7]. In general, PDZ domain-mediated binding plays an important role in regulating receptor and channel protein localization within membrane domains at synapses and in other cell–cell adhesion complexes and function to scaffold intracellular signaling protein complexes.

3 Unraveling Protein–Protein Interactions and Protein Complexes Related to ADGR by Screening Methods and Affinity Proteomics

Table 1 represents a comprehensive overview of proteins, which have been reported to interact with the 33 ADGRs. Since only a few binding proteins have been identified in the zebra fish *Danio rerio* or the frog *Xenopus laevis*, respectively, most of the binding partners listed in Table 1 originate from studies on mammalian ADGRs, namely, human and rodent ADGRs. Only for single ADGRs additional data are available from orthologues in *Drosophila melanogaster*, as, e.g., for Flamingo (CELSR) [65, 66] or *C. elegans* [67]. The overall protein network of vertebrate ADGRs and their interaction partners are shown in Fig. 2. It is obvious that the number of identified interaction partners largely varies between the ADGRs (Table 1, Fig. 2). Due to the lack of studies, no information about the binding partners for several ADGR proteins, e.g., ADGR3 (EMR3) or ADGRD2 (GPR144), has yet been reported, so far.

The intracellular binding partners of ADGRs listed in Table 1 have mainly been identified via yeast two-hybrid screens [8, 9], protein-peptide assays [29], and affinity proteomics methods, e.g., co-immunoprecipitation. Extracellular binding partners of ADGRs have been identified via cell adhesion assays [14–16], cell aggregation assays [10, 11], or affinity chromatography followed by mass spectrometry [10].

A large number of confirmed intracellular interacting proteins of ADGRs comprise scaffold proteins (Table 1). Only a few signaling molecules and putative components of downstream signaling pathways have been identified as binding proteins so far. The lack of identified downstream signaling molecules might be due to the screening approaches and assays implemented. These have in common that they require relatively high binding affinities and long periods of protein retention within assembled protein complexes. Such interaction characteristics are only fulfilled by binding of scaffold proteins to their interaction partners. In contrast, the interaction between components of signaling cascades are often low affine and transient. Nevertheless, screening approaches are valuable in order to get an impression of the general picture of ADGR networks. In this chapter, we only discuss interactions that have been reported in the literature. However, data mining in databases, e.g., *IntAct* (http://www.ebi.ac.uk/intact/), *BioGRID* (http://thebiogrid.

Table 1 Extracellular and intracellular ADGR interaction partners. (**a**) Comprehensive list of experimentally confirmed ADGR interaction partners of ADGR subfamily members. The protein nomenclature follows the HGNC guideline, and protein designations of human orthologues are used for all interaction partners. *Black lettering*: direct or indirect physical interaction. *Grey lettering*: evidence for indirect interaction. (**b**) Classification of ADGR binding protein types and ADGR binding regions. *Ptd-L-Ser* phosphatidylserine, *LPS* lipopolysaccharide

A

I	ADGRL1 (LPHN1)	ADGRL2 (LPHN2)	ADGRL3 (LPHN3)	ADGRL4 (ELTD1)
extra	*single TM receptors* NRXN1α,1β,2β,3β TENM2 FLRT1/3	FLTR3	FLTR1/3	
intra	*PDZ scaffold proteins* SHANK1/2/3 DLG3 DLG4 SCRIB MAGI1/2 *scaffold proteins w/o PDZ* ANKS1A *kinases* MAST2 ROS1 *Ion-channels* CACNA1A *G proteins* Gαo Gαs	*PDZ scaffold proteins* SHANK1/2 *others* CLN3		

II	ADGRE1 (EMR1)	ADGRE2 (EMR2)	ADGRE3 (EMR3)	ADGRE4 (EMR4)	ADGRE5 (CD97)
extra		*ECM proteins* Chrondroitin sulfate B			*single TM receptors* ITGA5/B1 ITGAV/B3 *ECM proteins* Chrondroitin sulfate B *others* CD55 THY1
intra		*G proteins* Gα15			*kinases* SRC *7TM-receptors* LPAR1 *small GTPases* RHOA *G proteins* Gα12/13

III	ADGRA1 (GPR123)	ADGRA2 (GPR124)	ADGRA3 (GPR125)
extra			
intra	*PDZ scaffold proteins* DLG4	*PDZ scaffold proteins* DLG1 *single TM kinases* RYK	*PDZ scaffold proteins* DLG1 DVL1 *single TM kinases* RYK

IV	ADGRC1 (CELSR1)	ADGRC2 (CELSR2)	ADGRC3 (CELSR3)
extra			
intra	*7TM-receptors* FZD6 *PDZ scaffold proteins* DVL2 *scaffold proteins w/o PDZ* DAAM1 *GEF s* ARHGEF11 *kinases* RYK *others* VANGL2 ATP6AP2	*single TM kinases* RYK *others* ATP6AP2	*single TM kinases* RYK *GEFs* SWAP70 *others* DST

V	ADGRD1 (GPR133)	ADGRD2 (GPR144)	
extra			
intra	*G proteins* Gαs Gαi		
extra			*others* SFTPD

(continued)

Table 1 (continued)

VI	ADGRF1 (GPR110)	ADGRF2 (GPR111)	ADGRF3 (GPR113)	ADGRF4 (GPR115)	ADGRF5 (GPR116)
intra	*G proteins* Gαq			*G proteins* Gα15	*G proteins* Gαq/11 *small GTPases* RHOA RAC1 *GEFs* ARHGEF25

VII	ADGRB1 (BAI1)		ADGRB2 (BAI2)	ADGRB3 (BAI3)
extra	*others* CD36 *single TM receptors* ITGAV/B3	*proteases* MMP-14 *lipids* Ptd-L-Ser LPS	*proteases* FURIN	*others* C1QL1-4
intra	*PDZ scaffold proteins* MAGI1/2/3 DLG 1/2/3/4 INADL LIN7A LRRC7 SNT1A,1B,2B,2G PDZD2 *GEF s* PARD3/ TIAM1 ELMO1/ DOCK1 *G proteins* Gα12/13 Gα16	*scaffold proteins w/o PDZ* BAIAP2 *7TM receptors* BAIAP3 *kinases* NEK4 MAPK1 *small GTPases* RAC1 RHOA *arrestins* ARRB2 *others* UBC PHYHIP	*PDZ scaffold proteins* SHANK1 *transcription factors* GABPB1/2 GABPA LRP2BP *arrestins* ARRB2 *GAPs* STARD13	*PDZ scaffold proteins* ELMO1/2/3 DOCK1 *arrestins* ARRB2

VIII	ADGRG1 (GPR56)	ADGRG2 (GPR64)	ADGRG3 (GPR97)	ADGRG4 (GPR112)	ADGRG5 (GPR114)	ADGRG6 (GPR126)	ADGRG7 (GPR128)
extra	*ECM proteins* COL3A1 *others* TGM2 CD9 CD81					*ECM proteins* COL4A4 LAMA2	
intra	*7TM receptors* ADRB2 *arrestins* ARRB2 *small GTPases* RHOA *G proteins* Gαq/11 Gα11/13	*transcription factors* SRF NFKB1 *small GTPases* RHOA *G proteins* Gα12/13 Gαq	*arrestins* ARRB2 *small GTPases* RHOA CDC42 *G proteins* Gαo		*G proteins* Gαs	*G proteins* Gαs Gαi	

IX	ADGRV1 (VLGR1)
extra	
intra	*scaffold proteins with PDZ* USH1C DFNB31 PDZD7 *single TM protein* VEZT *others* SNAP25 MYO7A ADCY6 *G proteins* Gαs Gαq Gαi

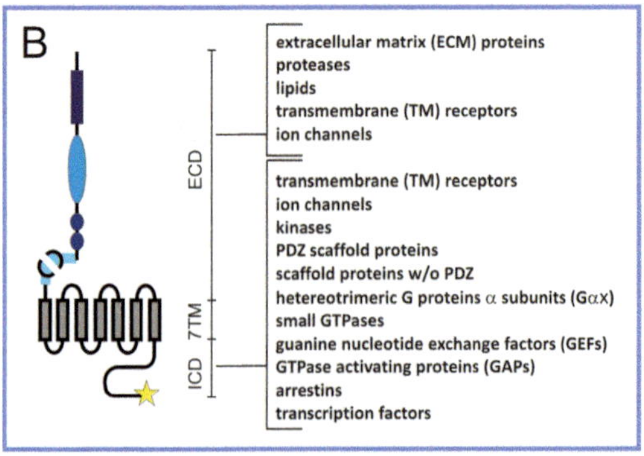

References: ADGRL1 [8–12], ADGRL2 [8, 10], ADGRL3 [10], ADGRE2 [13], ADGRE5 [14–16], ADGRA1 [17], ADGRA2 [18, 19], ADGRA3 [18, 20], ADGRC1 [21], ADGRC2 [22, 23], ADGRC3 [22, 24], ADGRD1 [25], ADGRF1 [26], ADGRF4 [26], ADGRF5 [27, 28], ADGRB1 [29–42], ADGRB2 [8, 38, 43, 44], ADGRB3 [45, 46], ADGRG1 [47–51], ADGRG2 [52], ADGRG3 [53], ADGRG5 [26], ADGRG6 [54–57], ADGRV1 [58–64].

org/), or *MINT* (http://mint.bio.uniroma2.it), provides additional information on putative complex partners for ADGRs, which have mainly been acquired by proteomic or genetic screens.

4 ADGR Protein Networks Define Functional Modules

Combining data on protein–protein interactions from experiments and literature and database searches for all ADGRs results in a complex overall protein network (Fig. 2). This has revealed interactions of ADGR members across different subfamilies with similar proteins and even with the same protein. These data indicate common mechanisms underlying the function of the diverse ADGRs and their integration in similar cellular contexts and molecular modules (see below Sect. 6). However, the overall ADGR interactome is still far from completion.

To highlight features of protein networks related to ADGRs, we have chosen three subfamilies of ADGRs for which reliable protein–protein interaction data and information on protein networks exist. We chose the following: (A) ADGRLs (LPHNs), also known as latrophilins; (B) ADGRBs, better known as brain-specific angiogenesis inhibitors (BAIs); and (C) the ADGR goliath ADGRV1, also known as very large G protein-coupled receptor (VLGR1), GPR98, or MASS1. In the following parts of this chapter, we will review these protein networks and discuss the impact of their function in cellular modules.

4.1 ADGRLs (LPHNs, Latrophilins) Are Organized by Scaffold Proteins at Synaptic Junctions

The group of ADGRLs is one of the most intensely studied subfamilies of ADGRs. Over the last two decades, several labs have analyzed ADGRLs in different organisms. Their traditional name "latrophilins" originates from their ligand α-latrotoxin, a component of the black widow spider venom. The binding of this poison activates ADGRLs and results in massive Ca^{2+}-independent exocytosis of neurotransmitters into the synaptic cleft [68, 69]. In vertebrates, the ADGRL subfamily is comprised of three paralogous genes (*ADGRL1* to *ADGRL3*) which each can be alternatively spliced. ADGRL1 and ADGRL3 are mainly expressed in the central nervous system, whereas ADGRL2 is expressed more ubiquitously [70]. Besides the typical secretin-like 7TM and the GAIN domain, all ADGRLs contain a characteristic intracellular domain with a C-terminal class I PBM (Fig. 3a). Their ECD displays a hormone receptor motif (HRM), an olfactomedin-like domain, and a rhamnose-binding lectin (RBL) motif. For both the ICD and ECD, several binding proteins have been identified which are part of protein networks related to cell–cell adhesion complexes predominantly found at synapses.

In the ECD of ADGRLs, two different protein-binding sites, namely, the olfactomedin-like domain and the RBL motif, have been described in vertebrates, so far (Fig. 3a). The olfactomedin-like domain of ADGRL1 is capable of binding

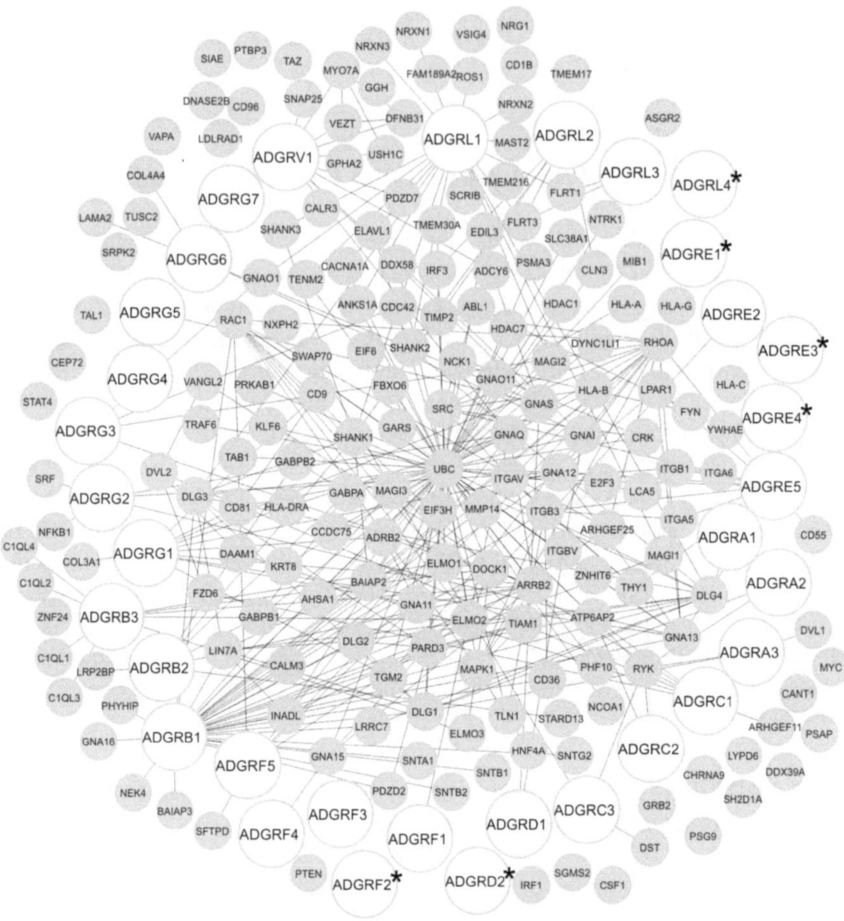

Fig. 2 The ADGR protein network. ADGRs are connected via their interaction partners and form complex networks. The present network is based on data from *STRING* (*Search Tool for the Retrieval of Interacting Genes/Proteins*, http://string-db.org/) and Cytoscape database searches (http://www.cytoscape.org/). The protein nomenclature follows the HGNC guideline, and protein designations of human orthologues are used for all interaction partners. *Black lines*, published experimental data; *green lines*, data from protein databases. Asterisk indicates ADGRs for which no known interaction partners are known so far

neurexins (NRXNs) (neurexins 1α, 1β, 2β, and 3β) [11]. Neurexins are single-pass transmembrane proteins and facilitate the formation of heterotypic intercellular transsynaptic junctions. Due to their frequent alternative splicing, a remarkable number of neurexin splice variants (~4000) is expressed [71]. The inclusion of a single exon at the splice site SS4 interrupts the sixth laminin/neurexin/sex hormone-binding globulin (LNS/LamG) domain of all neurexins [11]. This splicing event disrupts the neurexin–ADGRL1 interaction, indicating that the neurexin binding to ADGRL1 is mediated by this LNS domain. In neurons, neurexins are found at the pre-synapse and interact with neuroligin 1 (NLGN1), a neural cell adhesion molecule of the post-synapse. In this heterotypic transsynaptic junction, ADGRL1 competes with neuroligin in the binding of neurexins [11].

The RBL binding motif of ADGRL1 mediates binding of the single-pass transmembrane receptor teneurin-2 (TENM2), also known as Lasso [12]. Like ADGRLs, teneurins are highly enriched in the central nervous system where they localize to pre- and post-synapses. Apart from the full-length transmembrane protein, shorter isoforms of teneurins are expressed, so-called teneurin C-terminal-associated peptides (TCAPs) which lack the transmembrane domain. TCAPs are involved in paracrine signaling and have high structural homology with peptide ligands that activate GPCRs of the secretin family [72]. The Ushkaryov lab showed that the C-terminal fragment of teneurin-2 that corresponds mainly to the TCAP2 region is sufficient to activate ADGRL1 and ADGRL2, which results in presynaptic calcium release in hippocampal neuron cultures [12]. The ICD of teneurins contains a "teneurin"-like domain characterized by two Ca^{2+}-binding EF-hands and two Cap/ponsin sites that provide a link to the actin cytoskeleton [72]. Both, neurexins and teneurins, link ADGRLs to the dystroglycan complex (DGC) by binding to the α-subunit of dystroglycan (DAG1) [73–75]. The dystroglycan β-subunit is associated with dystrophin (DMD) and utrophin, which connect the DGC to the actin cytoskeleton. The DGC provides a crucial connection between the extracellular matrix and the cytoskeleton in muscle cells and is of importance in other cell types, namely, neurons. Thus, it is present at the neuromuscular junction [76] and participates in the maintenance of Schwann cell myelination and axon guidance [77].

A crucial role in axon guidance was also shown for the fibronectin leucine-rich repeat (LRR) transmembrane proteins (FLRTs) [10]. FLRT3 expression is essential for the chemotactic response of rostral thalamocortical neurons to the guidance cues Slit1 and Netrin1 in primary neuronal cultures [78]. Further, FLRTs have been demonstrated to bind to the RBL motif and the HRM domain of ADGRLs via their LRRs. Knockdown of FLRT3 and competitive binding of ADGRL3 extracellular domain fragments in hippocampal neuron cultures both reduce the density of glutamatergic synapses [10]. This suggests that ADGRLs and FLRTs act in concert to regulate the number of excitatory synapses.

The ICD of ADGRL1 binds via its PBM to a group of PDZ domain-containing scaffold proteins (Fig. 3a) which are key components of the postsynaptic density (PSD), including SH3 and multiple ankyrin repeat domains protein 1 (SHANK1), DLG4 (previously known as PSD95), and MAGUK protein membrane-associated guanylate kinase inverted-2 (MAGI2) [79, 80]. Although these scaffold proteins are characteristic for the PSD, they can also be found in the presynaptic compartment of some synapses (e.g., inhibitory synapses, photoreceptor synapses). Therefore, interactions of PSD scaffold proteins with ADGRLs do not strictly define a post-synaptic localization of ADGRLs.

At the synaptic membrane, scaffold proteins organize supramolecular adhesion complexes, which include, besides ADGRLs, some of their identified interaction partners, e.g., neurexins (Fig. 3a). In these complexes, ADGRLs are connected to

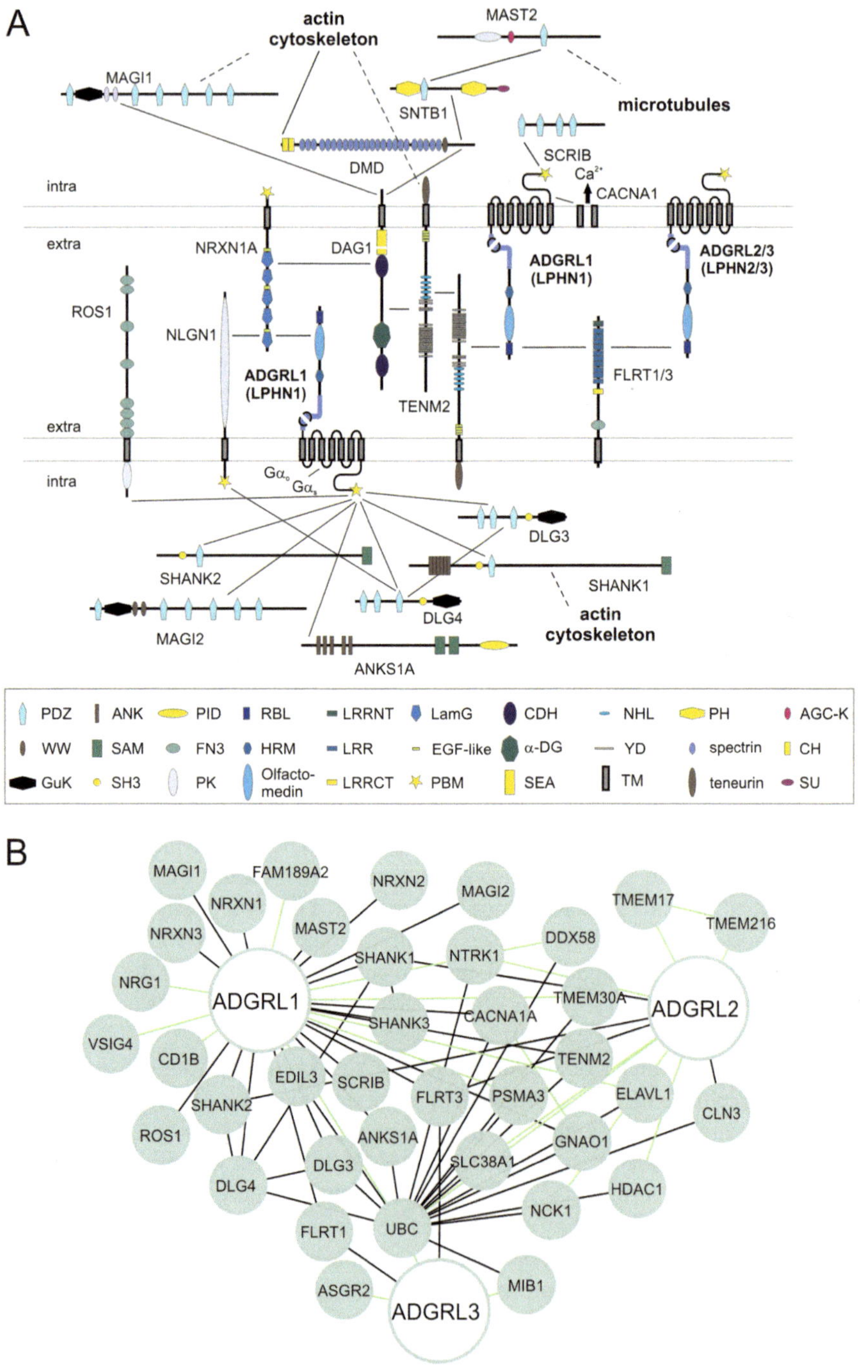

Fig. 3 The ADGRL (LPHN) network. (**a**) ADGRLs and their intracellular/extracellular interaction partners form a transmembrane-spanning protein network. *Black lines* depict experimental confirmed interactions. *Dotted lines* illustrate indirect connections. The protein nomenclature

actin filaments and the microtubule cytoskeleton via their interaction partners MAGI1/MAGI2 and the microtubule-associated serine/threonine 2 kinase (MAST2) [9]. Further, ADGRL1 binds to the planar cell protein (PCP) protein SCRIB [9], which is involved in cell migration, cell polarity, and cell proliferation [10, 81].

Insights into ADGRL signaling through heterotrimeric G proteins have been provided by two studies, so far. ADGRL1 interacts with Gαo, as demonstrated by affinity chromatography. In the same study an increase of cAMP and IP3 production in ADGRL1-transfected cells, but not in control cells, was observed [69]. A study in *C. elegans* on the ADGRL1 homologue LAT-1 showed coupling of LAT-1 to Gαs [67]. This Gαs-mediated cascade was essential for proper spindle orientation during cell division [67].

In summary, ADGRLs are mainly involved in synaptic protein networks, where they exert a role in axon guidance, synapse formation, the regulation of synaptic plasticity, and/or the control of the ratio between excitatory and inhibitory synapses. These functions are closely related to remodeling of the cytoskeleton and can be triggered by the reception of signaling cues from the extracellular space.

4.2 ADGRB (Brain-Specific Angiogenesis Inhibitor, BAI) Protein Networks Regulate Dendritic Spinogenesis, Synaptic Plasticity, and Phagocytosis

Intense studies conducted in the last decade have revealed a large number of molecules interacting with the three members of the ADGRB or BAI (brain-specific angiogenesis inhibitor) subfamily (Table 1, Fig. 4) [30]. ADGRB1–ADGRB3 are strongly expressed in neurons, astrocytes, and macrophages of the nervous system [31, 82, 83] but also found in non-neural tissues at lower levels of expression [32]. Their molecular structure is characterized by a variety of

Fig. 3 (continued) follows the HGNC guideline, and protein designations of human orthologues are used for all interaction partners. (**b**) Network of ADGRLs and their interactors. The visualized network is based on data from *STRING* (*Search Tool for the Retrieval of Interacting Genes/Proteins*, http://string-db.org/) and Cytoscape database search (http://www.cytoscape.org/). *Black lines*, published experimental data; *green lines*, data from protein databases. Abbreviations: *PDZ* (PSD95, Dlg, and ZO-1/ZO-2) domain; *WW* WW repeat; *GuK* guanylate kinase-like; *ANK* ankyrin repeats; *SAM* sterile alpha motif; *SH3* Src homology domain; *PID* phosphotyrosine interaction domain; *FN3* fibronectin-III type domain; *PK* protein kinase domain; *RBL* rhamnose-binding lectin-like domain; *HRM* hormone receptor motif; *Olfactomedin* olfactomedin domain; *LRRNT* leucine-rich repeat N-terminal; *LRR* leucine-rich repeat; *LRRCT* leucine-rich repeat C-terminal; *LamG* laminin G domain; *EGF* epidermal growth factor-like domain; *PBM* PDZ-binding motif; *CDH* cadherin-like repeat; *α-DG* α-dystroglycan domain; *SEA* sea urchin sperm protein, enterokinase and agrin domain; *NHL* NCL-1, HT2A, and Lin41 repeat; *YD* YD repeat; *TM* transmembrane domain; *PH* pleckstrin homology; *spectrin* spectrin repeats; *teneurin* teneurin intracellular domain; *AGC-K* AGC-kinase C-terminal domain; *CH* calponin homology domain; *SU* syntrophin unique domain

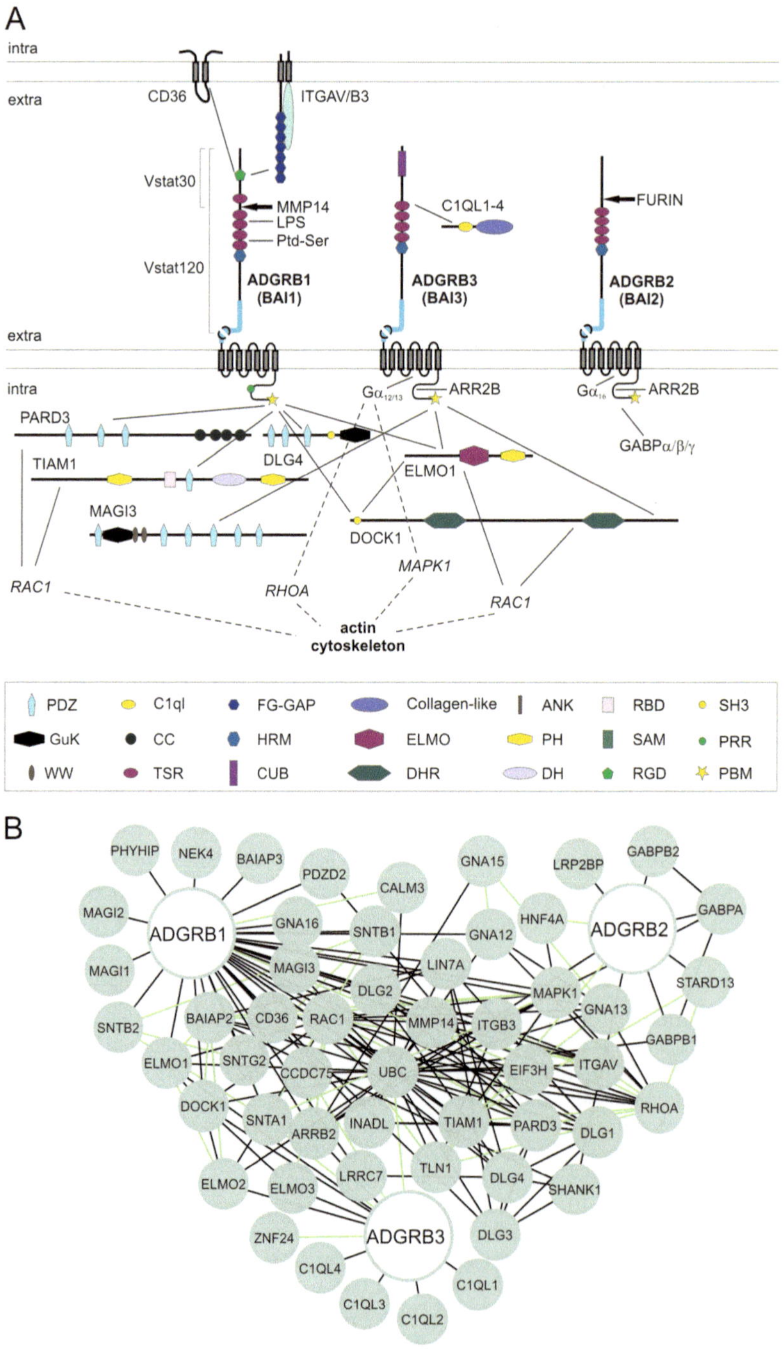

Fig. 4 The ADGRB (BAI) network. (**a**) ADGRBs form complex networks involved in actin cytoskeleton remodeling via activation of small RHO GTPases. The ECD of ADGRBs can be processed by proteases and interact with a heterogenous group of ligands. *Black lines* depict

conserved domains on both their ECD and ICD regions [30]. The ECDs of ADGRBs contain multiple thrombospondin type 1 repeats (TSRs) and a single hormone-binding domain (HBD). ADGRB1 additionally displays an Arg-Gly-Arg integrin-binding motif (RGD), whereas ADGRB3 possesses a (Cs1 and Csr/Uegf/BMP1) (CUB) domain on the ECD. Like most ADGRs, ADGRBs can be autoproteolytically cleaved at the GPS [3, 29, 84]. All ADGRBs possess a class I PBM at the very end of the C-terminus. The ICD of ADGRB1 also features a proline-rich region (PRR) suitable for binding to Src homology 3 (SH3) and WW domains.

The NTF of ADGRB1 was originally identified to regulate angiogenesis and thereby prevent tumorigenesis [30, 85]. The soluble N-terminal 120 kDa fragment, termed vasculostatin 120 (Vstat120), has a CTF-independent function [85]. Vstat120 inhibits endothelial cell migration in vitro and in vivo, and its antiangiogenic effect depends on its binding to CD36 which is expressed on the surface of endothelial cells [33]. The CLESH domain of CD36 and the antiangiogenesis TSR module of the ECD of ADGRB1 specifically mediate this in trans interaction [33]. TSRs are characteristic features of thrombospondins, proteins that are generally involved not only in angiogenesis, but also in the regulation of synaptogenesis [86].

In the CNS, ADGRB1 is highly enriched at the postsynaptic density (PSD) of dendrites [29, 34]. Similar to ADGRLs, ADGRB1 binds to various scaffold proteins that assemble complex networks at excitatory synapses as, e.g., MAGI1–MAGI3, DLG4 (PSD95), and DLG1 [29, 30, 35]. These protein–protein interactions are mediated by the binding of the C-terminal PBM of ADGRB1 to PDZ domains of the scaffold proteins [29].

ADGRB1 activates the RHO pathway via coupling to $G\alpha_{12/13}$, which concurrently leads to ERK (MAPK1) phosphorylation, which affects the actin cytoskeleton and thereby propagates synaptogenesis and dendritic spine formation [87]. Interestingly, the CTF alone activates the RHO pathway stronger than the full-length protein. Binding of the MAGUK family protein MAGI3 modulates the

Fig. 4 (continued) experimental confirmed interactions. *Dotted lines* illustrate indirect connections. The protein nomenclature follows the HGNC guideline, and protein designations of human orthologues are used for all interaction partners. (**b**) Network of ADGRBs and their interactors. The visualized network is based on data from *STRING* (*Search Tool for the Retrieval of Interacting Genes/Proteins*, http://string-db.org/) and Cytoscape database search (http://www.cytoscape.org/). *Black lines*, published experimental data; *green lines*, data from protein databases. Abbreviations: *PDZ* (PSD95, Dlg, and ZO-1/ZO-2) domain; *WW* WW repeat; *GuK* guanylate kinase-like; *C1ql* C1q-like domain; *CC* coiled-coiled domain; *TSR* thrombospondin type 1 repeats; *FG-GAP* FG-GAP repeat; *HRM* hormone receptor motif; *CUB* (C1r and C1s, uEGF, and bone morphogenetic protein) domain; *Collagen-like* collagen-like domain; *ELMO* ELMO domain; *DHR* dock homology region; *ANK* ankyrin repeats; *PH* pleckstrin homology domain; *DH* Dbl homology domain; *RBD* Raf-like Ras-binding domain; *SAM* sterile alpha motif; *RGD* Arg-Gly-Asp integrin-binding motif; *SH3* Src homology domain; *PRR* proline-rich region; *PBM* PDZ-binding motif activation of the ERK pathway [29]. For this, scaffold proteins may negatively regulate ß-arrestin 2 (ARRB2)-mediated receptor deactivation.

Further, a G protein-independent pathway of ADGRB1 signaling has been described in dendritic spines [34]. Here, ADGRB1 activates RAC1-mediated modulation of actin filament dynamics by the interaction of the two PDZ domain-containing proteins PARD3 and TIAM1 with the C-terminal PBM of ADGRB1. The induced PARD3/TIAM1 complex functions as a guanine exchange factor (GEF) for RAC1. Its recruitment to specific dendritic sites by ADGRB1 is crucial for proper spine and synapse formation [34]. Knockdown of ADGRB1 causes mislocalization of PARD3/TIAM1 and consequently loss of RAC1 and actin filaments from the spines [34]. Interestingly, the recruitment of the GEF by ADGRB1 seems to be regulated by a previously identified extracellular interaction partner of ADGRB1, integrin $\alpha v\beta 5$, that binds in trans to the RGD motif of ADGRB1 [34, 36].

Apart from its synaptic function, ADGRB1 also plays an important role in phagocytosis [31, 88] and myoblast fusion [89]. The engulfment of bacteria or apoptotic cells is induced by binding of lipopolysaccharides (LPS), endotoxins found in the outer membrane of Gram-negative bacteria, or phosphatidylserine (Ptd-L-Ser), a phospholipid of the plasma membrane enriched in neurons, respectively. For this, lipid compounds bind to the TSR domains of ADGRB1 and subsequently trigger a G protein-independent signaling pathway. Direct binding of ELMO1/DOCK1 to the ADGRB1 ICD activates the small GTPase RAC1 [31, 88]. Strikingly, the same signaling pathway is involved in the fusion of myoblasts forming syncytial muscle cells. During this process, a small subset of myoblasts undergo apoptosis, which then promotes the fusion of healthy myoblasts via the abovementioned ADGRB1-mediated RAC1 activation [89]. More recently, a comparable role in myoblast fusion has been demonstrated for ADGRB3 [45].

Secreted C1QL proteins, which bind to the TSR module of ADGRB3, can activate ADGRB3 [46]. C1QLs are unique among the C1Q/TNF superfamily of proteins [90], which are almost exclusively expressed in the brain during synaptogenesis [91, 92]. Together with ADGRB3 they regulate the density of excitatory synapses and thereby play an important role in synaptic plasticity [46]. During vertebrate brain development, C1QL1 expression controls climbing fiber synaptogenesis and the territory on Purkinje cells, thereby modulating Purkinje cell spinogenesis by activation of ADGRB3 [93].

Similar to the other two members of the ADGRB subfamily, ADGRB2 is also essential for proper neuronal cell development. Studies on ADGRB2-deficient mice reveal increased neurogenesis in hippocampal neurons [94]. The mechanism underlying the ADGRB2-dependent regulation of neurogenesis might be based on the binding of the transcription regulator GA-binding proteins α/γ and β/γ (GABPA, GABPB1, and GABPB2) to the ICD of ADGRB2. This association leads to the suppression of vascular endothelial growth factor (VEGF) expression, which is a stimulator of neurogenesis in the hippocampus [43].

In summary, ADGRBs regulate muscle cell development, neuronal function, and synaptogenesis through several different signaling pathways. This functional diversity is obtained via the variety of binding partners present in the interaction modules of the three ADGRBs (Fig. 4). First, the large number of PDZ domain-containing scaffold proteins provides multiple ways, which enable ADGRBs to integrate into membrane-associated protein networks. These modules are mostly associated with defined cell–cell contacts, in particular with synaptic junctions. Second, ADGRBs activate RAC1 via G protein-independent pathways by recruitment of the GEFs ELMO1/DOCK1 and TIAM1/PARD3, respectively. Third, ADGRBs can induce actin filament remodeling via a $G\alpha_{12/13}$-coupled pathway resulting in RHO activation. ADGRB signaling can be modulated by PDZ-proteins, β-arrestin 2 and potentially also via kinases such as NEK4 [37]. An additional mode of regulation of ADGRBs is provided by the NTF and CTF. Both fragments of ADGRBs can act independently or as a functional unit. Furthermore, the cleavage of the NTF by matrix metalloprotease-14 (MMP14) and furin can modulate ADGRB function [38, 95].

Interestingly, a couple of binding partners that are unusual for integral membrane proteins have been identified as ADGRB interactors, e.g., the GABP transcription factors [43] and PHYHIP [32], a component of peroxisomes. Together, this suggests a complex interplay between ADGRB regulation and its cellular function, depending on the cellular context.

4.3 ADGRV1 (VLGR1, GPR98), the Gigantic Receptor of Membrane Adhesion Networks in Sensory Cells and Synapses

ADGRV1, also known as very large G protein-coupled receptor (VLGR1), GPR98, or MASS1, is the largest member of the ADGR family, with a molecular weight of more than 700 kDa for the full-length protein [96]. To date, at least nine splice variants are known in vertebrates [97–100], but most probably many more splice variants remain uncharacterized [58]. All ADGRV1 splice variants contain varying numbers of Ca^{2+}-binding CalX-β repeats that resemble the aggregation factors of marine sponges [97]. ADGRV1b (VLGR1b), the largest isoform, contains 35 CalX-β domains, one laminin G/pentraxin domain (LamG/PTX), and seven epitempin/epilepsy-associated repeats (EPTP/EAR) on its NTF [96]. In contrast ADGRV1a (VLGR1a) exhibits a much shorter extracellular domain composed of only six CalX-β domains. Both ADGRV1a and ADGRV1b possess a CTF comprising the GAIN domain with the integral GPS, the 7TM domain, and an ICD with a class I PBM at the C-terminal end. However, similar to the vasculostatins of ADGRB1, most ADGRV1 isoforms lack the entire CTF. The function of these secreted soluble ECD polypeptides is unknown so far.

With the exception of Ca^{2+} ions there are no ligands known to bind to the ECD of ADGRV1 [59]. In contrast several proteins have been identified which bind to the ICD of ADGRV1 (Table 1, Fig. 5). The majority of these proteins are related to the human Usher syndrome (USH), the most common form of hereditary deaf-blindness [101, 102]. Mutations in the *ADGRV1* gene lead to the USH type 2C

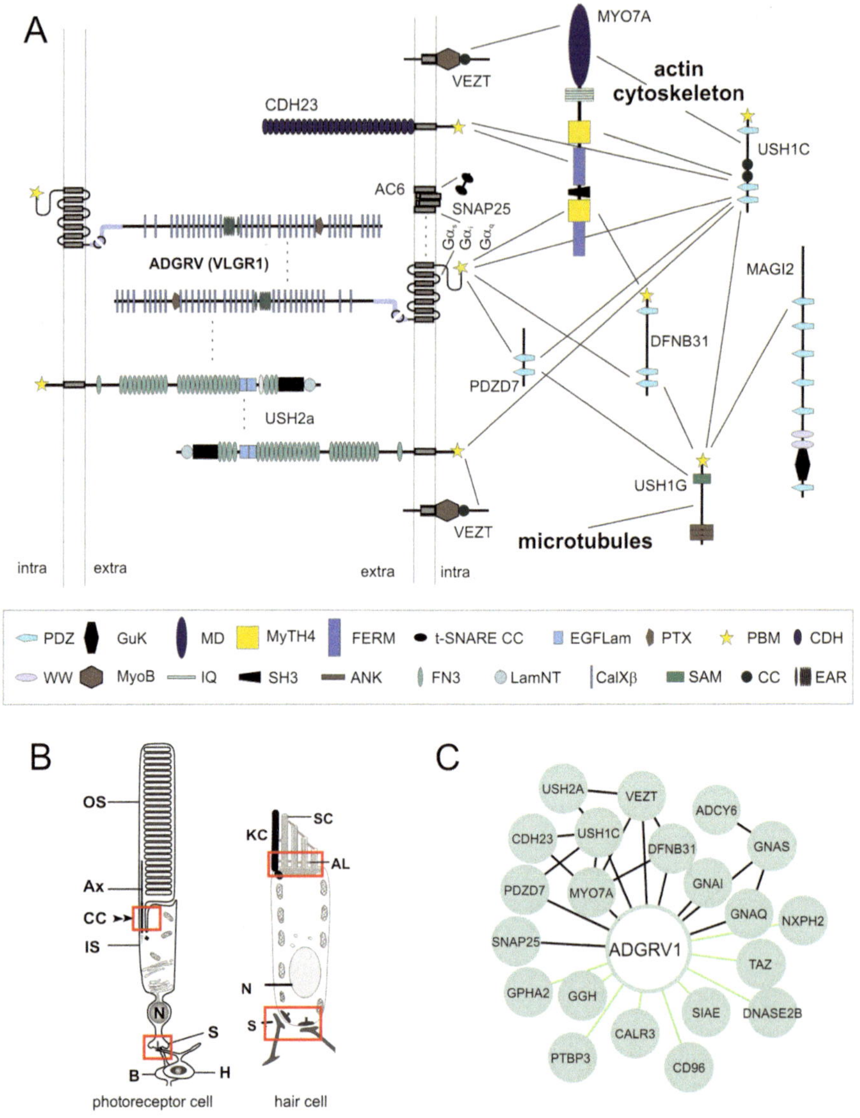

Fig. 5 The ADGRV1 (VLGR1) network. (**a**) ADGRV1 forms protein networks involving diverse PDZ scaffold proteins. The extracellular domain of ADGRV1 spans between membranes, potentially via homophilic interaction or by heterophilic interactions with USH2a (*dotted lines*). *Black lines* depict experimental confirmed interactions. *Dotted lines* illustrate indirect connections. The protein nomenclature follows the HGNC guideline, and protein designations of human orthologues are used for all interaction partners. (**b**) ADGRV1 networks are localized at the apical inner segment and the synapse of photoreceptor cells and at ankle-links between stereocilia and synapses of hair cells. (**c**) Network of ADGRV1 and its interactors. The visualized network is based on data from *STRING* (*Search Tool for the Retrieval of Interacting Genes/Proteins*, http://string-db.org/) and Cytoscape database search (http://www.cytoscape.org/). *Black lines*, published experimental data; *green lines*, data from protein databases. Abbreviations: *PDZ* (PSD95, Dlg, and ZO-1/ZO-2) domain; *WW* WW repeat; *GuK* guanylate kinase-like; *MyoB* myosin-binding domain; *MD* myosin

(USH2C), characterized by profound deafness and *retinitis pigmentosa* (retinal degeneration) (see below). ADGRV1 variants are mainly expressed in the affected neurons of the retina and the inner ear [96, 101], but they are also found in olfactory cells of the olfactory epithelium [103] and the brain [96].

Together with other USH proteins, ADGRV1 is localized at synaptic junctions, preferentially at the post-synapse of neurons (Fig. 5b) [60, 104], where they are thought to act in concert in the regulation of synaptic function [105]. The disruption of ADGRV1 function at CNS synapses most probably underlies the phenotype associated with audiogenic seizures and epilepsy in the *Frings* mouse [98]. However, in the sensory cells of the inner ear, the eye, and the nose, ADGRV1 is an essential component of membrane-membrane adhesion complexes at the proximal pole of the sensory neurons (Fig. 5b) [60, 61, 103].

In the mechanosensitive hair cells of the inner ear, the ECD of ADGRV1 is a core component of the ankle-links, fibers which span the membranes of neighboring stereocilia in the developing hair bundles (Fig. 5b) [62, 106, 107]. ADGRV1 deficient mice show splayed stereocilia which result in a auditory phenotype [62]. In these mice the absence of ADGRV1 leads to mislocalization or disappearance of other components of the Usher protein complex from hair bundles, e.g., USH2A, whirlin (DFNB31, USH2D), myosin VIIa (MYO7A, USH1B), and vezatin (VEZT) or specific isoforms of cadherin 23 (CDH23, USH1D) [62, 108, 109]. Furthermore, the adenylate cyclase 6 (ADCY6) is not only mislocalized, but its expression is also increased in the absence of ADGRV1 [62] arguing for a coupling of ADGRV1 to the Gαs/AC/cAMP pathway (see below). These findings support a signaling role of ADGRV1 in the ankle-links during hair bundle differentiation. More recently, PDZD7 has been identified as a binding partner of ADGRV1 [63, 110]. PDZD7 was previously identified as a deafness gene [110] and later as a genetic modifier of USH2 [63]. The PDZD7 protein co-localizes with whirlin and ADGRV1 and has been suggested to be a further component of the ankle-link complex [109].

Similar to the stereocilia of hair cells, whirlin (USH2D) and harmonin (USH1C), the scaffold proteins of the USH protein network, anchor and define the position of ADGRV1 in the photoreceptor cell membrane (Fig. 5a) [101]. In rod and cone photoreceptor cells, ADGRV1 is integrated in an USH protein network found at the periciliary ridge complex at the base of the photoreceptor cilium (Fig. 5b) [111]. Here, the ECD of ADGRV1 is essential for the formation of fibers linking the periciliary membrane of the inner segment with the membrane of the connecting

Fig. 5 (continued) motor domain; *IQ* IQ motif; *MyTH4* myosin tail homology 4 domain; *SH3* Src homology domain; *FERM*, 4.1 protein, ezrin, radixin moesin domain; *ANK* ankyrin repeats; *t-SNARE CC* t-SNARE coiled-coil homology; *FN3* fibronectin-III type domain; *EGFLam* laminin EGF-like 1; *LamNT* laminin N-terminal; *PTX* laminin G/pentraxin domain; *CalX-β* CalX-beta domain; *PBM* PDZ-binding motif; *SAM* sterile alpha motif; *CDH* cadherin-like repeat; *CC* coiled-coil; *EAR* epitempin/epilepsy-associated repeats; *OS* outer segment; *Ax* axoneme; *CC* connecting cilium; *IS* inner segment; *B* bipolar cell; *H* horizontal cell; *KC* kinocilium; *SC* stereocilia; *AL* ankle-links; nucleus cilium bridging the so-called ciliary pocket. We believe that this membrane-membrane adhesion complex and its intrinsic mechanical stabilization defines the target membrane for the vesicular transport and the import into photoreceptor cilia [111–113]. The relationship between ADGRV1 with intracellular vesicular transport is supported by more recent findings which indicate the presence of diverse ADGRV1 isoforms on transport vesicles and their association with the v-SNARE protein SNAP25 [58].

To date, only little is known about signaling pathways coupled to ADGRV1. Interestingly the two available studies on ADGRV1 signaling led to controversial results [59, 64]. First, Shin and coworkers (2013) reported the activation of PKC and PKA via $G\alpha_s$ and $G\alpha_q$, respectively, triggered by Ca^{2+}-binding to the ECD. This pathway seems to be involved in the stabilization of the myelin-associated glycoprotein, which is strongly expressed in oligodendrocytes. However, ADGRV1 mutant mice show no diminishment of myelination [59]. In contrast, Hu and coworkers (2014) demonstrated the coupling of ADGRV1 to a $G\alpha_i$-mediated pathway. Importantly both studies used different ADGRV1 fragments in their experiments. [59] created a "mini-ADGRV1" that contained an NTF composed of the first five CalX-β domains, the PTX domain and the EAR domain, which was fused to the CTF with an intact GAIN domain. In contrast, [64] applied only the CTF of ADGRV1. As shown similarly for other ADGRs, the absence of the NTF might abrogate its inhibiting effect on the activation of a $G\alpha_i$-mediated pathway [1]. However, the NTF might also provide a switch between the two different signaling pathways.

In any case, it remains elusive how ADGRV1 is activated, since no extracellular interaction partners or ligands have been identified so far. Furthermore, the role of the different ADGRV1 splice variants is unclear. Evidently, the elucidation of the ADGRV1 signaling network is still far from complete and seems to display a high degree of complexity, due to the large number of isoforms and potential modes of regulation.

4.4 Protein Networks Related to the Other Six ADGR Subfamilies

In comparison to the protein networks related to ADGRBs, ADGRLs, and ADGRV1 discussed above, the other six ADGR subfamilies have been less well studied so far. Nevertheless, at least for some of them, binding partners have been identified which are summarized in Table 1 and included in the overall network in Fig. 2.

For ADGRE5 (CD97) several ligands are known which activate the receptor via binding to its ECD [1]. Among these ligands is CD55, which is an inhibitor of the complement system and therefore of high clinical relevance [14]. In addition, THY1, an antigen present on the surface of immune cells, binds to the GAIN domain of ADGRE5 [14, 15]. ADGRE5 can also be activated by chondroitin sulfate B, a polysaccharide of the extracellular matrix, which can bind to the ADGRE5 paralogue ADGRE2 (EMR2) as well [13]. Further, the Arg-Gly-Asp

(RGD) motif downstream of the EGF-like domains in the ECD of ADGRE5 is recognized by integrins $\alpha5\beta1$ and $\alpha v\beta5$, mediating cell attachment [16]. ADGRE5 further dimerizes with the LPA receptor (LPAR1), which results in activation of RHO signaling [114].

All three members of ADGR subgroup III, ADGRA1–3 (GPR123–125), bind via their PBM to synaptic scaffold proteins of the MAGUK family, DLG4 and DLG1, respectively [17, 18]. ADGRA3 (GPR125) and ADGRA2 (GPR124) are additionally connected to Wnt signaling via their interaction with Dishevelled 1 (DVL1) [20] and the receptor tyrosine kinase RYK, respectively [22]. RYK is a receptor for Wnt molecules and involved in neuron differentiation and axon guidance [115]. Additionally, ADGRA2 functions as a Wnt7a-/Wnt7b-specific co-stimulator of β-catenin signaling [116].

Like ADGRA2 and ADGRA3, ADGRCs (CELSR1–3) are coupled to the Wnt signaling pathway. All three receptors interact with RYK [22]. ADGRC1 is additionally connected to Wnt signaling by further binding partners. Together with the Wnt receptor Frizzled 6 (FZD6), it forms a network that involves Dishevelled 2 (DVL2), DAAM1, and the GEF ARHGEF11. This network is localized at adherents junctions and is involved in RHO activation, which results in actomyosin-dependent contractions [21]. ADGRC1, FZD6, and DVL2 are key proteins of the planar cell polarity (PCP) pathway [117], and a further key component of the PCP pathway, VANGL2, has also been identified to bind ADGRC1. Both proteins work in concert with FZD6 and are indispensable for proper hair follicle orientation [118]. ADGRC3, however, is connected to the cytoskeleton by the linker protein DST and the GEF SWAP70, which both bind to F-actin [119, 120].

ADGRF5 (GPR116) is involved in lamellipodia and actin stress fiber formation, by $G\alpha_q$-ARHGEF25-mediated activation of RHOA and RAC1 [27]. For ADGRF5 the extracellular matrix surfactant protein D (SFTPD) has been described as the only ECD ligand so far [28]. It is a collagen-containing C-type lectin secreted from alveolar type 2 epithelial cells in the lung. ADGRF5 monitors the level of SFTPD and its deletion results in an emphysema-like pathology.

Two members of the VIII subfamily, ADGRG1 (GPR56) and ADGRG6 (GPR126), bind to fiber-forming collagen subtype 3 (COL3A1) and collagen subtype 4 (COL4A4) which does not assemble into fibers, respectively [47, 48]. COL4A4 interacts with the ECD of ADGRG6 that contains the CUB and PTX domains [54]. Furthermore, the interaction of ADGRG6 with a second ECM protein, laminin-211 (LAMA2), has been demonstrated. Both, COL4A4 and LAMA2, activate $G\alpha_s$-mediated signaling, which is important for the myelination of Schwann cells [54]. ADGRG1, however, does not contain such domains and therefore interacts via a different binding mode.

ADGRG1 (GPR56) and ADGRG3 (GPR97) can be regulated by β-arrestin 2 [48, 121] and have been shown to activate small RHO GTPases [47, 53], leading to remodeling of the actin cytoskeleton. ADGRG3 knockdown results in a change of the ratio of activated CDC42 vs. activated RHO leading to the redistribution of F-actin and paxillin. Binding of collagen subtype 3 (COL3A1) to ADGRG1 introduces the activation of the RHO pathway, which plays a crucial role in the regulation of proper lamination in the cerebral cortex [47].

Other ADGR groups are not well studied, although a growing set of data from public databases are available, e.g., for ADGRE1 (EMR1) and ADGRE2 (EMR2), which originate from proteomic screens but have not yet been validated [122, 123]. Pharmacological analyses have provided additional information on the heterotrimeric G proteins coupled to ADGR, which are described in [124, 125]. In conclusion, to date only a limited number of binding partners for ADGRs are known, and in most cases these interactions are poorly validated or uncharacterized. The number of valid binding partners of a specific ADGR is certainly highly dependent on the experimental efforts, which have been focused on that particular protein network.

5 Diseases Associated with ADGR Network Disturbances

Coherent protein networks require the smooth interplay of all network partners to provide proper cellular function. There are several examples in the ADGR field that the disruption of a single network component can have severe consequences for cell function and in turn for the affected organism. Malfunction of compounds within the same network often results in identical or similar phenotypes. The integration of molecules in the same protein network or pathway may provide potential common targets for pharmacological therapeutic interventions.

Mutations in *ADGRV1* (VLGR1) cause sensory neuronal degeneration, namely, the human Usher syndrome (USH) [102, 127] (see above Sect. 4.3). USH is a complex autosomal recessive disorder and the most common form of hereditary deaf-blindness. Based on the age of onset and the progression of the symptoms, three USH types (USH1–USH3) can be clinically differentiated. So far, ten different USH-causing genes and three genetic modifiers have been identified, and recent research revealed that all proteins related to USH are integrated into common protein networks [101, 102]. It is thought that defects in one USH protein can cause the disruption of the entire USH protein complexes in inner ear hair cells and retinal photoreceptor cells in the eye, leading to the sensory neuronal degeneration, which manifests in deaf-blindness. Furthermore, defects in the murine *Adgrv1* (*Vlgr1*) are associated with audiogenic seizures and epilepsy [98], which has also been discussed for human patients [24, 128].

Another severe disorder associated with an ADGR is bilateral frontoparietal polymicrogyria (BFPP), which is caused by mutations in ADGRG1 (GPR56) [129, 130]. Like USH, BFPP is an autosomal recessive disorder [131, 132]. Patients with BFPP suffer from cognitive interference, a delay in development of motor neurons, susceptibility to seizures, and ataxia. The most striking phenotype of BFPP-affected brains is the cobblestone-like cortical malformation [133, 134]. Interestingly, defects in collagen III (COL3A1), a validated ligand of ADGRG1 (see above, Sect. 5), develop an equivalent phenotype in mice [135] indicating that they interact in the same functional module.

A similar connection has been observed between ADGRB1 and its interaction partner BAIAP2 (IRS53). Both participate in the regulation of dendritic spine formation [34, 136], and their defects are suggested to be related to autism disorders [137, 138]. Interestingly, the expression of another interactor of ADGRB1, DLG4 (PSD95) [39], is increased in ADGRB1-deficient mice [34], and there are hints that the regulation of DLG4 is a general feature for genes involved in autism [139].

SNPs in *ADGRB3* lead to development of schizophrenia and probably participate in susceptibility to addiction [140, 141]. Further, ADGRB3 deficient mice show a depression-resistant phenotype [94].

For mutations in *ADGRL1* and *ADGRL3*, a connection to the attention deficit hyperactivity disorder (ADHD) [142, 143], hearing defects, brain malformations, and retardation has been reported [142]. One group of ADGRL1's extracellular interaction partners, neurexins, are associated with mental retardation and autism as well [144, 145].

ADGRCs are also mandatory for proper brain development. In humans, mutations in the corresponding genes cause neural tube defects and are associated with caudal agenesis [146, 147]. In addition, ADGRCs are involved in the proper development of ependymal cilia and disruptions of the network result in hydrocephalus. These effects occur due to disturbances of PCP and Wnt signaling pathways that are connected to ADGRCs [148].

Moreover, some ADGRs act as tumor suppressors and therefore gene defects in ADGRs can cause cancer and tumorigenesis. One of these receptors is ADGRE5 (CD97), which was found to be significantly upregulated in diverse cancer cells [149–151]. Others are ADGRG1 (GPR56), which is up- or downregulated, respectively, in a cell-specific manner in cancer cells [49, 152]. Specifically ADGRF5 is associated with breast cancer [27], and ADGRB1, is absent or downregulated in various cancer types [153–155]. The relation between cancer genesis and defects in ADGR networks might be due their role in cytoskeleton regulation via small GTPases.

Several additional ADGRs have been associated with severe human diseases but no proteomics and interactomics data are available for these so far. Further decipherment of protein networks related to diseases associated with ADGRs would not only enlighten the pathomechanisms leading to the disease but also elucidate potential targets for treatment and cure.

6 Common Features of ADGR-Related Protein Networks and Signaling Pathways

The data acquired for ADGRs so far indicate a high degree of similarity within the protein family. Since similar or identical interactions have been reported for ADGRs of various subgroups, it is probable that ADGRs are integrated in common networks and may even act in concert. ADGRs are authentic G protein-coupled receptors, and for several receptors heterotrimeric G proteins have been identified [156] (Fig. 6). Furthermore, the canonical desensitizing mechanism of GPCRs

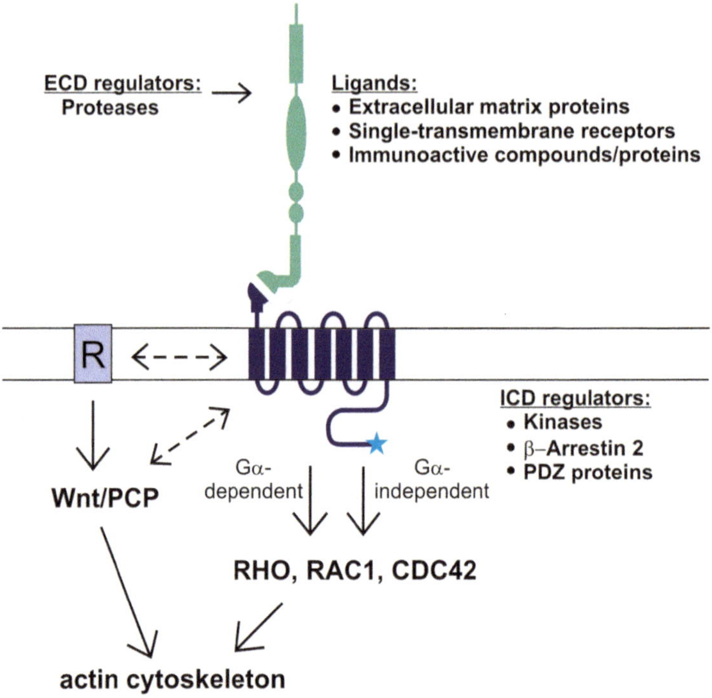

Fig. 6 Common features of ADGRs. ADGRs show similarities concerning ligand binding, signaling, and regulation. Many ADGRs activate RHO/RAC1/CDC42-mediated pathways. Some ADGRs may cross talk to Wnt/PCP signaling through other transmembrane receptors (R) or other components. Extracellular matrix proteins, single-transmembrane receptors, and immunoactive proteins/compounds mainly activate ADGRs. Regulation of ADGRs has been demonstrated for β-arrestin 2, PDZ scaffold proteins, proteases, and kinases

through β-arrestin 2 binding has also been identified in some ADGRs (Fig. 6), namely, ADGRBs (BAIs), ADGRG1 (GPR56), and ADGRG3 (GPR97) [48, 121].

Although ADGRs signal through diverse Gα-subunits to a variety of the downstream signaling pathways [1], the activation of small GTPases (RHO, RAC1, or CDC42) is a common theme (see above, Fig. 6). However, some ADGRs can also activate small GTPases independent from heterotrimeric G proteins as, e.g., ADGRB1 through PARD3/TIAM1 or ELMO/DOCK1, respectively [31, 34]. The downstream target of the small GTPases pathways is the actin cytoskeleton, and remodeling the cytoskeleton is a common process during neurite outgrowth or dendritic spine formation, which are known to be regulated by several ADGRs (see above). For at least five of the ADGRs, namely, the three ADGRCs (CELSRs) and ADGRAs (GPR125, GPR124), there is evidence for cross talk with the Wnt signaling pathway [20, 21]. They either are coupled to the receptor tyrosine kinase RYK, the Wnt receptor Frizzled 6 (FZD6) or interact with the FZD mediator Dishevelled (DVL) [22]. Interestingly, DVL can mediate the activation of the RHO pathway regulating the actin cytoskeleton [21].

Besides their obvious links to signaling networks, almost all ADGRs examined thus far are integrated in membrane-associated protein networks organized by scaffold proteins. This is usually mediated by the interaction of the C-terminal PDZ-binding motif (PBM) often found at the C-terminal end of ICD of ADGRs and PDZ domains of the diverse scaffold proteins. Most of these scaffold proteins belong to the membrane-associated guanylate kinase (MAGUK) protein family, e.g., DLG1–4 and MAGI1–MAGI3 (Figs. 2–4). These scaffold proteins facilitate the organization of ADGR-related networks at specific domains of the cell membrane, but may additionally regulate the signaling of the CTF of ADGRs. Strikingly, numerous ADGRs are integral components of adhesion complexes at synapses and dendritic spines showing a highly regulated expression profile during development (e.g., ADGRBs, ADGRLs, ADGRCs, and ADGRV1) [30, 96, 157, 158].

In cell adhesion complexes, the long ECD of ADGRs mediates cell–cell and cell–matrix interactions, which have been shown to promote ADGR signaling. Collagens or collagen-like proteins were identified as common interaction components from the extracellular matrix binding to different adhesion domains of the ECD of ADGRs [46, 47, 55]. Beside homophilic trans interactions, various adhesion domains in the ECDs of ADGRs are capable to interact with the ECDs of single-span transmembrane proteins or immunoactive proteins and compounds [10–12, 14–16].

Analysis of the protein interactome related to ADGRs presented in Fig. 2 shows that ten ADGRs interact with ubiquitin C (UBC). These interactions have been recently identified in proteomic screens targeting ubiquitinated proteins [159, 160]. So far the ubiquitination of ADGRs was not systematically studied, but it may have functional implications. As frequently reported for other signaling, ubiquitination of cell surface receptors may regulate the availability of ADGRs to interact with their extracellular ligands [161, 162].

Although ADGR families differ mainly in their ECD structure, several ADGRs from different families share the same domains, e.g., ADGRC1–ADGRC3 (CELSR1–CELSR3), ADGRD1 (GPR133), ADGRD2 (GPR144), ADGRG6 (GPR126), ADGRG4 (GPR112), and ADGRV1 (Vlgr1) share a LamG/PTX domain, whereas ADGRLs (LPHNs), ADGRBs (BAIs), ADGRA2 (GPR124), ADGRA3 (GPR125), and ADGRF3 (GPR113) have hormone receptor motifs (HRMs) [1]. The common ECD features of these ADGRs indicate that they may also be targets for similar or identical binding proteins and ligands, which are not described so far. Nevertheless, the available data indicate common mechanisms for ADGRs (see Fig. 6) in their integration and function in cellular contexts and molecular modules. Since similar or identical interactions have been reported for ADGRs of various subgroups, ADGRs seem to be integrated in common networks and may even act in concert.

7 Conclusions

To date, comprehensive approaches toward the systematic identification of valid interaction partners for ADGRs are still lacking. From future (larger) interactome data sets and systematic data mining, we expect to gain further insights into ADGR networks and a more holistic understanding of ADGR functional roles in cellular adhesion and signaling. The application of proteomic screening methods provides potential to learn more about these unique types of receptors. However, these approaches have to be complemented by functional assays and studies in animal models. Understanding the molecular mechanisms behind ADGR function will provide novel strategies for the cure of diseases that have been linked to malfunction of ADGRs, particularly various types of cancer, BFPP, and the human Usher syndrome.

Acknowledgments We thank Drs. Helen May-Simera and Kerstin Nagel-Wolfrum for critical reading of the manuscript and skillful language corrections. We also thank Dr. Kirsten A. Wunderlich and Fabian Möller for their help. Funding from the following sources is gratefully acknowledged: BMBF "HOPE2" (01GM1108D), DFG FOR 2149 Project 6 (WO54878-1), European Union FP7/2009/241955 (SYSCILIA) and FP7/2009/242013 (TREATRUSH), FAUN-Stiftung, Nuremberg, and the Foundation Fighting Blindness (FFB).

References

1. Hamann J, Aust G, Arac D, Engel FB, Formstone C et al (2015) International Union of Basic and Clinical Pharmacology. XCIV. Adhesion G protein-coupled receptors. Pharmacol Rev 67:338–367
2. Krishnan A, Nijmeijer S, de Graaf C, Schiöth HB (2016) Classification, nomenclature and structural aspects of adhesion GPCRs. In: Langenhan T, Schöneberg T (eds) Adhesion G protein-coupled receptors: molecular, physiological and pharmacological principles in health and disease. Springer, Heidelberg
3. Arac D, Boucard AA, Bolliger MF, Nguyen J, Soltis SM et al (2012) A novel evolutionarily conserved domain of cell-adhesion GPCRs mediates autoproteolysis. EMBO J 31:1364–1378
4. Araç D, Sträter N, Seiradake E (2016) Understanding the structural basis of adhesion GPCR functions. In: Langenhan T, Schöneberg T (eds) Adhesion G protein-coupled receptors: molecular, physiological and pharmacological principles in health and disease. Springer, Heidelberg
5. Nieberler M, Kittel RJ, Petrenko AG, Lin H-H, Langenhan T (2016) Control of adhesion GPCR function through proteolytic processing. In: Langenhan T, Schöneberg T (eds) Adhesion G protein-coupled receptors: molecular, physiological and pharmacological principles in health and disease. Springer, Heidelberg
6. Luck K, Charbonnier S, Trave G (2012) The emerging contribution of sequence context to the specificity of protein interactions mediated by PDZ domains. FEBS Lett 586:2648–2661
7. Dunn HA, Ferguson SS (2015) PDZ protein regulation of G protein-coupled receptor trafficking and signaling pathways. Mol Pharmacol 88:624–639
8. Kreienkamp HJ, Zitzer H, Gundelfinger ED, Richter D, Bockers TM (2000) The calcium-independent receptor for alpha-latrotoxin from human and rodent brains interacts with members of the ProSAP/SSTRIP/Shank family of multidomain proteins. J Biol Chem 275:32387–32390

9. Tobaben S, Sudhof TC, Stahl B (2000) The G protein-coupled receptor CL1 interacts directly with proteins of the Shank family. J Biol Chem 275:36204–36210
10. O'Sullivan ML, de Wit J, Savas JN, Comoletti D, Otto-Hitt S et al (2012) FLRT proteins are endogenous latrophilin ligands and regulate excitatory synapse development. Neuron 73:903–910
11. Boucard AA, Ko J, Sudhof TC (2012) High affinity neurexin binding to cell adhesion G-protein-coupled receptor CIRL1/latrophilin-1 produces an intercellular adhesion complex. J Biol Chem 287:9399–9413
12. Silva JP, Lelianova VG, Ermolyuk YS, Vysokov N, Hitchen PG et al (2011) Latrophilin 1 and its endogenous ligand Lasso/teneurin-2 form a high-affinity transsynaptic receptor pair with signaling capabilities. Proc Natl Acad Sci U S A 108:12113–12118
13. Stacey M, Chang GW, Davies JQ, Kwakkenbos MJ, Sanderson RD et al (2003) The epidermal growth factor-like domains of the human EMR2 receptor mediate cell attachment through chondroitin sulfate glycosaminoglycans. Blood 102:2916–2924
14. Hamann J, Vogel B, van Schijndel GM, van Lier RA (1996) The seven-span transmembrane receptor CD97 has a cellular ligand (CD55, DAF). J Exp Med 184:1185–1189
15. Wandel E, Saalbach A, Sittig D, Gebhardt C, Aust G (2012) Thy-1 (CD90) is an interacting partner for CD97 on activated endothelial cells. J Immunol 188:1442–1450
16. Wang T, Ward Y, Tian L, Lake R, Guedez L et al (2005) CD97, an adhesion receptor on inflammatory cells, stimulates angiogenesis through binding integrin counterreceptors on endothelial cells. Blood 105:2836–2844
17. Fernandez E, Collins MO, Uren RT, Kopanitsa MV, Komiyama NH et al (2009) Targeted tandem affinity purification of PSD-95 recovers core postsynaptic complexes and schizophrenia susceptibility proteins. Mol Syst Biol 5:269
18. Yamamoto Y, Irie K, Asada M, Mino A, Mandai K et al (2004) Direct binding of the human homologue of the Drosophila disc large tumor suppressor gene to seven-pass transmembrane proteins, tumor endothelial marker 5 (TEM5), and a novel TEM5-like protein. Oncogene 23:3889–3897
19. Wu G, Feng X, Stein L (2010) A human functional protein interaction network and its application to cancer data analysis. Genome Biol 11:R53
20. Li X, Roszko I, Sepich DS, Ni M, Hamm HE et al (2013) Gpr125 modulates Dishevelled distribution and planar cell polarity signaling. Development 140:3028–3039
21. Nishimura T, Honda H, Takeichi M (2012) Planar cell polarity links axes of spatial dynamics in neural-tube closure. Cell 149:1084–1097
22. Berndt JD, Aoyagi A, Yang P, Anastas JN, Tang L et al (2011) Mindbomb 1, an E3 ubiquitin ligase, forms a complex with RYK to activate Wnt/beta-catenin signaling. J Cell Biol 194:737–750
23. Hermle T, Guida MC, Beck S, Helmstadter S, Simons M (2013) Drosophila ATP6AP2/VhaPRR functions both as a novel planar cell polarity core protein and a regulator of endosomal trafficking. EMBO J 32:245–259
24. Nakayama M, Kikuno R, Ohara O (2002) Protein-protein interactions between large proteins: two-hybrid screening using a functionally classified library composed of long cDNAs. Genome Res 12:1773–1784
25. Bohnekamp J, Schoneberg T (2011) Cell adhesion receptor GPR133 couples to Gs protein. J Biol Chem 286:41912–41916
26. Gupte J, Swaminath G, Danao J, Tian H, Li Y et al (2012) Signaling property study of adhesion G-protein-coupled receptors. FEBS Lett 586:1214–1219
27. Tang X, Jin R, Qu G, Wang X, Li Z et al (2013) GPR116, an adhesion G-protein-coupled receptor, promotes breast cancer metastasis via the Galphaq-p63RhoGEF-Rho GTPase pathway. Cancer Res 73:6206–6218
28. Fukuzawa T, Ishida J, Kato A, Ichinose T, Ariestanti DM et al (2013) Lung surfactant levels are regulated by Ig-Hepta/GPR116 by monitoring surfactant protein D. PLoS One 8, e69451

29. Stephenson JR, Paavola KJ, Schaefer SA, Kaur B, Van Meir EG et al (2013) Brain-specific angiogenesis inhibitor-1 signaling, regulation, and enrichment in the postsynaptic density. J Biol Chem 288:22248–22256
30. Stephenson JR, Purcell RH, Hall RA (2014) The BAI subfamily of adhesion GPCRs: synaptic regulation and beyond. Trends Pharmacol Sci 35:208–215
31. Park D, Tosello-Trampont AC, Elliott MR, Lu M, Haney LB et al (2007) BAI1 is an engulfment receptor for apoptotic cells upstream of the ELMO/Dock180/Rac module. Nature 450:430–434
32. Koh JT, Lee ZH, Ahn KY, Kim JK, Bae CS et al (2001) Characterization of mouse brain-specific angiogenesis inhibitor 1 (BAI1) and phytanoyl-CoA alpha-hydroxylase-associated protein 1, a novel BAI1-binding protein. Brain Res Mol Brain Res 87:223–237
33. Kaur B, Cork SM, Sandberg EM, Devi NS, Zhang Z et al (2009) Vasculostatin inhibits intracranial glioma growth and negatively regulates in vivo angiogenesis through a CD36-dependent mechanism. Cancer Res 69:1212–1220
34. Duman JG, Tzeng CP, Tu YK, Munjal T, Schwechter B et al (2013) The adhesion-GPCR BAI1 regulates synaptogenesis by controlling the recruitment of the Par3/Tiam1 polarity complex to synaptic sites. J Neurosci 33:6964–6978
35. Wu Y, Dowbenko D, Spencer S, Laura R, Lee J et al (2000) Interaction of the tumor suppressor PTEN/MMAC with a PDZ domain of MAGI3, a novel membrane-associated guanylate kinase. J Biol Chem 275:21477–21485
36. Koh JT, Kook H, Kee HJ, Seo YW, Jeong BC et al (2004) Extracellular fragment of brain-specific angiogenesis inhibitor 1 suppresses endothelial cell proliferation by blocking alphavbeta5 integrin. Exp Cell Res 294:172–184
37. Basei FL, Meirelles GV, Righetto GL, Dos Santos Migueleti DL, Smetana JH et al (2015) New interaction partners for Nek4.1 and Nek4.2 isoforms: from the DNA damage response to RNA splicing. Proteome Sci 13:11
38. Cork SM, Kaur B, Devi NS, Cooper L, Saltz JH et al (2012) A proprotein convertase/MMP-14 proteolytic cascade releases a novel 40 kDa vasculostatin from tumor suppressor BAI1. Oncogene 31:5144–5152
39. Lim IA, Hall DD, Hell JW (2002) Selectivity and promiscuity of the first and second PDZ domains of PSD-95 and synapse-associated protein 102. J Biol Chem 277:21697–21711
40. Shiratsuchi A, Osada S, Kanazawa S, Nakanishi Y (1998) Essential role of phosphatidylserine externalization in apoptosing cell phagocytosis by macrophages. Biochem Biophys Res Commun 246:549–555
41. Shiratsuchi T, Futamura M, Oda K, Nishimori H, Nakamura Y et al (1998) Cloning and characterization of BAI-associated protein 1: a PDZ domain-containing protein that interacts with BAI1. Biochem Biophys Res Commun 247:597–604
42. Oda K, Shiratsuchi T, Nishimori H, Inazawa J, Yoshikawa H et al (1999) Identification of BAIAP2 (BAI-associated protein 2), a novel human homologue of hamster IRSp53, whose SH3 domain interacts with the cytoplasmic domain of BAI1. Cytogenet Cell Genet 84:75–82
43. Jeong BC, Kim MY, Lee JH, Kee HJ, Kho DH et al (2006) Brain-specific angiogenesis inhibitor 2 regulates VEGF through GABP that acts as a transcriptional repressor. FEBS Lett 580:669–676
44. Petersen HH, Hilpert J, Militz D, Zandler V, Jacobsen C et al (2003) Functional interaction of megalin with the megalin-binding protein (MegBP), a novel tetratrico peptide repeat-containing adaptor molecule. J Cell Sci 116:453–461
45. Hamoud N, Tran V, Croteau LP, Kania A, Cote JF (2014) G-protein coupled receptor BAI3 promotes myoblast fusion in vertebrates. Proc Natl Acad Sci U S A 111(10):3745–3750
46. Bolliger MF, Martinelli DC, Sudhof TC (2011) The cell-adhesion G protein-coupled receptor BAI3 is a high-affinity receptor for C1q-like proteins. Proc Natl Acad Sci U S A 108:2534–2539
47. Luo R, Jeong SJ, Jin Z, Strokes N, Li S, Piao X (2011) G protein-coupled receptor 56 and collagen III, a receptor-ligand pair, regulates cortical development and lamination. Proc Natl

Acad Sci U S A 108(31):12925–12930. doi:10.1073/pnas.1104821108, PubMed PMID: 21768377, PubMed Central PMCID: PMC3150909, Epub 2011/07/20
48. Paavola KJ, Stephenson JR, Ritter SL, Alter SP, Hall RA (2011) The N terminus of the adhesion G protein-coupled receptor GPR56 controls receptor signaling activity. J Biol Chem 286:28914–28921
49. Xu L, Begum S, Hearn JD, Hynes RO (2006) GPR56, an atypical G protein-coupled receptor, binds tissue transglutaminase, TG2, and inhibits melanoma tumor growth and metastasis. Proc Natl Acad Sci U S A 103:9023–9028
50. Little KD, Hemler ME, Stipp CS (2004) Dynamic regulation of a GPCR-tetraspanin-G protein complex on intact cells: central role of CD81 in facilitating GPR56-Galpha q/11 association. Mol Biol Cell 15:2375–2387
51. Iguchi T, Sakata K, Yoshizaki K, Tago K, Mizuno N et al (2008) Orphan G protein-coupled receptor GPR56 regulates neural progenitor cell migration via a G alpha 12/13 and Rho pathway. J Biol Chem 283:14469–14478
52. Peeters MC, Fokkelman M, Boogaard B, Egerod KL, van de Water B et al (2015) The adhesion G protein-coupled receptor G2 (ADGRG2/GPR64) constitutively activates SRE and NFkappaB and is involved in cell adhesion and migration. Cell Signal 27:2579–2588
53. Valtcheva N, Primorac A, Jurisic G, Hollmen M, Detmar M (2013) The orphan adhesion G protein-coupled receptor GPR97 regulates migration of lymphatic endothelial cells via the small GTPases RhoA and Cdc42. J Biol Chem 288:35736–35748
54. Monk KR, Naylor SG, Glenn TD, Mercurio S, Perlin JR et al (2009) A G protein-coupled receptor is essential for Schwann cells to initiate myelination. Science 325:1402–1405
55. Paavola KJ, Sidik H, Zuchero JB, Eckart M, Talbot WS (2014) Type IV collagen is an activating ligand for the adhesion G protein-coupled receptor GPR126. Sci Signal 7:ra76
56. Mogha A, Benesh AE, Patra C, Engel FB, Schoneberg T et al (2013) Gpr126 functions in Schwann cells to control differentiation and myelination via G-protein activation. J Neurosci 33:17976–17985
57. Petersen SC, Luo R, Liebscher I, Giera S, Jeong SJ et al (2015) The adhesion GPCR GPR126 has distinct, domain-dependent functions in Schwann cell development mediated by interaction with laminin-211. Neuron 85:755–769
58. Zallocchi M, Delimont D, Meehan DT, Cosgrove D (2012) Regulated vesicular trafficking of specific PCDH15 and VLGR1 variants in auditory hair cells. J Neurosci 32:13841–13859
59. Shin D, Lin ST, Fu YH, Ptacek LJ (2013) Very large G protein-coupled receptor 1 regulates myelin-associated glycoprotein via Galphas/Galphaq-mediated protein kinases A/C. Proc Natl Acad Sci U S A 110:19101–19106
60. Reiners J, van Wijk E, Marker T, Zimmermann U, Jurgens K et al (2005) Scaffold protein harmonin (USH1C) provides molecular links between Usher syndrome type 1 and type 2. Hum Mol Genet 14:3933–3943
61. van Wijk E, van der Zwaag B, Peters T, Zimmermann U, te Brinke H et al (2006) The DFNB31 gene product whirlin connects to the Usher protein network in the cochlea and retina by direct association with USH2A and VLGR1. Hum Mol Genet 15:751–765
62. Michalski N, Michel V, Bahloul A, Lefevre G, Barral J et al (2007) Molecular characterization of the ankle-link complex in cochlear hair cells and its role in the hair bundle functioning. J Neurosci 27:6478–6488
63. Ebermann I, Phillips JB, Liebau MC, Koenekoop RK, Schermer B et al (2010) PDZD7 is a modifier of retinal disease and a contributor to digenic Usher syndrome. J Clin Invest 120:1812–1823
64. Hu QX, Dong JH, Du HB, Zhang DL, Ren HZ et al (2014) Constitutive Galphai coupling activity of very large G protein-coupled receptor 1 (VLGR1) and its regulation by PDZD7 protein. J Biol Chem 289:24215–24225
65. Das G, Reynolds-Kenneally J, Mlodzik M (2002) The atypical cadherin Flamingo links Frizzled and Notch signaling in planar polarity establishment in the Drosophila eye. Dev Cell 2:655–666

66. Rawls AS, Wolff T (2003) Strabismus requires Flamingo and Prickle function to regulate tissue polarity in the Drosophila eye. Development 130:1877–1887
67. Muller A, Winkler J, Fiedler F, Sastradihardja T, Binder C, Schnabel R et al (2015) Oriented cell division in the C. elegans embryo is coordinated by G-protein signaling dependent on the adhesion GPCR LAT-1. PLoS Genet 11(10), e1005624
68. Krasnoperov VG, Bittner MA, Beavis R, Kuang Y, Salnikow KV et al (1997) Alpha-Latrotoxin stimulates exocytosis by the interaction with a neuronal G-protein-coupled receptor. Neuron 18:925–937
69. Lelianova VG, Davletov BA, Sterling A, Rahman MA, Grishin EV et al (1997) Alpha-latrotoxin receptor, latrophilin, is a novel member of the secretin family of G protein-coupled receptors. J Biol Chem 272:21504–21508
70. Sugita S, Ichtchenko K, Khvotchev M, Sudhof TC (1998) Alpha-Latrotoxin receptor CIRL/latrophilin 1 (CL1) defines an unusual family of ubiquitous G-protein-linked receptors. G-protein coupling not required for triggering exocytosis. J Biol Chem 273:32715–32724
71. Tabuchi K, Sudhof TC (2002) Structure and evolution of neurexin genes: insight into the mechanism of alternative splicing. Genomics 79:849–859
72. Woelfle R, D'Aquila AL, Pavlovic T, Husic M, Lovejoy DA (2015) Ancient interaction between the teneurin C-terminal associated peptides (TCAP) and latrophilin ligand-receptor coupling: a role in behavior. Front Neurosci 9:146
73. Sugita S (2001) A stoichiometric complex of neurexins and dystroglycan in brain. J Cell Biol 154:435–446
74. Trzebiatowska A, Topf U, Sauder U, Drabikowski K, Chiquet-Ehrismann R (2008) Caenorhabditis elegans teneurin, -1, is required for gonadal and pharyngeal basement membrane integrity and acts redundantly with integrin ina-1 and dystroglycan dgn-1. Mol Biol Cell 19:3898–3908
75. Chand D, Colacci M, Dixon K, Kollara A, Brown TJ et al (2014) C-terminal region of teneurin-1 co-localizes with the dystroglycan complex in adult mouse testes and regulates testicular size and testosterone production. Histochem Cell Biol 141:191–211
76. Banks GB, Fuhrer C, Adams ME, Froehner SC (2003) The postsynaptic submembrane machinery at the neuromuscular junction: requirement for rapsyn and the utrophin/dystrophin-associated complex. J Neurocytol 32:709–726
77. Masaki T, Matsumura K (2010) Biological role of dystroglycan in Schwann cell function and its implications in peripheral nervous system diseases. J Biomed Biotechnol 2010:740403
78. Leyva-Diaz E, del Toro D, Menal MJ, Cambray S, Susin R et al (2014) FLRT3 is a Robo1-interacting protein that determines Netrin-1 attraction in developing axons. Curr Biol 24:494–508
79. Feng W, Zhang M (2009) Organization and dynamics of PDZ-domain-related supramodules in the postsynaptic density. Nat Rev Neurosci 10:87–99
80. Kim E, Sheng M (2004) PDZ domain proteins of synapses. Nat Rev Neurosci 5:771–781
81. Yamanaka T, Ohno S (2008) Role of Lgl/Dlg/Scribble in the regulation of epithelial junction, polarity and growth. Front Biosci 13:6693–6707
82. Mori K, Kanemura Y, Fujikawa H, Nakano A, Ikemoto H et al (2002) Brain-specific angiogenesis inhibitor 1 (BAI1) is expressed in human cerebral neuronal cells. Neurosci Res 43:69–74
83. Sokolowski JD, Nobles SL, Heffron DS, Park D, Ravichandran KS et al (2011) Brain-specific angiogenesis inhibitor-1 expression in astrocytes and neurons: implications for its dual function as an apoptotic engulfment receptor. Brain Behav Immun 25:915–921
84. Park D, Ravichandran KS (2010) Emerging roles of brain-specific angiogenesis inhibitor 1. Adv Exp Med Biol 706:167–178
85. Kaur B, Brat DJ, Devi NS, Van Meir EG (2005) Vasculostatin, a proteolytic fragment of brain angiogenesis inhibitor 1, is an antiangiogenic and antitumorigenic factor. Oncogene 24:3632–3642

86. Adams JC, Tucker RP (2000) The thrombospondin type 1 repeat (TSR) superfamily: diverse proteins with related roles in neuronal development. Dev Dyn 218:280–299
87. Duman JG, Mulherkar S, Tu YK, X Cheng J, Tolias KF (2015) Mechanisms for spatiotemporal regulation of Rho-GTPase signaling at synapses. Neurosci Lett 601:4–10
88. Das S, Owen KA, Ly KT, Park D, Black SG et al (2011) Brain angiogenesis inhibitor 1 (BAI1) is a pattern recognition receptor that mediates macrophage binding and engulfment of Gram-negative bacteria. Proc Natl Acad Sci U S A 108:2136–2141
89. Hochreiter-Hufford AE, Lee CS, Kinchen JM, Sokolowski JD, Arandjelovic S et al (2013) Phosphatidylserine receptor BAI1 and apoptotic cells as new promoters of myoblast fusion. Nature 497:263–267
90. Ressl S, Vu BK, Vivona S, Martinelli DC, Sudhof TC et al (2015) Structures of C1q-like proteins reveal unique features among the C1q/TNF superfamily. Structure 23:688–699
91. Berube NG, Swanson XH, Bertram MJ, Kittle JD, Didenko V et al (1999) Cloning and characterization of CRF, a novel C1q-related factor, expressed in areas of the brain involved in motor function. Brain Res Mol Brain Res 63:233–240
92. Iijima T, Miura E, Watanabe M, Yuzaki M (2010) Distinct expression of C1q-like family mRNAs in mouse brain and biochemical characterization of their encoded proteins. Eur J Neurosci 31:1606–1615
93. Sigoillot SM, Iyer K, Binda F, Gonzalez-Calvo I, Talleur M, et al (2015) The secreted protein C1QL1 and its receptor BAI3 control the synaptic connectivity of excitatory inputs converging on cerebellar Purkinje cells. Cell Rep. pii: S2211-1247(15)00059-5
94. Okajima D, Kudo G, Yokota H (2011) Antidepressant-like behavior in brain-specific angiogenesis inhibitor 2-deficient mice. J Physiol Sci 61:47–54
95. Okajima D, Kudo G, Yokota H (2010) Brain-specific angiogenesis inhibitor 2 (BAI2) may be activated by proteolytic processing. J Recept Signal Transduct Res 30:143–153
96. McMillan DR, White PC, Grati M, Shin JB, Weston MD et al (2010) Studies on the very large G protein-coupled receptor: from initial discovery to determining its role in sensorineural deafness in higher animals. Adv Exp Med Biol 706:76–86
97. Nikkila H, McMillan DR, Nunez BS, Pascoe L, Curnow KM et al (2000) Sequence similarities between a novel putative G protein-coupled receptor and Na+/Ca2+ exchangers define a cation binding domain. Mol Endocrinol 14:1351–1364
98. Skradski SL, Clark AM, Jiang H, White HS, Fu YH et al (2001) A novel gene causing a mendelian audiogenic mouse epilepsy. Neuron 31:537–544
99. McMillan DR, Kayes-Wandover KM, Richardson JA, White PC (2002) Very large G protein-coupled receptor-1, the largest known cell surface protein, is highly expressed in the developing central nervous system. J Biol Chem 277:785–792
100. Weston MD, Luijendijk MW, Humphrey KD, Moller C, Kimberling WJ (2004) Mutations in the VLGR1 gene implicate G-protein signaling in the pathogenesis of Usher syndrome type II. Am J Hum Genet 74:357–366
101. Reiners J, Nagel-Wolfrum K, Jurgens K, Märker T, Wolfrum U (2006) Molecular basis of human Usher syndrome: deciphering the meshes of the Usher protein network provides insights into the pathomechanisms of the Usher disease. Exp Eye Res 83:97–119
102. Mathur P, Yang J (1852) Usher syndrome: hearing loss, retinal degeneration and associated abnormalities. Biochim Biophys Acta 2015:406–420
103. Jansen F, Kalbe B, Scholz P, Mikosz M, Wunderlich KA et al (2016) Impact of the Usher syndrome on olfaction. Hum Mol Genet 25(3):524–533
104. Specht D, Wu SB, Turner P, Dearden P, Koentgen F et al (2009) Effects of presynaptic mutations on a postsynaptic Cacna1s calcium channel colocalized with mGluR6 at mouse photoreceptor ribbon synapses. Invest Ophthalmol Vis Sci 50:505–515
105. Gregory FD, Bryan KE, Pangrsic T, Calin-Jageman IE, Moser T et al (2011) Harmonin inhibits presynaptic Cav1.3 Ca(2)(+) channels in mouse inner hair cells. Nat Neurosci 14:1109–1111

106. McGee J, Goodyear RJ, McMillan DR, Stauffer EA, Holt JR et al (2006) The very large G-protein-coupled receptor VLGR1: a component of the ankle link complex required for the normal development of auditory hair bundles. J Neurosci 26:6543–6553
107. Lefevre G, Michel V, Weil D, Lepelletier L, Bizard E et al (2008) A core cochlear phenotype in USH1 mouse mutants implicates fibrous links of the hair bundle in its cohesion, orientation and differential growth. Development 135:1427–1437
108. Yang J, Liu X, Zhao Y, Adamian M, Pawlyk B et al (2010) Ablation of whirlin long isoform disrupts the USH2 protein complex and causes vision and hearing loss. PLoS Genet 6, e1000955
109. Grati M, Shin JB, Weston MD, Green J, Bhat MA et al (2012) Localization of PDZD7 to the stereocilia ankle-link associates this scaffolding protein with the Usher syndrome protein network. J Neurosci 32:14288–14293
110. Schneider E, Marker T, Daser A, Frey-Mahn G, Beyer V et al (2009) Homozygous disruption of PDZD7 by reciprocal translocation in a consanguineous family: a new member of the Usher syndrome protein interactome causing congenital hearing impairment. Hum Mol Genet 18:655–666
111. Maerker T, van Wijk E, Overlack N, Kersten FF, McGee J et al (2008) A novel Usher protein network at the periciliary reloading point between molecular transport machineries in vertebrate photoreceptor cells. Hum Mol Genet 17:71–86
112. Sorusch N, Wunderlich K, Bauá K, Nagel-Wolfrum K, Wolfrum U (2014) Usher syndrome protein network functions in the retina and their relation to other retinal ciliopathies. In: Ash J, Hollyfield JG, LaVail MM, Anderson RE, Grimm C et al (eds) Retinal degenerative diseases. Springer, New York, NY
113. Bauss K, Knapp B, Jores P, Roepman R, Kremer H et al (2014) Phosphorylation of the Usher syndrome 1G protein SANS controls Magi2-mediated endocytosis. Hum Mol Genet 23:3923–3942
114. Ward Y, Lake R, Yin JJ, Heger CD, Raffeld M et al (2011) LPA receptor heterodimerizes with CD97 to amplify LPA-initiated RHO-dependent signaling and invasion in prostate cancer cells. Cancer Res 71:7301–7311
115. Keeble TR, Halford MM, Seaman C, Kee N, Macheda M et al (2006) The Wnt receptor Ryk is required for Wnt5a-mediated axon guidance on the contralateral side of the corpus callosum. J Neurosci 26:5840–5848
116. Posokhova E, Shukla A, Seaman S, Volate S, Hilton MB et al (2015) GPR124 functions as a WNT7-specific coactivator of canonical beta-catenin signaling. Cell Rep 10:123–130
117. Feng J, Han Q, Zhou L (2012) Planar cell polarity genes, Celsr1-3, in neural development. Neurosci Bull 28:309–315
118. Devenport D, Fuchs E (2008) Planar polarization in embryonic epidermis orchestrates global asymmetric morphogenesis of hair follicles. Nat Cell Biol 10:1257–1268
119. Ihara S, Oka T, Fukui Y (2006) Direct binding of SWAP-70 to non-muscle actin is required for membrane ruffling. J Cell Sci 119:500–507
120. Ferrier A, Boyer JG, Kothary R (2013) Cellular and molecular biology of neuronal dystonin. Int Rev Cell Mol Biol 300:85–120
121. Southern C, Cook JM, Neetoo-Isseljee Z, Taylor DL, Kettleborough CA et al (2013) Screening beta-arrestin recruitment for the identification of natural ligands for orphan G-protein-coupled receptors. J Biomol Screen 18:599–609
122. Dyer MD, Neff C, Dufford M, Rivera CG, Shattuck D et al (2010) The human-bacterial pathogen protein interaction networks of Bacillus anthracis, Francisella tularensis, and Yersinia pestis. PLoS One 5, e12089
123. Hein MY, Hubner NC, Poser I, Cox J, Nagaraj N et al (2015) A human interactome in three quantitative dimensions organized by stoichiometries and abundances. Cell 163:712–723
124. Liebscher I, Schöneberg T (2016) Tethered agonism: a common activation mechanism of adhesion GPCRs. In: Langenhan T, Schöneberg T (eds) Adhesion G protein-coupled

receptors: molecular, physiological and pharmacological principles in health and disease. Springer, Heidelberg

125. Kishore A, Hall RA (2016) Versatile signaling activity of adhesion GPCRs. In: Langenhan T, Schöneberg T (eds) Adhesion G protein-coupled receptors: molecular, physiological and pharmacological principles in health and disease. Springer, Heidelberg
126. Kovacs P, Schöneberg T (2016) The relevance of genomic signatures at adhesion GPCR loci in humans. In: Langenhan T, Schöneberg T (eds) Adhesion G protein-coupled receptors: molecular, physiological and pharmacological principles in health and disease. Springer, Heidelberg
127. Wolfrum U (2011) Protein networks related to the Usher syndrome gain insights in the molecular basis of the disease. In: Satpal A (ed) Usher syndrome: pathogenesis, diagnosis and therapy. Nova Science, New York, NY, pp 51–73
128. Scheel H, Tomiuk S, Hofmann K (2002) A common protein interaction domain links two recently identified epilepsy genes. Hum Mol Genet 11:1757–1762
129. Piao X, Hill RS, Bodell A, Chang BS, Basel-Vanagaite L et al (2004) G protein-coupled receptor-dependent development of human frontal cortex. Science 303:2033–2036
130. Piao X, Chang BS, Bodell A, Woods K, Benzeev B et al (2005) Genotype-phenotype analysis of human frontoparietal polymicrogyria syndromes. Ann Neurol 58:680–687
131. Piao X, Basel-Vanagaite L, Straussberg R, Grant PE, Pugh EW et al (2002) An autosomal recessive form of bilateral frontoparietal polymicrogyria maps to chromosome 16q12.2-21. Am J Hum Genet 70:1028–1033
132. Chang BS, Piao X, Bodell A, Basel-Vanagaite L, Straussberg R et al (2003) Bilateral frontoparietal polymicrogyria: clinical and radiological features in 10 families with linkage to chromosome 16. Ann Neurol 53:596–606
133. Li S, Jin Z, Koirala S, Bu L, Xu L et al (2008) GPR56 regulates pial basement membrane integrity and cortical lamination. J Neurosci 28:5817–5826
134. Bahi-Buisson N, Poirier K, Boddaert N, Fallet-Bianco C, Specchio N et al (2010) GPR56-related bilateral frontoparietal polymicrogyria: further evidence for an overlap with the cobblestone complex. Brain 133:3194–3209
135. Jeong SJ, Luo R, Li S, Strokes N, Piao X (2012) Characterization of G protein-coupled receptor 56 protein expression in the mouse developing neocortex. J Comp Neurol 520:2930–2940
136. Choi J, Ko J, Racz B, Burette A, Lee JR et al (2005) Regulation of dendritic spine morphogenesis by insulin receptor substrate 53, a downstream effector of Rac1 and Cdc42 small GTPases. J Neurosci 25:869–879
137. Toma C, Hervas A, Balmana N, Vilella E, Aguilera F et al (2011) Association study of six candidate genes asymmetrically expressed in the two cerebral hemispheres suggests the involvement of BAIAP2 in autism. J Psychiatr Res 45:280–282
138. Michaelson JJ, Shi Y, Gujral M, Zheng H, Malhotra D et al (2012) Whole-genome sequencing in autism identifies hot spots for de novo germline mutation. Cell 151:1431–1442
139. Tsai NP, Wilkerson JR, Guo W, Maksimova MA, DeMartino GN et al (2012) Multiple autism-linked genes mediate synapse elimination via proteasomal degradation of a synaptic scaffold PSD-95. Cell 151:1581–1594
140. Liu QR, Drgon T, Johnson C, Walther D, Hess J et al (2006) Addiction molecular genetics: 639,401 SNP whole genome association identifies many “cell adhesion” genes. Am J Med Genet B Neuropsychiatr Genet 141B:918–925
141. DeRosse P, Lencz T, Burdick KE, Siris SG, Kane JM et al (2008) The genetics of symptom-based phenotypes: toward a molecular classification of schizophrenia. Schizophr Bull 34:1047–1053
142. Bonaglia MC, Marelli S, Novara F, Commodaro S, Borgatti R et al (2010) Genotype-phenotype relationship in three cases with overlapping 19p13.12 microdeletions. Eur J Hum Genet 18:1302–1309

143. Arcos-Burgos M, Muenke M (2010) Toward a better understanding of ADHD: LPHN3 gene variants and the susceptibility to develop ADHD. Atten Defic Hyperact Disord 2:139–147
144. Zweier C, de Jong EK, Zweier M, Orrico A, Ousager LB et al (2009) CNTNAP2 and NRXN1 are mutated in autosomal-recessive Pitt-Hopkins-like mental retardation and determine the level of a common synaptic protein in Drosophila. Am J Hum Genet 85:655–666
145. Sudhof TC (2008) Neuroligins and neurexins link synaptic function to cognitive disease. Nature 455:903–911
146. Allache R, De Marco P, Merello E, Capra V, Kibar Z (2012) Role of the planar cell polarity gene CELSR1 in neural tube defects and caudal agenesis. Birth Defects Res A Clin Mol Teratol 94:176–181
147. Robinson A, Escuin S, Doudney K, Vekemans M, Stevenson RE et al (2012) Mutations in the planar cell polarity genes CELSR1 and SCRIB are associated with the severe neural tube defect craniorachischisis. Hum Mutat 33:440–447
148. Tissir F, Qu Y, Montcouquiol M, Zhou L, Komatsu K et al (2010) Lack of cadherins Celsr2 and Celsr3 impairs ependymal ciliogenesis, leading to fatal hydrocephalus. Nat Neurosci 13:700–707
149. Aust G, Eichler W, Laue S, Lehmann I, Heldin NE et al (1997) CD97: a dedifferentiation marker in human thyroid carcinomas. Cancer Res 57:1798–1806
150. Steinert M, Wobus M, Boltze C, Schütz A, Wahlbuhl M et al (2002) Expression and regulation of CD97 in colorectal carcinoma cell lines and tumor tissues. Am J Pathol 161:1657–1667
151. Galle J, Sittig D, Hanisch I, Wobus M, Wandel E et al (2006) Individual cell-based models of tumor-environment interactions: multiple effects of CD97 on tumor invasion. Am J Pathol 169:1802–1811
152. Zendman AJ, Cornelissen IM, Weidle UH, Ruiter DJ, van Muijen GN (1999) TM7XN1, a novel human EGF-TM7-like cDNA, detected with mRNA differential display using human melanoma cell lines with different metastatic potential. FEBS Lett 446:292–298
153. Kaur B, Brat DJ, Calkins CC, Van Meir EG (2003) Brain angiogenesis inhibitor 1 is differentially expressed in normal brain and glioblastoma independently of p53 expression. Am J Pathol 162:19–27
154. Fukushima Y, Oshika Y, Tsuchida T, Tokunaga T, Hatanaka H et al (1998) Brain-specific angiogenesis inhibitor 1 expression is inversely correlated with vascularity and distant metastasis of colorectal cancer. Int J Oncol 13:967–970
155. Zohrabian VM, Nandu H, Gulati N, Khitrov G, Zhao C et al (2007) Gene expression profiling of metastatic brain cancer. Oncol Rep 18:321–328
156. Paavola KJ, Hall RA (2012) Adhesion G protein-coupled receptors: signaling, pharmacology, and mechanisms of activation. Mol Pharmacol 82(5):777–783. doi:10.1124/mol.112.080309, PubMed PMID: 22821233, PubMed Central PMCID: PMC3477231, Epub 2012/07/24
157. Langenhan T, Promel S, Mestek L, Esmaeili B, Waller-Evans H et al (2009) Latrophilin signaling links anterior-posterior tissue polarity and oriented cell divisions in the C. elegans embryo. Dev Cell 17:494–504
158. Tissir F, Goffinet AM (2013) Atypical cadherins Celsr1-3 and planar cell polarity in vertebrates. Prog Mol Biol Transl Sci 116:193–214
159. Na CH, Jones DR, Yang Y, Wang X, Xu Y et al (2012) Synaptic protein ubiquitination in rat brain revealed by antibody-based ubiquitome analysis. J Proteome Res 11:4722–4732
160. Udeshi ND, Mani DR, Eisenhaure T, Mertins P, Jaffe JD et al (2012) Methods for quantification of in vivo changes in protein ubiquitination following proteasome and deubiquitinase inhibition. Mol Cell Proteomics 11:148–159
161. Lin AW, Man HY (2013) Ubiquitination of neurotransmitter receptors and postsynaptic scaffolding proteins. Neural Plast 2013:432057
162. Huangfu WC, Fuchs SY (2010) Ubiquitination-dependent regulation of signaling receptors in cancer. Genes Cancer 1:725–734

The Relevance of Genomic Signatures at Adhesion GPCR Loci in Humans

Peter Kovacs and Torsten Schöneberg

Graphical Abstract

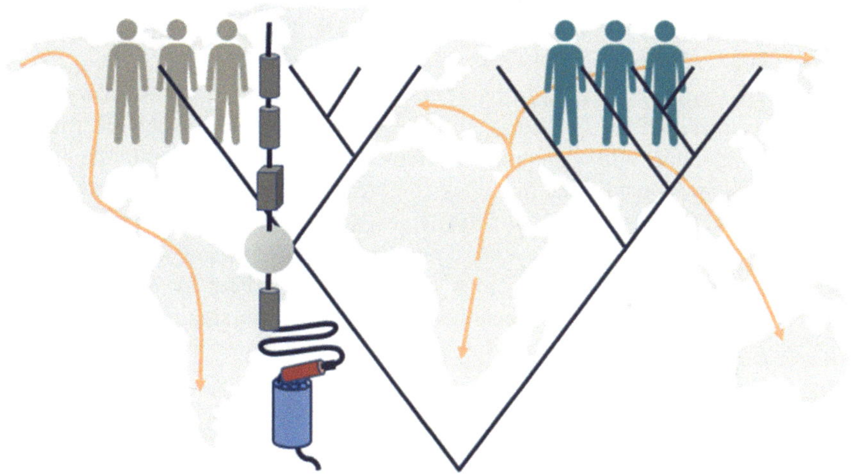

P. Kovacs (✉)
Integrated Research and Treatment Center (IFB) AdiposityDiseases, Medical Faculty, University of Leipzig, Liebigstr. 21, Leipzig 04103, Germany
e-mail: peter.kovacs@medizin.uni-leipzig.de

T. Schöneberg (✉)
Institute of Biochemistry, Medical Faculty, University of Leipzig, Johannisallee 30, Leipzig 04103, Germany
e-mail: schoberg@medizin.uni-leipzig.de

T. Langenhan, T. Schöneberg (eds.), *Adhesion G Protein-coupled Receptors*, Handbook of Experimental Pharmacology 234, DOI 10.1007/978-3-319-41523-9_9

Contents

Abstract

Adhesion G protein-coupled receptors (aGPCRs) have a long evolutionary history dating back to very basal unicellular eukaryotes. Almost every vertebrate is equipped with a set of different aGPCRs. Genomic sequence data of several hundred extinct and extant species allows for reconstruction of aGPCR phylogeny in vertebrates and non-vertebrates in general but also provides a detailed view into the recent evolutionary history of human aGPCRs. Mining these sequence sources with bioinformatic tools can unveil many facets of formerly unappreciated aGPCR functions. In this review, we extracted such information from the literature and open public sources and provide insights into the history of aGPCR in humans. This includes comprehensive analyses of signatures of selection, variability of human aGPCR genes, and quantitative traits at human aGPCR loci. As indicated by a large number of genome-wide genotype-phenotype association studies, variations in aGPCR contribute to specific human phenotypes. Our survey demonstrates that aGPCRs are significantly involved in adaptation processes, phenotype variations, and diseases in humans.

Keywords

Adhesion G protein-coupled receptors • Population genetics • Natural selection • Genome-wide association studies • Mutation • Disease

1 Introduction

Since their first availability in the mid-2000s, the next-generation sequencing (NGS) technologies have been proven to be powerful tools in many fields of genetics, evolutionary biology, and biomedicine. Whole-genome sequencing and whole-transcriptome sequencing are efficient to provide sequence data not only for ortho- and paralog genes and transcripts of gene families but also for variants of genes involved in adaptation processes and responsible for diseases. The trove of genetic data yielded by NGS has made a significant impact on understanding the origin of gene families, their evolutionary expansion and contraction, and their structural variability. By identifying rare variants in known genes, NGS contributes to today's clinical diagnoses and helps in the discovery of molecular pathomechanisms underlying human diseases. In parallel to this explosion in sequence information, an armada of bioinformatics tools and pipelines were developed to handle and meaningfully explore the complexity of these data. Since sequencing has become faster and more affordable, the number of available complete genomic sequences is exponentially increasing. As a result, the cost to process, analyze, transfer, and store the data is becoming the bottleneck for research and medical applications.

The wealth of sequence information comes from large sequencing consortia. For example, the whole genomes of over 2500 human individuals of 26 populations have been sequenced with low-coverage yielding 86 million variants [1]. A program has been launched in 2009 sequencing the whole genomes of 10,000 vertebrate species suitable for comparative genomic analyses [2]. Since then over 200 vertebrate species have been sequenced or completed [3]. Data of those multiple species and population genome ventures are used to construct the evolutionary tree of vertebrate clades, e.g., birds [4], and to reconstruct of migration waves of humans [5], respectively.

The advent of such large and diverse genomic data provides not only a valuable source to study the evolutionary history of species but also the evolution of specific gene families. This usually requires high-quality genomes of individual species and previously unavailable or unappreciated bioinformatics methods. Thus, the history of a protein family, the numeric gain and loss of family members, and their structural evolution by comparing orthologs and paralogs can be studied in the light of evolution. The comparison of orthologous sequences between species can shed light on evolutionary constrains and inventions which are of advantage in a given ecologic niche [6].

Evolutionary methods based on NGS data have already found numerous applications in biomedical research. Sequence differences in human populations or between extinct and recent hominoids are analyzed for signatures of selection providing clues for adaptation processes to specific environmental factors during human evolution. However, those differences may also influence or even cause human phenotypes and diseases. A great number of genome-wide genotype-phenotype association studies (GWAS) revealed previously unknown traits and pathways involving individual genes or gene families.

In this review we summarize applications of evolutionary methods specifically on the family of aGPCRs. Information that can be drawn from the currently available sequence data on the evolutionary history, selective forces acting on aGPCRs, and human phenotypes associated with aGPCR variants are summarized in this review.

2 The Evolutionary Origin and Development of aGPCRs

It has been estimated that GPCR prototypes were present already in the last common ancestor of eukaryotes [7], and most probably glutamate and cAMP receptors evolved in eukaryotic evolution about 1400 million years ago [8]. aGPCR apparently appeared later in eukaryotic life (1275 million years ago) since aGPCR-specific sequence signatures in 7TM segments were found in the genomes of *Amoebozoa* (*D. discoideum*) and *Alveolata* (*P. tetraurelia*) [8]. In these basal eukaryotes, aGPCRs present with short N termini. The lack of a large ectodomain is apparent also in fungi aGPCRs where this receptor family started to expand. The unicellular non-metazoa choanoflagellate *Monosiga brevicollis* and filasterea *Capsaspora owczarzaki* contain several genes for aGPCRs already equipped with GPS site [8]. Further structural diversification of the large ectodomain occurred in pre-bilaterian species [9]. One can speculate that structural diversification and numeric expansion of aGPCRs soon after the origin of metazoans were driven by cell-cell and cell-matrix contacts which are major factors of multicellularity. Being present in metazoan genomes, aGPCRs became apparently very important as suggested by the structural conservation and numerical expansion of aGPCR genes between invertebrate and vertebrate species [10]. However, the number of aGPCRs varies between vertebrate species [11]. In the human genome, there are 34 genes encoding aGPCRs. Out of these genes, 33 aGPCR genes contain an open reading frame, and one, *EMR4/ADGRE4,* is a pseudogene presenting a deletion-caused disruption of the open reading frame. The deletion is not present in great apes suggesting that EMR4 became nonfunctional only after human speciation [12]. Interestingly, available databases do not contain mRNA sequence of the human GPR144/ADGRD2 and only rare transcripts in other vertebrate species (own observations). In mouse GPR144 is a pseudogene. It is therefore unclear whether GPR144/ADGRD2 is functional in humans. There are several clusters where aGPCRs are genomically organized in a tandem-like structure. One clusters at chromosome X containing *GPR110/ADGRF1, GPR111/ADGRF2, GPR115/ADGRF4,* and *GPR116/ADGRF5,* and another group of aGPCRs (EMRs/ADGRE1–4, CD97/ADGRE5) clusters in two sections on human chromosome 19p13, suggesting common evolutionary origins, respectively [12, 13].

3 Signature of Selections in aGPCR Genes

The recent increase in human sequence and polymorphism data, together with the availability of genome sequences from several closely related primate species, provides an unprecedented opportunity to investigate how natural selection has shaped the human genome. A central aim in evolutionary biology is to understand the genetic mechanisms of how organisms adapt to their ecological niche. Typically, adaptation is viewed as a process where beneficial alleles are driven from low to high frequency in a population, reducing the variability in a genomic segment—so-called selective sweeps. On a genome-wide scale, immune-related genes and genes associated with fitness, reproduction, and fertility appear to be major targets of positive selection [14]. Lactase persistence is one of the classical examples where a line of genetic evidences suggests recent selection within the past 5000–10,000 years at the lactase locus in European-derived populations [15].

Positive selection can yield an excess of non-synonymous fixed differences (mainly between species), extended haplotype homozygosity (EHH), modified allele frequency distributions, and long-branch length in evolutionary trees. A comprehensive overview on the methods used and their limitations is given elsewhere [16, 17]. In brief, faster evolving genes can be detected between species by comparing the number of non-synonymous substitutions (dN) to the number of synonymous substitutions (dS). A likelihood method considering the dN/dS ratios across sites is implemented in phylogenetic analysis using maximum likelihood (PAML) [18, 19]. The McDonald-Kreitman (MK) test uses polymorphism data from a single species and fixed differences between multiple species to detect regions under selection [20]. The often used Tajima's *D* test compares the number of variants in a genomic region with the average number of pairwise differences [21]. The first scans for selection between populations that took advantage of differentiation across populations focused on single-marker F_{ST} [22]. F_{ST} statistic is useful to identify genomic regions that exhibit high variation in allelic frequency between groups, which is a characteristic of genomic regions that have gone through differential selection. Haplotype statistics, such as long-range haplotype (LRH) test [23], EHH [24], and integrated haplotype score (iHS) [25], use data from linked sites to identify past targets of natural selection. These methods are based on the expectation that recent positive sweeps yield extended haplotypes that have not had enough time to break down by recombination. An additional approach to identifying signals of selection through population comparison is the cross population extended haplotype heterozygosity (XP-EHH) test, which was designed to detect ongoing or nearly fixed selective sweeps by comparing haplotypes from two populations [26, 27].

Numerous studies have used the different methods to identify signature of selection and of convergent evolution between species and between or within populations of a single species producing long lists of genomic loci and candidate genes. Screening of those lists frequently reveals aGPCR loci with significant signatures of selection. For example, *GPR110* has been identified with such signatures linked to convergent evolution in echolocating mammals [28]. However, most studies focus on signals of selection during human evolution. Thus, sequence differences between primates, as closest relatives of humans, are used to identify obvious lineage-specific genomic events that are significantly different from random changes between species. Again, differences at aGPCR loci are frequently extracted from whole-genome comparison approaches (Table 1). For example, *GPR111* and *GPR123/ADGRA1* were found with signatures of selection in the human lineage in several studies. However, all these studies can neither provide the driving force of selection nor the selected phenotype but provide a rational to dig deeper into the physiology of the respective aGPCR also considering cross species differences in phenotypes. Significant cross species differences in the mRNA expression profile and level can also be helpful to elucidate the physiological significance of aGPCR. For example, a human-specific expression profile in brain prefrontal cortex and cerebellum compared to chimpanzees and rhesus macaques was found for *GPR116*, *CELSR2/ADGRC2*, *LPHN2/ADGRL2*, *BAI2/ADGRB2*, and

Table 1 Signatures of selection at genomic aGPCR loci in the human lineage

Gene name	New name	Human vs. species	Methods	Reference
LPHN1	*ADGRL1*	Chimpanzee	MK	[29]
LPHN3	*ADGRL3*	Chimpanzee	MK	[29]
EMR1	*ADGRE1*	Chimpanzee	MK	[29]
GPR123	*ADGRA1*	Chimpanzee	MK	[29]
		Primates	MK	[30]
CELSR1	*ADGRC1*	Chimpanzee	MK	[29]
GPR133	*ADGRD1*	Chimpanzee	MK	[29]
GPR110	*ADGRF1*	Chimpanzee	MK	[29]
GPR111	*ADGRF2*	Mammals	PAML	[31]
		Chimpanzee	PAML	[32]
BAI1	*ADGRB1*	Chimpanzee	MK	[29]
GPR56	*ADGRG1*	Chimpanzee	MK	[29]
GPR64	*ADGRG2*	Chimpanzee	MK	[29]
GPR126	*ADGRG6*	Chimpanzee	MK	[29]

BAI3/ADGRB3. Together with numerous other differentially expressed genes, they are considered to contribute to human-specific brain functions [33].

Recent advances in DNA extraction and amplification from fossil remains [34] and the new sequencing technologies for short DNA fragments have made it possible to retrieve substantial amounts of ancient DNA sequences and even almost complete genomes dating back close to 1 million years ago [35, 36]. Genome sequences of extinct organisms contain valuable information about past functions and gene evolution also of GPCRs [37, 38]. In a selective sweep analysis of the Neanderthal genome, a genomic locus containing *CELSR3/ADGRC3* was identified presenting signatures of selection in the Neanderthal lineage [39]. However, there are no specific genomic signals that aGPCRs may be involved in evolution of modern humans when compared to ancient hominids [40–44]. In contrast, analyses between different populations yielded a number of aGPCR loci with significant signatures of selection. Here, the clustered aGPCRs *GPR110*, *GPR111*, *GPR115*, and *GPR116* showed multiple signatures of recent selection among different populations (Table 2). Comparison of populations which majorly differ in respect to their environment revealed, for example, that the *GPR111* locus harbors significant signatures of selection in populations adapted to high altitude levels as found in Tibet, Mongolia, and Ethiopia, suggesting convergent evolution [45, 54]. The impact of different *GPR111* variants on this physiological trait is unsolved yet. Although analyses fall short in providing explanations for their unique genomic features and most probably many signals found at those loci are not related to the function of the specific gene product, they are indicative for a significant genomic dynamics at those loci in the recent human history.

Table 2 Signatures of selection at aGPCR loci in human genomes

Gene name	New name	Population	Methods	Reference
LPHN2	*ADGRL2*	Tibetan and Mongolian	iHS	[45]
LPHN3	*ADGRL3*	Oceanian vs. European	F_{ST}	[46]
EMR1	*ADGRE1*	South Asian vs. European	F_{ST}	[46]
EMR3	*ADGRE3*	African vs. Oceanian	F_{ST}	[46]
GPR123	*ADGRA1*	European vs. African	F_{ST}	[47]
GPR124	*ADGRA2*	Siberian	XP-EHH	[48]
CELSR1	*ADGRC1*	Yoruba, Japan	LRH, iHS, XP-EHH	[27]
		HapMap	HH	[49]
		Middle East, South Asian	iHS, XP-EHH	[50]
		Asian vs. African	SNP frequency	[51]
GPR133	*ADGRD1*	HapMap	Entropy of LD	[52]
		South Asian vs. European	F_{ST}	[46]
GPR110	*ADGRF1*	American, East Asian	iHS, XP-EHH	[50]
		East Asian vs. European	F_{ST}	[46]
GPR111	*ADGRF2*	Tibetan, Mongolian	iHS	[45]
		Asian	LRH	[53]
		Ethiopians	F_{ST}	[54]
		European	Neutrality test	[55]
		American, East Asian	iHS, XP-EHH	[50]
GPR113	*ADGRF3*	Americans	iHS	[50]
GPR115	*ADGRF4*	Tibetan, Mongolian	iHS	[45]
		Asian	LRH	[53]
		American, East Asian	iHS, XP-EHH	[50]
		South Asian vs. European	F_{ST}	[46]
GPR116	*ADGRF5*	American, East Asian	iHS, XP-EHH	[50]
		East Asian vs. South Asian	F_{ST}	[46]
BAI3	*ADGRB3*	African	LRH tests	[53]
		HapMap	EHH	[26, 56]
		African vs. European vs. East Asian	F_{ST}	[46]
GPR64	*ADGRG2*	Middle East	iHS	[50]
		European	iHS	[53]
GPR126	*ADGRG6*	European vs. Asian, Bantu vs. Pygmy	F_{ST}	[50]
GPR98	*ADGRV1*	HapMap	HH	[49]
		European, Middle East, South Asia	XP-EHH	[50]

4 Genetic Variability in the Coding Sequence of Human aGPCRs

Deep sequencing data, e.g., from population genetic studies revealed an unexpected variability of genes in humans. Most variants are single nucleotide polymorphisms (SNPs) without obvious impact on gene function. However, some SNP can have functional relevance and can even reduce the repertoire of genes by inactivation when gene function no longer provides a selective advantage. This process is called pseudogenization. Pseudogenes are inheritable and characterized by homology to a known gene and lack of functionality [57]. As a result of their nonfunctionality, most pseudogenes are released from selective pressure and accumulate missense, frameshifting, and nonsense mutations. Pseudogenization by mutations is considered for odorant and taste receptors [58] because of the loss of environmental constraints or the gain of other senses, such as vision. However, loss-of-function mutations, as observed in chemokine receptors, may have also some advantage in respect to pathogens [59, 60]. As listed in Table 3, the gene for the chemokine receptor *CCR5* showed a significant higher number of missense, nonsense, and frameshifting variants in human populations compared to other GPCR genes. Similarly, a great number of variants are found in the melanocortin type 1 receptor (*MC1R*) gene (Table 3), a major regulator of pigmentation in humans and other mammals [61]. Depending on the exposition to sunlight, full function of the MC1R (equatorial region) but also loss-of-function (northern hemisphere) intact can have an advantage, although the MC1R is not essentially required for life. Analysis of the non-synonymous SNP number in aGPCR revealed some receptors with similar variability as *CCR5* and *MC1R*. For example, a relative high number of inactivating mutations are naturally found in *EMR1/ADGRE1* and *EMR2/ADGRE2* (Table 3). In line with these findings, *Emr1*-deficient mice were healthy and fertile, indicating that EMR1 deficiency is most probably compatible with life also in humans [62]. Similarly, the relative high number of nonsense and frameshifting mutations found in *GPR110* and *GPR111* suggests vitality in humans even without these aGPCRs. Indeed, mice lacking *GPR110* or *GPR111* are obviously healthy and fertile [13]. This, however, does not exclude phenotypical differences between receptor-positive and receptor-negative individuals. Quantitative trait locus (QTL) analyses and GWAS may help to identify the physiological impact of the presence, modification, and absence of individual aGPCRs.

5 QTL and GWAS Linking aGPCR to Human Phenotypes

Identification of genetic determinants contributing to susceptibility to diseases is not only crucial for understanding the complexity of human health and disease but can also play an important role in designing novel diagnostic and treatment strategies. The ultimate goal would be the so-called personalized medicine, which tailors treatment to individual patients according to their genetic background. Along with candidate gene strategies focusing on biologically plausible genes, hypothesis-free genome-wide studies are the driving force in identification of genetic variants contributing to the complex etiology of human diseases. Until 2005, most genome-wide studies relied on the technique of genetic linkage. For decades, this approach has been the dominant way in investigating the genetic basis of inherited diseases. Its design is based on searching for genomic regions which carry alleles shared by the affected family members, i.e., identifying those regions

Table 3 Natural mutations in coding regions of human aGPCR

Gene name	New name	Missense	Nonsense	Frameshift	Length (codons)	Sum	Mutations/codon	Inactive/codon
LPHN1	ADGRL1	176	1	1	1475	178	0.121	0.001
LPHN2	ADGRL2	135	1	2	1474	138	0.094	0.002
LPHN3	ADGRL3	158	0	0	1500	158	0.105	0.000
ELTD1	ADGRL4	84	3	0	739	87	0.118	0.004
EMR1	ADGRE1	175	7	9	886	191	**0.216**	**0.018**
EMR2	ADGRE2	175	12	9	823	196	**0.238**	**0.026**
EMR3	ADGRE3	81	4	3	652	88	0.135	**0.011**
CD97	ADGRE5	159	4	2	835	165	0.198	0.007
GPR123	ADGRA1	276	11	7	1280	294	**0.230**	**0.014**
GPR124	ADGRA2	187	3	4	1338	194	0.145	0.005
GPR125	ADGRA3	184	2	2	1321	188	0.142	0.003
CELSR1	ADGRC1	552	2	1	3019	555	0.184	0.001
CELSR2	ADGRC2	395	2	0	2924	397	0.136	0.001
CELSR3	ADGRC3	402	4	7	3312	413	0.125	0.003
GPR133	ADGRD1	170	1	1	874	172	0.197	0.002
GPR144	ADGRD2	95	5	5	963	105	0.109	**0.010**
GPR110	ADGRF1	138	4	6	910	148	0.163	**0.011**
GPR111	ADGRF2	117	8	6	708	131	0.185	**0.020**
GPR113	ADGRF3	156	8	2	1079	166	0.154	0.009
GPR115	ADGRF4	103	2	3	752	108	0.144	0.007
GPR116	ADGRF5	163	1	0	1346	164	0.122	0.001
BAI1	ADGRB1	158	0	1	1584	159	0.100	0.001
BAI2	ADGRB2	170	1	2	1585	173	0.109	0.002
BAI3	ADGRB3	135	2	0	1522	137	0.090	0.001

(continued)

Table 3 (continued)

Gene name	New name	Missense	Nonsense	Frameshift	Length (codons)	Sum	Mutations/codon	Inactive/codon
GPR56	ADGRG1	90	4	1	693	95	0.137	0.007
GPR64	ADGRG2	111	0	1	1017	112	0.110	0.001
GPR97	ADGRG3	101	3	3	549	107	0.195	**0.011**
GPR112	ADGRG4	446	15	11	3080	472	0.153	0.008
GPR114	ADGRG5	124	1	1	528	126	**0.239**	0.004
GPR126	ADGRG6	145	2	1	1250	148	0.118	0.002
GPR128	ADGRG7	126	4	3	797	133	0.167	0.009
GPR98	ADGRV1	860	20	7	6306	887	0.141	0.004
TSHR		104	4	3	764	111	0.145	0.009
CaSR		102	1	0	1078	103	0.096	0.001
V2R		70	1	0	371	71	0.191	0.003
GPR34		31	0	1	381	32	0.084	0.003
GPR82		33	1	4	336	38	0.113	**0.015**
LHCGR		77	1	0	699	78	0.112	0.001
MC4R		44	2	1	332	47	0.142	0.009
MC1R		123	6	6	317	135	**0.426**	**0.038**
HTR1A		44	0	0	422	44	0.104	0.000
CCR5		69	5	8	352	82	**0.233**	**0.037**
GRM1		101	2	0	1194	103	**0.086**	**0.002**
GABBR1		109	3	0	961	112	**0.117**	**0.003**

The numbers of missense, nonsense, and frameshifting mutations in the coding regions of human aGPCRs were taken from dbSNP (www.ncbi.nlm.nih.gov/snp/) (last check: 2015/12/18). Only those variants were counted which were validated by frequency. The total number was referred to the maximum length of the proteins (www.ncbi.nlm.nih.gov/protein/). There was no significant difference ($P = 0.4$) in the variant/length ratio between aGPCR and non-aGPCR (in italics). Genes, one standard deviation above the mean (0.151 ± 0.060), are in bold. The ratio of obviously inactivating mutations (nonsense, frameshift) to the receptor length is given in the right row (although missense mutations can also inactivate a GPCR). All ratios >0.010 are in bold which segregate with the disease. Linkage studies turned out to be most powerful in identification of high-risk disease variants, such as those in *CFTR* (Cystic Fibrosis Transmembrane Conductance Regulator) gene responsible for cystic fibrosis [63]. Although genome-wide linkage analyses were successfully applied for a number of other single-gene Mendelian diseases (e.g., Huntington disease) [64], they did not appear pivotal for common diseases such as type 2 diabetes, cardiovascular diseases, or various cancer forms. However, with advances in high-throughput genotyping and sequencing technologies, GWAS became feasible and made a major contribution toward better understanding the genetics of complex traits and diseases. Based on the common disease-common variant hypothesis, GWAS allows detection of common disease variants with modest disease risks. In contrast to linkage studies, modest-risk allele sharing among unrelated affected patients is more striking than between affected subjects within families [65]. One of the major advantages of GWAS is, therefore, the relatively easy recruitment of unrelated affected individuals and healthy controls. However, since the region flanking the genetic marker in GWAS is much smaller than shared regions in related individuals in families, thousands of markers are required for a sufficient coverage of the genome and, so, successful identification of a potentially interesting genetic variant. With the availability of high-throughput array-based genotyping technologies, millions of SNPs can be determined easily and taken forward for association analyses with the clinical outcome of interest. Indeed, GWAS became the major tool in identification of genetic factors for common diseases such as obesity and type 2 diabetes in the last decade [66, 67]. GWAS did not only provide a proof of principle for many plausible candidate genes but also highlighted novel pathways and mechanisms underlying complex pathophysiologies of various diseases (e.g., *CDKN2A* (cyclin-dependent kinase inhibitor 2A) and *CDKAL1* (CDK5 regulatory subunit associated protein 1-like 1) in regeneration of the beta cell in type 2 diabetes) [66].

In respect to aGPCRs, numerous loci carrying the respective aGPCR genes have been shown to co-segregate with complex traits and diseases. In various family-based analyses, QTLs harboring aGPCRs have been identified for phenotypes of the immune system, skeleton (joint/bone inflammation) [68], renal/urinary system [69], and respiratory system (chronic and reversible airflow obstruction) [70, 71]. Most strikingly, majority of the aGPCR loci have been shown to be linked to traits related to homeostasis/metabolism (body weight/BMI, body fat, lipid levels, glucose levels, leptin) [72–74] and tumorigenesis (mammary and prostate tumor susceptibility) [75–78] (Table 4). Yet, like for most of the linkage studies, the genomic intervals with QTLs are very large to track down the respective causal genes. In this respect, numerous GWAS helped to refine data from linkage analyses and pointed to several interesting aGPCR candidates. For instance, the *LPHN1/ADGRL1* locus on chromosome 19 has previously been linked to obesity, lipid, and glucose levels [72–74] (Table 4). Consistently, recent GWAS identified rs4528684 within the locus to be associated with increased mortality in heart failure [189] (Table 5). Also for the *CELSR2* locus, consistent data have been obtained from both linkage and association studies. Whereas previous linkage analyses pointed to the

Table 4 Human aGPCR genes within QTLs

Receptor		QTL		
Gene name	New name	Chr.	Trait	Reference
LPHN1	*ADGRL1*	19	Urinary albumin excretion	[69]
			Chronic airflow obstruction	[70, 71]
			Multiple sclerosis susceptibility	[79]
			Lipid level	[73, 80–82]
			Body weight	[72, 83]
			Body fat	[84]
			Prostate tumor susceptibility	[76–78, 85]
			Glucose level	[74, 86]
LPHN2 *ELTD1*	*ADGRL2* *ADGRL4*	1	Joint/bone inflammation	[68]
			Chronic airflow obstruction	[87]
			Reversible airflow obstruction	[88]
			Leptin level	[89]
			Apolipoprotein B level	[90]
			Blood pressure	[91]
LPHN3	*ADGRL3*	4	Joint/bone inflammation	[92, 93]
			Chronic airflow obstruction	[70, 71, 94]
			Body fat	[84]
			Body weight	[95, 96]
			Mammary tumor susceptibility	[75]
			Glucose level	[74, 97]
EMR1 *EMR4*	*ADGRE1* *ADGRE4*	19	Urinary albumin excretion	[69]
			Blood pressure	[98]
			Chronic airflow obstruction	[71, 99]
			Multiple sclerosis susceptibility	[79]
			Lipid level	[73, 80–82]
			Birth weight	[83]
			Prostate tumor susceptibility	[76–78, 85]
			Glucose level	[74, 86]
EMR2 *EMR3* *CD97*	*ADGRE2* *ADGRE3* *ADGRE5*	19	Urinary albumin excretion	[69]
			Chronic airflow obstruction	[70, 71]
			Multiple sclerosis susceptibility	[79]
			Lipid level	[73, 80–82]
			Body weight	[72, 83]
			Body fat	[84]
			Prostate tumor susceptibility	[76–78, 85]
			Glucose level	[74, 86, 100]
GPR123	*ADGRA1*	10	Body weight	[101, 102]
			Body fat	[103]
			Prostate tumor susceptibility	[78, 104, 105]
			Glucose level	[74, 86, 106]

(continued)

Table 4 (continued)

Receptor		QTL		
Gene name	New name	Chr.	Trait	Reference
GPR124	*ADGRA2*	8	Blood pressure	[98]
			Body weight	[107]
			Multiple sclerosis susceptibility	[79]
			STNFR hormone level	[89]
			Lipid level	[108]
			Waist to hip ratio	[109]
			Adiponectin level	[110]
			Prostate tumor susceptibility	[111–113]
			Glucose level	[114]
GPR125	*ADGRA3*	4	Blood pressure	[115]
			Body weight/BMI	[116, 117]
			Serum IgE titer	[118]
			LDL particle size	[119]
			Body fat	[84]
			Prostate tumor susceptibility	[112, 120]
			Glucose level	[74, 100]
CELSR1	*ADGRC1*	22	Blood pressure	[98, 121]
			Chronic airflow obstruction	[71, 99]
			Body weight	[109]
			Prostate tumor susceptibility	[78, 105, 112, 113, 122, 123]
			Mammary tumor susceptibility	[124, 125]
CELSR2	*ADGRC2*	1	Blood pressure	[91, 98]
			Chronic airflow obstruction	[70, 71, 87]
			Body weight/BMI	[126, 127]
			Lipid level	[128]
			Prostate tumor susceptibility	[76, 129, 130]
			Glucose level	[86]
CELSR3	*ADGRC3*	3	Body fluid distribution	[131]
			Joint/bone inflammation	[132]
			Body weight/BMI	[133]
			Chronic airflow obstruction	[70, 71]
			Glucose level	[74, 134]
GPR133	*ADGRD1*	12	Joint/bone inflammation	[68]
			Reversible airflow obstruction	[135]
			Apolipoprotein A1 level	[136]
			Myocardial infarction	[137]
			Body weight/BMI	[95, 101]
			Body fat (abdominal)	[109, 138]
			Lipid level	[139]
			Leptin level	[140]
			Prostate tumor susceptibility	[104, 141]
			Glucose level	[142, 143]

(continued)

Table 4 (continued)

Receptor		QTL		
Gene name	New name	Chr.	Trait	Reference
GPR144	*ADGRD2*	9	Joint/bone inflammation	[132, 144]
			Multiple sclerosis	[145]
			Body weight/BMI	[101, 103, 126]
			Lipid level	[73]
			Glucose level	[74, 86, 146, 147]
			Prostate tumor susceptibility	[78, 129, 148]
			Insulin level	[149]
GPR110 *GPR111* *GPR115* *GPR116*	*ADGRF1* *ADGRF2* *ADGRF4* *ADGRF5*	6	Joint/bone inflammation	[93, 132]
			Blood pressure	[115]
			Reversible airflow obstruction	[150]
			Lipid level	[151, 152]
			LDL particle size	[153]
			Body fat	[109, 154]
			Prostate tumor susceptibility	[155]
GPR113	*ADGRF3*	2	Body weight/BMI	[107, 156, 157]
			Leptin level	[158–161]
			Lipid level	[73, 80]
			Glucose level	[86, 142, 146]
			Insulin level	[157]
			Body fat	[157]
			Prostate tumor susceptibility	[123]
BAI1	*ADGRB1*	8	Joint/bone inflammation	[132]
			Body weight/BMI	[131]
			Blood pressure	[162]
			Chronic airflow obstruction	[70]
			Body fat	[96]
			Waist to hip ratio	[163]
			Prostate tumor susceptibility	[78, 123]
			Glucose level	[164]
BAI2	*ADGRB2*	1	Body weight/BMI	[165]
			Multiple sclerosis	[145]
			Myocardial infarction	[137]
			Lipid level	[81]
			Blood pressure	[91]
			Body fat (abdominal)	[109]
			Ghrelin level	[166]
			Prostate tumor susceptibility	[78, 105, 123, 129, 141, 148, 155]
			Glucose level	[142, 167]

(continued)

Table 4 (continued)

Receptor		QTL		
Gene name	New name	Chr.	Trait	Reference
BAI3	*ADGRB3*	6	Joint/bone inflammation	[93, 168]
			Blood pressure	[115]
			Lipid level	[151]
			Body fat	[109]
			Body weight	[83]
			Glucose level	[74]
GPR56 *GPR97* *GPR114*	*ADGRG1* *ADGRG3* *ADGRG5*	16	Body weight/BMI	[72, 101, 140, 169]
			Lipid level	[170, 171]
			Leptin level	[140]
			Body fat	[140]
			Prostate tumor susceptibility	[78, 105, 112]
			Glucose level	[172]
GPR64	*ADGRG2*	X	Body weight	[117]
			Prostate tumor susceptibility	[78, 104, 105]
GPR112	*ADGRG4*	X	Prostate tumor susceptibility	[77, 111, 112, 129, 173, 174]
			Glucose level	[175]
GPR126	*ADGRG6*	6	LDL particle size	[176]
			Lipid level	[177–179]
			Body weight/BMI	[72, 179, 180]
			Waist	[140]
			Waist to hip ratio	[109]
			Glucose level	[146, 181–184]
			Leptin level	[140]
			Insulin level	[182]
GPR128	*ADGRG7*	3	Blood pressure	[98]
			Chronic airflow obstruction	[70, 71]
			Coronary artery disease	[185]
			Lipid level	[170]
			Adiponectin level	[110]
			Body weight/BMI	[186]
			Prostate tumor susceptibility	[141]
GPR98	*ADGRV1*	5	Joint/bone inflammation	[132]
			Reversible airflow obstruction	[88]
			Body weight/BMI	[84, 101, 126, 180, 187, 188]
			Apolipoprotein A-II level	[136]
			LDL particle size	[119]
			Prostate tumor susceptibility	[76, 123]
			Myocardial infarction	[137]

Table 5 Association of genetic variants in GPCRs identified in GWAS

Receptor		Association of genetic variants in GWAS			
Gene name	New name	SNP	Associated trait	*P*-value	Reference
LPHN1	*ADGRL1*	rs4528684-T	Mortality in heart failure	1×10^{-6}	[189]
LPHN2	*ADGRL2*	rs17498581-T	Toenail selenium levels	9×10^{-6}	[191]
		rs11163372-T	Temperament	3×10^{-6}	[192]
		rs6681544-C	Metabolite levels	6×10^{-6}	[193]
		rs1770678-G	TGF-β1	7×10^{-6}	[194]
		rs480745	Preterm birth	7×10^{-7}	[195]
		–	Climate	<0.05	[196]
LPHN3	*ADGRL3*	rs6856328	Response to antipsychotic treatment in schizophrenia	8×10^{-7}	[197]
			Partial epilepsies	3×10^{-6}	[198]
		rs2172802	HDL cholesterol	2×10^{-6}	[199]
		rs4599440-A	Subsistence	<0.05	[196]
		–	Climate	<0.05	[196]
ELTD1	*ADGRL4*	rs11162963-T	Aging (time to event)	4×10^{-6}	[200]
		rs10873998-T	Autism spectrum disorder, attention deficit-hyperactivity disorder, bipolar disorder, major depressive disorder, and schizophrenia (combined)	4×10^{-6}	[201]
EMR1	*ADGRE1*	rs3826782	Periodontitis	8×10^{-7}	[202]
EMR2	*ADGRE2*	–	Subsistence	<0.05	[196]
CD97	*ADGRE5*	rs4528684-T	Mortality in heart failure	1×10^{-6}	[189]
GPR124	*ADGRA2*	rs12676965-T	Axial length	3×10^{-3}	[203]
GPR125	*ADGRA3*	rs4697263 rs2141322	Age-related hearing impairment	9×10^{-9}	[204]
		rs6448119	Rheumatoid arthritis	7×10^{-6}	[205]
CELSR1	*ADGRC1*	rs5767218-C	Classic bladder exstrophy	3×10^{-6}	[206]
		rs9615362	Ischemic stroke	<0.05	[207]
CELSR2	*ADGRC2*	rs646776-G	Progranulin levels	2×10^{-30}	[208]
		rs646776-C	Response to statin therapy	4×10^{-6}	[209]
		rs629301-G	Total cholesterol	2×10^{-170}	[210, 211]
		rs646776-G	Total cholesterol	9×10^{-22}	[212]
		rs3902354-A	Total cholesterol	1×10^{-8}	[213]
		rs646776-C	LDL cholesterol	3×10^{-29}	[214]
		rs660240-A	LDL cholesterol	1×10^{-26}	[215]
		rs646776-G	LDL cholesterol	8×10^{-23}	[212, 216]
		rs12740374-T	LDL cholesterol	2×10^{-42}	[217]
		rs629301-G	LDL cholesterol	1×10^{-170}	[211]

(continued)

Table 5 (continued)

Receptor		Association of genetic variants in GWAS			
Gene name	New name	SNP	Associated trait	*P*-value	Reference
		rs646776	LDL cholesterol (fasting/whole)	2×10^{-20}	[218]
		rs660240-T	LDL cholesterol	2×10^{-22}	[219]
		rs12740374-T	LDL cholesterol	9×10^{-29}	[220]
		rs599839-G	LDL cholesterol	1×10^{-33}	[221]
		rs629301-G	LDL cholesterol	5×10^{-241}	[210]
		rs599839-A	LDL cholesterol	6×10^{-33}	[222]
		rs599839-G	LDL cholesterol	1×10^{-7}	[223, 224]
		rs602633-T	Coronary artery disease	1×10^{-8}	[225]
		rs646776-T	Coronary artery disease	8×10^{-11}	[226, 227]
		rs646776-T	Myocardial infarction (early onset)	8×10^{-12}	[228]
		rs7528419-A	Lipoprotein-associated phospholipase A2 activity and mass	1×10^{-17}	[229]
		rs12740374-T	Lipoprotein-associated phospholipase A2 activity and mass	2×10^{-22}	[230]
		rs646776	APOB (whole/fasting)	2×10^{-53}	[218]
GPR133	*ADGRD1*	rs1569019	Height	5×10^{-8}	[231]
		rs3847687	Longevity	$\leq 1 \times 10^{-6}$	[232]
		rs10466868	Erythropoietin	1×10^{-6}	[233]
		rs885389-A	RR interval (heart rate)	4×10^{-8}	[234]
		rs11061269	Amyotrophic lateral sclerosis (sporadic)	8×10^{-10}	[235]
		–	Pathogens	<0.05	[196]
		–	Subsistence	<0.05	[196]
		–	Climate	<0.05	[196]
GPR110	*ADGRF1*	–	Climate	<0.05	-
GPR111	*ADGRF2*	–	Subsistence	<0.05	[196]
		–	Climate	<0.05	[196]
GPR113	*ADGRF3*	rs6753473-G	Non-small cell lung cancer	4×10^{-6}	[236]
GPR116	*ADGRF5*	–	Climate	<0.05	[196]
BAI3	*ADGRB3*	rs2585617-A	Phospholipid levels (EPA)	6×10^{-7}	[237]
		rs3757057-T	Response to statin therapy (triglyceride sum)	4×10^{-6}	[209]
		rs9351730	Prostate cancer	5×10^{-6}	[238]
		rs9364554			
		rs9363918-A	Pancreatic cancer	1×10^{-6}	[239]
		rs78536982-T	Triglycerides	7×10^{-9}	[213]
		–	Pathogens	<0.05	[196]

(continued)

Table 5 (continued)

Receptor		Association of genetic variants in GWAS			
Gene name	New name	SNP	Associated trait	*P*-value	Reference
GPR126	*ADGRG6*	rs6570507	Scoliosis	1×10^{-14}	[240]
		rs6570507-A	Height	2×10^{-11}	[241]
		rs3748069-A	Height	5×10^{-14}	[242]
		rs4896582-A	Height	2×10^{-18}	[243]
		rs6570507-G	Height	2×10^{-7}	[244]
		rs6570507	Height	4×10^{-11}	[245]
		rs7763064-A	Height	1×10^{-33}	[246]
		rs225694	Height	1×10^{-7}	[246]
		rs7741741-T	Height	1×10^{-20}	[247]
		rs4896582-A	Height	3×10^{-55}	[248]
		rs3817928-A	Pulmonary function	1×10^{-9}	[249]
		rs3817928	Pulmonary function	3×10^{-12}	[250]
		rs1329705	Airflow obstruction	3×10^{-6}	[251]
		rs225675-G	Thiazide-induced adverse metabolic effects in hypertensive patients	1×10^{-7}	[252]
		rs7763064-A	Birth length	6×10^{-6}	[253]
GPR128	*ADGRG7*	–	Subsistence	<0.05	[196]
		–	Climate	<0.05	[196]
GPR98	*ADGRV1*	rs10514345	Hip geometry (Neck Z1)	2×10^{-7}	[254]
		rs1967256	Response to antipsychotic treatment	3×10^{-8}	[255]
		rs10074525-G	Obesity-related traits (weight z-score)	9×10^{-6}	[194]
		–	Pathogens	<0.05	[196]
		–	Climate	<0.05	[196]

GWAS signals according to the GWAS Catalog [190]

segregation of this locus with lipid levels and blood pressure- and obesity-related traits [98, 126, 128], it was the GWAS which ultimately identified polymorphisms within the *CELSR2* gene which were significantly associated with total and LDL cholesterol, clinical outcomes of coronary artery disease, and myocardial infarction but also response to statin therapy [209, 225, 228] (Table 5). Worth mentioning are also further traits shown to be associated with SNPs in aGPCR genes. In particular, genetic variants in *GPR123* and *GPR133/ADGRD1* have been shown to be associated with height [231, 246] and SNPs in *GPR125/ADGRA3* with rheumatoid arthritis and age-related impaired hearing [204, 205] (Table 5).

Despite the rapid and tremendous progress in the field of genetics of complex diseases by recovering hundreds of genetic variants in GWAS for numerous traits and diseases, the biological impact of these associated variants remains largely unknown. As mentioned above, common genetic variants such as SNPs are in strong contrast to rare high-risk alleles responsible for Mendelian genetic disorders such as cystic fibrosis [256]. Most of the identified SNPs associated with the trait of interest map to inter- or intragenic regions, thus suggesting their regulatory role in gene transcription and function. Therefore, one of the crucial steps following GWAS is to identify and validate target genes potentially affected by the associated SNPs. eQTL (expression QTL) analyses allow testing for correlations between genotype and tissue-specific gene expression levels and, so, help to identify chromosomal regions that influence whether and, if so, how much a gene is expressed. Thus, integrative approach combining GWAS and eQTL studies allows clarifying whether the genotype-phenotype association uncovered in a GWAS might be mediated by the regulation of a gene transcription by the corresponding SNP. According to the GTEx (Genotype-Tissue Expression) portal, a unique resource to study human gene expression and its relation to genetic variation [257], SNPs within *CELSR2* locus (e.g., rs660240) associated with lipid traits in recent GWAS [215] (Table 5), has also been shown to correlate with *CELSR2* mRNA expression in liver, skeletal muscle, and esophagus mucosa. Using GTEx, *GPR126/ADGRG6* SNPs (e.g., rs6570507, rs4896582, rs7763064) related to height in GWAS [241, 243–246] (Table 5) are associated with *GPR126* mRNA expression in subcutaneous adipose tissue. These findings strongly support the GWAS data and the potential functional consequences of the associated variants.

6 Monogenic Diseases Caused by Mutations in aGPCR

As shown in Table 5, there are numerous pathological traits in humans associated to genomic variations in aGPCR loci. *GPR133* was among the first aGPCRs identified in GWAS where SNPs were associated with human phenotypes [231, 258]. Usually, the SNPs linked or associated with phenotypes are not causal for the phenotypical trait but rather genetically linked (linkage disequilibrium, LD) to additional, likely causal variants at such loci. Thus, locus sequencing is required to identify those variants which are in LD with the indicative SNP (tag SNP). Database analysis of more than 1000 sequenced human genomes revealed over 5300 SNPs in the human *GPR133* gene [259]. Out of those 5 % SNPs are located in exons leading to 170 amino acid changes (Table 3). Compared to other GPCR genes, the number of non-synonymous SNPs/1000 bp of the coding region is not significantly different in the *GPR133* gene [260]. Interestingly, some non-synonymous SNP showed a functional relevance including complete loss of function when experimentally tested [259]. However, a direct causal link between individual SNPs and differences in the phenotypes linked to the *GPR133* locus needs to be provided.

Sequence variants can modulate GPCR activity in two directions—activation (gain-of-function) and inactivation (loss-of-function). In humans activating and inactivating mutations in GPCR can cause diseases with contrary phenotypes but also convergent phenotypes. For example, the rhodopsin-like GPCR V2 vasopressin receptor (V2R) is a key component in regulating renal water reabsorption.

Inactivating mutations of the V2R cause reduced water reabsorption (congenital nephrogenic diabetes insipidus), whereas activating mutations cause increased water reabsorption with serum hyposmolarity and high urinary sodium levels (nephrogenic syndrome of inappropriate antidiuresis) [261]. In contrast activating and inactivating mutations of rhodopsin cause retinitis pigmentosa [262]. Disease-causing mutations are rather rare in humans, and only 30–40 monogenic diseases caused by mutations in GPCR have been described so far [263–266].

Database entries from large-scale sequencing programs show that functionally relevant aGPCR variants exist in recent human populations (Table 3), and some are responsible for human diseases. There are several inherited diseases caused by loss-of-function mutations in humans. Inactivating mutations in *CELSR1/ADGRC1* have been found responsible for craniorachischisis and further neural tube defects [267] and hereditary lymphedema [268]. One inherited form of bilateral frontoparietal polymicrogyria is caused by inactivating mutations in the *GPR56/ADGRG1* gene [269] (see [270]) and the Usher syndrome type IIC by mutation in the *GPR98/ADGRV1* gene [271] (see [272]). Recently, the lethal congenital contracture syndrome-9 and arthrogryposis multiplex congenital have been linked to inactivation mutation in *GPR126*; however, final confirmation is pending [273]. Vibratory urticaria has been linked to a missense mutation in EMR2 [274] (see [275–277]). No disease-causing activating mutations have been described for aGPCR yet, but this is most probably just a matter of time.

7 Summary and Future Perspectives

Large-scale sequencing projects have provided a wealth of information to the aGPCR field in respect to evolution and sequence variability in humans and another species. Now, this information can be mined for studying the impact of aGPCR function on speciation, phenotypes, and susceptibility to diseases. Phenotypic and environmental comparisons of species which differ in the number of aGPCRs may provide clues for the loss or gain of individual aGPCRs. We are just at the beginning of utilizing this source to generate hypotheses which then can be tested in suitable model systems. For example, QTL studies suggest the involvement of CELSR2 and CELSR3 in the regulation of glucose homeostasis (Table 4). Although both aGPCRs were initially identified as highly relevant for planar cell polarity signaling in ependymal cilia development [278], detailed analyses revealed that loss of function of *CELSR2* and *CELSR3* alters differentiation of endocrine cells from polarized pancreatic progenitors, with a prevalent effect on insulin-producing β-cells and, therefore, glucose tolerance [279]. Numerous GWAS suggest that CELSR2 function is required for cholesterol homeostasis (see Table 5). However, sortilin, encoded by the *SORT1* gene which is in close genomic proximity to *CELSR2*, is an intracellular sorting receptor for apolipoprotein B100 and regulates plasma low-density lipoprotein (LDL) cholesterol [280]. It is therefore likely that functionally relevant SNPs in *SORT1* but not in *CELSR2* account for the strong GWAS signals. This example illustrates the need for experimental testing sequence data-derived hypothesis in the future. On the other hand, the field will majorly profit from an iterative process of linking-defined phenotypes to genomic variants and detailed characterizing gene-deficient animal models in order to unveil the functional relevance of the individual aGPCR in specific tissues. As an ultimate result, new human syndromes will be discovered taking advantage of well-defined sub-phenotypes identified in animal models. But not only inherited human diseases should be in the focus. There is mounting evidence that acquired diseases such as tumors harbor functionally relevant aGPCR variants. For example, melanomas and other skin cancers carry mutated aGPCR genes such as *GPR116*, *BAI3*, and *VLGR1/ ADGRV1* [281, 282]. Also, changes in expression levels of different aGPCRs are highly associated with specific tumors and may serve as prognostic markers [283, 284]. It will be a future task to clarify whether such aGPCR variants promote cancer proliferation and invasion.

Acknowledgments P.K. is supported by the Federal Ministry of Education and Research (BMBF), Germany, FKZ, 01EO1501 (AD2-060E), and by the German Research Foundation (CRC 1052; B03). T.S. is supported by the German Research Foundation (FOR 2149, P4, P5).

References

1. Genomes Project C, Auton A, Brooks LD, Durbin RM, Garrison EP, Kang HM et al (2015) A global reference for human genetic variation. Nature 526(7571):68–74. doi:10.1038/nature15393
2. Genome KCoS (2009) Genome 10K: a proposal to obtain whole-genome sequence for 10,000 vertebrate species. J Hered 100(6):659–674. doi:10.1093/jhered/esp086, PubMed PMID: 19892720; PubMed Central PMCID: PMCPMC2877544
3. Koepfli KP, Paten B, Genome KCoS, O'Brien SJ (2015) The Genome 10K Project: a way forward. Annu Rev Anim Biosci 3:57–111. doi:10.1146/annurev-animal-090414-014900
4. Jarvis ED, Mirarab S, Aberer AJ, Li B, Houde P, Li C et al (2014) Whole-genome analyses resolve early branches in the tree of life of modern birds. Science 346(6215):1320–1331. doi:10.1126/science.1253451, PubMed PMID: 25504713; PubMed Central PMCID: PMCPMC4405904
5. Pickrell JK, Reich D (2014) Toward a new history and geography of human genes informed by ancient DNA. Trends Genet 30(9):377–389. doi:10.1016/j.tig.2014.07.007, PubMed PMID: 25168683; PubMed Central PMCID: PMCPMC4163019
6. Le Duc D, Renaud G, Krishnan A, Almen MS, Huynen L, Prohaska SJ et al (2015) Kiwi genome provides insights into evolution of a nocturnal lifestyle. Genome Biol 16:147. doi:10.1186/s13059-015-0711-4, PubMed PMID: 26201466; PubMed Central PMCID: PMCPMC4511969
7. de Mendoza A, Sebe-Pedros A, Ruiz-Trillo I (2014) The evolution of the GPCR signaling system in eukaryotes: modularity, conservation, and the transition to metazoan multicellularity. Genome Biol Evol 6(3):606–619. doi:10.1093/gbe/evu038, PubMed PMID: 24567306; PubMed Central PMCID: PMCPMC3971589
8. Krishnan A, Almen MS, Fredriksson R, Schioth HB (2012) The origin of GPCRs: identification of mammalian like Rhodopsin, Adhesion, Glutamate and Frizzled GPCRs in fungi. PLoS One 7(1), e29817. doi:10.1371/journal.pone.0029817, PubMed PMID: 22238661; PubMed Central PMCID: PMCPMC3251606

9. Krishnan A, Schioth HB (2015) The role of G protein-coupled receptors in the early evolution of neurotransmission and the nervous system. J Exp Biol 218(Pt 4):562–571. doi:10.1242/jeb.110312
10. Nordstrom KJ, Lagerstrom MC, Waller LM, Fredriksson R, Schioth HB (2009) The Secretin GPCRs descended from the family of Adhesion GPCRs. Mol Biol Evol 26(1):71–84
11. Schioth HB, Nordstrom KJ, Fredriksson R (2010) The adhesion GPCRs; gene repertoire, phylogeny and evolution. Adv Exp Med Biol 706:1–13
12. Hamann J, Kwakkenbos MJ, de Jong EC, Heus H, Olsen AS, van Lier RA (2003) Inactivation of the EGF-TM7 receptor EMR4 after the Pan-Homo divergence. Eur J Immunol 33 (5):1365–1371. doi:10.1002/eji.200323881
13. Promel S, Waller-Evans H, Dixon J, Zahn D, Colledge WH, Doran J et al (2012) Characterization and functional study of a cluster of four highly conserved orphan adhesion-GPCR in mouse. Dev Dyn 241(10):1591–1602. doi:10.1002/dvdy.23841, Epub 2012/07/28
14. Lachance J, Tishkoff SA (2013) Population genomics of human adaptation. Annu Rev Ecol Evol Syst 44:123–143. doi:10.1146/annurev-ecolsys-110512-135833, PubMed PMID: 25383060; PubMed Central PMCID: PMCPMC4221232
15. Bersaglieri T, Sabeti PC, Patterson N, Vanderploeg T, Schaffner SF, Drake JA et al (2004) Genetic signatures of strong recent positive selection at the lactase gene. Am J Hum Genet 74 (6):1111–1120. doi:10.1086/421051, PubMed PMID: 15114531; PubMed Central PMCID: PMCPMC1182075
16. Vitti JJ, Grossman SR, Sabeti PC (2013) Detecting natural selection in genomic data. Annu Rev Genet 47:97–120. doi:10.1146/annurev-genet-111212-133526
17. Scheinfeldt LB, Tishkoff SA (2013) Recent human adaptation: genomic approaches, interpretation and insights. Nat Rev Genet 14(10):692–702. doi:10.1038/nrg3604
18. Yang Z (1997) PAML: a program package for phylogenetic analysis by maximum likelihood. Comput Appl Biosci 13(5):555–556
19. Yang Z (2007) PAML 4: phylogenetic analysis by maximum likelihood. Mol Biol Evol 24 (8):1586–1591
20. McDonald JH, Kreitman M (1991) Adaptive protein evolution at the Adh locus in Drosophila. Nature 351(6328):652–654. doi:10.1038/351652a0
21. Tajima F (1989) Statistical method for testing the neutral mutation hypothesis by DNA polymorphism. Genetics 123(3):585–595
22. Akey JM, Zhang G, Zhang K, Jin L, Shriver MD (2002) Interrogating a high-density SNP map for signatures of natural selection. Genome Res 12(12):1805–1814. doi:10.1101/gr.631202, PubMed PMID: 12466284; PubMed Central PMCID: PMCPMC187574
23. Sabeti PC, Schaffner SF, Fry B, Lohmueller J, Varilly P, Shamovsky O et al (2006) Positive natural selection in the human lineage. Science 312(5780):1614–1620
24. Sabeti PC, Reich DE, Higgins JM, Levine HZ, Richter DJ, Schaffner SF et al (2002) Detecting recent positive selection in the human genome from haplotype structure. Nature 419(6909):832–837. doi:10.1038/nature01140
25. Voight BF, Kudaravalli S, Wen X, Pritchard JK (2006) A map of recent positive selection in the human genome. PLoS Biol 4(3), e72
26. Tang K, Thornton KR, Stoneking M (2007) A new approach for using genome scans to detect recent positive selection in the human genome. PLoS Biol 5(7), e171
27. Sabeti PC, Varilly P, Fry B, Lohmueller J, Hostetter E, Cotsapas C et al (2007) Genome-wide detection and characterization of positive selection in human populations. Nature 449 (7164):913–918
28. Parker J, Tsagkogeorga G, Cotton JA, Liu Y, Provero P, Stupka E et al (2013) Genome-wide signatures of convergent evolution in echolocating mammals. Nature 502(7470):228–231. doi:10.1038/nature12511, PubMed PMID: 24005325; PubMed Central PMCID: PMCPMC3836225
29. Bustamante CD, Fledel-Alon A, Williamson S, Nielsen R, Hubisz MT, Glanowski S et al (2005) Natural selection on protein-coding genes in the human genome. Nature 437 (7062):1153–1157

30. Gaya-Vidal M, Alba MM (2014) Uncovering adaptive evolution in the human lineage. BMC Genomics 15:599. doi:10.1186/1471-2164-15-599, PubMed PMID: 25030307; PubMed Central PMCID: PMCPMC4124166
31. Kosiol C, Vinar T, da Fonseca RR, Hubisz MJ, Bustamante CD, Nielsen R et al (2008) Patterns of positive selection in six Mammalian genomes. PLoS Genet 4(8), e1000144. doi:10.1371/journal.pgen.1000144, PubMed PMID: 18670650; PubMed Central PMCID: PMCPMC2483296
32. Bakewell MA, Shi P, Zhang J (2007) More genes underwent positive selection in chimpanzee evolution than in human evolution. Proc Natl Acad Sci U S A 104(18):7489–7494. doi:10.1073/pnas.0701705104, PubMed PMID: 17449636; PubMed Central PMCID: PMCPMC1863478
33. Liu X, Somel M, Tang L, Yan Z, Jiang X, Guo S et al (2012) Extension of cortical synaptic development distinguishes humans from chimpanzees and macaques. Genome Res 22 (4):611–622. doi:10.1101/gr.127324.111, PubMed PMID: 22300767; PubMed Central PMCID: PMCPMC3317144
34. Rompler H, Dear PH, Krause J, Meyer M, Rohland N, Schoneberg T et al (2006) Multiplex amplification of ancient DNA. Nat Protoc 1(2):720–728
35. Miller W, Drautz DI, Ratan A, Pusey B, Qi J, Lesk AM et al (2008) Sequencing the nuclear genome of the extinct woolly mammoth. Nature 456(7220):387–390
36. Willerslev E, Cappellini E, Boomsma W, Nielsen R, Hebsgaard MB, Brand TB et al (2007) Ancient biomolecules from deep ice cores reveal a forested southern Greenland. Science 317 (5834):111–114
37. Lalueza-Fox C, Rompler H, Caramelli D, Staubert C, Catalano G, Hughes D et al (2007) A melanocortin 1 receptor allele suggests varying pigmentation among Neanderthals. Science 318(5855):1453–1455
38. Rompler H, Rohland N, Lalueza-Fox C, Willerslev E, Kuznetsova T, Rabeder G et al (2006) Nuclear gene indicates coat-color polymorphism in mammoths. Science 313(5783):62
39. Green RE, Krause J, Briggs AW, Maricic T, Stenzel U, Kircher M et al (2010) A draft sequence of the Neandertal genome. Science 328(5979):710–722. doi:10.1126/science.1188021
40. Prufer K, Racimo F, Patterson N, Jay F, Sankararaman S, Sawyer S et al (2014) The complete genome sequence of a Neanderthal from the Altai Mountains. Nature 505(7481):43–49. doi:10.1038/nature12886, PubMed PMID: 24352235; PubMed Central PMCID: PMCPMC4031459
41. Khrameeva EE, Bozek K, He L, Yan Z, Jiang X, Wei Y et al (2014) Neanderthal ancestry drives evolution of lipid catabolism in contemporary Europeans. Nat Commun 5:3584. doi:10.1038/ncomms4584, PubMed PMID: 24690587; PubMed Central PMCID: PMCPMC3988804
42. Castellano S, Parra G, Sanchez-Quinto FA, Racimo F, Kuhlwilm M, Kircher M et al (2014) Patterns of coding variation in the complete exomes of three Neandertals. Proc Natl Acad Sci U S A 111(18):6666–6671. doi:10.1073/pnas.1405138111, PubMed PMID: 24753607; PubMed Central PMCID: PMCPMC4020111
43. Meyer M, Kircher M, Gansauge MT, Li H, Racimo F, Mallick S et al (2012) A high-coverage genome sequence from an archaic Denisovan individual. Science 338(6104):222–226. doi:10.1126/science.1224344, PubMed PMID: 22936568; PubMed Central PMCID: PMCPMC3617501
44. Burbano HA, Hodges E, Green RE, Briggs AW, Krause J, Meyer M et al (2010) Targeted investigation of the Neandertal genome by array-based sequence capture. Science 328 (5979):723–725. doi:10.1126/science.1188046, PubMed PMID: 20448179; PubMed Central PMCID: PMCPMC3140021
45. Xing J, Wuren T, Simonson TS, Watkins WS, Witherspoon DJ, Wu W et al (2013) Genomic analysis of natural selection and phenotypic variation in high-altitude mongolians. PLoS

Genet 9(7), e1003634. doi:10.1371/journal.pgen.1003634, PubMed PMID: 23874230; PubMed Central PMCID: PMCPMC3715426
46. Sudmant PH, Mallick S, Nelson BJ, Hormozdiari F, Krumm N, Huddleston J et al (2015) Global diversity, population stratification, and selection of human copy-number variation. Science 349(6253):aab3761. doi:10.1126/science.aab3761, PubMed PMID: 26249230; PubMed Central PMCID: PMCPMC4568308
47. Oleksyk TK, Zhao K, De La Vega FM, Gilbert DA, O'Brien SJ, Smith MW (2008) Identifying selected regions from heterozygosity and divergence using a light-coverage genomic dataset from two human populations. PLoS One 3(3), e1712. doi:10.1371/journal.pone.0001712, PubMed PMID: 18320033; PubMed Central PMCID: PMCPMC2248624
48. Cardona A, Pagani L, Antao T, Lawson DJ, Eichstaedt CA, Yngvadottir B et al (2014) Genome-wide analysis of cold adaptation in indigenous Siberian populations. PLoS One 9(5), e98076. doi:10.1371/journal.pone.0098076, PubMed PMID: 24847810; PubMed Central PMCID: PMCPMC4029955
49. Kimura R, Fujimoto A, Tokunaga K, Ohashi J (2007) A practical genome scan for population-specific strong selective sweeps that have reached fixation. PLoS One 2(3), e286. doi:10.1371/journal.pone.0000286, PubMed PMID: 17356696; PubMed Central PMCID: PMCPMC1805687
50. Pickrell JK, Coop G, Novembre J, Kudaravalli S, Li JZ, Absher D et al (2009) Signals of recent positive selection in a worldwide sample of human populations. Genome Res 19 (5):826–837. doi:10.1101/gr.087577.108, PubMed PMID: 19307593; PubMed Central PMCID: PMCPMC2675971
51. Genomes Project C, Abecasis GR, Altshuler D, Auton A, Brooks LD, Durbin RM et al (2010) A map of human genome variation from population-scale sequencing. Nature 467 (7319):1061–1073. doi:10.1038/nature09534, PubMed PMID: 20981092; PubMed Central PMCID: PMCPMC3042601
52. Mizuno H, Atwal G, Wang H, Levine AJ, Vazquez A (2010) Fine-scale detection of population-specific linkage disequilibrium using haplotype entropy in the human genome. BMC Genet 11:27. doi:10.1186/1471-2156-11-27, PubMed PMID: 20416085; PubMed Central PMCID: PMCPMC2873552
53. International HapMap C, Frazer KA, Ballinger DG, Cox DR, Hinds DA, Stuve LL et al (2007) A second generation human haplotype map of over 3.1 million SNPs. Nature 449(7164):851–861. doi:10.1038/nature06258, PubMed PMID: 17943122; PubMed Central PMCID: PMCPMC2689609
54. Huerta-Sanchez E, Degiorgio M, Pagani L, Tarekegn A, Ekong R, Antao T et al (2013) Genetic signatures reveal high-altitude adaptation in a set of ethiopian populations. Mol Biol Evol 30(8):1877–1888. doi:10.1093/molbev/mst089, PubMed PMID: 23666210; PubMed Central PMCID: PMCPMC3708501
55. Andres AM, Hubisz MJ, Indap A, Torgerson DG, Degenhardt JD, Boyko AR et al (2009) Targets of balancing selection in the human genome. Mol Biol Evol 26(12):2755–2764. doi:10.1093/molbev/msp190, PubMed PMID: 19713326; PubMed Central PMCID: PMCPMC2782326
56. Lopez Herraez D, Bauchet M, Tang K, Theunert C, Pugach I, Li J et al (2009) Genetic variation and recent positive selection in worldwide human populations: evidence from nearly 1 million SNPs. PLoS One 4(11), e7888. doi:10.1371/journal.pone.0007888, PubMed PMID: 19924308; PubMed Central PMCID: PMCPMC2775638
57. Balakirev ES, Ayala FJ (2003) Pseudogenes: are they "junk" or functional DNA? Annu Rev Genet 37:123–151
58. Jiang P, Josue J, Li X, Glaser D, Li W, Brand JG et al (2012) Major taste loss in carnivorous mammals. Proc Natl Acad Sci U S A 109(13):4956–4961. doi:10.1073/pnas.1118360109, PubMed PMID: 22411809, PubMed Central PMCID: PMC3324019, Epub 2012/03/141118360109 [pii]

59. Rompler H, Schulz A, Pitra C, Coop G, Przeworski M, Paabo S et al (2005) The rise and fall of the chemoattractant receptor GPR33. J Biol Chem 280(35):31068–31075
60. Samson M, Libert F, Doranz BJ, Rucker J, Liesnard C, Farber CM et al (1996) Resistance to HIV-1 infection in caucasian individuals bearing mutant alleles of the CCR-5 chemokine receptor gene. Nature 382(6593):722–725
61. Hofreiter M, Schoneberg T (2010) The genetic and evolutionary basis of colour variation in vertebrates. Cell Mol Life Sci 67(15):2591–2603. doi:10.1007/s00018-010-0333-7
62. Schaller E, Macfarlane AJ, Rupec RA, Gordon S, McKnight AJ, Pfeffer K (2002) Inactivation of the F4/80 glycoprotein in the mouse germ line. Mol Cell Biol 22(22):8035–8043, PubMed PMID: 12391169; PubMed Central PMCID: PMCPMC134735
63. Kerem B, Rommens JM, Buchanan JA, Markiewicz D, Cox TK, Chakravarti A et al (1989) Identification of the cystic fibrosis gene: genetic analysis. Science 245(4922):1073–1080
64. MacDonald ME, Novelletto A, Lin C, Tagle D, Barnes G, Bates G et al (1992) The Huntington's disease candidate region exhibits many different haplotypes. Nat Genet 1 (2):99–103. doi:10.1038/ng0592-99
65. Carlson CS, Eberle MA, Kruglyak L, Nickerson DA (2004) Mapping complex disease loci in whole-genome association studies. Nature 429(6990):446–452. doi:10.1038/nature02623
66. Zeggini E, Weedon MN, Lindgren CM, Frayling TM, Elliott KS, Lango H et al (2007) Replication of genome-wide association signals in UK samples reveals risk loci for type 2 diabetes. Science 316(5829):1336–1341. doi:10.1126/science.1142364, PubMed PMID: 17463249; PubMed Central PMCID: PMCPMC3772310
67. Frayling TM, Timpson NJ, Weedon MN, Zeggini E, Freathy RM, Lindgren CM et al (2007) A common variant in the FTO gene is associated with body mass index and predisposes to childhood and adult obesity. Science 316(5826):889–894
68. Demissie S, Cupples LA, Myers R, Aliabadi P, Levy D, Felson DT (2002) Genome scan for quantity of hand osteoarthritis: the Framingham Study. Arthritis Rheum 46(4):946–952
69. Freedman BI, Beck SR, Rich SS, Heiss G, Lewis CE, Turner S et al (2003) A genome-wide scan for urinary albumin excretion in hypertensive families. Hypertension 42(3):291–296. doi:10.1161/01.HYP.0000087890.33245.41
70. Palmer LJ, Celedon JC, Chapman HA, Speizer FE, Weiss ST, Silverman EK (2003) Genome-wide linkage analysis of bronchodilator responsiveness and post-bronchodilator spirometric phenotypes in chronic obstructive pulmonary disease. Hum Mol Genet 12(10):1199–1210
71. Spira A, Beane J, Pinto-Plata V, Kadar A, Liu G, Shah V et al (2004) Gene expression profiling of human lung tissue from smokers with severe emphysema. Am J Respir Cell Mol Biol 31(6):601–610. doi:10.1165/rcmb.2004-0273OC
72. Meyre D, Lecoeur C, Delplanque J, Francke S, Vatin V, Durand E et al (2004) A genome-wide scan for childhood obesity-associated traits in French families shows significant linkage on chromosome 6q22.31-q23.2. Diabetes 53(3):803–811
73. Pollin TI, Hsueh WC, Steinle NI, Snitker S, Shuldiner AR, Mitchell BD (2004) A genome-wide scan of serum lipid levels in the Old Order Amish. Atherosclerosis 173(1):89–96. doi:10.1016/j.atherosclerosis.2003.11.012
74. Aulchenko YS, Vaessen N, Heutink P, Pullen J, Snijders PJ, Hofman A et al (2003) A genome-wide search for genes involved in type 2 diabetes in a recently genetically isolated population from the Netherlands. Diabetes 52(12):3001–3004
75. Smith P, McGuffog L, Easton DF, Mann GJ, Pupo GM, Newman B et al (2006) A genome wide linkage search for breast cancer susceptibility genes. Genes Chromosomes Cancer 45 (7):646–655. doi:10.1002/gcc.20330, PubMed PMID: 16575876; PubMed Central PMCID: PMCPMC2714969
76. Maier C, Herkommer K, Hoegel J, Vogel W, Paiss T (2005) A genomewide linkage analysis for prostate cancer susceptibility genes in families from Germany. Eur J Hum Genet 13 (3):352–360. doi:10.1038/sj.ejhg.5201333
77. Hsieh CL, Oakley-Girvan I, Balise RR, Halpern J, Gallagher RP, Wu AH et al (2001) A genome screen of families with multiple cases of prostate cancer: evidence of genetic

heterogeneity. Am J Hum Genet 69(1):148–158. doi:10.1086/321281, PubMed PMID: 11404817; PubMed Central PMCID: PMCPMC1226029
78. Schleutker J, Baffoe-Bonnie AB, Gillanders E, Kainu T, Jones MP, Freas-Lutz D et al (2003) Genome-wide scan for linkage in finnish hereditary prostate cancer (HPC) families identifies novel susceptibility loci at 11q14 and 3p25-26. Prostate 57(4):280–289. doi:10.1002/pros.10302
79. Eraksoy M, Kurtuncu M, Akman-Demir G, Kilinc M, Gedizlioglu M, Mirza M et al (2003) A whole genome screen for linkage in Turkish multiple sclerosis. J Neuroimmunol 143 (1–2):17–24
80. Imperatore G, Knowler WC, Pettitt DJ, Kobes S, Fuller JH, Bennett PH et al (2000) A locus influencing total serum cholesterol on chromosome 19p: results from an autosomal genomic scan of serum lipid concentrations in Pima Indians. Arterioscler Thromb Vasc Biol 20 (12):2651–2656
81. Elbein SC, Hasstedt SJ (2002) Quantitative trait linkage analysis of lipid-related traits in familial type 2 diabetes: evidence for linkage of triglyceride levels to chromosome 19q. Diabetes 51(2):528–535
82. Shmulewitz D, Heath SC, Blundell ML, Han Z, Sharma R, Salit J et al (2006) Linkage analysis of quantitative traits for obesity, diabetes, hypertension, and dyslipidemia on the island of Kosrae, Federated States of Micronesia. Proc Natl Acad Sci U S A 103 (10):3502–3509. doi:10.1073/pnas.0510156103, PubMed PMID: 16537441; PubMed Central PMCID: PMCPMC1533774
83. Arya R, Demerath E, Jenkinson CP, Goring HH, Puppala S, Farook V et al (2006) A quantitative trait locus (QTL) on chromosome 6q influences birth weight in two independent family studies. Hum Mol Genet 15(10):1569–1579. doi:10.1093/hmg/ddl076
84. Norman RA, Thompson DB, Foroud T, Garvey WT, Bennett PH, Bogardus C et al (1997) Genomewide search for genes influencing percent body fat in Pima Indians: suggestive linkage at chromosome 11q21-q22. Pima Diabetes Gene Group. Am J Hum Genet 60 (1):166–173, PubMed PMID: 8981960; PubMed Central PMCID: PMCPMC1712570
85. Wiklund F, Gillanders EM, Albertus JA, Bergh A, Damber JE, Emanuelsson M et al (2003) Genome-wide scan of Swedish families with hereditary prostate cancer: suggestive evidence of linkage at 5q11.2 and 19p13.3. Prostate 57(4):290–297. doi:10.1002/pros.10303
86. Vionnet N, Hani EH, Dupont S, Gallina S, Francke S, Dotte S et al (2000) Genomewide search for type 2 diabetes-susceptibility genes in French whites: evidence for a novel susceptibility locus for early-onset diabetes on chromosome 3q27-qter and independent replication of a type 2-diabetes locus on chromosome 1q21-q24. Am J Hum Genet 67 (6):1470–1480. doi:10.1086/316887, PubMed PMID: 11067779; PubMed Central PMCID: PMCPMC1287924
87. Silverman EK, Palmer LJ, Mosley JD, Barth M, Senter JM, Brown A et al (2002) Genomewide linkage analysis of quantitative spirometric phenotypes in severe early-onset chronic obstructive pulmonary disease. Am J Hum Genet 70(5):1229–1239. doi:10.1086/340316, PubMed PMID: 11914989; PubMed Central PMCID: PMCPMC447597
88. Colilla S, Nicolae D, Pluzhnikov A, Blumenthal MN, Beaty TH, Bleecker ER et al (2003) Evidence for gene-environment interactions in a linkage study of asthma and smoking exposure. J Allergy Clin Immunol 111(4):840–846
89. van der Kallen CJ, Cantor RM, van Greevenbroek MM, Geurts JM, Bouwman FG, Aouizerat BE et al (2000) Genome scan for adiposity in Dutch dyslipidemic families reveals novel quantitative trait loci for leptin, body mass index and soluble tumor necrosis factor receptor superfamily 1A. Int J Obes Relat Metab Disord 24(11):1381–1391
90. Cantor RM, de Bruin T, Kono N, Napier S, van Nas A, Allayee H et al (2004) Quantitative trait loci for apolipoprotein B, cholesterol, and triglycerides in familial combined hyperlipidemia pedigrees. Arterioscler Thromb Vasc Biol 24(10):1935–1941. doi:10.1161/01.ATV.0000142358.46276.a7

91. Ge D, Huang J, Yang W, Zhao J, Shen Y, Qiang B et al (2005) Linkage analysis of chromosome 1 with essential hypertension and blood pressure quantitative traits in Chinese families. Ann Hum Genet 69(Pt 1):45–54. doi:10.1046/j.1529-8817.2004.00136.x
92. Forster T, Chapman K, Marcelline L, Mustafa Z, Southam L, Loughlin J (2004) Finer linkage mapping of primary osteoarthritis susceptibility loci on chromosomes 4 and 16 in families with affected women. Arthritis Rheum 50(1):98–102. doi:10.1002/art.11427
93. Loughlin J, Mustafa Z, Irven C, Smith A, Carr AJ, Sykes B et al (1999) Stratification analysis of an osteoarthritis genome screen-suggestive linkage to chromosomes 4, 6, and 16. Am J Hum Genet 65(6):1795–1798. doi:10.1086/302685, PubMed PMID: 10577938; PubMed Central PMCID: PMCPMC1288394
94. Joost O, Wilk JB, Cupples LA, Harmon M, Shearman AM, Baldwin CT et al (2002) Genetic loci influencing lung function: a genome-wide scan in the Framingham Study. Am J Respir Crit Care Med 165(6):795–799. doi:10.1164/ajrccm.165.6.2102057
95. Kraja AT, Rao DC, Weder AB, Cooper R, Curb JD, Hanis CL et al (2005) Two major QTLs and several others relate to factors of metabolic syndrome in the family blood pressure program. Hypertension 46(4):751–757. doi:10.1161/01.HYP.0000184249.20016.bb
96. Liu YJ, Xu FH, Shen H, Liu YZ, Deng HY, Zhao LJ et al (2004) A follow-up linkage study for quantitative trait loci contributing to obesity-related phenotypes. J Clin Endocrinol Metab 89(2):875–882. doi:10.1210/jc.2003-030774
97. Wiltshire S, Frayling TM, Groves CJ, Levy JC, Hitman GA, Sampson M et al (2004) Evidence from a large U.K. family collection that genes influencing age of onset of type 2 diabetes map to chromosome 12p and to the MODY3/NIDDM2 locus on 12q24. Diabetes 53(3):855–860
98. Rice T, Rankinen T, Province MA, Chagnon YC, Perusse L, Borecki IB et al (2000) Genome-wide linkage analysis of systolic and diastolic blood pressure: the Quebec Family Study. Circulation 102(16):1956–1963
99. Silverman EK, Mosley JD, Palmer LJ, Barth M, Senter JM, Brown A et al (2002) Genome-wide linkage analysis of severe, early-onset chronic obstructive pulmonary disease: airflow obstruction and chronic bronchitis phenotypes. Hum Mol Genet 11(6):623–632
100. Hsueh WC, St Jean PL, Mitchell BD, Pollin TI, Knowler WC, Ehm MG et al (2003) Genome-wide and fine-mapping linkage studies of type 2 diabetes and glucose traits in the Old Order Amish: evidence for a new diabetes locus on chromosome 14q11 and confirmation of a locus on chromosome 1q21-q24. Diabetes 52(2):550–557
101. Ohman M, Oksanen L, Kaprio J, Koskenvuo M, Mustajoki P, Rissanen A et al (2000) Genome-wide scan of obesity in Finnish sibpairs reveals linkage to chromosome Xq24. J Clin Endocrinol Metab 85(9):3183–3190. doi:10.1210/jcem.85.9.6797
102. Almasy L, Goring HH, Diego V, Cole S, Laston S, Dyke B et al (2007) A novel obesity locus on chromosome 4q: the Strong Heart Family Study. Obesity (Silver Spring) 15 (7):1741–1748. doi:10.1038/oby.2007.207
103. Lee JH, Reed DR, Li WD, Xu W, Joo EJ, Kilker RL et al (1999) Genome scan for human obesity and linkage to markers in 20q13. Am J Hum Genet 64(1):196–209. doi:10.1086/302195, PubMed PMID: 9915959; PubMed Central PMCID: PMCPMC1377718
104. Gibbs M, Stanford JL, Jarvik GP, Janer M, Badzioch M, Peters MA et al (2000) A genomic scan of families with prostate cancer identifies multiple regions of interest. Am J Hum Genet 67(1):100–109. doi:10.1086/302969, PubMed PMID: 10820127; PubMed Central PMCID: PMCPMC1287067
105. Janer M, Friedrichsen DM, Stanford JL, Badzioch MD, Kolb S, Deutsch K et al (2003) Genomic scan of 254 hereditary prostate cancer families. Prostate 57(4):309–319. doi:10.1002/pros.10305
106. An P, Hong Y, Weisnagel SJ, Rice T, Rankinen T, Leon AS et al (2003) Genomic scan of glucose and insulin metabolism phenotypes: the HERITAGE Family Study. Metabolism 52 (2):246–253. doi:10.1053/meta.2003.50030

107. Mitchell BD, Cole SA, Comuzzie AG, Almasy L, Blangero J, MacCluer JW et al (1999) A quantitative trait locus influencing BMI maps to the region of the beta-3 adrenergic receptor. Diabetes 48(9):1863–1867
108. Arya R, Duggirala R, Almasy L, Rainwater DL, Mahaney MC, Cole S et al (2002) Linkage of high-density lipoprotein-cholesterol concentrations to a locus on chromosome 9p in Mexican Americans. Nat Genet 30(1):102–105. doi:10.1038/ng810
109. Wilson SG, Adam G, Langdown M, Reneland R, Braun A, Andrew T et al (2006) Linkage and potential association of obesity-related phenotypes with two genes on chromosome 12q24 in a female dizygous twin cohort. Eur J Hum Genet 14(3):340–348. doi:10.1038/sj.ejhg.5201551
110. Tejero ME, Cai G, Goring HH, Diego V, Cole SA, Bacino CA et al (2007) Linkage analysis of circulating levels of adiponectin in Hispanic children. Int J Obes (Lond) 31(3):535–542. doi:10.1038/sj.ijo.0803436
111. Gillanders EM, Xu J, Chang BL, Lange EM, Wiklund F, Bailey-Wilson JE et al (2004) Combined genome-wide scan for prostate cancer susceptibility genes. J Natl Cancer Inst 96 (16):1240–1247. doi:10.1093/jnci/djh228
112. Chang BL, Lange EM, Dimitrov L, Valis CJ, Gillanders EM, Lange LA et al (2006) Two-locus genome-wide linkage scan for prostate cancer susceptibility genes with an interaction effect. Hum Genet 118(6):716–724. doi:10.1007/s00439-005-0099-4
113. Xu J, Dimitrov L, Chang BL, Adams TS, Turner AR, Meyers DA et al (2005) A combined genomewide linkage scan of 1,233 families for prostate cancer-susceptibility genes conducted by the international consortium for prostate cancer genetics. Am J Hum Genet 77(2):219–229. doi:10.1086/432377, PubMed PMID: 15988677; PubMed Central PMCID: PMCPMC1224525
114. Elbein SC, Hoffman MD, Teng K, Leppert MF, Hasstedt SJ (1999) A genome-wide search for type 2 diabetes susceptibility genes in Utah Caucasians. Diabetes 48(5):1175–1182
115. Allayee H, de Bruin TW, Michelle Dominguez K, Cheng LS, Ipp E, Cantor RM et al (2001) Genome scan for blood pressure in Dutch dyslipidemic families reveals linkage to a locus on chromosome 4p. Hypertension 38(4):773–778
116. Arya R, Duggirala R, Jenkinson CP, Almasy L, Blangero J, O'Connell P et al (2004) Evidence of a novel quantitative-trait locus for obesity on chromosome 4p in Mexican Americans. Am J Hum Genet 74(2):272–282. doi:10.1086/381717, PubMed PMID: 14740316; PubMed Central PMCID: PMCPMC1181925
117. Stone S, Abkevich V, Hunt SC, Gutin A, Russell DL, Neff CD et al (2002) A major predisposition locus for severe obesity, at 4p15-p14. Am J Hum Genet 70(6):1459–1468, PubMed PMID: 11957135; PubMed Central PMCID: PMCPMC379132
118. Bethony J, Williams JT, Almasy L, Correa-Oliveira R, Blangero JC, Williams-Blangero S (2001) Genetic analysis of quantitative traits in highly ascertained samples: total serum IgE in families with asthma. Genet Epidemiol 21(Suppl 1):S174–S179
119. Bosse Y, Perusse L, Despres JP, Lamarche B, Chagnon YC, Rice T et al (2003) Evidence for a major quantitative trait locus on chromosome 17q21 affecting low-density lipoprotein peak particle diameter. Circulation 107(18):2361–2368. doi:10.1161/01.CIR.0000065577.60129.F5
120. Slager SL, Schaid DJ, Cunningham JM, McDonnell SK, Marks AF, Peterson BJ et al (2003) Confirmation of linkage of prostate cancer aggressiveness with chromosome 19q. Am J Hum Genet 72(3):759–762. doi:10.1086/368230, PubMed PMID: 12563560; PubMed Central PMCID: PMCPMC1180252
121. Perola M, Kainulainen K, Pajukanta P, Terwilliger JD, Hiekkalinna T, Ellonen P et al (2000) Genome-wide scan of predisposing loci for increased diastolic blood pressure in Finnish siblings. J Hypertens 18(11):1579–1585
122. Camp NJ, Cannon-Albright LA, Farnham JM, Baffoe-Bonnie AB, George A, Powell I et al (2007) Compelling evidence for a prostate cancer gene at 22q12.3 by the International

Consortium for Prostate Cancer Genetics. Hum Mol Genet 16(11):1271–1278. doi:10.1093/hmg/ddm075, PubMed PMID: 17478474; PubMed Central PMCID: PMCPMC2653215
123. Camp NJ, Farnham JM, Cannon Albright LA (2005) Genomic search for prostate cancer predisposition loci in Utah pedigrees. Prostate 65(4):365–374. doi:10.1002/pros.20287
124. Hartikainen JM, Tuhkanen H, Kataja V, Eskelinen M, Uusitupa M, Kosma VM et al (2006) Refinement of the 22q12-q13 breast cancer--associated region: evidence of TMPRSS6 as a candidate gene in an eastern Finnish population. Clin Cancer Res 12(5):1454–1462. doi:10.1158/1078-0432.CCR-05-1417
125. Hartikainen JM, Tuhkanen H, Kataja V, Dunning AM, Antoniou A, Smith P et al (2005) An autosome-wide scan for linkage disequilibrium-based association in sporadic breast cancer cases in eastern Finland: three candidate regions found. Cancer Epidemiol Biomarkers Prev 14(1):75–80
126. Chagnon YC, Rice T, Perusse L, Borecki IB, Ho-Kim MA, Lacaille M et al (2001) Genomic scan for genes affecting body composition before and after training in Caucasians from HERITAGE. J Appl Physiol (1985) 90(5):1777–1787
127. Adeyemo A, Luke A, Cooper R, Wu X, Tayo B, Zhu X et al (2003) A genome-wide scan for body mass index among Nigerian families. Obes Res 11(2):266–273. doi:10.1038/oby.2003.40
128. Feitosa MF, Borecki IB, Rankinen T, Rice T, Despres JP, Chagnon YC et al (2005) Evidence of QTLs on chromosomes 1q42 and 8q24 for LDL-cholesterol and apoB levels in the HERITAGE family study. J Lipid Res 46(2):281–286. doi:10.1194/jlr.M400252-JLR200
129. Xu J, Gillanders EM, Isaacs SD, Chang BL, Wiley KE, Zheng SL et al (2003) Genome-wide scan for prostate cancer susceptibility genes in the Johns Hopkins hereditary prostate cancer families. Prostate 57(4):320–325. doi:10.1002/pros.10306
130. Xu J, Zheng SL, Chang B, Smith JR, Carpten JD, Stine OC et al (2001) Linkage of prostate cancer susceptibility loci to chromosome 1. Hum Genet 108(4):335–345
131. Kotchen TA, Broeckel U, Grim CE, Hamet P, Jacob H, Kaldunski ML et al (2002) Identification of hypertension-related QTLs in African American sib pairs. Hypertension 40 (5):634–639
132. Jawaheer D, Seldin MF, Amos CI, Chen WV, Shigeta R, Monteiro J et al (2001) A genomewide screen in multiplex rheumatoid arthritis families suggests genetic overlap with other autoimmune diseases. Am J Hum Genet 68(4):927–936, PubMed PMID: 11254450; PubMed Central PMCID: PMCPMC1275647
133. Zhu X, Cooper RS, Luke A, Chen G, Wu X, Kan D et al (2002) A genome-wide scan for obesity in African-Americans. Diabetes 51(2):541–544
134. Lindgren CM, Mahtani MM, Widen E, McCarthy MI, Daly MJ, Kirby A et al (2002) Genomewide search for type 2 diabetes mellitus susceptibility loci in Finnish families: the Botnia study. Am J Hum Genet 70(2):509–516. doi:10.1086/338629, PubMed PMID: 11791216; PubMed Central PMCID: PMCPMC384923
135. Wilkinson J, Grimley S, Collins A, Thomas NS, Holgate ST, Morton N (1998) Linkage of asthma to markers on chromosome 12 in a sample of 240 families using quantitative phenotype scores. Genomics 53(3):251–259
136. Klos KL, Kardia SL, Ferrell RE, Turner ST, Boerwinkle E, Sing CF (2001) Genome-wide linkage analysis reveals evidence of multiple regions that influence variation in plasma lipid and apolipoprotein levels associated with risk of coronary heart disease. Arterioscler Thromb Vasc Biol 21(6):971–978
137. Wang Q, Rao S, Shen GQ, Li L, Moliterno DJ, Newby LK et al (2004) Premature myocardial infarction novel susceptibility locus on chromosome 1P34-36 identified by genomewide linkage analysis. Am J Hum Genet 74(2):262–271. doi:10.1086/381560, PubMed PMID: 14732905; PubMed Central PMCID: PMCPMC1181924
138. Perusse L, Rice T, Chagnon YC, Despres JP, Lemieux S, Roy S et al (2001) A genome-wide scan for abdominal fat assessed by computed tomography in the Quebec Family Study. Diabetes 50(3):614–621

139. Feitosa MF, Rice T, Rankinen T, Province MA, Chagnon YC, Gagnon J et al (2005) Evidence of QTLs on chromosomes 13q and 14q for triglycerides before and after 20 weeks of exercise training: the HERITAGE Family Study. Atherosclerosis 182(2):349–360. doi:10.1016/j.atherosclerosis.2005.01.052
140. Dai F, Keighley ED, Sun G, Indugula SR, Roberts ST, Aberg K et al (2007) Genome-wide scan for adiposity-related phenotypes in adults from American Samoa. Int J Obes (Lond) 31 (12):1832–1842. doi:10.1038/sj.ijo.0803675
141. Suarez BK, Lin J, Burmester JK, Broman KW, Weber JL, Banerjee TK et al (2000) A genome screen of multiplex sibships with prostate cancer. Am J Hum Genet 66(3):933–944. doi:10.1086/302818, PubMed PMID: 10712208; PubMed Central PMCID: PMCPMC1288174
142. An P, Teran-Garcia M, Rice T, Rankinen T, Weisnagel SJ, Bergman RN et al (2005) Genome-wide linkage scans for prediabetes phenotypes in response to 20 weeks of endurance exercise training in non-diabetic whites and blacks: the HERITAGE Family Study. Diabetologia 48(6):1142–1149. doi:10.1007/s00125-005-1769-4
143. Shaw JT, Lovelock PK, Kesting JB, Cardinal J, Duffy D, Wainwright B et al (1998) Novel susceptibility gene for late-onset NIDDM is localized to human chromosome 12q. Diabetes 47(11):1793–1796
144. Leppavuori J, Kujala U, Kinnunen J, Kaprio J, Nissila M, Heliovaara M et al (1999) Genome scan for predisposing loci for distal interphalangeal joint osteoarthritis: evidence for a locus on 2q. Am J Hum Genet 65(4):1060–1067. doi:10.1086/302569, PubMed PMID: 10486325; PubMed Central PMCID: PMCPMC1288239
145. Kenealy SJ, Babron MC, Bradford Y, Schnetz-Boutaud N, Haines JL, Rimmler JB et al (2004) A second-generation genomic screen for multiple sclerosis. Am J Hum Genet 75(6):1070–1078. doi:10.1086/426459, PubMed PMID: 15494893; PubMed Central PMCID: PMCPMC1182142
146. Iwasaki N, Cox NJ, Wang YQ, Schwarz PE, Bell GI, Honda M et al (2003) Mapping genes influencing type 2 diabetes risk and BMI in Japanese subjects. Diabetes 52(1):209–213
147. An P, Freedman BI, Rich SS, Mandel SA, Arnett DK, Myers RH et al (2006) Quantitative trait loci on chromosome 8q24 for pancreatic beta-cell function and 7q11 for insulin sensitivity in obese nondiabetic white and black families: evidence from genome-wide linkage scans in the NHLBI Hypertension Genetic Epidemiology Network (HyperGEN) study. Diabetes 55(2):551–558
148. Witte JS, Goddard KA, Conti DV, Elston RC, Lin J, Suarez BK et al (2000) Genomewide scan for prostate cancer-aggressiveness loci. Am J Hum Genet 67(1):92–99. doi:10.1086/302960, PubMed PMID: 10825281; PubMed Central PMCID: PMCPMC1287106
149. Chen G, Adeyemo A, Zhou J, Chen Y, Huang H, Doumatey A et al (2007) Genome-wide search for susceptibility genes to type 2 diabetes in West Africans: potential role of C-peptide. Diabetes Res Clin Pract 78(3):e1–e6. doi:10.1016/j.diabres.2007.04.010
150. Strauch K, Bogdanow M, Fimmers R, Baur MP, Wienker TF (2001) Linkage analysis of asthma and atopy including models with genomic imprinting. Genet Epidemiol 21(Suppl 1): S204–S209
151. North KE, Miller MB, Coon H, Martin LJ, Peacock JM, Arnett D et al (2005) Evidence for a gene influencing fasting LDL cholesterol and triglyceride levels on chromosome 21q. Atherosclerosis 179(1):119–125. doi:10.1016/j.atherosclerosis.2004.09.009
152. Knoblauch H, Muller-Myhsok B, Busjahn A, Ben Avi L, Bahring S, Baron H et al (2000) A cholesterol-lowering gene maps to chromosome 13q. Am J Hum Genet 66(1):157–166, PubMed PMID: 10631147; PubMed Central PMCID: PMCPMC1288321
153. Hixson JE, Blangero J (2000) Genomic searches for genes that influence atherosclerosis and its risk factors. Ann N Y Acad Sci 902:1–7
154. Norman RA, Bogardus C, Ravussin E (1995) Linkage between obesity and a marker near the tumor necrosis factor-alpha locus in Pima Indians. J Clin Invest 96(1):158–162. doi:10.1172/JCI118016, PubMed PMID: 7615786; PubMed Central PMCID: PMCPMC185184

155. Christensen GB, Camp NJ, Farnham JM, Cannon-Albright LA (2007) Genome-wide linkage analysis for aggressive prostate cancer in Utah high-risk pedigrees. Prostate 67(6):605–613. doi:10.1002/pros.20554
156. Dong C, Li WD, Li D, Price RA (2005) Interaction between obesity-susceptibility loci in chromosome regions 2p25-p24 and 13q13-q21. Eur J Hum Genet 13(1):102–108. doi:10.1038/sj.ejhg.5201292
157. Diego VP, Goring HH, Cole SA, Almasy L, Dyer TD, Blangero J et al (2006) Fasting insulin and obesity-related phenotypes are linked to chromosome 2p: the Strong Heart Family Study. Diabetes 55(6):1874–1878. doi:10.2337/db05-0668
158. Comuzzie AG, Hixson JE, Almasy L, Mitchell BD, Mahaney MC, Dyer TD et al (1997) A major quantitative trait locus determining serum leptin levels and fat mass is located on human chromosome 2. Nat Genet 15(3):273–276. doi:10.1038/ng0397-273
159. Rotimi CN, Comuzzie AG, Lowe WL, Luke A, Blangero J, Cooper RS (1999) The quantitative trait locus on chromosome 2 for serum leptin levels is confirmed in African-Americans. Diabetes 48(3):643–644
160. Delplanque J, Barat-Houari M, Dina C, Gallina P, Clement K, Guy-Grand B et al (2000) Linkage and association studies between the proopiomelanocortin (POMC) gene and obesity in caucasian families. Diabetologia 43(12):1554–1557. doi:10.1007/s001250051568
161. Hager J, Dina C, Francke S, Dubois S, Houari M, Vatin V et al (1998) A genome-wide scan for human obesity genes reveals a major susceptibility locus on chromosome 10. Nat Genet 20(3):304–308. doi:10.1038/3123
162. Atwood LD, Samollow PB, Hixson JE, Stern MP, MacCluer JW (2001) Genome-wide linkage analysis of blood pressure in Mexican Americans. Genet Epidemiol 20(3):373–382. doi:10.1002/gepi.7
163. Norris JM, Langefeld CD, Scherzinger AL, Rich SS, Bookman E, Beck SR et al (2005) Quantitative trait loci for abdominal fat and BMI in Hispanic-Americans and African-Americans: the IRAS Family study. Int J Obes (Lond) 29(1):67–77. doi:10.1038/sj.ijo.0802793
164. Frayling TM, Wiltshire S, Hitman GA, Walker M, Levy JC, Sampson M et al (2003) Young-onset type 2 diabetes families are the major contributors to genetic loci in the Diabetes UK Warren 2 genome scan and identify putative novel loci on chromosomes 8q21, 21q22, and 22q11. Diabetes 52(7):1857–1863
165. Platte P, Papanicolaou GJ, Johnston J, Klein CM, Doheny KF, Pugh EW et al (2003) A study of linkage and association of body mass index in the Old Order Amish. Am J Med Genet C Semin Med Genet 121C(1):71–80. doi:10.1002/ajmg.c.20005
166. Voruganti VS, Goring HH, Diego VP, Cai G, Mehta NR, Haack K et al (2007) Genome-wide scan for serum ghrelin detects linkage on chromosome 1p36 in Hispanic children: results from the Viva La Familia study. Pediatr Res 62(4):445–450. doi:10.1203/PDR.0b013e31813cbf02
167. Zhao JY, Xiong MM, Huang W, Wang H, Zuo J, Wu GD et al (2005) An autosomal genomic scan for loci linked to type 2 diabetes in northern Han Chinese. J Mol Med (Berl) 83(3):209–215. doi:10.1007/s00109-004-0587-3
168. Mustafa Z, Chapman K, Irven C, Carr AJ, Clipsham K, Chitnavis J et al (2000) Linkage analysis of candidate genes as susceptibility loci for osteoarthritis-suggestive linkage of COL9A1 to female hip osteoarthritis. Rheumatology (Oxford) 39(3):299–306
169. Wu X, Cooper RS, Borecki I, Hanis C, Bray M, Lewis CE et al (2002) A combined analysis of genomewide linkage scans for body mass index from the National Heart, Lung, and Blood Institute Family Blood Pressure Program. Am J Hum Genet 70(5):1247–1256. doi:10.1086/340362, PubMed PMID: 11923912; PubMed Central PMCID: PMCPMC447599
170. Shearman AM, Ordovas JM, Cupples LA, Schaefer EJ, Harmon MD, Shao Y et al (2000) Evidence for a gene influencing the TG/HDL-C ratio on chromosome 7q32.3-qter: a genome-wide scan in the Framingham study. Hum Mol Genet 9(9):1315–1320

171. Li WD, Dong C, Li D, Garrigan C, Price RA (2005) A genome scan for serum triglyceride in obese nuclear families. J Lipid Res 46(3):432–438. doi:10.1194/jlr.M400391-JLR200
172. Francke S, Manraj M, Lacquemant C, Lecoeur C, Lepretre F, Passa P et al (2001) A genome-wide scan for coronary heart disease suggests in Indo-Mauritians a susceptibility locus on chromosome 16p13 and replicates linkage with the metabolic syndrome on 3q27. Hum Mol Genet 10(24):2751–2765
173. Xu J, Meyers D, Freije D, Isaacs S, Wiley K, Nusskern D et al (1998) Evidence for a prostate cancer susceptibility locus on the X chromosome. Nat Genet 20(2):175–179. doi:10.1038/2477
174. Bochum S, Paiss T, Vogel W, Herkommer K, Hautmann R, Haeussler J (2002) Confirmation of the prostate cancer susceptibility locus HPCX in a set of 104 German prostate cancer families. Prostate 52(1):12–19. doi:10.1002/pros.10078
175. Silander K, Scott LJ, Valle TT, Mohlke KL, Stringham HM, Wiles KR et al (2004) A large set of Finnish affected sibling pair families with type 2 diabetes suggests susceptibility loci on chromosomes 6, 11, and 14. Diabetes 53(3):821–829
176. Austin MA, Edwards KL, Monks SA, Koprowicz KM, Brunzell JD, Motulsky AG et al (2003) Genome-wide scan for quantitative trait loci influencing LDL size and plasma triglyceride in familial hypertriglyceridemia. J Lipid Res 44(11):2161–2168. doi:10.1194/jlr.M300272-JLR200
177. Lin JP (2003) Genome-wide scan on plasma triglyceride and high density lipoprotein cholesterol levels, accounting for the effects of correlated quantitative phenotypes. BMC Genet 4(Suppl 1):S47. doi:10.1186/1471-2156-4-S1-S47, PubMed PMID: 14975115; PubMed Central PMCID: PMCPMC1866483
178. Yang Q, Lai CQ, Parnell L, Cupples LA, Adiconis X, Zhu Y et al (2005) Genome-wide linkage analyses and candidate gene fine mapping for HDL3 cholesterol: the Framingham Study. J Lipid Res 46(7):1416–1425. doi:10.1194/jlr.M400382-JLR200
179. Arya R, Lehman D, Hunt KJ, Schneider J, Almasy L, Blangero J et al (2003) Evidence for bivariate linkage of obesity and HDL-C levels in the Framingham Heart Study. BMC Genet 4 (Suppl 1):S52. doi:10.1186/1471-2156-4-S1-S52, PubMed PMID: 14975120; PubMed Central PMCID: PMCPMC1866489
180. Bell CG, Benzinou M, Siddiq A, Lecoeur C, Dina C, Lemainque A et al (2004) Genome-wide linkage analysis for severe obesity in French caucasians finds significant susceptibility locus on chromosome 19q. Diabetes 53(7):1857–1865
181. Wiltshire S, Hattersley AT, Hitman GA, Walker M, Levy JC, Sampson M et al (2001) A genomewide scan for loci predisposing to type 2 diabetes in a U.K. population (the Diabetes UK Warren 2 Repository): analysis of 573 pedigrees provides independent replication of a susceptibility locus on chromosome 1q. Am J Hum Genet 69(3):553–569, PubMed PMID: 11484155; PubMed Central PMCID: PMCPMC1235485
182. Duggirala R, Blangero J, Almasy L, Arya R, Dyer TD, Williams KL et al (2001) A major locus for fasting insulin concentrations and insulin resistance on chromosome 6q with strong pleiotropic effects on obesity-related phenotypes in nondiabetic Mexican Americans. Am J Hum Genet 68(5):1149–1164. doi:10.1086/320100, PubMed PMID: 11283790; PubMed Central PMCID: PMCPMC1226096
183. Xiang K, Wang Y, Zheng T, Jia W, Li J, Chen L et al (2004) Genome-wide search for type 2 diabetes/impaired glucose homeostasis susceptibility genes in the Chinese: significant linkage to chromosome 6q21-q23 and chromosome 1q21-q24. Diabetes 53(1):228–234
184. Shtir C, Nagakawa IS, Duren WL, Conneely KN, Scott LJ, Silander K et al (2007) Subsets of Finns with high HDL to total cholesterol ratio show evidence for linkage to type 2 diabetes on chromosome 6q. Hum Hered 63(1):17–25. doi:10.1159/000097927, PubMed PMID: 17179727; PubMed Central PMCID: PMCPMC2923439
185. Hauser ER, Crossman DC, Granger CB, Haines JL, Jones CJ, Mooser V et al (2004) A genomewide scan for early-onset coronary artery disease in 438 families: the GENECARD Study. Am J Hum Genet 75(3):436–447. doi:10.1086/423900, PubMed PMID: 15272420; PubMed Central PMCID: PMCPMC1182022

186. Dong C, Li WD, Geller F, Lei L, Li D, Gorlova OY et al (2005) Possible genomic imprinting of three human obesity-related genetic loci. Am J Hum Genet 76(3):427–437. doi:10.1086/428438, PubMed PMID: 15647995; PubMed Central PMCID: PMCPMC1196395
187. van Tilburg JH, Sandkuijl LA, Strengman E, Pearson PL, van Haeften TW, Wijmenga C (2003) Variance-component analysis of obesity in type 2 diabetes confirms loci on chromosomes 1q and 11q. Obes Res 11(11):1290–1294. doi:10.1038/oby.2003.175
188. Chen G, Adeyemo AA, Johnson T, Zhou J, Amoah A, Owusu S et al (2005) A genome-wide scan for quantitative trait loci linked to obesity phenotypes among West Africans. Int J Obes (Lond) 29(3):255–259. doi:10.1038/sj.ijo.0802873
189. Morrison AC, Felix JF, Cupples LA, Glazer NL, Loehr LR, Dehghan A et al (2010) Genomic variation associated with mortality among adults of European and African ancestry with heart failure: the cohorts for heart and aging research in genomic epidemiology consortium. Circ Cardiovasc Genet 3(3):248–255. doi:10.1161/CIRCGENETICS.109.895995, PubMed PMID: 20400778; PubMed Central PMCID: PMCPMC3033765
190. Welter D, MacArthur J, Morales J, Burdett T, Hall P, Junkins H et al (2014) The NHGRI GWAS Catalog, a curated resource of SNP-trait associations. Nucleic Acids Res 42(Database issue):D1001–D1006. doi:10.1093/nar/gkt1229, PubMed PMID: 24316577; PubMed Central PMCID: PMCPMC3965119
191. Cornelis MC, Fornage M, Foy M, Xun P, Gladyshev VN, Morris S et al (2015) Genome-wide association study of selenium concentrations. Hum Mol Genet 24(5):1469–1477. doi:10.1093/hmg/ddu546, PubMed PMID: 25343990; PubMed Central PMCID: PMCPMC4321444
192. Service SK, Verweij KJ, Lahti J, Congdon E, Ekelund J, Hintsanen M et al (2012) A genome-wide meta-analysis of association studies of Cloninger's Temperament Scales. Transl Psychiatry 2, e116. doi:10.1038/tp.2012.37, PubMed PMID: 22832960; PubMed Central PMCID: PMCPMC3365256
193. Yu B, Zheng Y, Alexander D, Manolio TA, Alonso A, Nettleton JA et al (2013) Genome-wide association study of a heart failure related metabolomic profile among African Americans in the Atherosclerosis Risk in Communities (ARIC) study. Genet Epidemiol 37 (8):840–845. doi:10.1002/gepi.21752, PubMed PMID: 23934736; PubMed Central PMCID: PMCPMC4079107
194. Comuzzie AG, Cole SA, Laston SL, Voruganti VS, Haack K, Gibbs RA et al (2012) Novel genetic loci identified for the pathophysiology of childhood obesity in the Hispanic population. PLoS One 7(12), e51954. doi:10.1371/journal.pone.0051954, PubMed PMID: 23251661; PubMed Central PMCID: PMCPMC3522587
195. Zhang H, Baldwin DA, Bukowski RK, Parry S, Xu Y, Song C et al (2015) A genome-wide association study of early spontaneous preterm delivery. Genet Epidemiol 39(3):217–226. doi:10.1002/gepi.21887, PubMed PMID: 25599974; PubMed Central PMCID: PMCPMC4366311
196. Fumagalli M, Sironi M, Pozzoli U, Ferrer-Admetlla A, Pattini L, Nielsen R (2011) Signatures of environmental genetic adaptation pinpoint pathogens as the main selective pressure through human evolution. PLoS Genet 7(11), e1002355. doi:10.1371/journal.pgen.1002355, PubMed PMID: 22072984; PubMed Central PMCID: PMCPMC3207877
197. McClay JL, Adkins DE, Aberg K, Bukszar J, Khachane AN, Keefe RS et al (2011) Genome-wide pharmacogenomic study of neurocognition as an indicator of antipsychotic treatment response in schizophrenia. Neuropsychopharmacology 36(3):616–626. doi:10.1038/npp.2010.193, PubMed PMID: 21107309; PubMed Central PMCID: PMCPMC3055694
198. Kasperaviciute D, Catarino CB, Heinzen EL, Depondt C, Cavalleri GL, Caboclo LO et al (2010) Common genetic variation and susceptibility to partial epilepsies: a genome-wide association study. Brain 133(Pt 7):2136–2147. doi:10.1093/brain/awq130, PubMed PMID: 20522523; PubMed Central PMCID: PMCPMC2892941
199. Zemunik T, Boban M, Lauc G, Jankovic S, Rotim K, Vatavuk Z et al (2009) Genome-wide association study of biochemical traits in Korcula Island, Croatia. Croat Med J 50(1):23–33, PubMed PMID: 19260141; PubMed Central PMCID: PMCPMC2657564

200. Walter S, Atzmon G, Demerath EW, Garcia ME, Kaplan RC, Kumari M et al (2011) A genome-wide association study of aging. Neurobiol Aging 32(11):2109e15–2109e28. doi:10.1016/j.neurobiolaging.2011.05.026, PubMed PMID: 21782286; PubMed Central PMCID: PMCPMC3193030
201. Cross-Disorder Group of the Psychiatric Genomics C (2013) Identification of risk loci with shared effects on five major psychiatric disorders: a genome-wide analysis. Lancet 381 (9875):1371–1379. doi:10.1016/S0140-6736(12)62129-1, PubMed PMID: 23453885; PubMed Central PMCID: PMCPMC3714010
202. Divaris K, Monda KL, North KE, Olshan AF, Reynolds LM, Hsueh WC et al (2013) Exploring the genetic basis of chronic periodontitis: a genome-wide association study. Hum Mol Genet 22(11):2312–2324. doi:10.1093/hmg/ddt065, PubMed PMID: 23459936; PubMed Central PMCID: PMCPMC3652417
203. Cheng CY, Schache M, Ikram MK, Young TL, Guggenheim JA, Vitart V et al (2013) Nine loci for ocular axial length identified through genome-wide association studies, including shared loci with refractive error. Am J Hum Genet 93(2):264–277. doi:10.1016/j.ajhg.2013.06.016, PubMed PMID: 24144296; PubMed Central PMCID: PMCPMC3772747
204. Fransen E, Bonneux S, Corneveaux JJ, Schrauwen I, Di Berardino F, White CH et al (2015) Genome-wide association analysis demonstrates the highly polygenic character of age-related hearing impairment. Eur J Hum Genet 23(1):110–115. doi:10.1038/ejhg.2014.56, PubMed PMID: 24939585; PubMed Central PMCID: PMCPMC4266741
205. Padyukov L, Seielstad M, Ong RT, Ding B, Ronnelid J, Seddighzadeh M et al (2011) A genome-wide association study suggests contrasting associations in ACPA-positive versus ACPA-negative rheumatoid arthritis. Ann Rheum Dis 70(2):259–265. doi:10.1136/ard.2009.126821, PubMed PMID: 21156761; PubMed Central PMCID: PMCPMC3015094
206. Draaken M, Knapp M, Pennimpede T, Schmidt JM, Ebert AK, Rosch W et al (2015) Genome-wide association study and meta-analysis identify ISL1 as genome-wide significant susceptibility gene for bladder exstrophy. PLoS Genet 11(3), e1005024. doi:10.1371/journal.pgen.1005024, PubMed PMID: 25763902; PubMed Central PMCID: PMCPMC4357422
207. Yamada Y, Fuku N, Tanaka M, Aoyagi Y, Sawabe M, Metoki N et al (2009) Identification of CELSR1 as a susceptibility gene for ischemic stroke in Japanese individuals by a genome-wide association study. Atherosclerosis 207(1):144–149. doi:10.1016/j.atherosclerosis.2009.03.038
208. Carrasquillo MM, Nicholson AM, Finch N, Gibbs JR, Baker M, Rutherford NJ et al (2010) Genome-wide screen identifies rs646776 near sortilin as a regulator of progranulin levels in human plasma. Am J Hum Genet 87(6):890–897. doi:10.1016/j.ajhg.2010.11.002, PubMed PMID: 21087763; PubMed Central PMCID: PMCPMC2997361
209. Barber MJ, Mangravite LM, Hyde CL, Chasman DI, Smith JD, McCarty CA et al (2010) Genome-wide association of lipid-lowering response to statins in combined study populations. PLoS One 5(3), e9763. doi:10.1371/journal.pone.0009763, PubMed PMID: 20339536; PubMed Central PMCID: PMCPMC2842298
210. Global Lipids Genetics C, Willer CJ, Schmidt EM, Sengupta S, Peloso GM, Gustafsson S et al (2013) Discovery and refinement of loci associated with lipid levels. Nat Genet 45 (11):1274–1283. doi:10.1038/ng.2797, PubMed PMID: 24097068; PubMed Central PMCID: PMCPMC3838666
211. Teslovich TM, Musunuru K, Smith AV, Edmondson AC, Stylianou IM, Koseki M et al (2010) Biological, clinical and population relevance of 95 loci for blood lipids. Nature 466(7307):707–713. doi:10.1038/nature09270, PubMed PMID: 20686565; PubMed Central PMCID: PMCPMC3039276
212. Aulchenko YS, Ripatti S, Lindqvist I, Boomsma D, Heid IM, Pramstaller PP et al (2009) Loci influencing lipid levels and coronary heart disease risk in 16 European population cohorts. Nat Genet 41(1):47–55. doi:10.1038/ng.269, PubMed PMID: 19060911; PubMed Central PMCID: PMCPMC2687074

213. Ko A, Cantor RM, Weissglas-Volkov D, Nikkola E, Reddy PM, Sinsheimer JS et al (2014) Amerindian-specific regions under positive selection harbour new lipid variants in Latinos. Nat Commun 5:3983. doi:10.1038/ncomms4983, PubMed PMID: 24886709; PubMed Central PMCID: PMCPMC4062071
214. Kathiresan S, Melander O, Guiducci C, Surti A, Burtt NP, Rieder MJ et al (2008) Six new loci associated with blood low-density lipoprotein cholesterol, high-density lipoprotein cholesterol or triglycerides in humans. Nat Genet 40(2):189–197. doi:10.1038/ng.75, PubMed PMID: 18193044; PubMed Central PMCID: PMCPMC2682493
215. Waterworth DM, Ricketts SL, Song K, Chen L, Zhao JH, Ripatti S et al (2010) Genetic variants influencing circulating lipid levels and risk of coronary artery disease. Arterioscler Thromb Vasc Biol 30(11):2264–2276. doi:10.1161/ATVBAHA.109.201020, PubMed PMID: 20864672; PubMed Central PMCID: PMCPMC3891568
216. Sabatti C, Service SK, Hartikainen AL, Pouta A, Ripatti S, Brodsky J et al (2009) Genome-wide association analysis of metabolic traits in a birth cohort from a founder population. Nat Genet 41(1):35–46. doi:10.1038/ng.271, PubMed PMID: 19060910; PubMed Central PMCID: PMCPMC2687077
217. Kathiresan S, Willer CJ, Peloso GM, Demissie S, Musunuru K, Schadt EE et al (2009) Common variants at 30 loci contribute to polygenic dyslipidemia. Nat Genet 41(1):56–65. doi:10.1038/ng.291, PubMed PMID: 19060906; PubMed Central PMCID: PMCPMC2881676
218. Chasman DI, Pare G, Mora S, Hopewell JC, Peloso G, Clarke R et al (2009) Forty-three loci associated with plasma lipoprotein size, concentration, and cholesterol content in genome-wide analysis. PLoS Genet 5(11), e1000730. doi:10.1371/journal.pgen.1000730, PubMed PMID: 19936222; PubMed Central PMCID: PMCPMC2777390
219. Middelberg RP, Ferreira MA, Henders AK, Heath AC, Madden PA, Montgomery GW et al (2011) Genetic variants in LPL, OASL and TOMM40/APOE-C1-C2-C4 genes are associated with multiple cardiovascular-related traits. BMC Med Genet 12:123. doi:10.1186/1471-2350-12-123, PubMed PMID: 21943158; PubMed Central PMCID: PMCPMC3189113
220. Lettre G, Palmer CD, Young T, Ejebe KG, Allayee H, Benjamin EJ et al (2011) Genome-wide association study of coronary heart disease and its risk factors in 8,090 African Americans: the NHLBI CARe Project. PLoS Genet 7(2), e1001300. doi:10.1371/journal.pgen.1001300, PubMed PMID: 21347282; PubMed Central PMCID: PMCPMC3037413
221. Sandhu MS, Waterworth DM, Debenham SL, Wheeler E, Papadakis K, Zhao JH et al (2008) LDL-cholesterol concentrations: a genome-wide association study. Lancet 371 (9611):483–491. doi:10.1016/S0140-6736(08)60208-1, PubMed PMID: 18262040; PubMed Central PMCID: PMCPMC2292820
222. Willer CJ, Sanna S, Jackson AU, Scuteri A, Bonnycastle LL, Clarke R et al (2008) Newly identified loci that influence lipid concentrations and risk of coronary artery disease. Nat Genet 40(2):161–169. doi:10.1038/ng.76
223. Wallace C, Newhouse SJ, Braund P, Zhang F, Tobin M, Falchi M et al (2008) Genome-wide association study identifies genes for biomarkers of cardiovascular disease: serum urate and dyslipidemia. Am J Hum Genet 82(1):139–149. doi:10.1016/j.ajhg.2007.11.001, PubMed PMID: 18179892; PubMed Central PMCID: PMCPMC2253977
224. Kim YJ, Go MJ, Hu C, Hong CB, Kim YK, Lee JY et al (2011) Large-scale genome-wide association studies in East Asians identify new genetic loci influencing metabolic traits. Nat Genet 43(10):990–995. doi:10.1038/ng.939
225. Dichgans M, Malik R, Konig IR, Rosand J, Clarke R, Gretarsdottir S et al (2014) Shared genetic susceptibility to ischemic stroke and coronary artery disease: a genome-wide analysis of common variants. Stroke 45(1):24–36. doi:10.1161/STROKEAHA.113.002707, PubMed PMID: 24262325; PubMed Central PMCID: PMCPMC4112102
226. Reilly MP, Li M, He J, Ferguson JF, Stylianou IM, Mehta NN et al (2011) Identification of ADAMTS7 as a novel locus for coronary atherosclerosis and association of ABO with

myocardial infarction in the presence of coronary atherosclerosis: two genome-wide association studies. Lancet 377(9763):383–392. doi:10.1016/S0140-6736(10)61996-4, PubMed PMID: 21239051; PubMed Central PMCID: PMCPMC3297116
227. Coronary Artery Disease Genetics C (2011) A genome-wide association study in Europeans and South Asians identifies five new loci for coronary artery disease. Nat Genet 43 (4):339–344. doi:10.1038/ng.782
228. Myocardial Infarction Genetics C, Kathiresan S, Voight BF, Purcell S, Musunuru K, Ardissino D et al (2009) Genome-wide association of early-onset myocardial infarction with single nucleotide polymorphisms and copy number variants. Nat Genet 41 (3):334–341. doi:10.1038/ng.327, PubMed PMID: 19198609; PubMed Central PMCID: PMCPMC2681011
229. Grallert H, Dupuis J, Bis JC, Dehghan A, Barbalic M, Baumert J et al (2012) Eight genetic loci associated with variation in lipoprotein-associated phospholipase A2 mass and activity and coronary heart disease: meta-analysis of genome-wide association studies from five community-based studies. Eur Heart J 33(2):238–251. doi:10.1093/eurheartj/ehr372, PubMed PMID: 22003152; PubMed Central PMCID: PMCPMC3258449
230. Chu AY, Guilianini F, Grallert H, Dupuis J, Ballantyne CM, Barratt BJ et al (2012) Genome-wide association study evaluating lipoprotein-associated phospholipase A2 mass and activity at baseline and after rosuvastatin therapy. Circ Cardiovasc Genet 5(6):676–685. doi:10.1161/CIRCGENETICS.112.963314
231. Tonjes A, Koriath M, Schleinitz D, Dietrich K, Bottcher Y, Rayner NW et al (2009) Genetic variation in GPR133 is associated with height: genome wide association study in the self-contained population of Sorbs. Hum Mol Genet 18(23):4662–4668
232. Yashin AI, Wu D, Arbeev KG, Ukraintseva SV (2010) Joint influence of small-effect genetic variants on human longevity. Aging (Albany, NY) 2(9):612–620, PubMed PMID: 20834067; PubMed Central PMCID: PMCPMC2984609
233. Melzer D, Perry JR, Hernandez D, Corsi AM, Stevens K, Rafferty I et al (2008) A genome-wide association study identifies protein quantitative trait loci (pQTLs). PLoS Genet 4(5), e1000072. doi:10.1371/journal.pgen.1000072, PubMed PMID: 18464913; PubMed Central PMCID: PMCPMC2362067
234. Marroni F, Pfeufer A, Aulchenko YS, Franklin CS, Isaacs A, Pichler I et al (2009) A genome-wide association scan of RR and QT interval duration in 3 European genetically isolated populations: the EUROSPAN project. Circ Cardiovasc Genet 2(4):322–328. doi:10.1161/CIRCGENETICS.108.833806, PubMed PMID: 20031603; PubMed Central PMCID: PMCPMC2760953
235. Xie T, Deng L, Mei P, Zhou Y, Wang B, Zhang J et al (2014) Genome-wide association study combining pathway analysis for typical sporadic amyotrophic lateral sclerosis in Chinese Han populations. Neurobiol Aging 35(7):1778e9–e23. doi:10.1016/j.neurobiolaging.2014.01.014
236. Lee Y, Yoon KA, Joo J, Lee D, Bae K, Han JY et al (2013) Prognostic implications of genetic variants in advanced non-small cell lung cancer: a genome-wide association study. Carcinogenesis 34(2):307–313. doi:10.1093/carcin/bgs356
237. Lemaitre RN, Tanaka T, Tang W, Manichaikul A, Foy M, Kabagambe EK et al (2011) Genetic loci associated with plasma phospholipid n-3 fatty acids: a meta-analysis of genome-wide association studies from the CHARGE Consortium. PLoS Genet 7(7), e1002193. doi:10.1371/journal.pgen.1002193, PubMed PMID: 21829377; PubMed Central PMCID: PMCPMC3145614
238. Tao S, Wang Z, Feng J, Hsu FC, Jin G, Kim ST et al (2012) A genome-wide search for loci interacting with known prostate cancer risk-associated genetic variants. Carcinogenesis 33 (3):598–603. doi:10.1093/carcin/bgr316, PubMed PMID: 22219177; PubMed Central PMCID: PMCPMC3291863

239. Wu C, Miao X, Huang L, Che X, Jiang G, Yu D et al (2012) Genome-wide association study identifies five loci associated with susceptibility to pancreatic cancer in Chinese populations. Nat Genet 44(1):62–66. doi:10.1038/ng.1020
240. Kou I, Takahashi Y, Johnson TA, Takahashi A, Guo L, Dai J et al (2013) Genetic variants in GPR126 are associated with adolescent idiopathic scoliosis. Nat Genet 45(6):676–679. doi:10.1038/ng.2639
241. He M, Xu M, Zhang B, Liang J, Chen P, Lee JY et al (2015) Meta-analysis of genome-wide association studies of adult height in East Asians identifies 17 novel loci. Hum Mol Genet 24 (6):1791–1800. doi:10.1093/hmg/ddu583, PubMed PMID: 25429064; PubMed Central PMCID: PMCPMC4351379
242. Gudbjartsson DF, Walters GB, Thorleifsson G, Stefansson H, Halldorsson BV, Zusmanovich P et al (2008) Many sequence variants affecting diversity of adult human height. Nat Genet 40 (5):609–615. doi:10.1038/ng.122
243. Lettre G, Jackson AU, Gieger C, Schumacher FR, Berndt SI, Sanna S et al (2008) Identification of ten loci associated with height highlights new biological pathways in human growth. Nat Genet 40(5):584–591. doi:10.1038/ng.125, PubMed PMID: 18391950; PubMed Central PMCID: PMCPMC2687076
244. Liu JZ, Medland SE, Wright MJ, Henders AK, Heath AC, Madden PA et al (2010) Genome-wide association study of height and body mass index in Australian twin families. Twin Res Hum Genet 13(2):179–193. doi:10.1375/twin.13.2.179, PubMed PMID: 20397748; PubMed Central PMCID: PMCPMC3232006
245. Soranzo N, Rivadeneira F, Chinappen-Horsley U, Malkina I, Richards JB, Hammond N et al (2009) Meta-analysis of genome-wide scans for human adult stature identifies novel Loci and associations with measures of skeletal frame size. PLoS Genet 5(4), e1000445. doi:10.1371/journal.pgen.1000445, PubMed PMID: 19343178; PubMed Central PMCID: PMCPMC2661236
246. Lango Allen H, Estrada K, Lettre G, Berndt SI, Weedon MN, Rivadeneira F et al (2010) Hundreds of variants clustered in genomic loci and biological pathways affect human height. Nature 467(7317):832–838. doi:10.1038/nature09410, PubMed PMID: 20881960; PubMed Central PMCID: PMCPMC2955183
247. Berndt SI, Gustafsson S, Magi R, Ganna A, Wheeler E, Feitosa MF et al (2013) Genome-wide meta-analysis identifies 11 new loci for anthropometric traits and provides insights into genetic architecture. Nat Genet 45(5):501–512. doi:10.1038/ng.2606, PubMed PMID: 23563607; PubMed Central PMCID: PMCPMC3973018
248. Wood AR, Esko T, Yang J, Vedantam S, Pers TH, Gustafsson S et al (2014) Defining the role of common variation in the genomic and biological architecture of adult human height. Nat Genet 46(11):1173–1186. doi:10.1038/ng.3097, PubMed PMID: 25282103; PubMed Central PMCID: PMCPMC4250049
249. Hancock DB, Eijgelsheim M, Wilk JB, Gharib SA, Loehr LR, Marciante KD et al (2010) Meta-analyses of genome-wide association studies identify multiple loci associated with pulmonary function. Nat Genet 42(1):45–52. doi:10.1038/ng.500, PubMed PMID: 20010835; PubMed Central PMCID: PMCPMC2832852
250. Hancock DB, Artigas MS, Gharib SA, Henry A, Manichaikul A, Ramasamy A et al (2012) Genome-wide joint meta-analysis of SNP and SNP-by-smoking interaction identifies novel loci for pulmonary function. PLoS Genet 8(12), e1003098. doi:10.1371/journal.pgen.1003098, PubMed PMID: 23284291; PubMed Central PMCID: PMCPMC3527213
251. Wilk JB, Shrine NR, Loehr LR, Zhao JH, Manichaikul A, Lopez LM et al (2012) Genome-wide association studies identify CHRNA5/3 and HTR4 in the development of airflow obstruction. Am J Respir Crit Care Med 186(7):622–632. doi:10.1164/rccm.201202-0366OC, PubMed PMID: 22837378; PubMed Central PMCID: PMCPMC3480517
252. Del-Aguila JL, Beitelshees AL, Cooper-Dehoff RM, Chapman AB, Gums JG, Bailey K et al (2014) Genome-wide association analyses suggest NELL1 influences adverse metabolic

response to HCTZ in African Americans. Pharmacogenomics J 14(1):35–40. doi:10.1038/tpj.2013.3, PubMed PMID: 23400010; PubMed Central PMCID: PMCPMC3812324
253. van der Valk RJ, Kreiner-Moller E, Kooijman MN, Guxens M, Stergiakouli E, Saaf A et al (2015) A novel common variant in DCST2 is associated with length in early life and height in adulthood. Hum Mol Genet 24(4):1155–1168. doi:10.1093/hmg/ddu510, PubMed PMID: 25281659; PubMed Central PMCID: PMCPMC4447786
254. Kiel DP, Demissie S, Dupuis J, Lunetta KL, Murabito JM, Karasik D (2007) Genome-wide association with bone mass and geometry in the Framingham Heart Study. BMC Med Genet 8 (Suppl 1):S14. doi:10.1186/1471-2350-8-S1-S14, PubMed PMID: 17903296; PubMed Central PMCID: PMCPMC1995606
255. Adkins DE, Aberg K, McClay JL, Bukszar J, Zhao Z, Jia P et al (2011) Genomewide pharmacogenomic study of metabolic side effects to antipsychotic drugs. Mol Psychiatry 16(3):321–332. doi:10.1038/mp.2010.14, PubMed PMID: 20195266; PubMed Central PMCID: PMCPMC2891163
256. Bush WS, Moore JH (2012) Chapter 11: Genome-wide association studies. PLoS Comput Biol 8(12), e1002822. doi:10.1371/journal.pcbi.1002822, PubMed PMID: 23300413; PubMed Central PMCID: PMCPMC3531285
257. Consortium GT (2013) The Genotype-Tissue Expression (GTEx) project. Nat Genet 45 (6):580–585. doi:10.1038/ng.2653, PubMed PMID: 23715323; PubMed Central PMCID: PMCPMC4010069
258. Kim JJ, Park YM, Baik KH, Choi HY, Yang GS, Koh I et al (2012) Exome sequencing and subsequent association studies identify five amino acid-altering variants influencing human height. Hum Genet 131(3):471–478. doi:10.1007/s00439-011-1096-4
259. Fischer L, Wilde C, Schöneberg T, Liebscher I (2016) Functional relevance of naturally occurring mutations in adhesion G protein-coupled receptor ADGRD1 (GPR133). BMC Genomics (in press)
260. Stäubert C, Le Duc D, Schöneberg T (2014) Examining the dynamic evolution of G protein-coupled receptors. In: Stevens CW (ed) G protein-coupled receptor genetics: research and methods in the post-genomic era. Springer, New York, NY, pp 23–44
261. Rosenthal SM, Feldman BJ, Vargas GA, Gitelman SE (2006) Nephrogenic syndrome of inappropriate antidiuresis (NSIAD): a paradigm for activating mutations causing endocrine dysfunction. Pediatr Endocrinol Rev 4(Suppl 1):66–70
262. Mendes HF, van der Spuy J, Chapple JP, Cheetham ME (2005) Mechanisms of cell death in rhodopsin retinitis pigmentosa: implications for therapy. Trends Mol Med 11(4):177–185. doi:10.1016/j.molmed.2005.02.007
263. Schoneberg T, Schulz A, Biebermann H, Hermsdorf T, Rompler H, Sangkuhl K (2004) Mutant G-protein-coupled receptors as a cause of human diseases. Pharmacol Ther 104 (3):173–206
264. Spiegel AM (2007) Inherited endocrine diseases involving G proteins and G protein-coupled receptors. Endocr Dev 11:133–144. doi:10.1159/0000111069
265. Thompson MD, Hendy GN, Percy ME, Bichet DG, Cole DE (2014) G protein-coupled receptor mutations and human genetic disease. Methods Mol Biol 1175:153–187. doi:10.1007/978-1-4939-0956-8_8
266. Vassart G, Costagliola S (2011) G protein-coupled receptors: mutations and endocrine diseases. Nat Rev Endocrinol 7(6):362–372. doi:10.1038/nrendo.2011.20
267. Robinson A, Escuin S, Doudney K, Vekemans M, Stevenson RE, Greene ND et al (2012) Mutations in the planar cell polarity genes CELSR1 and SCRIB are associated with the severe neural tube defect craniorachischisis. Hum Mutat 33(2):440–447. doi:10.1002/humu.21662
268. Gonzalez-Garay ML, Aldrich MB, Rasmussen JC, Guilliod R, Lapinski PE, King PD et al (2016) A novel mutation in CELSR1 is associated with hereditary lymphedema. Vasc Cell 8:1. doi:10.1186/s13221-016-0035-5, PubMed PMID: 26855770; PubMed Central PMCID: PMCPMC4743364
269. Piao X, Hill RS, Bodell A, Chang BS, Basel-Vanagaite L, Straussberg R et al (2004) G protein-coupled receptor-dependent development of human frontal cortex. Science 303 (5666):2033–2036

270. Sigoillot SM, Monk KR, Piao X, Selimi F, Harty BL (2016) Adhesion G protein-coupled receptors in the nervous system: from synapse and dendrite morphogenesis to myelination. In: Langenhan T, Schöneberg T (eds) Adhesion G protein-coupled receptors: molecular, physiological and pharmacological principles in health and disease. Springer, Heidelberg
271. Weston MD, Luijendijk MW, Humphrey KD, Moller C, Kimberling WJ (2004) Mutations in the VLGR1 gene implicate G-protein signaling in the pathogenesis of Usher syndrome type II. Am J Hum Genet 74(2):357–366
272. Knapp B, Wolfrum U (2016) Adhesion G protein-coupled receptor-related protein networks. In: Langenhan T, Schöneberg T (eds) Adhesion G protein-coupled receptors: molecular, physiological and pharmacological principles in health and disease. Springer, Heidelberg
273. Ravenscroft G, Nolent F, Rajagopalan S, Meireles AM, Paavola KJ, Gaillard D et al (2015) Mutations of GPR126 are responsible for severe arthrogryposis multiplex congenita. Am J Hum Genet 96(6):955–961. doi:10.1016/j.ajhg.2015.04.014, PubMed PMID: 26004201; PubMed Central PMCID: PMCPMC4457946
274. Boyden SE, Desai A, Cruse G, Young ML, Bolan HC, Scott LM et al (2016) Vibratory urticaria associated with a missense variant in ADGRE2. N Engl J Med 374(7):656–663. doi:10.1056/NEJMoa1500611, PubMed PMID: 26841242; PubMed Central PMCID: PMCPMC4782791
275. Nieberler M, Kittel RJ, Petrenko AG, Lin H-H, Langenhan T (2016) Control of adhesion GPCR function through proteolytic processing. In: Langenhan T, Schöneberg T (eds) Adhesion G protein-coupled receptors: molecular, physiological and pharmacological principles in health and disease. Springer, Heidelberg
276. Scholz N, Monk KR, Kittel RJ, Langenhan T (2016) Adhesion GPCRs as a putative class of metabotropic mechanosensors. In: Langenhan T, Schöneberg T (eds) Adhesion G protein-coupled receptors: molecular, physiological and pharmacological principles in health and disease. Springer, Heidelberg
277. Hamann J, Hsiao C-C, Lee CS, Ravichandran KS, Lin H-H (2016) Adhesion GPCRs as modulators of immune cell function. In: Langenhan T, Schöneberg T (eds) Adhesion G protein-coupled receptors: molecular, physiological and pharmacological principles in health and disease. Springer, Heidelberg
278. Tissir F, Qu Y, Montcouquiol M, Zhou L, Komatsu K, Shi D et al (2010) Lack of cadherins Celsr2 and Celsr3 impairs ependymal ciliogenesis, leading to fatal hydrocephalus. Nat Neurosci 13(6):700–707. doi:10.1038/nn.2555
279. Cortijo C, Gouzi M, Tissir F, Grapin-Botton A (2012) Planar cell polarity controls pancreatic beta cell differentiation and glucose homeostasis. Cell Rep 2(6):1593–1606. doi:10.1016/j.celrep.2012.10.016, PubMed PMID: 23177622; PubMed Central PMCID: PMCPMC3606931
280. Kjolby M, Andersen OM, Breiderhoff T, Fjorback AW, Pedersen KM, Madsen P et al (2010) Sort1, encoded by the cardiovascular risk locus 1p13.3, is a regulator of hepatic lipoprotein export. Cell Metab 12(3):213–223. doi:10.1016/j.cmet.2010.08.006
281. Parker SC, Gartner J, Cardenas-Navia I, Wei X, Ozel Abaan H, Ajay SS et al (2012) Mutational signatures of de-differentiation in functional non-coding regions of melanoma genomes. PLoS Genet 8(8), e1002871. doi:10.1371/journal.pgen.1002871, PubMed PMID: 22912592; PubMed Central PMCID: PMCPMC3415438
282. Martincorena I, Roshan A, Gerstung M, Ellis P, Van Loo P, McLaren S et al (2015) Tumor evolution. High burden and pervasive positive selection of somatic mutations in normal human skin. Science 348(6237):880–886. doi:10.1126/science.aaa6806, PubMed PMID: 25999502; PubMed Central PMCID: PMCPMC4471149
283. Pabst C, Bergeron A, Lavallee VP, Yeh J, Gendron P, Norddahl GL et al (2016) GPR56 identifies primary human acute myeloid leukemia cells with high repopulating potential in vivo. Blood 127(16):2018–2027. doi:10.1182/blood-2015-11-683649, Epub 2016 Feb 1
284. Tang X, Jin R, Qu G, Wang X, Li Z, Yuan Z et al (2013) GPR116, an adhesion G-protein-coupled receptor, promotes breast cancer metastasis via the Galphaq-p63RhoGEF-Rho GTPase pathway. Cancer Res 73(20):6206–6218. doi:10.1158/0008-5472.CAN-13-1049

Part II

Adhesion GPCRs as Pharmakotargets in Organ Function and Development

Adhesion GPCRs as a Putative Class of Metabotropic Mechanosensors

Nicole Scholz, Kelly R. Monk, Robert J. Kittel, and Tobias Langenhan

Graphical Abstract

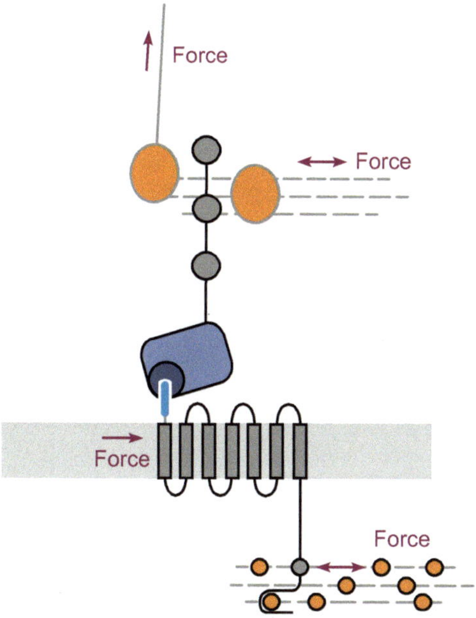

Adhesion GPCRs as mechanosensors. Different aGPCR homologs and their cognate ligands have been described in settings, which suggest that they function in a mechanosensory capacity. For details, see text

N. Scholz (✉) • R.J. Kittel • T. Langenhan (✉)
Department of Neurophysiology, Institute of Physiology, University of Würzburg, Röntgenring 9, Würzburg 97070, Germany
e-mail: nicole.scholz@uni-wuerzburg.de; tobias.langenhan@gmail.com

K.R. Monk
Department of Developmental Biology, Hope Center for Neurologic Disorders, Washington University School of Medicine, St. Louis 63110, MO, USA

T. Langenhan, T. Schöneberg (eds.), *Adhesion G Protein-coupled Receptors*,
Handbook of Experimental Pharmacology 234, DOI 10.1007/978-3-319-41523-9_10

Contents

Abstract

G protein-coupled receptors (GPCRs) constitute the most versatile superfamily of biosensors. This group of receptors is formed by hundreds of GPCRs, each of which is tuned to the perception of a specific set of stimuli a cell may encounter emanating from the outside world or from internal sources. Most GPCRs are receptive for chemical compounds such as peptides, proteins, lipids, nucleotides, sugars, and other organic compounds, and this capacity is utilized in several sensory organs to initiate visual, olfactory, gustatory, or endocrine signals. In contrast, GPCRs have only anecdotally been implicated in the perception of mechanical stimuli. Recent studies, however, show that the family of adhesion GPCRs (aGPCRs), which represents a large panel of over 30 homologs within the GPCR superfamily, displays molecular design and expression patterns that are compatible with receptivity toward mechanical cues (Fig. 1). Here, we review physiological and molecular principles of established mechanosensors, discuss their relevance for current research of the mechanosensory function of aGPCRs, and survey the current state of knowledge on aGPCRs as mechanosensing molecules.

Keywords

Adhesion GPCR • Mechanosensor • Mechanosensation

1 General Overview of Mechanosensation

Each cell in our body is constantly exposed to a multitude of mechanical cues emanating from cell movements, hydrostatic pressure, shear stress exerted by fluids, or compressive and tensile forces of various origins. In order to react to mechanical conditions, these need to be (1) conveyed to mechanosensitive molecules (a process termed *mechanotransmission*; Fig. 2). (2) These molecular mechanosensors perceive force application through conformational changes (*mechanoreception* or *mechanosensing*). (3) Force-dependent changes in mechanosensor structure are finally handed over to intracellular signaling cascades that are not necessarily force dependent per se, including metabotropic effectors and transcriptional pathways, which finally shape the cellular response to the mechanical change (*mechanoresponse*) [1]. Here, we will concentrate on the role of aGPCRs in the process of mechanoreception.

Mechanosensors frequently constitute integral membrane proteins, and it is commonly accepted that mechanical forces that impinge onto the cell surface affect local and global membrane tension. Thus far, a number of molecules have been attributed with a function in mechanoreception, among them ion channels, cell surface receptors (e.g., integrins, cadherins/selectins), and, more recently, G protein-coupled receptors (GPCRs).

2 Established Mechanosensors

Mechanosensors evolved different receptive strategies to reliably perceive, integrate, and convey mechanical information into the cell. Even though detailed molecular knowledge of the mechanotransduction cascade is, for many mechanosensors, still unknown, an understanding of the receptive strategies of these molecules will help define to what extent aGPCRs serve as metabotropic mechanosensitive devices. Thus, the following section delineates potential mechanisms that underlie mechanotransduction and provides an overview of established mechanosensors.

2.1 Molecular Models of Mechanosensation

Two models currently encompass how mechanical strain is transmitted onto mechanosensors. The "tethered" model is based on the concept that surface molecules are anchored to intra- and/or extracellular components via molecular springs, which convey mechanical forces to membrane integral mechanosensors (e.g., TMC, extracellular; see Sect. 2.2.5 NOMPC, intracellular; see Sect. 2.2.3). Alternatively, in the "membrane" model, mechanosensitivity is shaped by the lateral pressure profile [2] at the receptor/bilayer interface promoting conformational changes of mechanoreceptive proteins. Given that a cell's response to force is signified by its mechanical context by means of the surrounding extracellular matrix (ECM), neighboring cells, and the mechanical properties of the cell itself, it is reasonable to assume that both mechanisms are employed in parallel to ensure reliable integration of physical forces.

2.2 Ionotropic Mechanosensors

Guharay and Sachs were the first to discover that skeletal muscles in chick express an ion channel that can be activated by membrane stretch [3, 4]. After that, several years passed until the first mechanosensitive channel genes, *mscL* and *mscS*, were identified and cloned from prokaryotes [5–7]. In *Bacteria* and *Archaea*, mechanosensitive channels are primarily involved in detecting osmotically induced cell swelling to regulate turgor [6, 8].

A large number of mechanosensitive channels were identified in eukaryotic cells [9] and, based on their ion permeability, can be divided into two classes: excitatory cation-selective ion channels and inhibitory depolarization-gated ion channels. We will briefly discuss examples for both groups below. Furthermore, a novel ion channel family—transmembrane channel-like 1 and 2 (TMC1/TMC2)—was recently identified.

2.2.1 Piezo

Vertebrate Piezo1 and Piezo2 (or Fam38A and Fam38B) channel subunits comprise large membrane proteins with 14 predicted membrane-spanning segments [10]. Heterologous expression of Piezo conveys mechanosensitivity to a mechanically insensitive cell [11]. Furthermore, functional ion channels could be reconstituted from Piezo proteins in liposomes [12]. Thus, ectopically expressed Piezo appears to form mechanosensitive ion channels independent from cytoskeletal elements implying that the channel activating tension originates from the membrane [11–13].

Mammalian *Piezo 1*/*Piezo 2* was detected in the lung, bladder, colon, and skin. *Piezo 2* was additionally enriched in dorsal root ganglia (DRGs) [11], where it acts as a mechanotransducer to regulate the perception of tactile stimuli [14, 15]. Similarly, the *piezo2b* homolog plays a role in zebrafish touch sensation [16]. Moreover, Piezo1 was assigned roles in the regulation of erythrocyte cell volume and vascular development in zebrafish and mouse, respectively [17–19]. The sole Piezo homolog in *Drosophila* (*dmpiezo*) produces, similar to its vertebrate counterpart, mechano-dependent ion currents and is required for the perception of noxious mechanical stimuli in vivo. Interestingly, Piezo appears to act in parallel to Pickpocket (PPK), which belongs to the (DEG)/ENaC family, to regulate mechanical nociception in *Drosophila* larvae [20].

2.2.2 DEG/ENaC Channel

The DEG/ENaC channel family was named after their first members, degenerins (*mec-4 and deg-1*) and epithelial sodium channel (ENaC) from *C. elegans* and mammals, respectively [21–24]. DEG/ENaC channels are expressed in different tissues across a large range of phyla, are usually Na^+-selective, and appear to be activated by a wide spectrum of stimuli including mechanical force [25]. Two transmembrane domains (N- and C-termini intracellular) connected by a large extracellular loop define the protomer architecture of DEG/ENaC proteins, which can form both homo- and heteromeric ion channels [26–28].

The first identified eukaryotic genes involved in touch sensation, *mec-4* and *mec-10 (mec*hanosensory abnormal) [29], were uncovered through forward genetic screens in *C. elegans*, in which mutants were probed for their ability to respond to gentle body touch conveyed through a defined set of touch receptor neurons [29–31]. Genetic, biochemical, functional, and morphological analyses implicate four MEC proteins (MEC-4, MEC-10, MEC-2, MEC-6) to compose mechanosensory channel complexes [32–34]. The pore-forming subunits are presumably contributed by MEC-4 and MEC-10 [23, 32, 35, 36]. Interestingly, the local lipid milieu appears to influence the function of this mechanosensory complex [33, 37–39], suggestive of a membrane stretch-mediated gating mechanism.

2.2.3 TRP Channels

Transient receptor potential (TRP) channels are involved in recognizing a broad range of mechanical and chemical stimuli [40]. Each TRP channel gene belongs to one of seven subfamilies [TRPC (classical), TRPV (vanilloid), TRPM (melastatin), TRPN (NOMPC), TRPA (ANKTM1), TRPML (mucolipin), and TRPP (polycystin)] [41] and encodes a channel subunit with at least six transmembrane-spanning segments and intracellular N- and C-terminal regions with subfamily-specific domain layout [42]. Four subunits assemble into functional homo- or heteromeric ion channels, commonly nonselective for cations and permeable to Ca^{2+} [41, 43].

Mechanosensitivity has been reported for a number of TRP channel subunits; however, to what extent these proteins are directly involved in mechanotransduction remains to be resolved for many family members. Interestingly, TRP channels were also identified as partners of GPCRs and aGPCRs in this context.

For example, TRPC6 is a downstream target of a mechanosensitive GPCR responsible for mediating myogenic vasoconstriction in response to elevated intraluminal pressure [44]. Further, mouse sensory neurons express TRPC1, TRPC3, and TRPC6 [45]. *Trpc3/Trpc6* double knockout mice display altered responses to tactile stimuli. These mutants also exhibit pronounced hearing impairments and vestibular defects consistent with TRPC3 and TRPC6 expression in cochlear and vestibular hair cells [45]. Recent work indicated that TRPC1, TRPC3, TRPC6, and TRPC5 heteromultimerize to contribute to cutaneous and auditory mechanosensation. Moreover, these TRP channel subunits were proposed to contribute indirectly to cochlear mechanotransduction [46].

Further, members of the TRPN family have been associated with the perception of mechanical stimuli in *Drosophila* (no receptor potential; NOMPC) [47, 48], hearing in zebrafish (TRPN1) [49], and proprioception in *C. elegans* (TRP-4) [50]. NOMPC localizes to the ciliary tips of mechanosensory neurons of the Johnston's organ and chordotonal organs in *Drosophila* [51, 52]. NOMPC can be directly mechanically activated and confers mechanosensitivity to otherwise insensitive cells, rendering NOMPC a bona fide mechanotransduction channel [53, 54]. Recent data indicate that the 29 ankyrin repeats of NOMPC's N-terminus constitute a molecular spring that conveys forces generated by the cytoskeleton to modulate ion channel gating [55, 56]. However, loss of *nompC* does not completely abolish fly hearing [47], suggesting the existence of additional transducer molecules [57]. Alternatively, NOMPC may sensitize or adjust the mechanosensory complex for mechanical input [58]. Interestingly, genetic analysis suggests that the aGPCR dCIRL/latrophilin modulates NOMPC activity in sensory chordotonal neurons [59] (see Sect. 3.2).

The *C. elegans* TRPV proteins OSM-9 and OCR-2 form heteromeric channel complexes in sensory neurons that respond to touch and hyperosmolarity [60–62]. Consequently, loss of function of *osm-9* results in worms that are less sensitive to osmotic and mechanical challenges [60, 62]. *Drosophila* encodes TRPV proteins, INACTIVE and NANCHUNG, that are mutually required for the assembly of heteromultimeric channels involved in fly hearing and possibly auditory mechanotransduction [57, 58, 63, 64]. Epistatic analysis implicated NANCHUNG function in a mechanosensory signaling pathway together with the aGPCR latrophilin/CIRL [59].

Finally, TRPP homologs TRPP2 (polycystin 2 or PC2) and TRPP3 (polycystic kidney disease 2-like 1, PKD2L2) have been associated with mechanosensation. TRPP2 forms a receptor-ion channel complex with polycystin 1 (PC1, TRPP1) [65, 66], which localizes to primary cilia of renal epithelial cells and endothelial cells [67–70]. PC1 does not belong to the TRP channel family; it contains 11 -transmembrane-spanning regions and a large extracellular N-terminal domain that promotes cell-cell and cell-matrix interactions [71, 72] (see also [73]). The TRPP2/PC1 complex mediates Ca^{2+} transients in response to ciliary deflections induced by luminal shear stress [67, 68, 74]. Intriguingly, filamin A cross-links TRPP2 and the actin cytoskeleton to regulate stretch-activated cation channels (SACs) involved in cardiovascular pressure sensing [70]. Furthermore, TRPP2 forms ciliary mechanosensitive sensors with TRPV4 to transduce mechanical stress [75].

2.2.4 K_{2P} Channels

K_{2P} channels are comprised of four transmembrane-passing segments, two pore-forming regions, and intracellular N- and C-terminal portions. They assemble into hetero- and homodimeric ion channels. TWIK-related K^+ channels (TREK) and TWIK-related arachidonic acid-stimulated K^+ channel (TRAAK) are membrane tension-gated channels characterized by low mechanical threshold and a broad range of tension activation [76].

TREK-1 and TREK-2 are widely expressed within the central and peripheral nervous systems as well as nonneuronal tissues such as the cardiovascular system, lung, colon, and kidney. By contrast, TRAAK expression appears to be confined to neuronal tissues [77]. In the nervous system, K_{2P} channels are physiologically relevant as they contribute considerably to cell hyperpolarization, which balances/counteracts depolarization-induced action potentials, thereby shaping the electrical response to mechanical force. For example, TRAAK activity counteracts activation of Piezo to curtail action current generation in cultured neuroblastoma cells [78]. Further, TRAAK/TREK and mechanosensitive cation channels are coexpressed in sensory neurons of DRGs [79]. Deletion of *Traak*, *Trek-1*, or

Trek-2 genes in mice renders them hypersensitive to mechanical challenge [80, 81], which is in line with the notion that K_{2P} channels set the mechanical threshold for action potential generation [76]. Evidence for the physiological relevance of TREK-1 in mechanosensation outside the nervous system has emerged as well. Polycystins play a role in force-dependent apoptosis of renal epithelial cells, a function that was shown to rely on the opening of TREK-1 [82]. TREK-1 is enriched in the bladder, uterus, and colon where it regulates stretch-induced contraction of smooth muscle cells [83].

2.2.5 Transmembrane Channel-Like 1 and 2 Proteins

Recently, the transmembrane channel-like 1 and 2 (TMC1 and TMC2) proteins emerged as novel components required for auditory and vestibular mechanosensation in mammals [84–86]. The *Tmc* gene family encodes proteins with at least six transmembrane regions flanked by intracellular N- and C-terminal portions [87] reminiscent of TRP channel proteins. However, TMC proteins do not share sequence homologies with known ion channels [85, 88], even though recent work indicated that *C. elegans tmc-1* assembles into nonselective cation channels when heterologously expressed [89]. Several lines of evidence implicate TMC proteins as key components of the hair cell transduction complex. First, consistent with the mRNA expression pattern [90], TMC1 and TMC2 fusion proteins localize to stereocilia of the inner ear hair cells [86]. Second, mutations of the human *Tmc1* gene have been associated with dominant and recessive nonsyndromic sensorineural hearing deficits [84]. Similarly, dominant and recessive mutations of mouse *Tmc1* result in animals that exhibit defective hearing (*Beethoven*) and deafness (*dn*), respectively [84, 91]. Third, in mice, loss of function of *Tmc1* or *Tmc2* deprives hair cells of mechanosensory responses [90, 92]. Fourth, robust apical FM1-43 dye or gentamicin uptake reported for wild-type hair cells was abolished in the absence of TMC1 and TMC2 [90, 93, 94]. Fifth, expression of either TMC1 or TMC2 rescues loss of hair cell mechanosensitivity displayed by homozygous null mutants [90]. Sixth, protocadherin-15 (PCDH15) together with cadherin-23 forms extracellular tethers (called tip links), which connect adjacent pairs of stereocilia and transmit hair cell bundle deflection-based mechanical forces to mechanosensors [95, 96]. Strikingly, recent work uncovered a direct interaction between PCDH15 and TMC1 as well as TMC2 in fish and mouse [97, 98].

Taken together, these results suggest that TMC proteins are strong candidates for the long sought-after mechanosensitive channel required for mechanotransduction in mammalian hair cells [85].

2.3 Integrins and Cadherins

Cell adhesion enables the generation of mechanical tissue cohesion as well as cell-cell and cell-matrix communication vital for a myriad of physiological processes. But how do cells sense the mechanical context of their microenvironment (e.g., extracellular matrix or adjacent cells), and how are these signals translated into cellular responses? Based on structural and functional properties, integrins and cadherins have been associated with mechanosensation in this vein.

2.3.1 Integrins

Integrins constitute a large family of glycoprotein receptors that bridge cell-cell contacts (cell adhesion) and interconnect intra- and extracellular matrices (ECMs) in metazoans. Integrins exist as heterodimers comprised of non-covalently bound α- and β-subunits, which are combined to form more than 20 distinct, functionally nonredundant receptor variants [99, 100].

Integrin heterodimers are inserted into the plasma membrane through a single transmembrane domain per protomer. The large extracellular domain (ECD) mediates interactions with ECM components (e.g., collagens, fibronectins, or laminins) and adjacent receptors [e.g., vascular cell adhesion molecule-1 (VCAM-1) and intercellular cell adhesion molecule (ICAM)] [100, 101]. The short intracellular domain (ICD) interacts, via cytoplasmic adaptors, with cytoskeletal elements to modulate local actin polymerization and global cytoskeletal dynamics [102, 103]. This molecular layout allows integrins to mechanically interlink the cell's exterior and interior providing the backbone for mechanical stress-induced assembly of focal adhesions (FA) [104–107], stimulation of signaling pathways, and gene transcription [108–110].

Importantly, integrins possess the capacity to confer signals in an "outside-in" or "inside-out" fashion associated with different biological ramifications. "Outside-in" signaling is based on the interaction of integrin ECDs with ECM molecules or opposing cellular receptors. These engagements provide information about ECM rigidity as well as adjacent cell/tissue geometry, which in turn regulate cellular growth, differentiation and apoptosis, cell polarity, and formation of focal adhesion complexes [105, 111–114]. During "inside-out" signaling, intracellular activators such as kindlins or talins associate with integrin ICDs to induce conformational changes that alter the affinity to extracellular ligands ("integrin activation"). Hence, this signaling mode affects cell adhesion and migration as well as ECM assembly [102]. Interestingly, the ionotropic Piezo1/Fam38A channel subunit was shown to induce integrin activation through an R-Ras-dependent mechanism [115].

In sum, integrin-based adhesion complexes are fit to perceive and integrate external forces into the preexisting cellular mechanical context to balance the ECM resistance according to tension forces generated by the cytoskeleton.

2.3.2 Cadherins

Cadherins are a large group of cell surface receptors that mediate Ca^{2+}-dependent cell adhesion [116]. The cadherin family includes a vast number of cell surface molecules; however, classical cadherins (E-, N-, P-cadherins) and the related desmosomal cadherins are, thus far, the best characterized. They feature cadherin repeat-containing ECDs, a single membrane-spanning region, and ICDs which interact with β-catenin (classical cadherins), γ-catenin (desmosomal cadherins), and components of the cytoskeleton [117].

Cadherins engage in ECD-dependent homo- and heterophilic interactions that may mediate adhesive and selectivity functions, respectively [118]. For example, force measurements uncovered that cadherins establish three spatially distinct bonds at different ECD regions that possess varying kinetic and mechanical properties [119, 120]. In contrast, ICDs associate with F-actin via catenin/vinculin interactions [121–124]. This structural setup allows cadherin complexes to act as mechanosensors that actively sense and transduce exogenous and endogenous mechanical fluctuations to trigger biochemical changes that direct cell responses [125]. Strikingly, the tensile force administered to an adhesion site is directly proportional to ECM stiffness [126].

Several lines of evidence support the role of cadherin-based adhesive junctions in mechanotransduction. First, cell-cell junctions undergo force-dependent remodeling not only in vitro but also in vivo [127, 128]. In *Drosophila* and *C. elegans*, cytoskeletal-provoked tension at cadherin junction coincides with a decline in junction expansion in various developmental processes. For example, the association of the actomyosin apparatus with cadherin junctions is a prerequisite for contraction-driven apical constriction during *C. elegans* gastrulation. A similar mechanism has been reported for ventral furrow formation in *Drosophila* [129]. Intriguingly, recent work demonstrated that during germ band extension in *Drosophila*, junction shrinkage appeared to result from tension-generated cadherin accumulation, which was counteracted by cadherin endocytosis [130]. Second, the mechanics of intercellular junctions (e.g., cell traction, adhesion strength, and junction rigidity) change according to the mechanical environment [122, 131–134]. In *Drosophila*, junctional mechanics are regulated in a forward loop that controls cadherin levels with respect to actomyosin density to enable adequate tissue morphogenesis [130, 135].

Taken together, cadherin complexes probe their environment for tensional changes to preserve the mechanical integrity of cellular junctions and to regulate morphogenesis and homeostasis [136–139].

2.4 Metabotropic Mechanosensors

GPCRs are biosensors vital for the transduction of light, olfactory and gustatory stimuli, as well as neurotransmitters and hormones into biochemical signals and thus shape a multitude of physiological processes [140]. Based on this perception profile, GPCRs were classically considered chemoreceptors. However, in recent years, mounting evidence suggests that mechanical forces can also be perceived through GPCRs. Therefore, GPCRs most likely constitute polymodal sensors with both chemoreceptive and mechanoreceptive properties.

The first mechanosensitive GPCR identified was the angiotensin II type 1 receptor (AT_1R), which mediates functions in preload-induced cardiac hypertrophy in vitro and in vivo. In contrast to autocrine-mediated vasoconstriction, activity of AT_1R is brought about independently of angiotensin II and facilitates intracellular signaling through $G_{q/11}$ proteins [141, 142]. In addition to this pathophysiological function, intraluminal pressure stimulates AT_1Rs located in the arterial blood vessels causing myogenic vasoconstriction, a mechanism referred to as the Bayliss effect [44]. Mounting evidence implicates PLC kinases in mechano-dependent signaling pathways, which may subsequently trigger activation of ionotropic TRPC channels (see Sect. 2.2.3) to set arterial myogenic tone [44, 141, 143, 144].

Strikingly, AT_1R undergoes mechanically induced conformational changes that stabilize the seventh transmembrane domain near the agonist-binding pocket, a molecular organization that is different from that induced by agonist and inverse agonist signaling [145–147], suggesting stimulus modality-dependent signaling modes of this GPCR.

Thus far, eight additional putative mechanosensitive $G_{q/11}$ protein-coupled receptors have been described [148]. For example, time-resolved fluorescence microscopy of endothelial cells unveiled that mechanical strain causes a ligand-independent rise in bradykinin 2 receptor (B_2R) activity [149]. The authors suggest that, similar to AT_1R, conformational alterations of B_2R are due to tension-based properties of the lipid bilayer. However, it remains unclear if mechanical stimulation of B_2R is causally linked to effective G protein coupling. Another interesting example of a potential mechanosensitive GPCR derives from type 1 parathyroid hormone receptor (PTH_1R) essential for Ca^{2+} homeostasis in osteoblasts [150, 151]. Bone formation and bone mass regulation rely on the availability of parathyroid hormone (PTH) and mechanical loading, signal modalities proposed to converge at PTH_1R [151]. Evidence for a role in mechanoreception was also reported for dopamine receptor type 5 (D_5R) and formyl peptide receptor (FPR_1), which are both, similar to PTH_1R, excitable through fluid flow-generated shear stress in endothelial cells and neutrophils, respectively [152, 153]. Interestingly, D_5R localizes to primary cilia in vitro and in vivo and regulates their length through cofilin and actin polymerization. In addition, the authors report that ciliary sensitivity to fluid shear stress can be modified through chemosensory properties of the receptor [152]. Furthermore, electrophysiological recordings unveiled hypotonicity-dependent responses of heterologously expressed vasopressor receptors, ET_A endothelin receptor (ET_AR) and V_{1A} receptor ($V_{1A}R$), as well as H_1 histamine receptor (H_1R) and M_5 muscarinic acetylcholine receptor (M_5R) [44].

3 Adhesion GPCRs in Mechanosensation

A novel intriguing line of evidence for the mechanoreceptive role of GPCRs derives from the large family of aGPCRs, which stand out from canonical GPCRs because of their structural and functional profile as well as the conspicuous lack of soluble ligands [154]. Thus, are aGPCRs particularly prone to sense mechanical cues?

Three members of the aGPCR class, GPR56/ADGRG1, GPR126/ADGRG6, and latrophilin/CIRL/ADGRL, have been recently suggested to be involved in mechanosensation (Fig. 1) [59, 155, 156]. Like canonical GPCRs, aGPCRs possess a seven-transmembrane helix region (7TM) that can intracellularly couple to

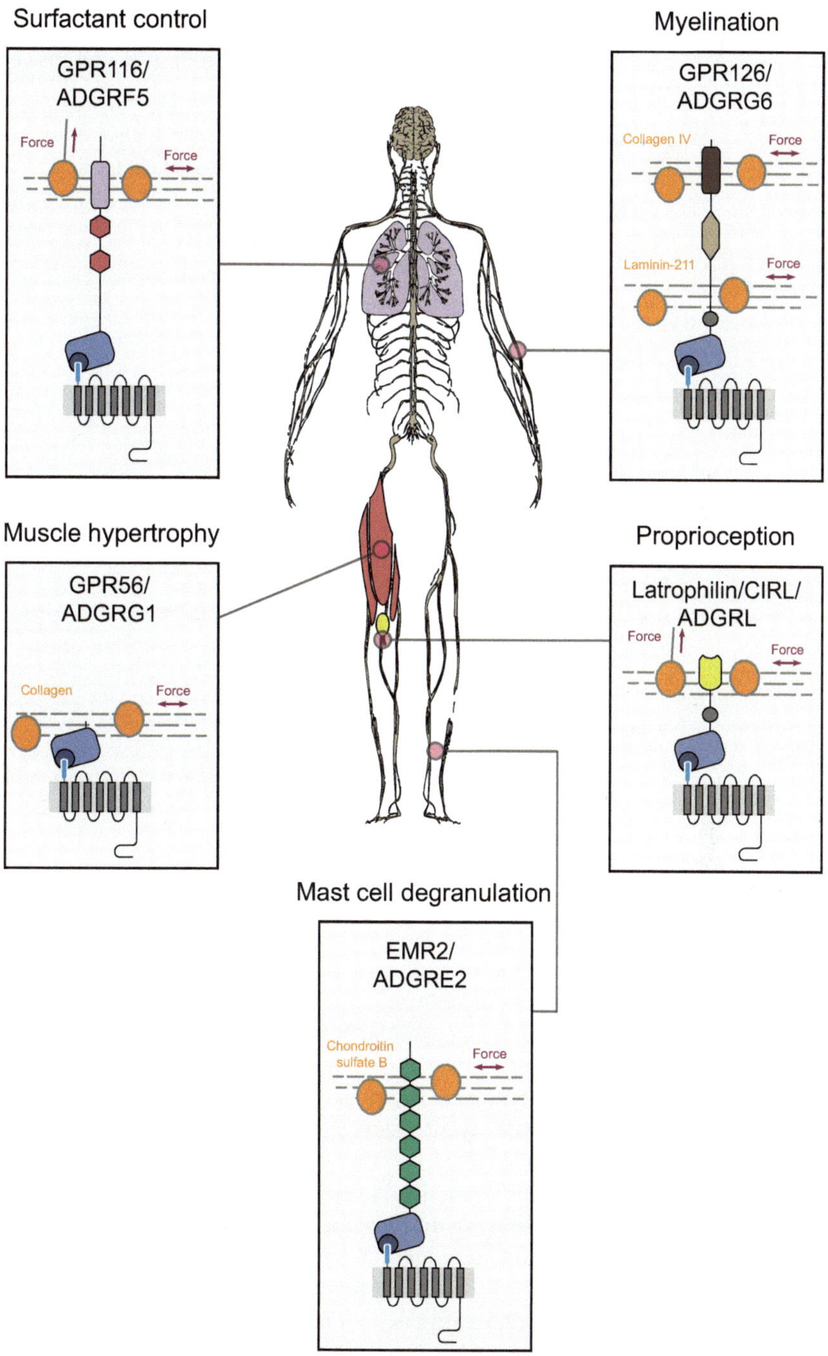

Fig. 1 Adhesion GPCRs as mechanosensors. Different aGPCR homologs and their cognate ligands have been described in settings, which suggest that they function in a mechanosensory capacity. For details, see text

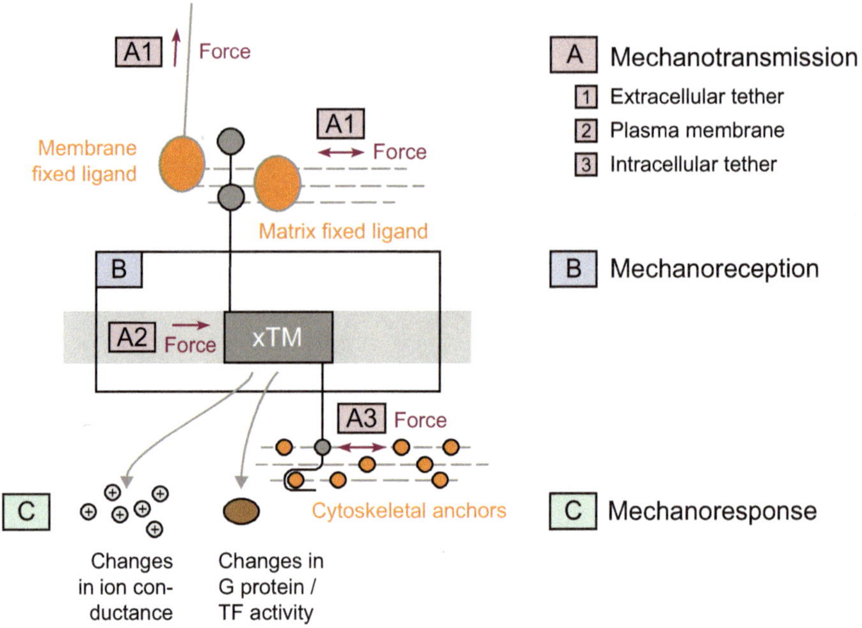

Fig. 2 Principal elements of mechanotransduction. The process of mechanotransduction can be grossly divided into three steps. (**a**) Mechanotransmission (indicated in *light pink*) that, according to current models, can be conveyed through extracellular or intracellular tethers and/or the plasma membrane. (**b**) Mechanosensation (*blue*), perceived through force-dependent conformational changes of membrane integral sensor proteins (*gray*). (**c**) Mechanoresponse (*green*), which induces changes in cell physiology (e.g., ion channel conductances, metabolic states, or modulation of transcriptional activity). xTM = variable number of transmembrane helices depending on the mechanosensor protein architecture (TRP channels, 6TM; aGPCR, 7TM; polycystin 1, 11TM; Piezo, 14TM)

heterotrimeric G proteins capable of triggering a multitude of signaling cascades that determine cellular responses (Fig. 2; see [157, 158]).

Further, aGPCRs possess large extracellular N-termini, which for many aGPCRs contain domains that are involved in cell-cell or cell-matrix adhesion [154] (see also [159]). Uniquely, the N-termini of many aGPCRs can be separated from the 7TM through an autoproteolytic event catalyzed by the GPCR proteolytic site (GPS) motif [160, 161], which is part of a much larger protein fold (GPCR autoproteolysis-inducing (GAIN) domain) [162]. Receptor cleavage is thought to occur early in the secretory pathway during protein maturation and generates an N-terminal fragment (NTF) and C-terminal fragment (CTF), which are predicted to traffic together to the cell surface where they are located as non-covalently bound heterodimers (Fig. 3; see also [73, 159]).

For several aGPCRs, a cryptic tethered ligand, termed the *Stachel* sequence, has been described, which acts as a potent agonist for downstream signaling [163–166] (Fig. 3; see also [157, 158]). Yet, crystal structures of aGPCR GAIN domains

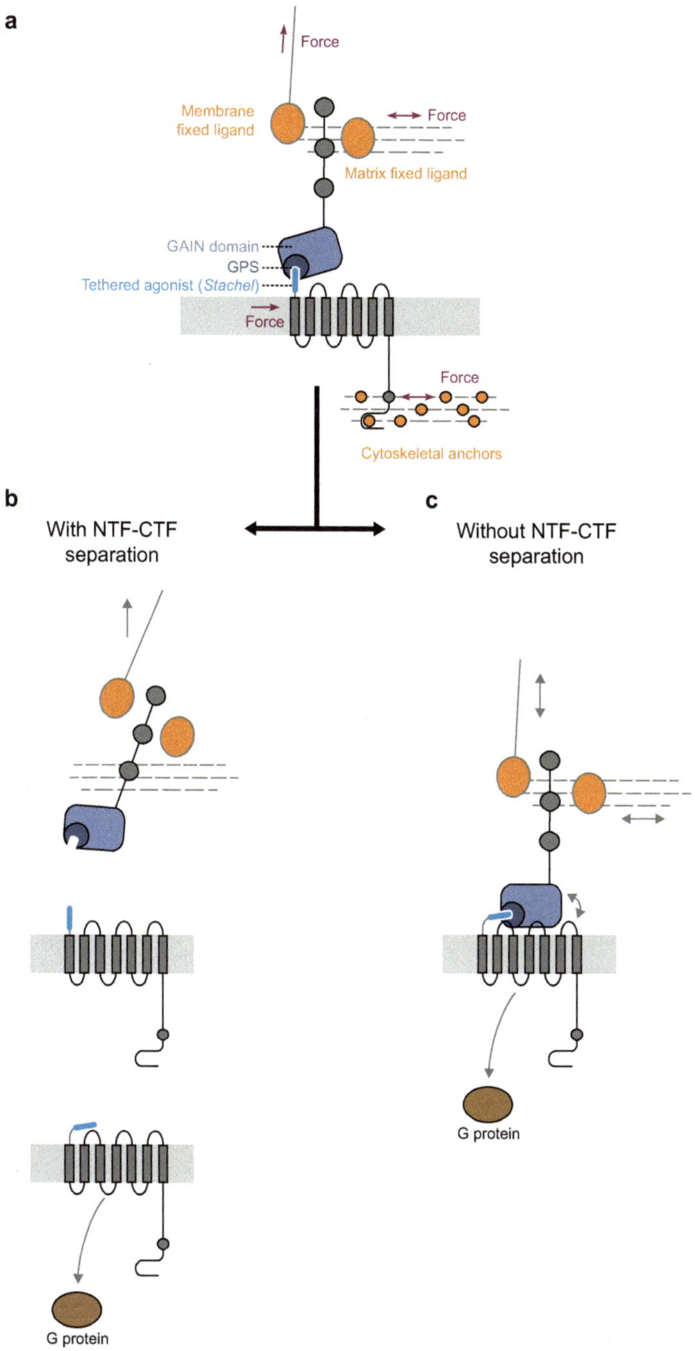

Fig. 3 Sequence of putative events during aGPCR-mediated mechanotransduction. (**a**) aGPCRs are subjected to mechanical force through their adhesive ECDs, at the level of the membrane, or through intracellular anchors. (**b**) Depending on whether the receptor is autoproteolyzed and

demonstrate that the *Stachel* sequence is deeply buried between two β-sheets [162], suggesting that the tethered agonist, at least for some aGPCRs, may not be readily exposed. How, then, might the *Stachel* sequence "escape" the ensconcement of the GAIN domain β-sheets in order to confer a signal to the receptor's 7TM stretch? Recent studies suggest that mechanical activation may be at play, which could induce conformational changes within the GAIN domain or complementary portions of the receptor to liberate the *Stachel* sequence and to allow receptor activation (Fig. 3).

3.1 GPR126/ADGRG6

As described in [167], myelin is the multilamellar sheath generated by specialized glial cells called Schwann cells in the peripheral nervous system and oligodendrocytes in the central nervous system. In the peripheral nervous system, GPR126 couples to $G\alpha_s$ and elevates cAMP in Schwann cells to initiate myelination (Fig. 1) [155, 163, 168–171]. The *Stachel* sequence of GPR126 is a potent agonist to trigger cAMP elevation, and analysis of zebrafish mutants has recently provided evidence that *Stachel*-mediated signaling is critically important in vivo. Liebscher and colleagues generated a new mutant allele of *gpr126*, $gpr126^{stl215}$, in which two amino acids essential for signaling were precisely deleted from the *Stachel* sequence. In vitro analysis demonstrated that this mutant receptor could be cleaved and that it trafficked appropriately to the cell surface, although it was incapable of signaling via cAMP. Accordingly, in vivo, $gpr126^{stl215}$ mutant Schwann cells could not initiate myelination [163]. This work suggested a model in which *Stachel*-mediated activation of Gpr126 is required for cAMP elevation and subsequent myelin initiation.

To determine how the *Stachel* sequence of GPR126 might be exposed during Schwann cell myelination, Petersen and colleagues investigated the relationship between one GPR126 binding partner, laminin-211, and GPR126 signaling [155] (discussed further in [167]). Laminin-211 is a heterotrimeric protein encoded by *Lama2*, *Lamb1*, and *Lamc1* genes [172], and like GPR126, laminin-211 is required for Schwann cell development and myelination [173]. Overexpression of *lama2* in zebrafish rescues myelination in *gpr126* hypomorphic mutants in a cAMP-dependent manner, suggesting that laminin-211 activates $G\alpha_s$ signaling.

Fig. 3 (continued) present as an NTF/CTF heterodimer at the cell surface, force transmission may separate the NTF from the CTF, thereby exposing the cryptic tethered agonist that interacts with the 7TM domain and triggers intracellular responses. (**c**) Several aGPCRs cannot be autoproteolytically cleaved, yet they appear to harbor a *Stachel* sequence that can stimulate receptor activity. This indicates that the tethered agonist can interact with the 7TM domain even if it remains associated with GAIN domain, possibly through conformational changes of the GAIN domain that render the agonist accessible. Note that this scheme illustrates putative intramolecular interactions between GAIN/*Stachel* and the 7TM region of aGPCRs; however, intermolecular interactions are conceivable as well

Intriguingly, addition of laminin-211 to GPR126 in heterologous cell systems suppressed cAMP elevation under standard culture conditions and elevated cAMP levels under dynamic conditions (vibration or shaking) [155]. These data can be reconciled by observations from related aGPCRs, noted above, that the *Stachel* sequence is buried, not exposed, between two β-sheets of the GAIN domain [162]. The *Stachel* sequence is so deeply buried that significant structural changes might be required for *Stachel*-mediated receptor activation to occur (Fig. 3). In vitro analyses of mutant receptors demonstrated that laminin-211 addition facilitates greater GPR126 activation than dynamic forces alone and that dynamic force signaling requires the *Stachel* sequence to activate GPR126.

These in vitro findings suggested that laminin-211 enhances GPR126 structural changes to allow for *Stachel*-mediated signaling, and the overexpression studies in zebrafish demonstrated that increased *lama2* levels can drive Gpr126-dependent myelination. What physical forces might laminin-211 impart upon GPR126 in vivo? Interestingly, laminin-211 polymerization is known to be essential for Schwann cell development; $Lama2^{dy2j}$ mouse mutants have a spontaneous mutation that leads to aberrant splicing and deletion of the laminin α2 polymerization domain, and these mice have defects in Schwann cell development and myelination [174–176]. To test if laminin α2 polymerization could be an activating force on Gpr126, reminiscent of physical forces in vitro, Petersen and colleagues engineered a non-polymerizable Lama2 overexpression construct to mimic the $Lama2^{dy2j}$ mouse mutation. Unlike wild-type *lama2*, the polymerization-defective *lama2 (dy2j)* was not sufficient to rescue myelination defects in hypomorphic *gpr126* zebrafish mutants [155].

In summary, the in vitro studies suggested the possibility that GPR126 might be sensitive to mechanical force, while the in vivo studies implicated laminin-211 polymerization as a potential source to transmit forces and stimulate mechanosensation in vivo.

3.2 Latrophilins/ADGRL1–ADGRL3

Contemporaneous work in *Drosophila* has even more strongly implicated an aGPCR in mechanosensation, as latrophilin/CIRL was shown to shape the response of chordotonal sensory neurons by determining the sensitivity of certain mechanosensors in a cell-autonomous manner [59]. Chordotonal neurons are peripheral, compound mechanosensory neurons in *Drosophila*, which perceive mechanical signals such as sound, touch, and muscle stretch [47, 52, 177]. Latrophilin/CIRL is robustly expressed in these cells (unpublished data) suggestive of a function in this cell type. Indeed, analyses of a newly engineered *dCirl* null allele revealed many interesting phenotypes consistent with a role for latrophilin/CIRL in mechanosensation. *dCirl* mutant larvae exhibited aberrant crawling patterns, and although locomotion is a complex behavior controlled by both motor outputs and peripheral sensory inputs, this defect could be partially rescued by chordotonal neuron-specific expression of *dCirl*. Moreover, *dCirl* mutant larvae were also less responsive to gentle touch (touch perception is a known function of the chordotonal organ), and this phenotype could only be suppressed by chordotonal neuron-specific re-expression of this aGPCR [59]. Interestingly, the morphology of chordotonal neurons is grossly normal in *dCirl* null mutants, suggesting that latrophilin/CIRL is required specifically for the function of these neurons and not for their development or morphology.

To elucidate the potential role of latrophilin/CIRL in mechanosensation, Scholz and colleagues directly applied mechanical stimulation (vibration stimuli) to the cap cells of the chordotonal organ and simultaneously recorded action currents from chordotonal neurons. While *dCirl* mutant neurons maintained spontaneous activity in the absence of vibration stimuli, the mutant neurons displayed significantly lower action current frequencies across the entire tested vibration spectrum. Compound mutant analyses also demonstrated that *dCirl* genetically interacts with TRP channels, known as mechanosensor molecules of *Drosophila* chordotonal neurons [59] (see also Sect. 2.2.3).

All in all, these studies establish that latrophilin/CIRL can modulate the sensitivity of neuronal mechanosensation (Fig. 1). Given that several cell-specific rescue experiments demonstrate that *dCirl* functions cell autonomously in chordotonal neurons, likely via CTF signaling, it will be interesting to determine if and how *Stachel* sequence signaling modulates these functions of latrophilin/CIRL in mechanosensation (Fig. 3).

3.3 GPR56/ADGRG1

The role of aGPCRs as mechanosensors also extends beyond the nervous system. Skeletal muscle can respond to multiple stimuli, including mechanical tension, which is a potent regulator of muscle mass. In myotubes, a splice isoform of the peroxisome proliferator-activated receptor gamma coactivator 1-alpha (PGC-1α) induces muscle hypertrophy. Interestingly, overexpression of PGC-1α increases *Gpr56* expression, and GPR56 overexpression also induces muscle hypertrophy dependent on $G\alpha_{12/13}$ signaling. Resistance exercise causes muscle hypertrophy, but this phenotype is attenuated in *Gpr56* mutant mice [156] (see also [178]). These results collectively highlight a potential role for Gpr56 in sensing muscle fiber size, which could be mediated by mechanical sensitivity to stretch (Fig. 1).

Furthermore, in the central nervous system, GPR56 is required for proper cortical development (i.e., cortical folding), which has been directly linked to mechanosensation [179]. GPR56 and other aGPCRs including GPR114/ADGRG5 and GPR97/ADGRG3 are also highly expressed on lymphocytes [180]; it is tempting to speculate that signaling via these aGPCRs might be affected during the rolling and extravasation behaviors characteristic of these cells.

3.4 Mechanosensation by Other Adhesion GPCRs

These recent studies implicating Gpr126, latrophilin/CIRL, and GPR56 as putative mechanosensitive receptors raise the interesting and obvious question as to whether other aGPCRs might share this mechanosensitive property.

Indeed, one splice variant of GPR114 is sensitive to mechanical stimulation [166], although it is not yet clear in what physiological context this receptor may partake in mechanoreception. In addition, recent reports implicate *EMR2/ADGRE2* in vibratory urticaria [181], an autosomal dominant disorder that is signified by hives and systemic manifestations in response to cutaneous vibration. Interestingly, mast cells express EMR2, and mechanosensing through this receptor appears to depend on the stability between the NTF and CTF heterodimer of EMR2. A human mutation in the GAIN domain destabilizes this interaction and conceivably results in increased receptor activity upon mechanical challenge, which ultimately leads to massive mast cell degranulation [181]. Moreover, observations from previous studies on several aGPCRs can be interpreted in the light of these new insights to support a unifying model of mechanical susceptibility of this receptor class.

In the murine lung, loss of *Gpr116/Adgrf5* causes reduced surfactant uptake, leading to massive and pathological accumulations of surfactant lipids and proteins in the alveolar space [182–184] (see also [185]). Given that surfactant is essential to temper lung surface tension to allow for effortless lung expansion during inhalation, it is enticing to speculate that GPR116 is involved in monitoring lung tension (Fig. 1).

Similar to the expression of dCIRL in mechanosensory cells, GPR98/VLGR/ADGRV1 localizes to inner ear hair cells [186], where it is involved in proper development of the cochlear organ of Corti [187]. In the future, it will be interesting to determine the exact role of GPR98 during auditory mechanotransduction in these cells.

4 Conclusions

In sum, ionotropic mechanosensors are distributed throughout various organ systems across different species where they assure mechanosensory responses with latencies on millisecond timescales, rather atypical for metabotropic mechanosensors. Interestingly, however, in some instances these responses appear to be adjusted through the activity of GPCRs, which begs the question if ionotropic and metabotropic mechanosensors form functional units to enable reliable mechanotransduction.

aGPCRs regulate a multitude of physiological processes signified by the capacity to sense mechanics in different cellular contexts, which could explain the participation of aGPCRs in various, seemingly unrelated, biological phenomena. The structural and functional layouts of aGPCRs potentially reflect their optimization for the perception of mechanical forces, rather than a general sensitivity to a range of multiple sensory stimuli known for other GPCR families. Another interesting aspect will be to unveil the signaling cascades through which aGPCRs influence ionotropic mechanotransduction processes and whether ion channels and GPCRs form functional mechanosensory units. Further research efforts will clarify the physiological and pharmacological properties underlying the mechanobiological roles of aGPCRs.

Acknowledgments The writing of this manuscript was supported by grants from the Deutsche Forschungsgemeinschaft to R. J. K. and T. L. (Research Unit FOR 2149, Projects P1 [LA 2861/4-1] and P3 [LA 2861/5-1, KI 1460/2-1]; SFB 1047, Project A5; SFB-TR 166 Projects B4 and C3; LA 2861/7-1). K. R. M. acknowledges support from the NIH (R01 NS079445, R01 HD080601).

References

1. Hoffman BD, Grashoff C, Schwartz MA (2011) Dynamic molecular processes mediate cellular mechanotransduction. Nature 475:316–323. doi:10.1038/nature10316
2. Cantor RS (1997) Lateral pressures in cell membranes: a mechanism for modulation of protein function. J Phys Chem B 101:1723–1725. doi:10.1021/jp963911x
3. Guharay F, Sachs F (1984) Stretch-activated single ion channel currents in tissue-cultured embryonic chick skeletal muscle. J Physiol 352:685–701
4. Guharay F, Sachs F (1985) Mechanotransducer ion channels in chick skeletal muscle: the effects of extracellular pH. J Physiol 363:119–134
5. Sukharev SI, Blount P, Martinac B, Blattner FR, Kung C (1994) A large-conductance mechanosensitive channel in E coli encoded by mscL alone. Nature 368:265–268. doi:10.1038/368265a0
6. Levina N, Tötemeyer S, Stokes NR, Louis P, Jones MA, Booth IR (1999) Protection of Escherichia coli cells against extreme turgor by activation of MscS and MscL mechanosensitive channels: identification of genes required for MscS activity. EMBO J 18:1730–1737. doi:10.1093/emboj/18.7.1730
7. Perozo E, Rees DC (2003) Structure and mechanism in prokaryotic mechanosensitive channels. Curr Opin Struct Biol 13:432–442
8. Ajouz B, Berrier C, Garrigues A, Besnard M, Ghazi A (1998) Release of thioredoxin via the mechanosensitive channel MscL during osmotic downshock of Escherichia coli cells. J Biol Chem 273:26670–26674
9. Kung C (2005) A possible unifying principle for mechanosensation. Nature 436:647–654. doi:10.1038/nature03896
10. Ge J, Li W, Zhao Q, Li N, Chen M, Zhi P et al (2015) Architecture of the mammalian mechanosensitive Piezo1 channel. Nature 527:64–69. doi:10.1038/nature15247
11. Coste B, Mathur J, Schmidt M, Earley TJ, Ranade S, Petrus MJ et al (2010) Piezo1 and Piezo2 are essential components of distinct mechanically activated cation channels. Science 330:55–60. doi:10.1126/science.1193270
12. Coste B, Xiao B, Santos JS, Syeda R, Grandl J, Spencer KS et al (2012) Piezo proteins are pore-forming subunits of mechanically activated channels. Nature 483:176–181. doi:10.1038/nature10812
13. Gottlieb PA, Sachs F (2011) Piezo1: properties of a cation selective mechanical channel. Channels (Austin) 6:214–219. doi:10.4161/chan.21050
14. Woo SH, Ranade S, Weyer AD, Dubin AE, Baba Y (2014) Piezo2 is required for Merkel-cell mechanotransduction. Nature 509:622–626
15. Coste B, Murthy SE, Mathur J, Schmidt M, Mechioukhi Y, Delmas P et al (2015) Piezo1 ion channel pore properties are dictated by C-terminal region. Nat Commun 6:7223. doi:10.1038/ncomms8223

16. Faucherre A, Nargeot J, Mangoni ME, Jopling C (2013) Piezo2b regulates vertebrate light touch response. J Neurosci 33:17089–17094. doi:10.1523/JNeurosci0522-13.2013
17. Faucherre A, Kissa K, Nargeot J, Mangoni ME, Jopling C (2014) Piezo1 plays a role in erythrocyte volume homeostasis. Haematologica 99:70–75. doi:10.3324/haematol.2013.086090
18. Li J, Hou B, Tumova S, Muraki K, Bruns A, Ludlow MJ et al (2014) Piezo1 integration of vascular architecture with physiological force. Nature 515:279–282. doi:10.1038/nature13701
19. Ranade SS, Qiu Z, Woo SH, Hur SS (2014) Piezo1, a mechanically activated ion channel, is required for vascular development in mice. Proc Natl Acad Sci U S A 111(28):10347–10352. doi:10.1073/pnas.1409233111
20. Kim SE, Coste B, Chadha A, Cook B, Patapoutian A (2012) The role of Drosophila Piezo in mechanical nociception. Nature 483:209–212. doi:10.1038/nature10801
21. Canessa CM, Horisberger JD, Rossier BC (1993) Epithelial sodium channel related to proteins involved in neurodegeneration. Nature 361:467–470. doi:10.1038/361467a0
22. Canessa CM, Schild L, Buell G, Thorens B, Gautschi I, Horisberger JD et al (1994) Amiloride-sensitive epithelial Na^+ channel is made of three homologous subunits. Nature 367:463–467. doi:10.1038/367463a0
23. Driscoll M, Chalfie M (1991) The mec-4 gene is a member of a family of Caenorhabditis elegans genes that can mutate to induce neuronal degeneration. Nature 349:588–593. doi:10.1038/349588a0
24. Lingueglia E, Voilley N, Waldmann R, Lazdunski M, Barbry P (1993) Expression cloning of an epithelial amiloride-sensitive Na^+ channel. A new channel type with homologies to Caenorhabditis elegans degenerins. FEBS Lett 318:95–99
25. Kellenberger S, Schild L (2002) Epithelial sodium channel/degenerin family of ion channels: a variety of functions for a shared structure. Physiol Rev 82:735–767. doi:10.1152/physrev.00007.2002
26. Askwith CC, Wemmie JA, Price MP, Rokhlina T, Welsh MJ (2004) Acid-sensing ion channel 2 (ASIC2) modulates ASIC1 H^+-activated currents in hippocampal neurons. J Biol Chem 279:18296–18305. doi:10.1074/jbc.M312145200
27. Baron A, Voilley N, Lazdunski M, Lingueglia E (2008) Acid sensing ion channels in dorsal spinal cord neurons. J Neurosci 28:1498–1508. doi:10.1523/JNEUROSCI.4975-07.2008
28. Benson CJ, Xie J, Wemmie JA, Price MP, Henss JM, Welsh MJ et al (2002) Heteromultimers of DEG/ENaC subunits form H^+-gated channels in mouse sensory neurons. Proc Natl Acad Sci U S A 99:2338–2343. doi:10.1073/pnas.032678399
29. Chalfie M, Sulston J (1981) Developmental genetics of the mechanosensory neurons of Caenorhabditis elegans. Dev Biol 82:358–370
30. Sulston J, Dew M, Brenner S (1975) Dopaminergic neurons in the nematode Caenorhabditis elegans. J Comp Neurol 163:215–226. doi:10.1002/cne.901630207
31. Chalfie M, Au M (1989) Genetic control of differentiation of the Caenorhabditis elegans touch receptor neurons. Science 243:1027–1033
32. Goodman MB, Ernstrom GG, Chelur DS, O'Hagan R, Yao CA, Chalfie M (2002) MEC-2 regulates C. elegans DEG/ENaC channels needed for mechanosensation. Nature 415:1039–1042. doi:10.1038/4151039a
33. Zhang S, Árnadóttir J, Keller C, Caldwell GA, Yao CA, Chalfie M (2004) MEC-2 is recruited to the putative mechanosensory complex in C. elegans touch receptor neurons through its stomatin-like domain. Curr Biol 14:1888–1896. doi:10.1016/j.cub.2004.10.030
34. Chelur DS, Ernstrom GG, Goodman MB, Yao CA, Chen L, O'Hagan R et al (2002) The mechanosensory protein MEC-6 is a subunit of the C. elegans touch-cell degenerin channel. Nature 420:669–673. doi:10.1038/nature01205
35. Huang M, Chalfie M (1994) Gene interactions affecting mechanosensory transduction in Caenorhabditis elegans. Nature 367:467–470. doi:10.1038/367467a0

36. O'Hagan R, Chalfie M, Goodman MB (2004) The MEC-4 DEG/ENaC channel of Caenorhabditis elegans touch receptor neurons transduces mechanical signals. Nat Neurosci 8:43–50. doi:10.1038/nn1362
37. Getz GS, Reardon CA (2004) Paraoxonase, a cardioprotective enzyme: continuing issues. Curr Opin Lipidol 15:261
38. Huber TB, Schermer B, Muller RU, Hohne M, Bartram M, Calixto A et al (2006) Podocin and MEC-2 bind cholesterol to regulate the activity of associated ion channels. Proc Natl Acad Sci U S A 103:17079–17086. doi:10.1073/pnas.0607465103
39. Barnes TM, Jin Y, Horvitz HR, Ruvkun G (1996) The Caenorhabditis elegans behavioral gene unc-24 encodes a novel bipartite protein similar to both erythrocyte band 7.2 (stomatin) and nonspecific lipid transfer protein. J Neurochem 67(1):46–57
40. Clapham DE (2003) TRP channels as cellular sensors. Nature 426:517–524. doi:10.1038/nature02196
41. Montell C (2005) The TRP superfamily of cation channels. Sci Signal 2005:re3. doi:10.1126/stke.2722005re3
42. Christensen AP, Corey DP (2007) TRP channels in mechanosensation: direct or indirect activation? Nat Rev Neurosci 8:510–521. doi:10.1038/nrn2149
43. Owsianik G, Talavera K, Voets T, Nilius B (2006) Permeation and selectivity of TRP channels. Annu Rev Physiol 68:685–717. doi:10.1146/annurev.physiol.68.040204.101406
44. Schnitzler MMY, Storch U, Meibers S, Nurwakagari P, Breit A, Essin K et al (2008) Gq-coupled receptors as mechanosensors mediating myogenic vasoconstriction. EMBO J 27:3092–3103. doi:10.1038/emboj.2008.233
45. Quick K, Zhao J, Eijkelkamp N, Linley JE, Rugiero F, Cox JJ et al (2012) TRPC3 and TRPC6 are essential for normal mechanotransduction in subsets of sensory neurons and cochlear hair cells. Open Biol 2:120068. doi:10.1098/rsob.120068
46. Sexton JE, Desmonds T, Quick K, Taylor R, Abramowitz J, Forge A et al (2015) The contribution of TRPC1, TRPC3, TRPC5 and TRPC6 to touch and hearing. Neurosci Lett 610:36–42. doi:10.1016/j.neulet.2015.10.052
47. Eberl DF, Hardy RW, Kernan MJ (2000) Genetically similar transduction mechanisms for touch and hearing in Drosophila. J Neurosci 20:5981–5988
48. Walker RG, Willingham AT, Zuker CS (2000) A Drosophila mechanosensory transduction channel. Science 287:2229–2234
49. Sidi S, NompC TRP (2003) Channel required for vertebrate sensory hair cell mechanotransduction. Science 301:96–99. doi:10.1126/science.1084370
50. Li W, Feng Z, Sternberg PW, Xu XZS (2006) A C. elegans stretch receptor neuron revealed by a mechanosensitive TRP channel homologue. Nature 440:684–687. doi:10.1038/nature04538
51. Lee J, Moon S, Cha Y, Chung YD (2010) Drosophila TRPN(=NOMPC) channel localizes to the distal end of mechanosensory cilia. PLoS One 5(6), e11012
52. Cheng LE, Song W, Looger LL, Jan LY, Jan YN (2010) The role of the TRP channel NompC in Drosophila larval and adult locomotion. Neuron 67:373–380. doi:10.1016/j.neuron.2010.07.004
53. Yan Z, Zhang W, He Y, Gorczyca D, Xiang Y, Cheng LE et al (2013) Nature 493:221–225. doi:10.1038/nature11685
54. Gong J, Wang Q, Wang Z (2013) NOMPC is likely a key component of Drosophila mechanotransduction channels. Eur J Neurosci 38:2057–2064. doi:10.1111/ejn.12214
55. Liang X, Madrid J, Gärtner R, Verbavatz J-M, Schiklenk C, Wilsch-Bräuninger M et al (2013) A NOMPC-dependent membrane-microtubule connector is a candidate for the gating spring in fly mechanoreceptors. Curr Biol 23:755–763. doi:10.1016/j.cub.2013.03.065
56. Zhang W, Cheng LE, Kittelmann M, Li J, Petkovic M, Cheng T et al (2015) Ankyrin repeats convey force to gate the NOMPC mechanotransduction channel. Cell 162:1391–1403. doi:10.1016/j.cell.2015.08.024

57. Göpfert MC, Albert JT, Nadrowski B, Kamikouchi A (2006) Specification of auditory sensitivity by Drosophila TRP channels. Nat Neurosci 9:999–1000. doi:10.1038/nn1735
58. Lehnert BP, Baker AE, Gaudry Q, Chiang A-S, Wilson RI (2013) Distinct roles of TRP channels in auditory transduction and amplification in Drosophila. Neuron 77:115–128. doi:10.1016/j.neuron.2012.11.030
59. Scholz N, Gehring J, Guan C, Ljaschenko D, Fischer R, Lakshmanan V et al (2015) The adhesion GPCR latrophilin/CIRL shapes mechanosensation. Cell Rep. doi:10.1016/j.celrep.2015.04.008
60. Tobin DM, Madsen DM, Kahn-Kirby A, Peckol EL (2002) Combinatorial expression of TRPV channel proteins defines their sensory functions and subcellular localization in C. elegans neurons. Neuron 35(2):307–318
61. Kaplan JM, Horvitz HR (1993) A dual mechanosensory and chemosensory neuron in Caenorhabditis elegans. Proc Natl Acad Sci U S A 90(6):2227–2231
62. Colbert HA, Smith TL, Bargmann CI (1997) OSM-9, a novel protein with structural similarity to channels, is required for olfaction, mechanosensation, and olfactory adaptation in Caenorhabditis elegans. J Neurosci 17:8259–8269
63. Kim J, Chung YD, Park D-Y, Choi S, Shin DW, Soh H et al (2003) A TRPV family ion channel required for hearing in Drosophila. Nature 424:81–84. doi:10.1038/nature01733
64. Gong Z, Son W, Chung YD, Kim J, Shin DW, McClung CA et al (2004) Two interdependent TRPV channel subunits, inactive and Nanchung, mediate hearing in Drosophila. J Neurosci 24:9059–9066. doi:10.1523/JNeurosci.1645-04.2004
65. Hanaoka K, Qian F, Boletta A, Bhunia AK, Piontek K, Tsiokas L et al (2000) Co-assembly of polycystin-1 and -2 produces unique cation-permeable currents. Nature 408:990–994. doi:10.1038/35050128
66. Yu Y, Ulbrich MH, Li M-H, Buraei Z, Chen X-Z, Ong ACM et al (2009) Structural and molecular basis of the assembly of the TRPP2/PKD1 complex. Proc Natl Acad Sci U S A 106:11558–11563. doi:10.1073/pnas.0903684106
67. Nauli SM, Alenghat FJ, Luo Y, Williams E, Vassilev P, Li X et al (2003) Polycystins 1 and 2 mediate mechanosensation in the primary cilium of kidney cells. Nat Genet 33:129–137. doi:10.1038/ng1076
68. Nauli SM, Kawanabe Y, Kaminski JJ, Pearce WJ, Ingber DE, Zhou J (2008) Endothelial cilia are fluid shear sensors that regulate calcium signaling and nitric oxide production through polycystin-1. Circulation 117:1161–1171. doi:10.1161/CIRCULATIONAHA.107.710111
69. AbouAlaiwi WA, Takahashi M, Mell BR, Jones TJ, Ratnam S, Kolb RJ et al (2009) Ciliary polycystin-2 is a mechanosensitive calcium channel involved in nitric oxide signaling cascades. Circ Res 104:860–869. doi:10.1161/CIRCRESAHA.108.192765
70. Sharif-Naeini R, Folgering JHA, Bichet D, Duprat F, Lauritzen I, Arhatte M et al (2009) Polycystin-1 and -2 dosage regulates pressure sensing. Cell 139:587–596. doi:10.1016/j.cell.2009.08.045
71. Wilson PD (2004) Polycystic kidney disease: new understanding in the pathogenesis. Int J Biochem Cell Biol 36:1868–1873. doi:10.1016/j.biocel.2004.03.012
72. Harris PC, Torres VE (2009) Polycystic kidney disease. Annu Rev Med 60:321–337. doi:10.1146/annurev.med.60.101707.125712
73. Nieberler M, Kittel RJ, Petrenko AG, Lin H-H, Langenhan T (2016) Control of adhesion GPCR function through proteolytic processing. In: Langenhan T, Schöneberg T (eds) Adhesion G protein-coupled receptors: molecular, physiological and pharmacological principles in health and disease. Springer, Heidelberg
74. Praetorius HA, Spring KR (2001) Bending the MDCK cell primary cilium increases intracellular calcium. J Membr Biol 184:71–79
75. Köttgen M, Buchholz B, Garcia-Gonzalez MA, Kotsis F, Fu X, Doerken M et al (2008) TRPP2 and TRPV4 form a polymodal sensory channel complex. J Cell Biol 182:437–447. doi:10.1083/jcb.200805124

76. Brohawn SG (2015) How ion channels sense mechanical force: insights from mechanosensitive K2P channels TRAAK, TREK1, and TREK2. Ann N Y Acad Sci 1352:20–32. doi:10.1111/nyas.12874
77. Noël J, Sandoz G, Lesage F (2014) Molecular regulations governing TREK and TRAAK channel functions. Channels (Austin) 5:402–409. doi:10.4161/chan.5.5.16469
78. Brohawn SG, Su Z, MacKinnon R (2014) Mechanosensitivity is mediated directly by the lipid membrane in TRAAK and TREK1 K^+ channels. Proc Natl Acad Sci U S A 111:3614–3619. doi:10.1073/pnas.1320768111
79. Enyedi P, Czirják G (2010) Molecular background of leak K^+ currents: two-pore domain potassium channels. Physiol Rev 90:559–605. doi:10.1152/physrev.00029.2009
80. Alloui A, Zimmermann K, Mamet J, Duprat F, Noël J, Chemin J et al (2006) TREK-1, a K+ channel involved in polymodal pain perception. EMBO J 25:2368–2376. doi:10.1038/sj.emboj.7601116
81. Heurteaux C, Guy N, Laigle C, Blondeau N, Duprat F, Mazzuca M et al (2004) TREK-1, a K+ channel involved in neuroprotection and general anesthesia. EMBO J 23:2684–2695. doi:10.1038/sj.emboj.7600234
82. Peyronnet R, Sharif-Naeini R, Folgering J, Arhatte M (2012) Mechanoprotection by polycystins against apoptosis is mediated through the opening of stretch-activated K2P channels. Cell Rep 1:241–250
83. Sanders KM, Don KS (2006) Two-pore-domain potassium channels in smooth muscles: new components of myogenic regulation. J Physiol 570:37–43. doi:10.1113/jphysiol.2005.098897
84. Kurima K, Peters LM, Yang Y, Riazuddin S, Ahmed ZM, Naz S et al (2002) Dominant and recessive deafness caused by mutations of a novel gene, TMC1, required for cochlear hair-cell function. Nat Genet 30:277–284. doi:10.1038/ng842
85. Pan B, Géléoc GS, Asai Y, Horwitz GC, Kurima K, Ishikawa K et al (2013) TMC1 and TMC2 are components of the mechanotransduction channel in hair cells of the mammalian inner ear. Neuron 79:504–515. doi:10.1016/j.neuron.2013.06.019
86. Kurima K, Ebrahim S, Pan B, Sedlacek M, Sengupta P, Millis BA et al (2015) TMC1 and TMC2 localize at the site of mechanotransduction in mammalian inner ear hair cell stereocilia. Cell Rep 12:1606–1617. doi:10.1016/j.celrep.2015.07.058
87. Keresztes G, Mutai H, Heller S (2003) TMC and EVER genes belong to a larger novel family, the TMC gene family encoding transmembrane proteins. BMC Genomics 4:24. doi:10.1186/1471-2164-4-24
88. Labay V, Weichert RM, Makishima T, Griffith AJ (2010) Topology of transmembrane channel-like gene 1 protein. Biochemistry 49:8592–8598. doi:10.1021/bi1004377
89. Chatzigeorgiou M, Bang S, Hwang SW, Schafer WR (2013) tmc-1 encodes a sodium-sensitive channel required for salt chemosensation in C. elegans. Nature 494:95–99. doi:10.1038/nature11845
90. Kawashima Y, Géléoc GSG, Kurima K, Labay V, Lelli A, Asai Y et al (2011) Mechanotransduction in mouse inner ear hair cells requires transmembrane channel-like genes. J Clin Invest 121:4796–4809. doi:10.1172/JCI60405
91. Vreugde S, Erven A, Kros CJ, Marcotti W, Fuchs H, Kurima K et al (2002) Beethoven, a mouse model for dominant, progressive hearing loss DFNA36. Nat Genet 30:257–258. doi:10.1038/ng848
92. Beurg M, Kim KX, Fettiplace R (2014) Conductance and block of hair-cell mechanotransducer channels in transmembrane channel-like protein mutants. J Gen Physiol 144:55–69. doi:10.1038/ng848
93. Gale JE, Marcotti W, Kennedy HJ, Kros CJ, Richardson GP (2001) FM1-43 dye behaves as a permeant blocker of the hair-cell mechanotransducer channel. J Neurosci 21:7013–7025
94. Marcotti W, Van Netten SM (2005) The aminoglycoside antibiotic dihydrostreptomycin rapidly enters mouse outer hair cells through the mechano-electrical transducer channels. J Physiol 567:505–521
95. Kazmierczak P, Sakaguchi H, Tokita J (2007) Cadherin 23 and protocadherin 15 interact to form tip-link filaments in sensory hair cells. Nature 449(7158):87–91. Abstract

96. Sakaguchi H, Tokita J, Müller U, Kachar B (2009) Tip links in hair cells: molecular composition and role in hearing loss. Curr Opin Otolaryngol Head Neck Surg 17:388. doi:10.1097/MOO.0b013e3283303472
97. Maeda R, Kindt KS, Mo W, Morgan CP, Erickson T, Zhao H et al (2014) Tip-link protein protocadherin 15 interacts with transmembrane channel-like proteins TMC1 and TMC2. Proc Natl Acad Sci U S A 111:12907–12912. doi:10.1073/pnas.1402152111
98. Beurg M, Xiong W, Zhao B, Müller U (2015) Subunit determination of the conductance of hair-cell mechanotransducer channels. 2015. Proc Natl Acad Sci U S A 112(5):1589–1594. doi:10.1073/pnas.1420906112
99. Hynes R (1987) Integrins: a family of cell surface receptors. Cell 48:549–554. doi:10.1016/0092-8674(87)90233-9
100. Hynes RO (2002) Integrins: bidirectional, allosteric signaling machines. Cell 110:673–687
101. Plow EF, Haas TA, Zhang L, Loftus J, Smith JW (2000) Ligand binding to integrins. J Biol Chem 275:21785–21788. doi:10.1074/jbc.R000003200
102. Shattil SJ, Kim C, Ginsberg MH (2010) The final steps of integrin activation: the end game. Nat Rev Mol Cell Biol 11:288–300. doi:10.1038/nrm2871
103. Campbell ID, Humphries MJ (2011) Integrin structure, activation, and interactions. Cold Spring Harb Perspect Biol 3(3):a004994
104. Balaban NQ, Schwarz US, Riveline D, Goichberg P, Tzur G, Sabanay I et al (2001) Force and focal adhesion assembly: a close relationship studied using elastic micropatterned substrates. Nat Cell Biol 3:466–472. doi:10.1038/35074532, Abstract
105. Galbraith CG, Yamada KM, Sheetz MP (2002) The relationship between force and focal complex development. J Cell Biol 159(4):695–705
106. Wang N, Butler J, Ingber D (1993) Mechanotransduction across the cell surface and through the cytoskeleton. Science 260:1124–1127. doi:10.1126/science.7684161
107. Choquet D, Felsenfeld DP, Sheetz MP (1997) Extracellular matrix rigidity causes strengthening of integrin–cytoskeleton linkages. Cell 88(1):39–48
108. Meyer CJ, Alenghat FJ, Rim P, Fong JH-J, Fabry B, Ingber DE (2000) Mechanical control of cyclic AMP signalling and gene transcription through integrins. Nat Cell Biol 2:666–668. doi:10.1038/35023621, Abstract
109. Chen J, Fabry B, Schiffrin EL, Wang N (2001) Twisting integrin receptors increases endothelin-1 gene expression in endothelial cells. Am J Physiol Cell Physiol 280:C1475–C1484
110. Riveline D, Zamir E, Balaban NQ, Schwarz US, Ishizaki T, Narumiya S et al (2001) Focal contacts as mechanosensors: externally applied local mechanical force induces growth of focal contacts by an mDia1-dependent and rock-independent mechanism. J Cell Biol 153:1175–1186. doi:10.1083/jcb.141.2.539
111. Lo CM, Wang HB, Dembo M, Wang Y (2000) Cell movement is guided by the rigidity of the substrate. Biophys J 79(1):144–152
112. Chen CS, Mrksich M, Huang S, Whitesides GM, Ingber DE (1997) Geometric control of cell life and death. Science 276(5317):1425–1428
113. Dike LE, Chen CS, Mrksich M, Tien J, Whitesides GM, Ingber DE (1999) Geometric control of switching between growth, apoptosis, and differentiation during angiogenesis using micropatterned substrates. Dev Biol 35:441–448
114. Parker KK, Brock AL, Brangwynne C, Mannix RJ, Wang N, Ostuni E et al (2002) Directional control of lamellipodia extension by constraining cell shape and orienting cell tractional forces. FASEB J 16(10):1195–1204
115. McHugh BJ, Buttery R, Lad Y, Banks S (2010) Integrin activation by Fam38A uses a novel mechanism of R-Ras targeting to the endoplasmic. J Cell Sci 123(1):51–61. doi:10.1242/jcs.056424
116. Nose A, Tsuji K, Takeichi M (1990) Localization of specificity determining sites in cadherin cell adhesion molecules. Cell 61:147–155

117. Shapiro L, Weis WI (2009) Structure and biochemistry of cadherins and catenins. Cold Spring Harb Perspect Biol 1:a003053. doi:10.1101/cshperspect.a003053
118. Prakasam AK, Maruthamuthu V, Leckband DE (2006) Similarities between heterophilic and homophilic cadherin adhesion. Proc Natl Acad Sci U S A 103:15434–15439. doi:10.1073/pnas.0606701103
119. Leckband D, Prakasam A (2006) Mechanism and dynamics of cadherin adhesion. Annu Rev Biomed Eng 8:259–287. doi:10.1146/annurev.bioeng.8.061505.095753
120. Shi Q, Maruthamuthu V, Li F, Leckband D (2010) Allosteric cross talk between cadherin extracellular domains. Biophys J 99:95–104. doi:10.1016/j.bpj.2010.03.062
121. Desai R, Sarpal R, Ishiyama N, Pellikka M, Ikura M, Tepass U (2013) Monomeric α-catenin links cadherin to the actin cytoskeleton. Nature 15:261–273. doi:10.1038/ncb2685
122. le Duc Q, Shi Q, Blonk I, Sonnenberg A, Wang N, Leckband D et al (2010) Vinculin potentiates E-cadherin mechanosensing and is recruited to actin-anchored sites within adherens junctions in a myosin II-dependent manner. J Cell Biol 189:1107–1115. doi:10.1083/jcb.201001149
123. Yonemura S, Wada Y, Watanabe T, Nagafuchi A, Shibata M (2010) alpha-Catenin as a tension transducer that induces adherens junction development. Nature 12:533–542. doi:10.1038/ncb2055
124. Yonemura S (2011) A mechanism of mechanotransduction at the cell-cell interface: emergence of α-catenin as the center of a force-balancing mechanism for morphogenesis in multicellular organisms. Bioessays 33:732–736. doi:10.1002/bies.201100064
125. Leckband DE, de Rooij J (2014) Cadherin adhesion and mechanotransduction. Annu Rev Cell Dev Biol 30:291–315. doi:10.1146/annurev-cellbio-100913-013212
126. Saez A, Buguin A, Silberzan P, Ladoux B (2005) Is the mechanical activity of epithelial cells controlled by deformations or forces? Biophys J 89:L52–L54. doi:10.1529/biophysj.105.071217
127. Liu Q-S, Zhang J, Liu M, Dong W-G (2010) Lentiviral-mediated miRNA against liver-intestine cadherin suppresses tumor growth and invasiveness of human gastric cancer. Cancer Sci 101:1807–1812. doi:10.1111/j.1349-7006.2010.01600.x
128. Barry AK, Wang N, Leckband DE (2015) Local VE-cadherin mechanotransduction triggers long-ranged remodeling of endothelial monolayers. J Cell Sci 128:1341–1351. doi:10.1242/jcs.159954
129. Roh-Johnson M, Shemer G, Higgins CD, McClellan JH, Werts AD, Tulu US et al (2012) Triggering a cell shape change by exploiting preexisting actomyosin contractions. Science 335:1232–1235. doi:10.1126/science.1217869
130. Levayer R, Lecuit T (2013) Oscillation and polarity of E-cadherin asymmetries control actomyosin flow patterns during morphogenesis. Dev Cell 26:162–175. doi:10.1016/j.devcel.2013.06.020
131. Tabdili H, Langer M, Shi Q, Poh Y-C, Wang N, Leckband D (2012) Cadherin-dependent mechanotransduction depends on ligand identity but not affinity. J Cell Sci 125:4362–4371. doi:10.1242/jcs.105775
132. Ladoux B, Anon E, Lambert M, Rabodzey A, Hersen P, Buguin A et al (2010) Strength dependence of cadherin-mediated adhesions. Biophys J 98:534–542. doi:10.1016/j.bpj.2009.10.044
133. Dufour S, Mège R-M, Thiery JP (2014) α-Catenin, vinculin, and F-actin in strengthening E-cadherin cell–cell adhesions and mechanosensing. Cell Adh Migr 7:345–350. doi:10.4161/cam.25139
134. Thomas WA, Boscher C, Chu Y-S, Cuvelier D, Martinez-Rico C, Seddiki R et al (2013) α-Catenin and vinculin cooperate to promote high E-cadherin-based adhesion strength. J Biol Chem 288:4957–4969. doi:10.1074/jbc.M112.403774
135. Lecuit T, Lenne P-F, Munro E (2010) Force generation, transmission, and integration during cell and tissue morphogenesis. Annu Rev Cell Dev Biol 27:157–184. doi:10.1146/annurev-cellbio-100109-104027

136. Maître J-L, Heisenberg C-P (2013) Three functions of cadherins in cell adhesion. Curr Biol 23:R626–R633. doi:10.1016/j.cub.2013.06.019
137. Andreeva AV, Han J, Kutuzov MA, Profirovic J, Tkachuk VA, Voyno-Yasenetskaya TA (2010) T-cadherin modulates endothelial barrier function. J Cell Physiol 223:94–102. doi:10.1002/jcp.22014
138. Groschwitz KR, Hogan SP (2009) Intestinal barrier function: molecular regulation and disease pathogenesis. J Allergy Clin Immunol 124:3–20. doi:10.1016/j.jaci.2009.05.038, quiz 21–2
139. Gumbiner BM (2005) Regulation of cadherin-mediated adhesion in morphogenesis. Nat Rev Mol Cell Biol 6:622–634. doi:10.1038/nrm1699
140. Bockaert J, Pin JP (1999) Molecular tinkering of G protein-coupled receptors: an evolutionary success. EMBO J 18:1723–1729. doi:10.1093/emboj/18.7.1723
141. Schnitzler MY, Storch U, Gudermann T (2011) AT1 receptors as mechanosensors. Curr Opin Pharmacol 11:112–116
142. Zou Y, Akazawa H, Qin Y, Sano M, Takano H, Minamino T et al (2004) Mechanical stress activates angiotensin II type 1 receptor without the involvement of angiotensin II. Nat Cell Biol 6:499–506. doi:10.1038/ncb1137
143. Jarajapu YPR, Knot HJ (2002) Role of phospholipase C in development of myogenic tone in rat posterior cerebral arteries. Am J Physiol Heart Circ Physiol 283:H2234–H2238. doi:10.1152/ajpheart.00624.2002
144. Osol G, Laher I, Kelley M (1993) Myogenic tone is coupled to phospholipase C and G protein activation in small cerebral arteries. Am J Physiol Heart Circ Physiol 265:H415–H420
145. Farrens DL, Altenbach C, Yang K, Hubbell WL (1996) Requirement of rigid-body motion of transmembrane helices for light activation of rhodopsin. Science 274(5288):768–770
146. Karnik S, Gogonea C, Patil S, Saad Y, Takezako T (2003) Activation of G-protein-coupled receptors: a common molecular mechanism. Trends Endocrinol Metab 14:431–437. doi:10.1016/j.tem.2003.09.007
147. Yasuda N, Miura S-I, Akazawa H, Tanaka T, Qin Y, Kiya Y et al (2008) Conformational switch of angiotensin II type 1 receptor underlying mechanical stress-induced activation. EMBO Rep 9:179–186. doi:10.1038/sj.embor.7401157
148. Storch U, Mederos y Schnitzler M, Gudermann T (2012) G protein-mediated stretch reception. Am J Physiol Heart Circ Physiol 302:H1241–H1249. doi:10.1152/ajpheart.00818.2011
149. Chachisvilis M, Zhang Y-L, Frangos JA (2006) G protein-coupled receptors sense fluid shear stress in endothelial cells. Proc Natl Acad Sci U S A 103:15463–15468. doi:10.1073/pnas.0607224103
150. Mannstadt M, Jüppner H, Gardella TJ (1999) Receptors for PTH and PTHrP: their biological importance and functional properties. Am J Physiol 277:F665–F675
151. Zhang Y-L, Frangos JA, Chachisvilis M (2009) Mechanical stimulus alters conformation of type 1 parathyroid hormone receptor in bone cells. Am J Physiol Cell Physiol 296:C1391–C1399. doi:10.1152/ajpcell.00549.2008
152. Abdul-Majeed S, Nauli SM (2011) Dopamine receptor type 5 in the primary cilia has dual chemo- and mechano-sensory roles. Hypertension 58:325–331
153. Makino A, Prossnitz ER, Bunemann M, Wang JM, Yao WJ, Schmid-Schoenbein GW (2006) G protein-coupled receptors serve as mechanosensors for fluid shear stress in neutrophils. Am J Physiol Cell Physiol 290:C1633–C1639. doi:10.1152/ajpcell.00576.2005
154. Langenhan T, Aust G, Hamann J (2013) Sticky signaling-adhesion class G protein-coupled receptors take the stage. Sci Signal 6:re3. doi:10.1126/scisignal.2003825
155. Petersen SC, Luo R, Liebscher I, Giera S, Jeong S-J, Mogha A et al (2015) The adhesion GPCR GPR126 has distinct, domain-dependent functions in Schwann cell development mediated by interaction with laminin-211. Neuron 85:755–769. doi:10.1016/j.neuron.2014.12.057

156. White JP, Wrann CD, Rao RR, Nair SK, Jedrychowski MP, You JS et al (2014) G protein-coupled receptor 56 regulates mechanical overload-induced muscle hypertrophy. Proc Natl Acad Sci U S A. doi:10.1073/pnas.1417898111
157. Liebscher I, Schöneberg T (2016) Tethered agonism: a common activation mechanism of adhesion GPCRs. In: Langenhan T, Schöneberg T (eds) Adhesion G protein-coupled receptors: molecular, physiological and pharmacological principles in health and disease. Springer, Heidelberg
158. Kishore A, Hall RA (2016) Versatile signaling activity of adhesion GPCRs. In: Langenhan T, Schöneberg T (eds) Adhesion G protein-coupled receptors: molecular, physiological and pharmacological principles in health and disease. Springer, Heidelberg
159. Araç D, Sträter N, Seiradake E (2016) Understanding the structural basis of adhesion GPCR functions. In: Langenhan T, Schöneberg T (eds) Adhesion G protein-coupled receptors: molecular, physiological and pharmacological principles in health and disease. Springer, Heidelberg
160. Lin H-H, Chang G-W, Davies JQ, Stacey M, Harris J, Gordon S (2004) Autocatalytic cleavage of the EMR2 receptor occurs at a conserved G protein-coupled receptor proteolytic site motif. J Biol Chem 279:31823–31832. doi:10.1074/jbc.M402974200
161. Krasnoperov VG, Bittner MA, Beavis R, Kuang YN, Salnikow KV, Chepurny OG et al (1997) alpha-Latrotoxin stimulates exocytosis by the interaction with a neuronal G-protein-coupled receptor. Neuron 18:925–937. doi:10.1016/S0896-6273(00)80332-3
162. Araç D, Boucard AA, Bolliger MF, Nguyen J, Soltis SM, Südhof TC et al (2012) A novel evolutionarily conserved domain of cell-adhesion GPCRs mediates autoproteolysis. EMBO J 31:1364–1378. doi:10.1038/emboj.2012.26
163. Liebscher I, Schön J, Petersen SC, Fischer L, Auerbach N, Demberg LM et al (2014) A tethered agonist within the ectodomain activates the adhesion G protein-coupled receptors GPR126 and GPR133. Cell Rep 9:2018–2026. doi:10.1016/j.celrep.2014.11.036
164. Stoveken HM, Hajduczok AG, Xu L, Tall GG (2015) Adhesion G protein-coupled receptors are activated by exposure of a cryptic tethered agonist. Proc Natl Acad Sci U S A. doi:10.1073/pnas.1421785112
165. Demberg LM, Rothemund S, Schöneberg T, Liebscher I (2015) Identification of the tethered peptide agonist of the adhesion G protein-coupled receptor GPR64/ADGRG2. Biochem Biophys Res Commun 464:743–747. doi:10.1016/j.bbrc.2015.07.020
166. Wilde C, Fischer L, Lede V, Kirchberger J, Rothemund S, Schöneberg T et al (2015) The constitutive activity of the adhesion GPCR GPR114/ADGRG5 is mediated by its tethered agonist. FASEB J. doi:10.1096/fj.15-276220
167. Sigoillot SM, Monk KR, Piao X, Selimi F, Harty BL (2016) Adhesion G protein-coupled receptors in the nervous system: from synapse and dendrite morphogenesis to myelination. In: Langenhan T, Schöneberg T (eds) Adhesion G protein-coupled receptors: molecular, physiological and pharmacological principles in health and disease. Springer, Heidelberg
168. Monk KR, Naylor SG, Glenn TD, Mercurio S, Perlin JR, Dominguez C et al (2009) A G protein-coupled receptor is essential for Schwann cells to initiate myelination. Science 325:1402–1405. doi:10.1126/science.1173474
169. Monk KR, Oshima K, Jörs S, Heller S, Talbot WS (2011) Gpr126 is essential for peripheral nerve development and myelination in mammals. Development 138:2673–2680. doi:10.1242/dev.062224
170. Glenn TD, Talbot WS (2013) Analysis of Gpr126 function defines distinct mechanisms controlling the initiation and maturation of myelin. Development 140:3167–3175. doi:10.1242/dev.093401
171. Mogha A, Benesh AE, Patra C, Engel FB, Schöneberg T, Liebscher I et al (2013) Gpr126 functions in Schwann cells to control differentiation and myelination via G-protein activation. J Neurosci 33:17976–17985. doi:10.1523/JNEUROSCI.1809-13.2013
172. Feltri ML, Wrabetz L (2005) Laminins and their receptors in Schwann cells and hereditary neuropathies. J Peripher Nerv Syst 10:128–143. doi:10.1111/j.1085-9489.2005.0010204.x

173. Monk KR, Feltri ML, Taveggia C (2015) New insights on Schwann cell development. Glia 63:1376–1393. doi:10.1002/glia.22852
174. Colognato H, Yurchenco PD (1999) The laminin alpha2 expressed by dystrophic dy(2J) mice is defective in its ability to form polymers. Curr Biol 9:1327–1330
175. Yang D (2005) Coordinate control of axon defasciculation and myelination by laminin-2 and -8. J Cell Biol 168:655–666. doi:10.1083/jcb.200411158
176. Xu H, Christmas P, Wu XR, Wewer UM, Engvall E (1994) Defective muscle basement membrane and lack of M-laminin in the dystrophic dy/dy mouse. Proc Natl Acad Sci U S A 91:5572–5576
177. Eberl DF (1999) Feeling the vibes: chordotonal mechanisms in insect hearing. Curr Opin Neurobiol 9:389–393. doi:10.1016/S0959-4388(99)80058-0
178. White JP (2016) Control of skeletal muscle cell growth and size through adhesion GPCRs. In: Langenhan T, Schöneberg T (eds) Adhesion G protein-coupled receptors: molecular, physiological and pharmacological principles in health and disease. Springer, Heidelberg
179. Budday S, Steinmann P, Kuhl E (2015) Physical biology of human brain development. Front Cell Neurosci 9:257. doi:10.3389/fncel.2015.00257
180. Peng Y-M, van de Garde MDB, Cheng K-F, Baars PA, Remmerswaal EBM, van Lier RAW et al (2011) Specific expression of GPR56 by human cytotoxic lymphocytes. J Leukoc Biol 90:735–740. doi:10.1189/jlb.0211092
181. Boyden SE, Desai MS, Cruse G, Young ML, Bolan HC, Scott LM et al (2016) Vibratory urticaria associated with a missense variant in ADGRE2. N Engl J Med 374(7):656–663. doi:10.1056/NEJMoa1500611
182. Bridges JP, Ludwig M-G, Mueller M, Kinzel B, Sato A, Xu Y et al (2013) Orphan G protein-coupled receptor GPR116 regulates pulmonary surfactant pool size. Am J Respir Cell Mol Biol 49:348–357. doi:10.1165/rcmb.2012-0439OC
183. Yang MY, Hilton MB, Seaman S, Haines DC, Nagashima K, Burks CM et al (2013) Essential regulation of lung surfactant homeostasis by the orphan G protein-coupled receptor GPR116. Cell Rep 3:1457–1464. doi:10.1016/j.celrep.2013.04.019
184. Fukuzawa T, Ishida J, Kato A, Ichinose T, Ariestanti DM, Takahashi T et al (2013) Lung surfactant levels are regulated by Ig-Hepta/GPR116 by monitoring surfactant protein D. PLoS One 8:e69451. doi:10.1371/journal.pone.0069451
185. Ludwig M-G, Seuwen K, Bridges JP (2016) Adhesion GPCR function in pulmonary development and disease. In: Langenhan T, Schöneberg T (eds) Adhesion G protein-coupled receptors: molecular, physiological and pharmacological principles in health and disease. Springer, Heidelberg
186. McMillan DR, Kayes-Wandover KM, Richardson JA, White PC (2002) Very large G protein-coupled receptor-1, the largest known cell surface protein, is highly expressed in the developing central nervous system. J Biol Chem 277:785–792. doi:10.1074/jbc.M108929200
187. McGee J, Goodyear RJ, McMillan DR, Stauffer EA, Holt JR, Locke KG et al (2006) The very large G-protein-coupled receptor VLGR1: a component of the ankle link complex required for the normal development of auditory hair bundles. J Neurosci 26:6543–6553. doi:10.1523/JNEUROSCI.0693-06.2006

Adhesion GPCRs Govern Polarity of Epithelia and Cell Migration

David Strutt, Ralf Schnabel, Franziska Fiedler, and Simone Prömel

Graphical Abstract

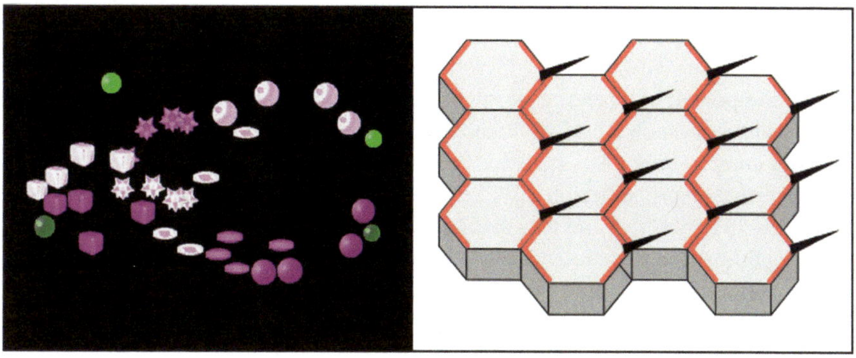

D. Strutt (✉)
Bateson Centre and Department of Biomedical Science, University of Sheffield, Sheffield, UK
e-mail: d.strutt@sheffield.ac.uk

R. Schnabel (✉)
Institute of Genetics, TU Braunschweig, Braunschweig, Germany
e-mail: r.schnabel@tu-bs.de

F. Fiedler • S. Prömel (✉)
Medical Faculty, Institute of Biochemistry, Leipzig University, Leipzig, Germany
e-mail: simone.proemel@medizin.uni-leipzig.de

T. Langenhan, T. Schöneberg (eds.), *Adhesion G Protein-coupled Receptors*,
Handbook of Experimental Pharmacology 234, DOI 10.1007/978-3-319-41523-9_11

Contents

Abstract

In multicellular organisms cells spatially arrange in a highly coordinated manner to form tissues and organs, which is essential for the function of an organism. The component cells and resulting structures are often polarised in one or more axes, and how such polarity is established and maintained correctly has been one of the major biological questions for many decades. Research progress has shown that many adhesion GPCRs (aGPCRs) are involved in several types of polarity. Members of the two evolutionarily oldest groups, Flamingo/Celsr and Latrophilins, are key molecules in planar cell polarity of epithelia or the propagation of cellular polarity in the early embryo, respectively. Other adhesion GPCRs play essential roles in cell migration, indicating that this receptor class includes essential molecules for the control of various levels of cellular organisation.

1 Introduction

Life is order, life is counteracting entropy. It has been a very interesting question for a long time in biology how the linear information of the DNA codes for the three-dimensional topology of cells, tissues and ultimately the whole organism. The spatial arrangement of cells as well as their structure, in general referred to as polarity, is an indispensable prerequisite to order cells and tissues in a multicellular organism. Various types of polarity are distinguished. Even cell migration can be considered as a form of polarity in a broader sense: migrating cells are polarised as they have a clear front and back [1]. In most cells, asymmetric organisation of cellular components and structures form the basis for the establishment and maintenance of polarity.

Polarity not only defines cellular function, it is also an essential part of various biological processes such as embryogenesis or the formation of tissues. Cell polarity is already evident very early in embryogenesis when cells divide asymmetrically or in an oriented manner, and it is a fundamental requirement for the organisation and patterning of the developing embryo, especially during gastrulation (reviewed in [2]). The formation of anterior-posterior, dorsal-ventral and left-right axes rely on polarisation. Later on in development, the collective alignment of cell polarity across a tissue plane mediates coordination and organisation of entire cell layers (reviewed in [3, 4]). This process is referred to as planar cell polarity (PCP) and is involved, for example, in the uniform orientation of hairs [5, 6] or branching morphogenesis in the lung [7].

Although various forms of polarity have been subject to intense study, and progress has been made in the past decades towards a detailed knowledge about the mechanisms underlying each type of polarity, our understanding on a molecular level of the cellular signals involved is still limited. Several major pathways have been identified to be essential for establishment and maintenance of polarity. Anterior-posterior tissue polarity cascades as well as the non-canonical Wnt/Frizzled (Fz) and Fat/Dachsous/Four-jointed (Fat/Ds/Fj) PCP pathways are involved in oriented cell division (reviewed in [8]), with PCP pathways also playing other major regulatory roles in animal development (reviewed in [3, 4]). However, these molecular cascades cover only certain aspects of polarity, and many components of the complex network of regulation remain elusive.

It has become increasingly clear that aGPCRs are not only involved in shaping polarity but are potentially major players in the coordination of various types of polarity such as oriented cell division, planar cell polarity and cell migration. Receptors of five of the nine subfamilies have been identified to play a role in these processes [9]. Besides the recently described Gpr125 the two most prominent and the best characterised ones are Latrophilins and Flamingo/Celsr (Fig. 1). This

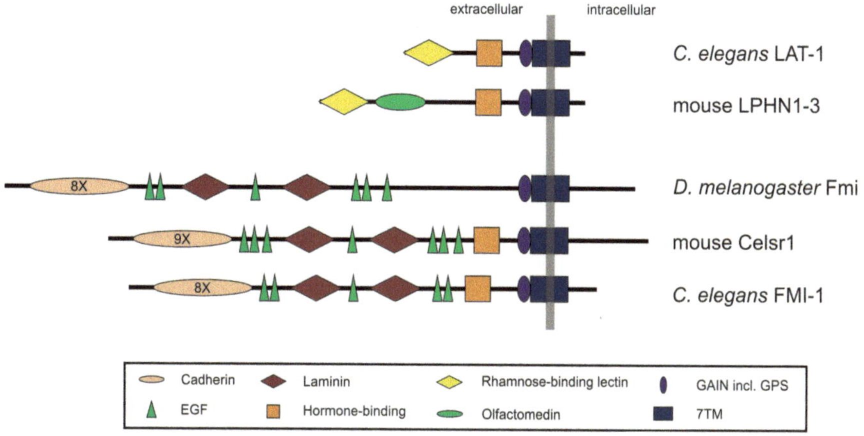

Fig. 1 Domain structure of invertebrate and mammalian Latrophilin and Flamingo/Celsr homologues. Both receptor families are conserved in invertebrates and vertebrates and have the GAIN (GPCR autoproteolysis inducing) domain containing the GPS (GPCR proteolysis site), which is a hallmark feature for aGPCRs. The Latrophilin family is determined by a Rhamnose-binding lectin domain, whereas Cadherins are a hallmark of the Flamingo group

book chapter will discuss the current state of knowledge on the physiological functions of aGPCRs in polarity of epithelia and cell migration as well as the molecular mechanisms they mediate.

2 Latrophilin Signals in Tissue Polarity of Embryogenesis

2.1 Embryonic Development of *C. elegans*

All life today is based on the division of cells, which in metazoans originates from the female gamete. Development is then just a modification of this cell. It is therefore an interesting question what the basal state of the original progenitor cell looks like. Does the embryo have to arrange the polarity of each cell division de novo? If so there should be specific cues or even forces to arrange the directions of mitotic spindles in space. Or is there an inherent polarity rule governing the directions of consecutive cell divisions? If so this may organise a basal arrangement of cells, which may only be modified in specific situations and one cue or force could suffice to do so. Two scenarios for the developmental programme of cells are conceivable. A cell can either get information through interaction with neighbouring cells or follows a preprogrammed series of divisions without any or much interaction with other cells (reviewed in [10, 11]).

The dogma that cells are solely specified autonomously in the roundworm *Caenorhabditis elegans* [12] was broken in the late 1980s [13, 14] suggesting that inductions via contacts with neighbouring cells also play a key role. The hermaphroditic *C. elegans* has always been an ideal organism to study polarity and cell division, especially after it was revealed that its cell lineage is invariant [12, 15]. The division planes and axes of blastomeric cleavages are highly reproducible and tightly controlled by various cell induction events. After fertilisation, the first division occurs in an anterior-posterior direction leading to the anterior AB and the posterior P1 blastomere (Fig. 2). In the second cleavage, P1 generates the two blastomeres EMS and P2, whereas AB divides into an anterior daughter ABa and a posterior daughter ABp. In the subsequent divisions up to the 12-cell stage embryo, the dorsal-ventral and left-right body axes of the embryo are formed through a set of controlled asymmetric and symmetric cell divisions. The

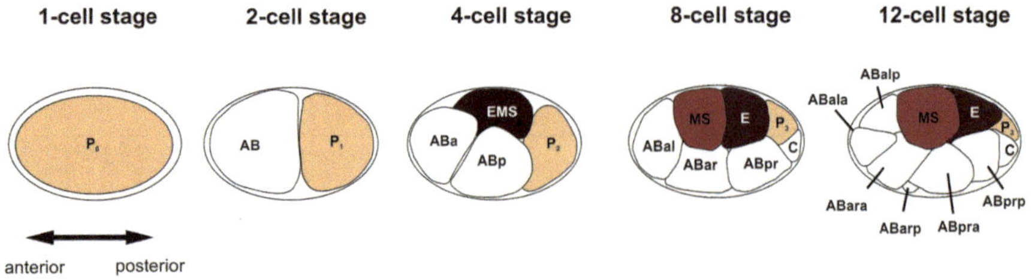

Fig. 2 Cell division in the early *C. elegans* embryo. Cell division in a *C. elegans* embryo occurs in a coordinated and unchangeable manner

mechanisms and pathways controlling these highly concerted and tightly regulated processes remained elusive. However, only 10 years after the first reports of the invariant cell lineage and cell inductions a sudden burst of studies established that the Wnt/Frizzled (Fz) pathway—or more accurately, pathway modules stitched together for different purposes—supplies the polarity required for anterior-posterior fate decisions and for controlling the spindle directions of the EMS and ABar blastomeres. In nematodes mutant for components of the Wnt pathway, spindles are tilted compared to spindles of wild-type embryos leading to severe defects including embryonic lethality [16–20]. This finally also disproved the notion that cell cleavage directions are specified merely autonomously [12]. The fact that Wnt/Fz also regulate spindle directions, which depend on an arrangement of cytoskeletal components, was a surprise as it had been shown before that the directions of early spindles depend on the *par* genes establishing the initial polarity of the zygote and the early blastomeres [21].

It is noteworthy that there has never been a *C. elegans* mutant identified, in which any of the other three AB-derived blastomeres of the eight-cell stage embryo (apart from ABar) cleaved specifically in an aberrant direction deranging the left-right inductions of anterior or even posterior blastomeres [22]. The exception is ABar where mutation of *ced-10/Rac* results in spindles turning in the same direction as in *wnt* mutants. It turned out to be the missing link between polarity conferred by the Wnt pathway and the modulation of the cell cytoskeleton during engulfment of cell corpses, spindle turning and cell migration [23].

In 2009, a novel player in the regulation of these highly complex processes of orienting spindles in the early embryo was discovered—the aGPCR Latrophilin [24]. Interestingly, the three Latrophilin homologues in mammals (ADGRL1–3/LPHN1–3) had so far only been associated with neuronal functions [25, 26]. They were first identified as targets of latrotoxin, a component of the black widow spider's venom [27, 28]. One of the two homologues in *C. elegans*, LAT-1 was now found to be essential for spindle orientation in certain AB descendants of the eight-cell stage of the early embryo (Fig. 3a).

2.2 LAT-1 Is Required for Spindle Orientation in the ABal Blastomere

Nematodes null for *lat-1* displayed several defects including embryonic and larval lethality yielding only about 25 % of the total brood to reach adulthood [24] which suggested an involvement of the receptor in cell division or similar processes. Subsequently, more detailed analyses revealed that indeed, *lat-1* mutant nematodes have a very distinct defect in spindle orientation in the early embryo (Fig. 3a). In embryos from homozygous *lat-1* mutant mothers, the ABal division direction is tilted from the normal $128° \pm 8°$ towards a direction of $90° \pm 18°$ (mean ± SD), which is almost perpendicular to the anterior-posterior axis [29]. Thus, the anterior daughter ABala is located closer to the centre of the embryo than it is in the wild type. This new position has fatal consequences for the embryo as now not only

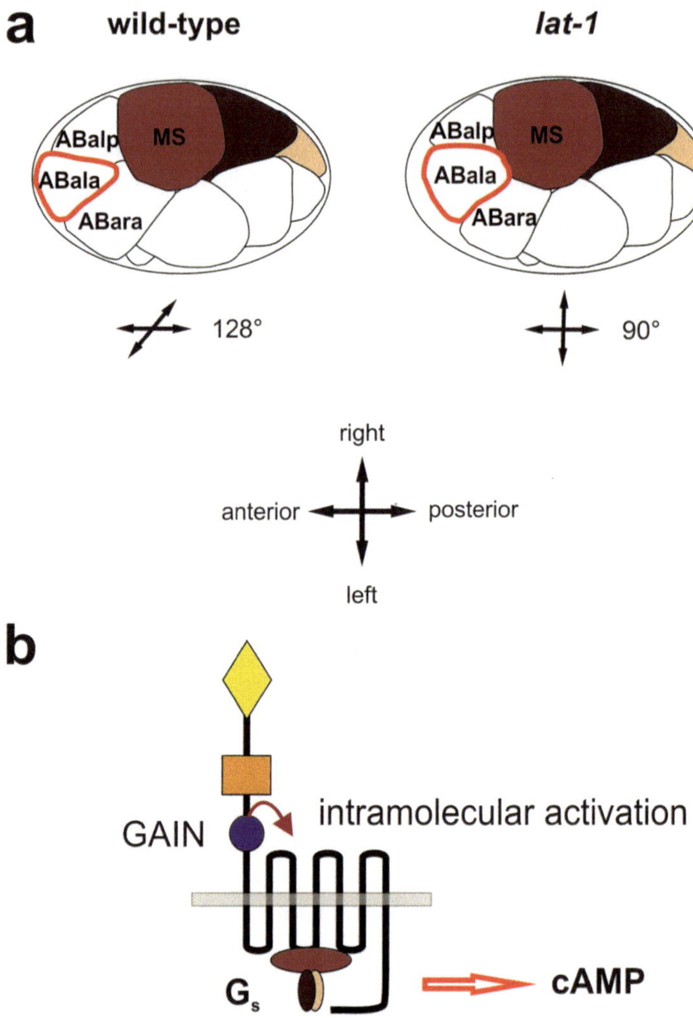

Fig. 3 Cleavage plane orientation defects in *lat-1* mutant embryos and LAT-1 signalling. (**a**) In wild-type *C. elegans* embryos, the mother blastomere of ABala and ABalp divides in an 128° angle to the anterior-posterior axis. This angle is shifted to 90° in *lat-1* mutant individuals. As a consequence, ABala is not displaced towards the anterior pole of the embryo as in the wild type but is located more towards the centre of the embryo. This leads to ABala contacting MS in *lat-1* mutants. (**b**) Schematic depiction of LAT-1 signalling. LAT-1 is activated by an intramolecular sequence within the GAIN domain which potentially interacts with the seven transmembrane region. Subsequently, the receptor triggers an elevation of cAMP levels via a G_s protein cascade leading to correct cell division plane orientation

ABalp, as in normal embryogenesis, but also ABala touches the MS blastomere (Fig. 3a). Through this contact, the MS blastomere generally induces left-right identity in the two cells ABara and ABalp. As in *lat-1* embryos, ABala is also in contact with MS; consequently, not only ABara is induced to execute the ABara blastomere identity but by mistake also ABala, resulting in two ABara fates in the embryo. This effect is lethal for the embryo as subsequent errors in the following cell divisions, and thus, wrong induction events occur. In the original description of the *lat-1* phenotype, only the deletion of the ABala fate was documented [24]. Interestingly, cleavage orientation in the other blastomeres at this stage does not seem to be controlled by LAT-1 function.

It is a key question in the contemplation of LAT-1's role whether or not there is a basal system guiding spindle orientations in consecutive cell divisions in the absence of any external polarity. Earlier observations including the analysis of the development of isolated AB blastomeres were consistent with the notion that such a basal pattern (0° along the anterior-posterior axis followed by 90° along the left-right and dorsal-ventral axis, respectively) may exist [30, 31]. New investigations using much more sophisticated methods show that this basal system coordinates spindle directions during the first cleavages, and it is expected that parallel cleavages of the four AB-derived blastomeres in the eight-cell stage embryo reflects a basal state of the cells (R. Schnabel, unpublished results). Modulation of such a system in a specific new direction may only require one cue or force. In this scenario LAT-1 may be sufficient to direct the spindle of ABal away from MS to avoid a left-right induction. The same is true for the turning of the ABar spindle by the Wnt pathway.

2.3 Genetics of *lat-1*

The function of the Latrophilin gene in *C. elegans* was not identified in a mutant screen as for many other regulators of embryonic polarity [15] but genetically defined in a reverse genetic approach by isolating an induced deletion in the gene [24]. This approach is not unproblematic. If not all exons are deleted, there may be still some gene activity left. The mutation *lat-1(ok1465)* passed all classical and modern genetic tests of excluding residual activity; thus, it most likely is a null allele. The mutation has a maternal effect, which may be expected concerning its very early function at the eight-cell stage embryo. Correspondingly the mRNA of *lat-1* is present in the very early embryo. The fact that the homozygous mutation causes lethality of L1 larvae after a maternal rescue indicates that the gene also has zygotic function(s). A hallmark of all observed phenotypes is that they are not fully penetrant. Classically, this is seen as an indication that the mutation is not amorphic. However, if the *lat-1* mutation is indeed a null mutation, another attractive interpretation is that the gene serves in a "failsafe" process limiting variability in critical situations.

2.4 Oriented Cell Division Is Controlled by G Protein Signalling

The molecular mechanisms underlying the establishment of polarity in the early *C. elegans* embryo are intensely studied, but due to the complexity of the processes involved, many are still poorly understood. Naturally, one central question that occurred after it became clear that LAT-1 is a novel component essential for the control of cell division and spindle orientation in the early *C. elegans* embryo was how the receptor mediated its function. While for the mammalian homologues of Latrophilins several ligands have been identified [32–34], LAT-1 was long considered to be an orphan receptor meaning that neither agonists nor signalling mechanisms were known. As this is the case for many aGPCRs, for very few a direct link between physiological function and signalling mechanisms has been demonstrated in vivo thus far. It was shown that for the regulation of spindle polarity, LAT-1 transduces a signal mostly via a G_s protein [29]. Biochemical as well as genetic analyses showed that via activation of this G_s protein, LAT-1 modulates intracellular cAMP levels (Fig. 3b). Consistently, cAMP concentrations in *lat-1* mutant embryos are lower than in wild-type individuals. Artificially raising cAMP levels in these *lat-1* embryos, for instance, by application of an adenylyl cyclase activator, leads to an amelioration of ABal blastomere cleavage defects showing that via LAT-1 a GPCR-dependent G protein signalling cascade is involved in oriented cell division. Further, it was shown that LAT-1 is activated by an intramolecular sequence within the GPS [29], possibly through an interaction with the seven transmembrane region (Fig. 3b) [35]. This concept of activation has already been shown for other aGPCRs [36–38] and might be a common principle for many members of this class (see also [39, 40] on this topic).

Interestingly, neither LAT-1 nor the signal immediately transduced by the receptor need to be distributed in a polarised manner to mediate this effect. Consistently, expression of a *lat-1::gfp* fusion shows no polarised distribution of the receptor in cells as seen for the components of the Wnt pathway [24]. However, it is possible that the signal is polarised further downstream of the cascade.

2.5 LAT-1 Functions in Later Embryonic Stages

There is evidence that LAT-1 function is not restricted to the ABal blastomere but also plays a role in later embryonic stages. *Lat-1* is expressed in various cells of the developing embryo [24]. Among others, this later expression is detected in pharyngeal cells. Initially it may be expressed at a lower level in most hypodermal (epidermal) cells but later shows a strong expression in the left ABarppa- and Caaa-derived hypodermal cells indicating LAT-1 function in these lineages. Several additional observations suggest additional LAT-1 functions. Phenotypes in *lat-1* mutants are not penetrant as evident by the observation that not all embryos display identical ABal cleavages and subsequent lineage fate deviations. Weak alterations of the ABal cleavage directions towards MS, as opposed to strong ones, may not cause ABala to touch MS significantly enough to cause a left-right induction [29]. Embryos also exhibit severe defects in the posterior far away from the anterior fate transformations caused by the turning of the ABal spindle. Interestingly, two embryos executed a wild-type ABal cleavage but nevertheless showed severe morphogenetic defects and developmental arrest, also supporting the hypothesis that LAT-1 has essential functions independent of its very early activity. These functions can be manifold such as directing cell fate decisions or cell migration.

The descendants of the ABarp blastomere undergo the most extensive morphogenetic movements of all AB-derived cells to form a y-shaped structure on the dorsal side of the embryo, and they are therefore ideally suited to search for the causes of the observed morphogenetic problems in *lat-1* mutant embryos (Fig. 4). Langenhan et al. already showed that cells from the ABarp lineage, which form eight large hypodermal cells on the left and right side, each fail to move to their proper place and that two cells H2L and H2R, which are normally located on the left or right side of the embryo, stay together on the right side [24]. As the specification of left or right is based on anterior-posterior cleavages (anterior specifies left, posterior right) and LAT-1 fails to direct the ABal spindle in its proper anterior-posterior direction, one explanation for the observed phenomenon is that LAT-1, like the Wnt pathway, directs anterior-posterior decisions [24]. In another scenario LAT-1 could be involved in cell migration and cell sorting. This hypothesis is supported by occasional *lat-1* mutant embryos in which only certain anterior-posterior decisions occur, making it rather improbable that anterior-posterior fate alterations were a secondary consequence of transverse divisions.

As reported, above a certain number of *lat-1* mutant embryos survive, most probably because the ABal spindle is still (possibly by chance) in a direction preventing a significant contact of ABala to MS. Despite these occasional correct cell division orientations in the absence of LAT-1, the function of this receptor is essential for cell division plane orientation and subsequently, for establishing correct cell-cell contacts. Therefore, LAT-1 may serve in a "failsafe" system limiting variability in critical situations. How it could be involved in coordinating cell migration is a challenging question to be addressed in the future.

3 Flamingo/Celsr Are Key Components in Planar Cell Polarity Signalling

3.1 Flamingo/Celsr Homologues Are Developmental Patterning Genes

Members of the family of Celsr homologues (ADGRC1–3/CELSR1–3) were first functionally characterised following their discovery in the fruit fly *Drosophila melanogaster* by two groups in the late 1990s [41, 42]. The gene product was found to have an unusual molecular structure consisting of an extraordinarily long extracellular N terminus followed by a seven-pass transmembrane domain with similarity to members of the secretin family of GPCRs and an intracellular C terminus [41, 42]. This architecture led to the coining of the name Flamingo for the *Drosophila* homologue [42]. The receptor structure was found to be highly conserved throughout almost the entire animal kingdom [43, 44]. As the extracellular domain contains 9 Cadherin repeats, 3–5 EGF-like domains and 2 laminin G

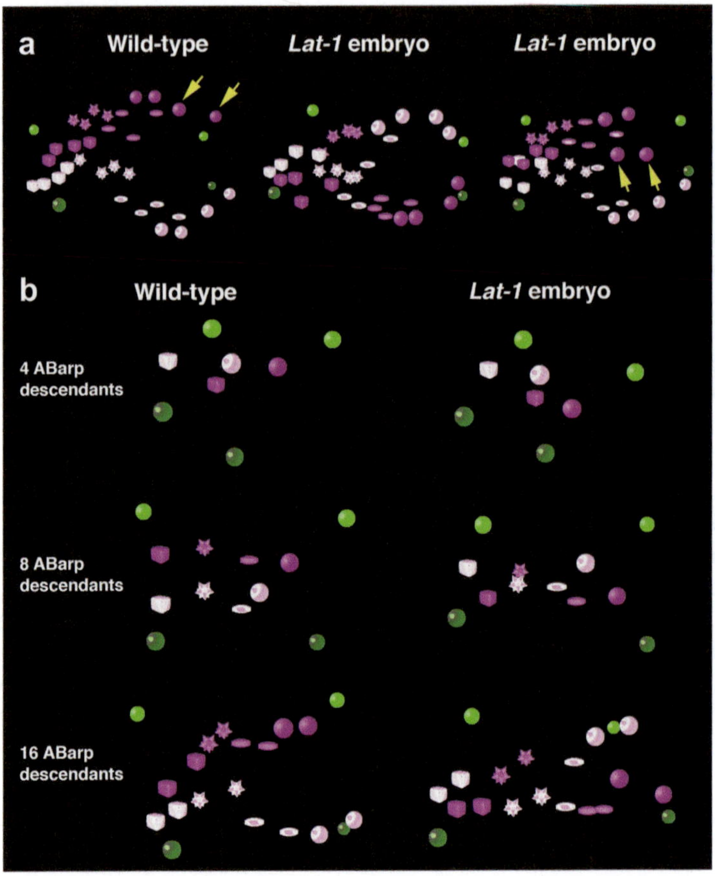

Fig. 4 LAT-1 functions in later embryonic stages. (**a**) In wild-type embryos, the descendants of ABarp form a y-shaped arrangement of 32 cells. The anterior (*left* in figure) contains some neurons, while the posterior part mostly consists of large hypodermal cells. The later left cells are derived from ABarpaa (a descends marked as *cubes*, p descendants marked as *spikes*) and ABarpap (a descends marked as *ufos*, p descendants marked as *spheres* all *white-purple* cells). The right cells are derived from ABarpap and ABarppp (*purple* cells), and the shapes follow the same system as the left cells. The *light green* cells are the most anterior and posterior descendants of ABpra, the *dark green* cells those of ABpla. These cells define the left-right axes of the embryos. In the *lat-1* embryo 19 cells of left or right identity exchanged their positions in a mosaic fashion. These may be explained by an exchange of the anterior-posterior polarities during the ABarpa and ABarpp cleavages. The *lat-1* embryo 41 shows a displacement of the two most posterior hypodermal cells on the *right side* (they moved up in a parallel position to their cousins *arrows*). These positions are consistent with the assumption that both mothers have an anterior fate. Since all cell divisions occurred properly on the anterior-posterior axis, one had to postulate that LAT-1 is like the Wnt pathway not only involved in specifying the cell cleavage directions but also in the specification of cell fates. (**b**) In wild-type embryos, descendants of ABarp with left identities are born to the right of their homologues with right identities but exchange immediately their positions to move into their final positions. In the *lat-1* embryo 19, only two cells exchanged their positions, while the others move to the wrong side. R. Schnabel, unpublished data

motifs, the name Cadherin EGF LAG seven-pass G-type Receptor (Celsr) was adopted in parallel to Flamingo for the human and mouse homologues [45–47] (Fig. 1; see also [48]).

The first clue as to the function of Fmi-Celsr homologues came through the analysis of *Drosophila* mutant phenotypes. Loss of *fmi* gene function leads to a disruption of the polarity of cuticular structures of the adult fly [41, 42, 49]. This phenotype is strikingly consistent with a loss of function of the Frizzled-dependent "core" planar polarity pathway and led to the idea of Flamingo/Celsr homologues being developmental patterning genes. Planar polarity (planar cell polarity, PCP) refers to the coordinated polarisation of cells in the plane of a tissue (commonly an epithelial sheet) orthogonal to the apicobasal axis of the tissue [50] (Fig. 5a, b). The core planar polarity pathway (or simply "core pathway") is one of the major developmental regulators of planar polarity throughout the animal kingdom

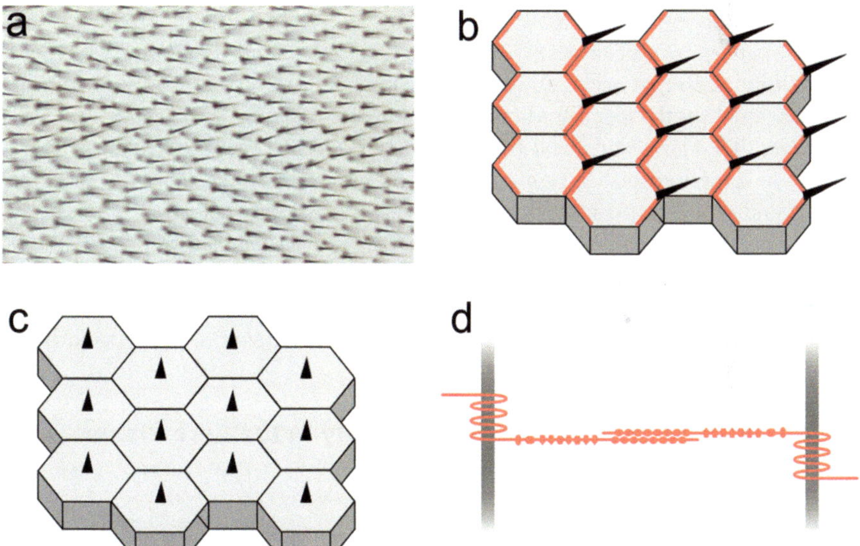

Fig. 5 *Drosophila* Flamingo and planar polarity specification in the wing. (**a**) Image of hairs (trichomes) on the surface of the adult wing of *Drosophila* (distal to the right). In wings of wild-type animals, each cell produces a single hair that projects distally, showing alignment of cellular polarity both with the overall proximodistal axis of the wing and with each neighbouring cell (D. Strutt, unpublished data). (**b**) Cartoon of cells in a wild-type wing at the time of trichome production, showing distally pointing trichomes (*black*) emerging from the distal vertex of each cell. Localisation of Flamingo protein is indicated in *red*, at the level of apicolateral junctions and enriched to both proximal and distal cell edges. (**c**) Cartoon of cells in a wing lacking Flamingo protein at the time of trichome production, showing production of a trichome in the centre of the apical surface of each cell, indicating loss of cell polarity cues. Note that subsequently in development, these mispolarised trichomes will adopt a swirling pattern. (**d**) Cartoon showing homophilic binding of two Flamingo molecules (*red*) between junctions of adjacent cells. Homophilic binding is dependent on the presence of the Cadherin repeats; however, the exact binding interfaces have not been mapped. See main text for respective references

[43, 50]. In *Drosophila* the loss of its activity leads to defects in the polarity of a variety of epidermal derivatives such as the hairs (also known as trichomes) on the surface of the wing (Fig. 5b, c), the orientation of bristles on the back and abdomen and the orientation of photoreceptor clusters in the compound eye.

It had long being recognised that the specification of planar polarity during tissue development requires the transmission of polarised signals between cells and that planar polarity genes were likely to encode proteins involved in such cell-cell signalling [51, 52]. However, little was known regarding the underlying mechanism. In this respect, the molecular characterisation of Fmi protein localisation and behaviour in *Drosophila* was a major step forward.

Generation of antibodies against Fmi revealed that during the process of cell polarisation in the fly wing, the Fmi protein became asymmetrically distributed within cells, being found at proximodistal cell junctions [42] (Fig. 5b). Further biochemical and genetic evidence indicated that Fmi bound homophilically to itself, with distally localised molecules binding to proximally localised molecules in the neighbouring cell (Fig. 5d).

This pattern of localisation generates a puzzle: Fmi is required for cell polarisation, such that each cell in the wing can distinguish its distal side from its proximal side, yet the localisation of the protein is equal at both distal and proximal cell edges. To address this, and based on elegant genetic and molecular analyses, Usui et al. proposed that through interactions with other planar polarity proteins, Fmi would exist in two forms with different activities on each side of the proximodistal cell boundaries and would be involved in feedback amplification of initially small differences in cell polarity in order to generate a robust polarity cue [42]. Furthermore, the ability of Fmi to link adjacent cells through homophilic binding provides an elegant mechanism for cell-cell coupling of planar polarity.

3.2 The Role of Flamingo/Celsr Activity in the Core Planar Polarity Pathway

Following the discovery of the asymmetric cellular localisation of Fmi in the developing *Drosophila* wing, a series of studies quickly followed showing that such asymmetric localisation was a hallmark of all the protein products of the genetically defined components of the core planar polarity pathway (reviewed in [53]). Furthermore, unlike Fmi which is found concentrated at both proximal and distal cell edges, each of the other components was found to preferentially localise either proximally or distally (Fig. 6a). Similar patterns of subcellular asymmetry for the core proteins were also found to occur in other *Drosophila* tissues such as the developing eye, suggesting that these polarised localisations are a general property of the core pathway components.

Consistent with the genetic requirement for activity of the core pathway components for correct planar polarisation of adult cuticular structures, the activity of each protein is also required for the asymmetric subcellular localisation of all the other core proteins (Fig. 6a–d). This supports the view that the core proteins all act

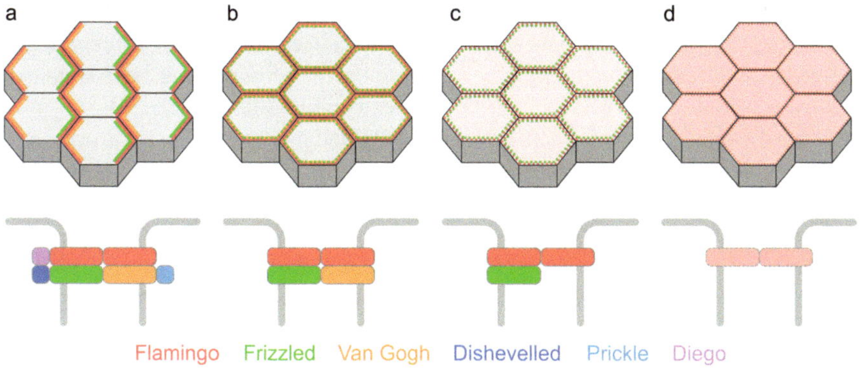

Fig. 6 Localisation of Flamingo and the core planar polarity proteins in the developing wing. (**a**) In a wild-type wing prior to trichome formation, the core planar polarity proteins are asymmetrically localised along the proximodistal axis in the apicolateral region of wing cells. Fz, Dsh and Dgo are localised at distal cell edges (all three represented in *green* in cartoon at *top*), where they associate with Fmi (*red*). Distally localised Fmi forms homophilic adhesions with Fmi molecules at the proximal edge of neighbouring cells, which are associated with proximally localised Vang and Pk (represented in *orange* in cartoon at *top*). (**b**) In the absence of the cytoplasmic core proteins, Fz, Vang and Fmi still tightly localise to the apicolateral junctional region but are no longer asymmetrically distributed on the proximodistal cell axis. (**c**) In the absence of Vang (with or without the cytoplasmic complex components), Fz and Fmi still localise largely to apicolateral cells junctions. Fz-Fmi preferentially bind homophilically to Fmi in neighbouring cells, although a small fraction of Fmi is also seen in apical cell membranes. The formation of Fz-Fmi:Fmi intercellular complexes appears to be the first symmetry breaking step in planar polarisation. (**d**) In the absence of Fz and Vang (with or without the cytoplasmic complex components), Fmi shows only weak homophilic junctional binding and is largely seen localised to apical cell membranes. See main text for respective references

together as part of a common molecular machinery for the planar polarisation of cells and suggests that it is the asymmetric localisation of the proteins themselves that signals the production of correctly polarised cuticular structures such as the trichomes of the wing. This latter hypothesis was validated by the subsequent discovery that a group of downstream "effector" proteins, which act on the cytoskeleton to produce polarised wing trichomes, are themselves asymmetrically localised within wing cells under the control of the core pathway components (reviewed in [53]).

Taking together the studies to date examining core protein behaviour, it is possible to construct a hierarchy of core protein function, with Fmi intercellular homodimers lying at the top, providing the central function of linking the localisation of core proteins in adjacent cells. As already noted, in common with most other cadherins, Fmi acts as a junctionally localised homophilic binding molecule, with Fmi in one cell binding to Fmi in adjacent cells, in a manner dependent on the presence of the extracellular cadherin repeats [42, 54]. Notably, this preferential homophilic junctional binding of Fmi is not an intrinsic property but depends on the presence of two of the other core proteins, the seven-pass transmembrane protein Frizzled (Fz) and the four-pass transmembrane protein

Van Gogh (Vang, also known as Strabismus [Stbm]) [55]. In the absence of both Fz and Vang, the majority of Fmi appears to be localised to apical cell membranes where it cannot bind homophilically to molecules in neighbouring cells (Fig. 6d) and from where it is subject to high rates of endocytic turnover [55, 56].

Mutual dependency between Fmi, Fz and Vang for junctional localisation is further evident in that in the absence of Fmi, Fz fails to show junctional localisation [57] and similarly Vang junctional localisation is noticeably compromised [58]. The promotion of these junctional localisations is believed to depend on direct protein-protein interactions between Fmi and Fz and Vang. Specific binding interfaces between Fmi and its putative partners have not been mapped, in part due to the inherent difficulties of carry out biochemical studies with large multipass transmembrane proteins. Nevertheless, the coincidence of their junctional colocalisation, their mutual dependence for such localisation and a number of cell culture and co-immunoprecipitation studies with both *Drosophila* and vertebrate homologues [55, 59, 60] lend support to the view that their interactions might be direct.

Interestingly, Fmi appears to preferentially form homophilic junctional complexes when Fz is only present in one of the two adjacent cells (Fig. 5c), consistent with the formation of a complex consisting of Fz-Fmi in one cell with Fmi in the neighbouring cell being the first symmetry breaking step in planar polarity complex formation [55]. Taken together with other evidence, this has led to the view that by association with different binding partners (in particular Fz and Vang), Fmi exists in two states, a "distal" form and a "proximal" form, which have different binding affinities both for each other in adjacent cells and for other partners in the same cell [55, 59].

Following this initial Fz-dependent symmetry breaking step that leads to formation of asymmetric complexes of the form Fz-Fmi:Fmi-Vang between the adjacent cells, the next key activity of the core proteins is the asymmetric distribution of these complexes at the cellular level (Fig. 6a). This requires the activity of a group of cytosolic core pathway components, notably Dishevelled (Dsh) which is recruited into the complex most likely by direct interactions with Fz [61, 62], Prickle (Pk) which interacts with Vang [58, 63] and Diego (Dgo) which is recruited in a Fz- and Dsh-dependent manner [64, 65]. The exact mechanism by which the core proteins mediate their own asymmetric distribution within cells is a subject of active investigation, but a large body of evidence supports the view that feedback interactions between core protein complexes lead to the amplification of initially small asymmetries to produce the final robust patterns of polarised subcellular localisation (reviewed in [4, 53, 66]) (Fig. 7a). The exact source of the initial small asymmetries has been (and continues to be) a subject of some controversy, although the likely situation is that both within and between different tissues, multiple cues play a role (reviewed in [4]).

This role of Fmi-Celsr proteins as key components of asymmetrically distributed intercellular protein complexes that mediate planar polarisation of cell sheets has been confirmed by numerous studies in vertebrates (reviewed in [4, 50]), and

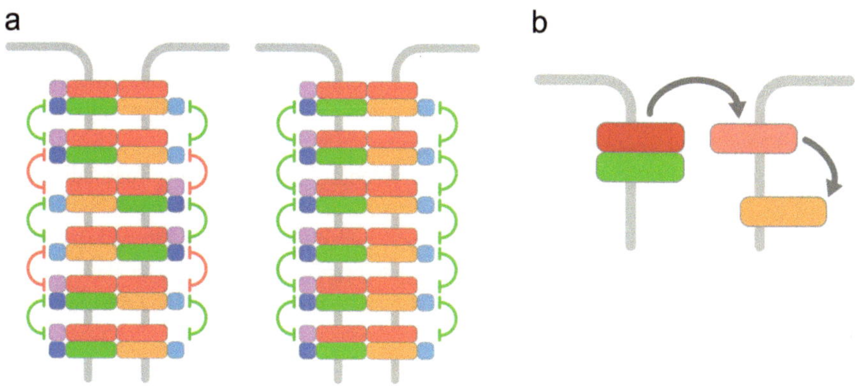

Fig. 7 Feedback models for amplification of asymmetry and possible mechanisms of Flamingo action. (**a**) Amplification of cellular asymmetry and the distribution of Fz and Vang containing complexes to opposite cell ends are believed to depend on feedback mechanisms that locally sort core protein complexes into polarised domains. On the *left* shows a situation where complexes are poorly sorted and asymmetry is low. It is believed that local interactions between complexes might lead to mutual stabilisation of complexes of the same orientation (*green* interactions) and destabilisation of complexes of opposite orientation (*red* interactions), leading to the situation shown on the *right* where complexes are locally sorted into domains of common orientation. Feedback interactions that promote stability might function by increasing cis multimerisation or transhomophilic binding between neighbouring Fmi molecules, whereas destabilising interactions might block cis- or trans-Fmi interactions or promote endocytosis of Fmi. (**b**) Model indicating that Fz-Fmi in one cell might act as a ligand for Fmi in a neighbouring cell, "activating" Fmi and causing it to interact with Vang, for instance, via a conformational change. See main text for respective references

functions in planar polarisation are also likely in other invertebrate groups (see [43]). In particular, studies in epithelia such as those of the inner ear, skin, airways and oviduct have revealed similar patterns of subcellular asymmetric localisation of Celsr homologues and other core planar polarity proteins and also evidence of mutual dependency of core protein function for this localisation and for planar polarisation of the tissue [60, 67–69].

To date, overlapping activities of the three Celsr homologues found in vertebrates have been implicated through genetic studies in knockout animals in a plethora of developmental processes (reviewed in [70, 71]). Based on either evidence for planar polarisation of the tissue, or the requirement for other components of the core planar polarity pathway, many of these seem likely to be genuine planar polarity functions, dependent on polarised cell-cell communication mediated by asymmetrically distributed core protein complexes: examples include polarised growth of hair follicles or polarised positioning of cilia in tissues such as the airways or oviduct (as noted above) but also polarised cell movements such as seen in gastrulation (including neural tube closure) [72], wound healing [73] and commissural axon guidance (reviewed in [74]).

However, it should be noted that in both *Drosophila* and vertebrates, there are examples of Fmi-Celsr functions that are either not evidently related to planar polarisation of tissues, which do not require the activity of the other core planar polarity proteins and/or which do not appear to require homophilic interactions between Fmi/Celsr molecules in neighbouring cells. In *Drosophila*, these include roles in dendritic tiling [54, 75] and photoreceptor axon spacing and targeting [76, 77]; and in vertebrates some roles in gastrulation movements [78, 79] or forebrain patterning [80] appear to be core pathway independent. While these alternative functions of Fmi/Celsr are of great interest, the existing lack of knowledge regarding the molecular mechanisms means that they currently add little to our overall understanding of Fmi/Celsr biology.

3.3 Molecular Mechanisms of Flamingo/Celsr Function

The large size and multidomain structure of Fmi-Celsr proteins hint at perhaps multiple functions for these molecules in various developmental processes, a view supported by the observation that mutations affect both planar polarity and non-planar polarity processes. Even regarding the better characterised functions in planar polarity, studies of the molecular mechanisms of Fmi-Celsr function are still in their infancy and in large part have been limited to simple structure-function analyses. Although belonging to the family of aGPCRs, no G protein-dependent function has been shown. Generally, very little is known about whether Fmi-Celsr proteins can signal via similar mechanisms to G protein-coupled receptors or whether alternatively (at the opposite extreme) they simply act as passive protein scaffolds onto which other signalling active proteins can assemble.

So far no specific functions in planar polarity have been assigned to the EGF-like domains, laminin G motifs or GPS region in the extracellular domain. However, in *Drosophila*, immunoblotting using antibodies against the Fmi protein shows that species of different molecular weights are present, consistent with the GPS providing a functional cleavage site [42].

Consistent with being aGPCRs, the extracellular cadherin repeats are required for homophilic adhesion interactions [42, 54], and this activity appears to be strongly modulated by the partner proteins Fz and Vang [55]. Intriguingly, similar to a construct lacking the cadherin repeats, a deletion construct lacking the hormone binding domain is non-functional when overexpressed [54], perhaps indicating a key role for the hormone binding domain in mediating cadherin-dependent homophilic adhesion. Overexpression of full-length Fmi in the *Drosophila* wing gives a phenotype similar to loss of Fz activity [42] in which cells repolarise towards the region of overexpression, and under these conditions, excess Fz is seen recruited to cell junctions [55] and Vang to the junctions of neighbouring cells [59]. Additionally, the intracellular domain of Fmi is not required for the overexpression phenotype to be seen [54, 55, 59], and if a form of Fmi lacking the intracellular domain is expressed at close to physiological levels, it preferentially localises at distal cell edges with Fz and appears to be compromised in its ability to interact with Vang [55]. In this context, it is notable that normally Vang only appears to have strong interactions with Fmi homodimers if they are already bound to Fz in the neighbouring cell [55]. Conversely, overexpression of a form of Fmi lacking almost all the extracellular regions gives a phenotype similar to loss of Vang activity whereby cells repolarise away from the region [54].

Taken together, these results suggest that Fmi has critical interactions with Fz that require either the extracellular or transmembrane regions, but not the intracellular domain. Furthermore, Vang interactions are promoted by (but not are not absolutely dependent on) an activity conferred by the intracellular tail of Fmi, and these interactions are stimulated by interactions with Fz-Fmi complexes in the neighbouring cell [55] (Fig. 7b). As it is generally assumed that the intracellular tail of GPCR-like molecules such as Fmi would be required for intracellular signalling activity, it is tempting to speculate that a Fz-Fmi complex in one cell acts as a ligand for a Fmi molecule in the neighbouring cell and induces a signalling event that leads to recruitment of Vang. The nature of such a signalling event is unclear, but by analogy to other GPCRs, it could involve a conformational change that exposes a binding site. Furthermore, in zebrafish, the intracellular domain of a Celsr homologue is able to affect planar polarised convergent extension movements in a manner dependent on a conserved cluster of potential serine-threonine phosphorylation sites, suggesting that an unknown kinase may be involved in Fmi/Celsr activation [78].

There is limited evidence that any signalling activity of Fmi-Celsr is dependent on these molecules acting as guanine nucleotide exchange factors (GEFs) for heterotrimeric G proteins in the normal manner of GPCRs. However, in rat brain slice culture, Celsr2 and Celsr3 can cause calcium release in phospholipase C-dependent manner in response to exogenously applied Celsr cadherin repeats bound to beads [81]. This supports the view that Fmi-Celsr molecules can act as homophilic ligands to activate their own signalling activity. Identification of a requirement for heterotrimeric G proteins would be strong evidence that such a mechanism was involved in planar polarity activity, but given the broad roles of such molecules in developmental and physiological signalling events, a specific function in planar polarity might be masked.

In addition to forming Fz-Fmi:Fmi-Vang intracellular complexes, planar polarisation requires the sorting of these complexes into locally polarised domains [56]. Plausible mechanisms underlying the feedback interactions that mediate such sorting include the modulation of dimerisation activity between complexes or regulation of endocytosis and trafficking of complex components (Fig. 7a). Interestingly, zebrafish Celsr2 has been shown to be able to dimerise in cis in a manner that does not require the complete seven-pass transmembrane region or intracellular domain [78], but how such dimerisation could be modulated to regulate planar polarisation is unclear. Similarly, mouse Celsr1 can be specifically internalised at the cell surface in response to cell cycle-dependent phosphorylation by Polo-like kinase 1 [82, 83], but again no role for this mechanism in sorting of planar polarity complexes is known.

3.4 Emerging Roles of Adhesion GPCRs in Cell Migration

The concept of polarity is not only essential for the development of epithelia and organs but is also a central element in cell migration. Migratory processes are essential for development, wound healing or immune responses and can lead to tumour formation when misregulated [84]. Naturally, cell migration belongs to the most important and tightly controlled biological mechanisms in multicellular organisms. In general, two types of migration can be distinguished: single cell migration and collective cell migration (Fig. 8a, b). However, the basic mechanisms underlying both processes are similar. Upon receiving a migration-promoting signal, a cell polarises by forming protrusions called lamellipodia or filopodia caused by actin polymerisation. By developing adhesion contacts to components of the cell's environment such as the extracellular matrix or surrounding cells, the lamellipodia or filopodia of the migrating cells get stabilised. Subsequently,

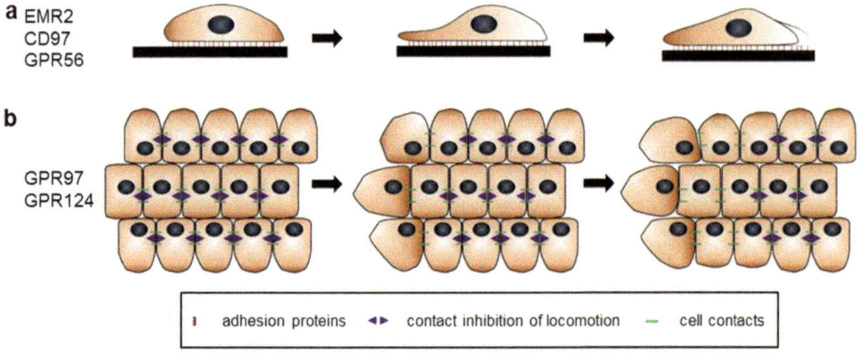

c

Adhesion-GPCR	tissue / cells involved in migration	Single cell/ Collective cell Migration	Reference
ADGRE2/EMR2	activated leukocytes	single	[86]
ADGRE5/CD97	activated leukocytes	single	[85]
ADGRG1/GPR56	radial glia cells and other neural progenitor cells	single	[90]
ADGRG3/GPR97	mouse lymphatic endothelium	collective	[95]
ADGRA2/GPR124	central nervous system	collective	[94]

Fig. 8 Schematic illustration of single cell and collective cell migration and the involvement of aGPCRs in these processes. (**a**) Single cell migration. After a migration-promoting signal, a cell forms protrusions generally driven by actin polymerisation. By forming new adhesion contacts to the environment of the cell, these protrusions are stabilised. A traction force translocates the cell over the substrate. In a cell layer migration is inhibited by contact inhibition of locomotion signals. After a migration-promoting signal, cells polarise at the edge of the free space. By exerting attractive force between neighbouring cells, the contact inhibition of locomotion is resolved and thus coordinates movement of the cell layer. (**b**) Collective cell migration displays the same underlying mechanisms as single cell migration but in a cellular environment. (**c**) Several aGPCRs are involved in cell migration

actin- and myosin-dependent contractions lead to a traction force which translocates the cell and retracts it from its contact (Fig. 8a, b) [85, 86].

Given that cell migration is highly dependent on the ability of cells to form adhesive contacts, it was not surprising to find that several aGPCRs are implicated in migratory processes. Group II was the first one to be linked with migration (Fig. 8c). Members of this group are very prominent in immune cells [87] and have been shown to be involved in single cell migration of various types of these cells. Single cell migration is an important mechanism in immunity, which allows immune cells to invade into inflamed tissue in a rapid manner and evoke an immune response [85, 88]. Expression of ADGRE2/EMR2 and ADGRE5/CD97 is tightly linked to activation of leukocytes [89, 90] and increased migration of these immune cells [90, 91]. While EMR2 shows a redistribution to the leading edge of activated leukocytes similarly to Rac, a redistribution of CD97 within the activated leukocytes was not observed [90, 92]. However, the details of the molecular mechanisms underlying the involvement of group II aGPCRs in immune cell migration remains to be determined. First steps towards an understanding of their signalling mechanisms revealed that C-terminally truncated versions of CD97 as well as EMR2 showed no significant migration to a chemotactic agent, whereas the full-length receptors showed migration towards a chemoattractant. This sheds light on possible signal transduction via the Rho family of small GTPases, which are a central player in migratory processes [85, 90, 93].

Besides immune cell migration, aGPCRs are also implicated in the migration of neurons. An example for inhibiting migratory processes of this cell type is ADGRG1/GPR56 which is expressed in migrating neurons [94, 95] (Fig. 8c). By interacting with collagen III, the receptor suppresses migration of neurons in the developing cerebral cortex [96]. In mice lacking *Gpr56*, neurons overmigrate beyond the pial basement membrane which leads to cerebral cortex malformation. This phenomenon is linked to bilateral frontoparietal polymicrogyria (BFPP) in humans, which is a recessive inherited development disorder of the cerebral cortex [94, 96, 97].

There is also evidence for aGPCRs playing roles not only in single cell but also in collective cell migration. The collective migration process, in which cells connected by cell-cell contacts move together as a tissue, often takes place in embryogenesis and wound healing [86]. Interestingly, Flamingo/Celsr are implicated in collective cell migratory processes such as gastrulation [72] and wound healing [73] as part of their PCP function (described above). Other prominent examples for the involvement of an aGPCR in collective cell migration are ADGRG3/GPR97 and ADGRA2/GPR124 (Fig. 8c). *Gpr124* is highly expressed in the endothelium of the central nervous system (CNS) [98]. Mice lacking *Gpr124* show severe CNS haemorrhage restricted to the forebrain but not in the mid- or hindbrain [98]. Further, it was shown that GPR124 regulates CNS angiogenesis in a cell-autonomous manner in vivo and in vitro. A brain endothelial cell line expressing *Gpr124* exhibits a directional migration towards gradients of conditioned medium of forebrain cortical cells in a Cdc42-dependent manner [98]. Again, the signals mediated by the receptor remain elusive. GPR97 is expressed in lymphatic endothelial cells (LEC) (Fig. 8c). Interestingly, an enhanced single cell migration but a decreased collective cell migration was observed in these cells in vitro upon silencing of *Gpr97* [99]. As single cell migration can progress into collective cell migration, it is very possible that GPR97 acts as a linchpin between both migratory systems. Further analyses indicate that GPR97 is directly involved in migration without an intermediary soluble factor. Changes in the actin cytoskeleton and focal adhesions are possibly caused by GPR97 through influencing the balance between the small G proteins RhoA and Cdc42 [99].

Taken together, several aGPCRs have a role in migratory processes. Although details of the signalling mechanisms underlying their functions are yet to be elucidated, preliminary data indicate that they probably are directly involved in mediating cell adhesion. Especially for group II aGPCRs, the EGF domains in their N termini are potentially distinctive for the formation of adhesion contacts [87, 100]. Although signalling via heterotrimeric G proteins has not been shown for any of these receptors and cannot be excluded, there is evidence that the aGPCRs involved in migratory processes signal via members of the Rho family of small GTPases. These proteins cause cytoskeletal changes and focal adhesion rearrangements [88, 99].

4 Conclusion

Polarity in multicellular organisms is a fundamental concept for the development of tissues and organs but also for maintaining biological processes such as immune responses or wound healing. Due to the highly complex and tightly regulated processes ensuring the correct establishment and maintenance of polarity, the details on how they are governed especially on a molecular level are not fully understood.

Over the past years, intense research has revealed that aGPCRs play essential roles in various types of polarity. Interestingly, the members of the two evolutionarily oldest groups, Latrophilins and Flamingo/Celsr, are involved in tissue polarity. While for Flamingo/Celsr it has been shown that vertebrate homologues of this aGPCR group have similar functions to *Drosophila* Fmi, for homologues of the Latrophilin group, this has not yet been the case. Vertebrate Latrophilins are implicated in synaptic functions [25, 26, 34], but no role in establishing or maintaining polarity has been assigned to any of the three mammalian Latrophilin homologues (ADGRL1–3). It can be speculated that the reason for this is that due to the complexity of the mammalian system together with possible redundancy regarding the function of the different homologues, analyses of these receptors are still incomplete and some of their functions have not yet been discovered. It is, however, also conceivable that due to the existence of so many more aGPCRs in mammals (33 in humans compared to 3 in *C. elegans* [101, 102]), the invertebrate receptors have different roles than their vertebrate homologues.

Similarly there are major gaps in our knowledge regarding the actual mechanisms of signalling by aGPCRs in polarity processes. It has been shown for several receptors of this family that they do have a function in these processes, but the details and the impact of their individual roles need further investigation. Especially the molecular details underlying their physiological functions remain elusive. It is quite possible that several if not all receptors follow similar mechanisms of signalling such as activation by an intramolecular agonist or transmission of signals through heterotrimeric or small G proteins. In particular for members of the group of Flamingo/Celsr, which have been characterised in great detail regarding their function in PCP, this information would be invaluable in understanding how they mediate polarity and also if and how their function in PCP is related to the one in migratory processes. Thus, future analyses need to focus on further clarifying the mechanisms of activation as well as signalling cascades of aGPCRs involved in the different types of polarity.

Acknowledgements Work in the Strutt laboratory is funded by a Wellcome Trust Senior Fellowship in Basic Biomedical Science. Work in the Prömel laboratory is funded by grants from the Deutsche Forschungsgemeinschaft (PR1534 1-1; FOR2149 Project 2 [PR 1534/2-1]) and the Medical Faculty of the University of Leipzig (Formel 1 junior research grant).

References

1. Wolpert L (2013) Cell polarity. Philos Trans R Soc Lond B Biol Sci 368:20130419
2. Hunter MV, Fernandez-Gonzalez R (2013) Gastrulation: cell polarity comes full circle. Curr Biol 23:R845–R848
3. Zallen JA (2007) Planar polarity and tissue morphogenesis. Cell 129:1051–1063
4. Devenport D (2014) The cell biology of planar cell polarity. J Cell Biol 207:171–179
5. Gubb D, Garcia-Bellido A (1982) A genetic analysis of the determination of cuticular polarity during development in Drosophila melanogaster. J Embryol Exp Morphol 68:37–57
6. Wong LL, Adler PN (1993) Tissue polarity genes of Drosophila regulate the subcellular location for prehair initiation in pupal wing cells. J Cell Biol 123:209–221
7. Yates LL, Schnatwinkel C, Murdoch JN, Bogani D, Formstone CJ et al (2010) The PCP genes Celsr1 and Vangl2 are required for normal lung branching morphogenesis. Hum Mol Genet 19:2251–2267
8. Segalen M, Bellaiche Y (2009) Cell division orientation and planar cell polarity pathways. Semin Cell Dev Biol 20:972–977
9. Li X, Roszko I, Sepich DS, Ni M, Hamm HE, Marlow FL, Solnica-Krezel L (2013) Gpr125 modulates dishevelled distribution and planar cell polarity signaling. Development 140:3028–3039
10. Maduro MF (2010) Cell fate specification in the C. elegans embryo. Dev Dyn 239:1315–1329
11. Schnabel R, Priess JR (1997) Specification of cell fates in the early embryo. In: Riddle DL, Blumenthal T, Meyer BJ, Priess JR (eds) C. elegans II, 2nd edn. Cold Spring Harbor, New York, NY
12. Sulston JE, Schierenberg E, White JG, Thomson JN (1983) The embryonic cell lineage of the nematode Caenorhabditis elegans. Dev Biol 100:64–119
13. Priess JR, Thomson JN (1987) Cellular interactions in early C. elegans embryos. Cell 48:241–250
14. Schierenberg E (1987) Reversal of cellular polarity and early cell-cell interaction in the embryos of Caenorhabditis elegans. Dev Biol 122:452–463
15. Gonczy P, Rose LS (2005) Asymmetric cell division and axis formation in the embryo. WormBook October:1–20

16. Kaletta T, Schnabel H, Schnabel R (1997) Binary specification of the embryonic lineage in Caenorhabditis elegans. Nature 390:294–298
17. Lin R, Hill RJ, Priess JR (1998) POP-1 and anterior-posterior fate decisions in C. elegans embryos. Cell 92:229–239
18. Rocheleau CE, Downs WD, Lin R, Wittmann C, Bei Y et al (1997) Wnt signaling and an APC-related gene specify endoderm in early C. elegans embryos. Cell 90:707–716
19. Thorpe CJ, Schlesinger A, Carter JC, Bowerman B (1997) Wnt signaling polarizes an early C. elegans blastomere to distinguish endoderm from mesoderm. Cell 90:695–705
20. Walston T, Tuskey C, Edgar L, Hawkins N, Ellis G et al (2004) Multiple Wnt signaling pathways converge to orient the mitotic spindle in early C. elegans embryos. Dev Cell 7:831–841
21. Nance J, Zallen JA (2011) Elaborating polarity: PAR proteins and the cytoskeleton. Development 138:799–809
22. Hutter H, Schnabel R (1995) Establishment of left-right asymmetry in the Caenorhabditis elegans embryo: a multistep process involving a series of inductive events. Development 121:3417–3424
23. Cabello J, Neukomm LJ, Gunesdogan U, Burkart K, Charette SJ et al (2010) The Wnt pathway controls cell death engulfment, spindle orientation, and migration through CED-10/Rac. PLoS Biol 8, e1000297
24. Langenhan T, Prömel S, Mestek L, Esmaeili B, Waller-Evans H et al (2009) Latrophilin signaling links anterior-posterior tissue polarity and oriented cell divisions in the C. elegans embryo. Dev Cell 17:494–504
25. Sudhof TC (2001) alpha-Latrotoxin and its receptors: neurexins and CIRL/latrophilins. Annu Rev Neurosci 24:933–962
26. Willson J, Amliwala K, Davis A, Cook A, Cuttle MF et al (2004) Latrotoxin receptor signaling engages the UNC-13-dependent vesicle-priming pathway in C. elegans. Curr Biol 14:1374–1379
27. Krasnoperov VG, Beavis R, Chepurny OG, Little AR, Plotnikov AN et al (1996) The calcium-independent receptor of alpha-latrotoxin is not a neurexin. Biochem Biophys Res Commun 227:868–875
28. Krasnoperov VG, Bittner MA, Beavis R, Kuang Y, Salnikow KV et al (1997) alpha-Latrotoxin stimulates exocytosis by the interaction with a neuronal G-protein-coupled receptor. Neuron 18:925–937
29. Müller A, Winkler J, Fiedler F, Sastradihardja T, Binder C et al (2015) Oriented cell division in the C. elegans embryo is coordinated by G-protein signaling dependent on the adhesion GPCR LAT-1. PLoS Genet 11, e1005624
30. Bischoff M, Schnabel R (2006) A posterior centre establishes and maintains polarity of the Caenorhabditis elegans embryo by a Wnt-dependent relay mechanism. PLoS Biol 4, e396
31. Hyman AA, White JG (1987) Determination of cell division axes in the early embryogenesis of Caenorhabditis elegans. J Cell Biol 105:2123–2135
32. Boucard AA, Ko J, Sudhof TC (2012) High-affinity neurexin binding to the cell-adhesion G-protein coupled receptor CIRL1/latrophilin-1 produces an intercellular adhesion complex. J Biol Chem 287(12):9399–9413
33. O'Sullivan ML, de Wit J, Savas JN, Comoletti D, Otto-Hitt S et al (2012) FLRT proteins are endogenous latrophilin ligands and regulate excitatory synapse development. Neuron 73:903–910
34. Silva JP, Lelianova VG, Ermolyuk YS, Vysokov N, Hitchen PG et al (2011) Latrophilin 1 and its endogenous ligand Lasso/teneurin-2 form a high-affinity transsynaptic receptor pair with signaling capabilities. Proc Natl Acad Sci U S A 108:12113–12118
35. Prömel S, Frickenhaus M, Hughes S, Mestek L, Staunton D et al (2012) The GPS motif is a molecular switch for bimodal activities of adhesion class G protein-coupled receptors. Cell Rep 2:321–331
36. Demberg LM, Rothemund S, Schoneberg T, Liebscher I (2015) Identification of the tethered peptide agonist of the adhesion G protein-coupled receptor GPR64/ADGRG2. Biochem Biophys Res Commun 464(3):743–747

37. Liebscher I, Schon J, Petersen SC, Fischer L, Auerbach N et al (2014) A tethered agonist within the ectodomain activates the adhesion G protein-coupled receptors GPR126 and GPR133. Cell Rep 9:2018–2026
38. Stoveken HM, Hajduczok AG, Xu L, Tall GG (2015) Adhesion G protein-coupled receptors are activated by exposure of a cryptic tethered agonist. Proc Natl Acad Sci U S A 112:6194–6199
39. Liebscher I, Schöneberg T (2016) Tethered agonism: a common activation mechanism of adhesion GPCRs. In: Langenhan T, Schöneberg T (eds) Adhesion G protein-coupled receptors: molecular, physiological and pharmacological principles in health and disease. Springer, Heidelberg
40. Kishore A, Hall RA (2016) Versatile signaling activity of adhesion GPCRs. In: Langenhan T, Schöneberg T (eds) Adhesion G protein-coupled receptors: molecular, physiological and pharmacological principles in health and disease. Springer, Heidelberg
41. Chae J, Kim MJ, Goo JH, Collier S, Gubb D et al (1999) The *Drosophila* tissue polarity gene *starry night* encodes a member of the protocadherin family. Development 126:5421–5429
42. Usui T, Shima Y, Shimada Y, Hirano S, Burgess RW et al (1999) Flamingo, a seven-pass transmembrane cadherin, regulates planar cell polarity under the control of Frizzled. Cell 98:585–595
43. Hale R, Strutt D (2015) Conservation of planar polarity pathway function across the animal kingdom. Annu Rev Genet 49:529–551
44. Nordström KJ, Lagerström MC, Wallér LM, Fredriksson R, Schiöth HB (2009) The Secretin GPCRs descended from the family of Adhesion GPCRs. Mol Biol Evol 26:71–84
45. Hadjantonakis AK, Formstone CJ, Little PF (1998) mCelsr1 is an evolutionarily conserved seven-pass transmembrane receptor and is expressed during mouse embryonic development. Mech Dev 78:91–95
46. Hadjantonakis AK, Sheward WJ, Harmar AJ, de Galan L, Hoovers JM et al (1997) Celsr1, a neural-specific gene encoding an unusual seven-pass transmembrane receptor, maps to mouse chromosome 15 and human chromosome 22qter. Genomics 45:97–104
47. Nakayama M, Nakajima D, Nagase T, Nomura N, Seki N et al (1998) Identification of high-molecular-weight proteins with multiple EGF-like motifs by motif-trap screening. Genomics 51:27–34
48. Araç D, Sträter N, Seiradake E (2016) Understanding the structural basis of adhesion GPCR functions. In: Langenhan T, Schöneberg T (eds) Adhesion G protein-coupled receptors: molecular, physiological and pharmacological principles in health and disease. Springer, Heidelberg
49. Lu B, Usui T, Uemura T, Jan L, Jan Y-N (1999) Flamingo controls the planar polarity of sensory bristles and asymmetric division of sensory organ precursors in *Drosophila*. Curr Biol 9:1247–1250
50. Goodrich LV, Strutt D (2011) Principles of planar polarity in animal development. Development 138:1877–1892
51. Vinson CR, Adler PN (1987) Directional non-cell autonomy and the transmission of polarity information by the *frizzled* gene of *Drosophila*. Nature 329:549–551
52. Vinson CR, Conover S, Adler PN (1989) A *Drosophila* tissue polarity locus encodes a protein containing seven potential transmembrane domains. Nature 338:262–264
53. Strutt H, Strutt D (2009) Asymmetric localisation of planar polarity proteins: mechanisms and consequences. Semin Cell Dev Biol 20:957–963
54. Kimura H, Usui T, Tsubouchi A, Uemura T (2006) Potential dual molecular interaction of the Drosophila 7-pass transmembrane cadherin Flamingo in dendritic morphogenesis. J Cell Sci 119:1118–1129
55. Strutt H, Strutt D (2008) Differential stability of flamingo protein complexes underlies the establishment of planar polarity. Curr Biol 18:1555–1564
56. Strutt H, Warrington SJ, Strutt D (2011) Dynamics of core planar polarity protein turnover and stable assembly into discrete membrane subdomains. Dev Cell 20:511–525

57. Strutt DI (2001) Asymmetric localization of Frizzled and the establishment of cell polarity in the *Drosophila* wing. Mol Cell 7:367–375
58. Bastock R, Strutt H, Strutt D (2003) Strabismus is asymmetrically localised and binds to Prickle and Dishevelled during *Drosophila* planar polarity patterning. Development 130:3007–3014
59. Chen WS, Antic D, Matis M, Logan CY, Povelones M et al (2008) Asymmetric homotypic interactions of the atypical cadherin flamingo mediate intercellular polarity signaling. Cell 133:1093–1105
60. Devenport D, Fuchs E (2008) Planar polarization in embryonic epidermis orchestrates global asymmetric morphogenesis of hair follicles. Nat Cell Biol 10:1257–1268
61. Axelrod JD (2001) Unipolar membrane association of Dishevelled mediates Frizzled planar cell polarity signaling. Genes Dev 15:1182–1187
62. Shimada Y, Usui T, Yanagawa S, Takeichi M, Uemura T (2001) Asymmetric colocalization of Flamingo, a seven-pass transmembrane cadherin, and Dishevelled in planar cell polarization. Curr Biol 11:859–863
63. Jenny A, Darken RS, Wilson PA, Mlodzik M (2003) Prickle and Strabismus form a functional complex to generate a correct axis during planar cell polarity signaling. EMBO J 22:4409–4420
64. Das G, Jenny A, Klein TJ, Eaton S, Mlodzik M (2004) Diego interacts with Prickle and Strabismus/Van Gogh to localize planar cell polarity complexes. Development 131:4467–4476
65. Feiguin F, Hannus M, Mlodzik M, Eaton S (2001) The ankyrin repeat protein Diego mediates Frizzled-dependent planar polarization. Dev Cell 1:93–101
66. Tree DR, Ma D, Axelrod JD (2002) A three-tiered mechanism for regulation of planar cell polarity. Semin Cell Dev Biol 13:217–224
67. Davies A, Formstone C, Mason I, Lewis J (2005) Planar polarity of hair cells in the chick inner ear is correlated with polarized distribution of c-flamingo-1 protein. Dev Dyn 233:998–1005
68. Shi D, Komatsu K, Hirao M, Toyooka Y, Koyama H et al (2014) Celsr1 is required for the generation of polarity at multiple levels of the mouse oviduct. Development 141:4558–4568
69. Vladar EK, Bayly RD, Sangoram AM, Scott MP, Axelrod JD (2012) Microtubules enable the planar cell polarity of airway cilia. Curr Biol 22:2203–2212
70. Boutin C, Goffinet AM, Tissir F (2012) Celsr1-3 cadherins in PCP and brain development. Curr Top Dev Biol 101:161–183
71. Formstone CJ (2010) 7TM-Cadherins: developmental roles and future challenges. Adv Exp Med Biol 706:14–36
72. Curtin JA, Quint E, Tsipouri V, Arkell RM, Cattanach B et al (2003) Mutation of Celsr1 disrupts planar polarity of inner ear hair cells and causes severe neural tube defects in the mouse. Curr Biol 13:1129–1133
73. Caddy J, Wilanowski T, Darido C, Dworkin S, Ting SB et al (2010) Epidermal wound repair is regulated by the planar cell polarity signaling pathway. Dev Cell 19:138–147
74. Zou Y (2012) Does planar cell polarity signaling steer growth cones? Curr Top Dev Biol 101:141–160
75. Gao FB, Kohwi M, Brenman JE, Jan LY, Jan YN (2000) Control of dendritic field formation in *Drosophila*: the roles of Flamingo and competition between homologous neurons. Neuron 28:91–101
76. Lee RC, Clandinin TR, Lee CH, Chen PL, Meinertzhagen IA et al (2003) The protocadherin Flamingo is required for axon target selection in the *Drosophila* visual system. Nat Neurosci 6:557–563
77. Senti KA, Usui T, Boucke K, Greber U, Uemura T et al (2003) Flamingo regulates R8 axon-axon and axon-target interactions in the *Drosophila* visual system. Curr Biol 13:828–832

78. Carreira-Barbosa F, Kajita M, Morel V, Wada H, Okamoto H et al (2009) Flamingo regulates epiboly and convergence/extension movements through cell cohesive and signalling functions during zebrafish gastrulation. Development 136:383–392
79. Witzel S, Zimyanin V, Carreira-Barbosa F, Tada M, Heisenberg CP (2006) Wnt11 controls cell contact persistence by local accumulation of Frizzled 7 at the plasma membrane. J Cell Biol 175:791–802
80. Qu Y, Huang Y, Feng J, Alvarez-Bolado G, Grove EA et al (2014) Genetic evidence that Celsr3 and Celsr2, together with Fzd3, regulate forebrain wiring in a Vangl-independent manner. Proc Natl Acad Sci U S A 111:E2996–E3004
81. Shima Y, Kawaguchi SY, Kosaka K, Nakayama M, Hoshino M et al (2007) Opposing roles in neurite growth control by two seven-pass transmembrane cadherins. Nat Neurosci 10:963–969
82. Devenport D, Oristian D, Heller E, Fuchs E (2011) Mitotic internalization of planar cell polarity proteins preserves tissue polarity. Nat Cell Biol 13:893–902
83. Shrestha R, Little KA, Tamayo JV, Li W, Perlman DH et al (2015) Mitotic control of planar cell polarity by polo-like kinase 1. Dev Cell 33:522–534
84. Vedula SRK, Ravasio A, Lim CT, Ladoux B (2013) Collective cell migration: a mechanistic perspective. Physiology (Bethesda, MD) 28:370–379
85. Ridley AJ (2015) Rho GTPase signalling in cell migration. Curr Opin Cell Biol 36:103–112
86. Rørth P (2009) Collective cell migration. Annu Rev Cell Dev Biol 25:407–429
87. Stacey M, Lin H-H, Gordon S, McKnight AJ (2000) LNB-TM7, a group of seven-transmembrane proteins related to family-B G-protein-coupled receptors. Trends Biochem Sci 25:284–289
88. Ridley AJ, Schwartz MA, Burridge K, Firtel RA, Ginsberg MH et al (2003) Cell migration: integrating signals from front to back. Science (New York, NY) 302:1704–1709
89. Eichler W, Aust G, Hamann D (1994) Characterization of an early activation-dependent antigen on lymphocytes defined by the monoclonal antibody BL-Ac(F2). Scand J Immunol 39:111–115
90. Yona S, Lin H-H, Dri P, Davies JQ, Hayhoe RPG et al (2008) Ligation of the adhesion-GPCR EMR2 regulates human neutrophil function. FASEB J 22:741–751
91. Leemans JC, te Velde AA, Florquin S, Bennink RJ, Kd B et al (2004) The epidermal growth factor-seven transmembrane (EGF-TM7) receptor CD97 is required for neutrophil migration and host defense. J Immunol 172:1125–1131
92. Bokoch GM (2005) Regulation of innate immunity by Rho GTPases. Trends Cell Biol 15:163–171
93. Galle J, Sittig D, Hanisch I, Wobus M, Wandel E et al (2006) Individual cell-based models of tumor-environment interactions: multiple effects of CD97 on tumor invasion. Am J Pathol 169:1802–1811
94. Li S, Jin Z, Koirala S, Bu L, Xu L et al (2008) GPR56 regulates pial basement membrane integrity and cortical lamination. J Neurosci 28:5817–5826
95. Luo R, Jeong S-J, Jin Z, Strokes N, Li S et al (2011) G protein-coupled receptor 56 and collagen III, a receptor-ligand pair, regulates cortical development and lamination. Proc Natl Acad Sci U S A 108:12925–12930
96. Singer K, Luo R, Jeong S-J, Piao X (2013) GPR56 and the developing cerebral cortex: cells, matrix, and neuronal migration. Mol Neurobiol 47:186–196
97. Piao X, Basel-Vanagaite L, Straussberg R, Grant PE, Pugh EW et al (2002) An autosomal recessive form of bilateral frontoparietal polymicrogyria maps to chromosome 16q12.2-21. Am J Hum Genet 70:1028–1033
98. Kuhnert F, Mancuso MR, Shamloo A, Wang H-T, Choksi V et al (2010) Essential regulation of CNS angiogenesis by the orphan G protein-coupled receptor GPR124. Science (New York, NY) 330:985–989

99. Valtcheva N, Primorac A, Jurisic G, Hollmén M, Detmar M (2013) The orphan adhesion G protein-coupled receptor GPR97 regulates migration of lymphatic endothelial cells via the small GTPases RhoA and Cdc42. J Biol Chem 288:35736–35748
100. Fredriksson R, Lagerström MC, Lundin L-G, Schiöth HB (2003) The G-protein-coupled receptors in the human genome form five main families. Phylogenetic analysis, paralogon groups, and fingerprints. Mol Pharmacol 63:1256–1272
101. Fredriksson R, Gloriam DE, Hoglund PJ, Lagerstrom MC, Schioth HB (2003) There exist at least 30 human G-protein-coupled receptors with long Ser/Thr-rich N-termini. Biochem Biophys Res Commun 301:725–734
102. Schioth HB, Fredriksson R (2005) The GRAFS classification system of G-protein coupled receptors in comparative perspective. Gen Comp Endocrinol 142:94–101

Adhesion GPCRs as Novel Actors in Neural and Glial Cell Functions: From Synaptogenesis to Myelination

Séverine M. Sigoillot, Kelly R. Monk, Xianhua Piao, Fekrije Selimi, and Breanne L. Harty

Graphical Abstract

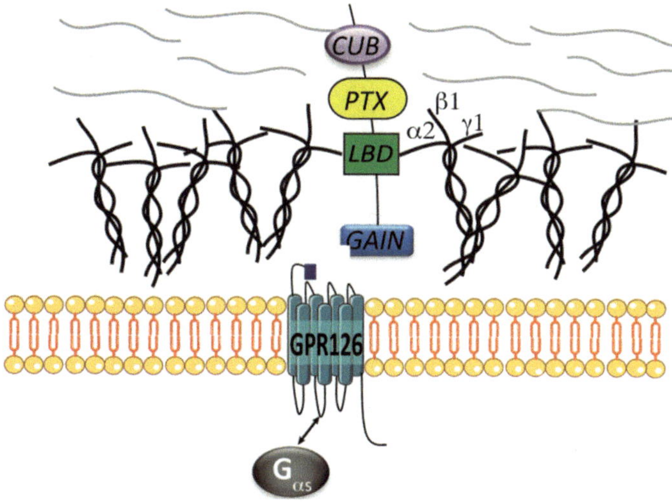

S.M. Sigoillot (✉) • F. Selimi
Center for Interdisciplinary Research in Biology (CIRB), Collège de France, CNRS UMR 7241, INSERM U1050, PSL Research University, Paris, France
e-mail: severine.sigoillot@college-de-france.fr

K.R. Monk • B.L. Harty (✉)
Department of Developmental Biology, Washington University School of Medicine, St. Louis, MO, USA
e-mail: blharty@wustl.edu

X. Piao
Division of Newborn Medicine, Boston Children's Hospital, Harvard Medical School, Boston, MA, USA

T. Langenhan, T. Schöneberg (eds.), *Adhesion G Protein-coupled Receptors*, Handbook of Experimental Pharmacology 234, DOI 10.1007/978-3-319-41523-9_12

Contents

Abstract

Adhesion G-protein-coupled receptors (aGPCRs) are emerging as key regulators of nervous system development and health. aGPCRs can regulate many aspects of neural development, including cell signaling, cell-cell and cell-matrix interactions, and, potentially, mechanosensation. Here, we specifically focus on the roles of several aGPCRs in synapse biology, dendritogenesis, and myelinating glial cell development. The lessons learned from these examples may be extrapolated to other contexts in the nervous system and beyond.

Keywords

Adhesion GPCR • Neurodevelopment • Synaptogenesis • Dendritogenesis • Myelination

1 Introduction

The nervous system contains a multitude of complex cell-cell and cell-environment interactions. Neural cells—neurons and glia—utilize a battery of electrical, chemical, and mechanical signal transduction methods to rapidly transmit information across the nervous system and to other cells throughout the organism, often over very long distances. To do this, neurons and glia maintain close associations with one another to allow constant communication within individual cells, between cells and their extracellular matrix (ECM).

The adhesion G-protein-coupled receptor (aGPCR) family regulates many cellular processes in a wide variety of cell types and contexts [1]. Despite being the second largest family of GPCRs, aGPCRs are relatively understudied, in part owing to their complex structure. aGPCRs are defined by a large extracellular N-terminus and a GPCR autoproteolysis-inducing (GAIN) domain that can cleave the receptor at the GPCR proteolysis site (GPS) into an N-terminal fragment (NTF) and a 7-transmembrane (7TM)-containing C-terminal fragment (CTF) during the maturation process [2, 3]. The NTF and CTF are thought to remain non-covalently attached at the cell surface [2, 4]. In addition to their characteristic autoproteolytic cleavage, aGPCRs are distinct from other GPCR families in that their N-termini often contain protein domains that are classically associated with cell-cell or cell-ECM adhesion, hence the name "adhesion GPCR" (see also [5, 6]). There are a total of 33 aGPCRs in humans, 31 in mice and rats [7], and over 50 in zebrafish [8]. In vertebrates, there are nine distinct subgroups of aGPCRs that can be separated based on the phylogeny of their 7TM regions [7] (see also [9, 10]).

This chapter will focus in detail on the roles of aGPCRs in two specific areas of neurobiology: synapto-/dendritogenesis and myelination. However, it is important to note that aGPCRs have proven to be critical for many different aspects of nervous system development including, but not limited to, cortical lamination and blood-brain barrier formation [11–13]. For an exhaustive overview of mammalian aGPCRs and their known functional roles, please refer to other chapters of this book and [1].

2 Adhesion GPCRs in Synaptogenesis and Dendritogenesis

Neurons communicate with each other at synapses, specialized junctions that ensure the transfer of neuronal information. Numerous types of neurons and synapses exist, characterized by different morphologies and properties, depending on their position in the network. Among the nine subfamilies of aGPCRs, three have come to light for their roles in synaptogenesis and/or dendrite morphogenesis: latrophilins (ADGRLs/LPHNs), Flamingo/cadherin, EGF-like, LAG-like, and seven-pass G-type receptors (ADGRCs/Fmi/CELSRs) and brain-specific angiogenesis inhibitors (ADGRBs/BAIs).

2.1 LPHN1 and LPHN3 Regulate Synapse Formation and Function

The LPHN subfamily is composed of three members, *LPHN1–3* [14]. Unlike LPHN2, LPHN1 and LPHN3 are highly enriched in the brains of zebrafish, rodents, and humans [8, 15, 16]. In rodents, expression of *Lphn1* and *Lphn3* is high until the third postnatal week and then decreases in older animals, suggesting a role during postnatal brain development [17]. Indeed, various studies have demonstrated that LPHN1 and LPHN3 regulate synaptic development and function, through their interactions with multiple partners.

2.1.1 LPHN1's Interaction with NRXs May Regulate Synapse Function

Originally identified as the endogenous receptor of α-latrotoxin, a component of black widow spider venom, LPHNs act as co-receptors for the neurotoxin together with the presynaptically localized single-pass transmembrane neurexins (NRX-1α, NRX-1β, NRX-2β, and NRX-3β) [15]. LPHNs bind to α-latrotoxin in a calcium-independent manner, whereas NRXs bind it in a calcium-dependent manner.

Disruption of LPHN1 and/or NRX-1α in mice confirmed that the two receptors bind α-latrotoxin independently but act cooperatively to regulate the effect of α-latrotoxin on neurotransmitter release [18]. Furthermore, LPHN1 interacts at high affinity with NRX-1α and NRX-1β, through its olfactomedin (Olf) domain, to form a transcellular adhesion complex in a calcium-dependent manner (Fig. 1). Together with the fact that LPHN1 co-enrich with postsynaptic markers in biochemical preparations [18], this suggests that LPHN1/NRX1 could produce a trans-synaptic contact between neurons [16].

2.1.2 LPHN1 Interacts with TEN2 to Stabilize Synapses and Induce Neurotransmitter Release

Other endogenous binding partners for LPHNs are the teneurins (TENs), single-pass transmembrane glycoproteins encoded by four different genes *(Ten1–4)*. In the rat brain, TEN2 binds with high affinity to LPHN1 and low affinity to LPHN2, but has no affinity for LPHN3. The interaction between LPHN1 and TEN2 was shown to mediate intercellular neuronal contact in a calcium-independent manner [16, 20]. In addition, in an artificial synapse formation assay, presynaptically enriched LPHN1 induced postsynaptic specialization in neurons, whereas the postsynaptically enriched TEN2 influenced axon-target interaction (Fig. 1) [20]. LPHN1 thus seems to regulate synapse stability and function through its interaction with TEN2. Similarly, TEN1 and TEN3 both interact with LPHN3 [21], although a specific role for these interactions in synapses has not yet been demonstrated.

2.1.3 LPHN/FLRT Trans-synaptic Complexes Regulate the Strength and Density of Synapses

Another group of LPHN interacting partners is the fibronectin leucine-rich transmembrane (FLRT) proteins, as shown by affinity purifications from rat synaptosome extracts using the LPHN3 or FLRT3 extracellular domain [21]. FLRT3 is a postsynaptic protein partially co-localizing with the postsynaptic scaffolding protein PSD95 at glutamatergic but not GABAergic synapses. While LPHN3, like LPHN1, is sufficient to induce postsynaptic differentiation, FLRT3 is not sufficient to induce presynaptic differentiation. However, *Lphn3* or *Flrt3* knockdown experiments in hippocampus, dentate gyrus, or cerebral cortex reveal a decrease in the strength and number of synapses, as well as a decrease in glutamatergic transmission, sometimes accompanied by a reduction in the number of dendritic spines [21, 23]. These results suggest that the LPHN3-FLRT3 complex positively regulates synaptic development and function.

In vertebrates, while NRXs are expressed ubiquitously, FLRTs and TENs are expressed in a cell-specific manner that is complementary and nonoverlapping [21]. Therefore, only certain combinations of LPHNs and FLRTs or TENs seem to be present at any given synapse, suggesting their role in the specific formation of different neuronal circuits. Indeed, *lphn3*-knockdown zebrafish larvae are reported to have defects in dopaminergic neurons, which control locomotor activity and impulsivity. This phenotype is reminiscent of *Lphn3* mutant mice, which are

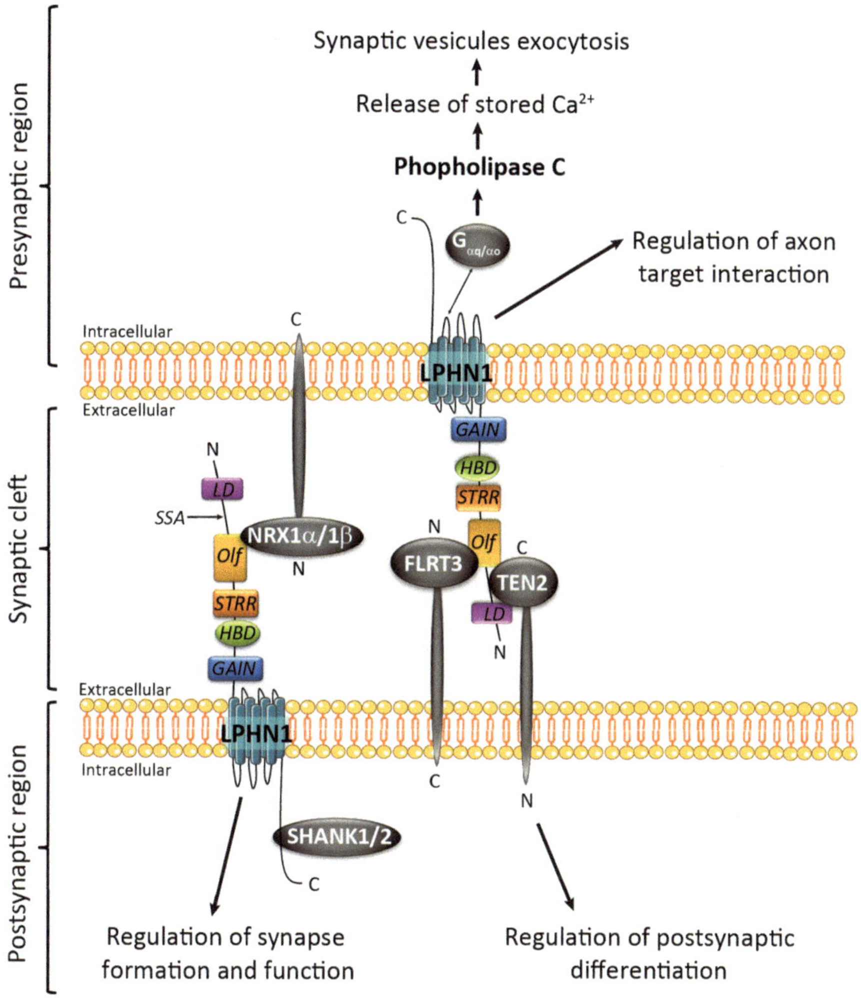

Fig. 1 Structure and functions of LPHN1 at the synapse. The extracellular N-terminal fragment (NTF) of latrophilins contains a rhamnose-binding lectin domain (RBL) [19], a central olfactomedin-like domain (Olf), a serine/threonine-rich region (STRR), and a hormone-binding domain (HBD), in addition to the characteristic GAIN domain containing the autoproteolytic GPS motif. The 7TM domain is followed by a long cytoplasmic tail where most of the alternative splicing occurs. One splicing site, referred as SSA, is also present between the lectin and olfactomedin domains in the N-terminal region. LPHN1 interacts extracellularly with neurexin1α and 1β (NRX1α/1β) [16], Teneurin 2 (TEN2) [20], and FLRT3 [21] and intracellularly with $G\alpha_q/\alpha_o$ [22] and the postsynaptic scaffolding proteins SHANK1/2 [17, 18]

hyperactive and display increased sensitivity to a locomotor stimulant. In both species, the defects can be reversed by ADHD medication [24].

Although LPHNs are conserved in invertebrates, their roles in synaptogenesis have not yet been demonstrated in these models. However, the *C. elegans* homolog of LPHN1, LAT1, was shown to be involved in the modulation of neurotransmitter release in the pharyngeal nervous system [25], so some functions may be conserved. Interestingly, invertebrate LPHN homologs lack the Olf domain, which is required for forming the LPHN-NRX complexes as well as the functions of LPHN/FLRT and LPHN/TEN complexes in synapses [16, 23]. The Olf domain is absent in LPHNs in invertebrates and thus may confer a specific role for LPHNs in the vertebrate brain. Additionally, in *C. elegans*, TEN1 and LAT1 are possibly engaged in cis interactions and thus might have different modes of action than their vertebrate counterparts [3] (see also [26] on this topic).

2.2 FMI/CELSRs Regulate Synaptogenesis and Dendritogenesis

The aGPCR Flamingo (Fmi) is found in invertebrates and is evolutionary conserved, as mammals have three *Fmi* homologs named *Celsr1–3*. In *C. elegans*, *fmi-1* is expressed in neurons and localized in axons during development and adulthood [27]. In *Drosophila, fmi* is expressed at the neuromuscular junction and in dorsal sensory neurons of the peripheral nervous system (PNS) as well as in the central nervous system (CNS), where it is mostly present on dendritic and axonal membranes [28, 29]. Similarly, in rodents, *Celsr1–3* are predominantly expressed in the brain. While *Celsr1* is mostly restricted to regions of neural stem cell proliferation, *Celsr3* is preferentially expressed in postmitotic neurons, and *Celsr2 is* expressed in both proliferating and postmitotic neurons of the postnatal brain [30, 31].

2.2.1 Fmi Regulates Axon Guidance and Synaptogenesis in Invertebrates

In the ventral nerve cord of *C. elegans*, pioneer axon growth allows the formation of a path that will be followed by later growing follower axons. In *fmi-1* mutants, the navigation of both pioneer and follower axons is perturbed. The navigation of pioneer axons requires the CTF of FMI-1, whereas the navigation of follower axons requires the NTF [27], suggesting that FMI-1 controls axon navigation through signal transduction for pioneer axons and through axon-axon adhesion and/or *trans* signaling for follower axons.

At the larval neuromuscular junction of *Drosophila fmi* mutants, Fmi prevents the formation of inappropriate ectopic synapses. *fmi* mutants also present defective synaptic responses in a subset of muscles, as well as an age-dependent loss of muscle innervation due to drastic degeneration of axons in the third larval stage. Importantly, Fmi restoration in neurons rescues these synaptic and axonal defects. Fmi is thus required in neurons for the selection of synaptic targets, for synaptogenesis, and for the maintenance of axons and synapses in adulthood (Fig. 2) [28].

In the sensory system of *Drosophila*, processes of dendritic avoidance or repulsion occur during early development and are required to obtain a specific dendritic branching pattern for each of the dorsal sensory neurons. *fmi* mutant embryos show

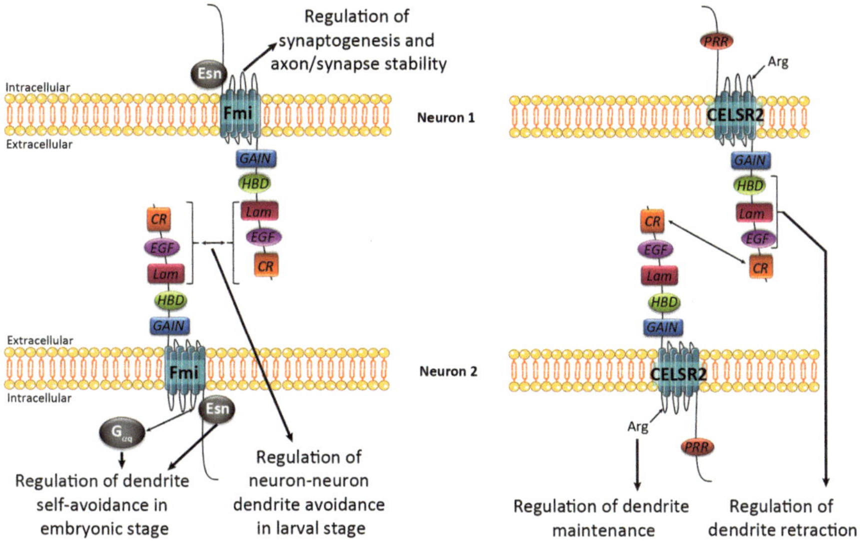

Fig. 2 Structure, partners, and functions of FMI and CELSR2 in neurons. In the NTF, Fmi/CELSR proteins contain nine cadherin repeats (CR), six EGF-like domains (EGF), two laminin globular-like domains (Lam), a hormone-binding domain (HBD), and the GPS-containing GAIN domain. The three CELSR proteins share 50 % sequence identity in their N-terminal and 7TM regions but have different C-terminal regions: 300 residues long for CELSR1–2 and 590 residues long for CELSR3. The C-terminal intracellular domain does not contain any conserved motif between invertebrates and vertebrates, except for a proline-rich region (PRR). In addition, CELSR2 has two splice variants, one long and one short, which differ based on splicing of exon 31, affecting 20 residues in the cytoplasmic tail. Fmi interacts with Espinas (Esn) [32] and Gα_q [33]. CELSR2 and CELSR3 are localized both pre- and postsynaptically and are engaged in homophilic interactions [34]

dendritic overgrowth and overlap of dorsal sensory neurons [29, 34]. These defects can be partially rescued by the specific expression of an *fmi* construct containing only the extracellular hormone-binding domain (HBD) that impairs Fmi homophilic binding (Fig. 2). Intracellularly, genetic experiments suggest a signaling pathway (Fig. 2) involving the proteins Fmi, Esn, and Van Gogh [32] that could be localized in dendrites and locally activate RhoA upon contact of two branches to redirect dendrites away from each other. In addition, neuronal G protein Gα_q may also function downstream of Fmi to inhibit dendritic growth [33]. At the larval stage, *fmi* mutants show dendritic outgrowth from the contralateral sides with overlapping of dendrites that can only be rescued by Fmi-containing cadherin repeats (Fig. 2). Thus, Fmi first restricts dendritic growth in embryos and then controls neuron-neuron dendritic avoidance through homophilic binding at the larval stage.

2.2.2 CELSRs Modulate Dendritic Growth in the Mammalian Brain

In rodents, CELSR2 is distributed in dendrites and axons of various embryonic and postnatal neuronal cells [35]. *Celsr2* loss-of-function manipulations, performed on organotypic brain slices from rat, caused simplification of the dendritic arborization of hippocampal and cortical pyramidal neurons and cerebellar Purkinje cells (PCs). This decrease in dendritic complexity is likely due to dendrite retraction, indicating that CELSR2 is required for dendritic maintenance [36]. Rescue experiments suggest a dual role of CELSR2: promotion of dendritic maintenance through homophilic interactions mediated by the extracellular cadherin repeats and decreasing dendritic complexity through the EGF-HBD region (Fig. 2). This suggests that different ligands of CELSR2, binding to different domains, could exert opposite effects on dendrites. This property of CELSR2 could be important for the control of dendritic complexity depending on environmental molecular cues.

In contrast to CELSR2, CELSR3 suppresses dendritic growth [37]. This functional difference may be determined by a single amino acid in the 7TM domain: *Celsr1*, *Celsr3*, and *Fmi* possess a histidine in the first intracellular loop of the 7TM domain, whereas *Celsr2* has an arginine. This residue switch is correlated with an inversion of the roles of the corresponding proteins in dendritogenesis, perhaps suggesting the presence of different second messengers initiating different signaling pathways. The role of Celsr3 in dendritogenesis in vivo remains to be confirmed since *Celsr3* knockout mice lack a dendritic phenotype, in contrast with the reported knockdown phenotype [37, 38].

2.3 BAIs Regulate Dendrite, Spine, and Synapse Formation

Three different BAI receptors are encoded by *Bai1–3* and are only present in vertebrates. Their expression increases in the brain after birth, with a peak in rodents at postnatal day 1 for *Bai3* and at postnatal day 10 for *Bai1* and *Bai2*. *Bai1* and *Bai3* are specifically expressed in the brain, whereas *Bai2* is also found in many other tissues [8, 39–42]. Although suggested by the antidepressant phenotype of the *Bai2* knockout mice [43], the involvement of BAI2 in neuronal circuit formation remains to be demonstrated. In contrast, roles for BAI1 and BAI3 in dendritogenesis and synaptogenesis have been described by several studies.

2.3.1 BAI1 Regulates Synaptogenesis and Synaptic Plasticity

BAI1 is enriched in the postsynaptic density (PSD) of excitatory synapses. Here, it co-localizes with the AMPA receptor GluR1 [44, 45] and potentially interacts with a variety of PDZ proteins such as PSD95, PSD93, the MAGUKs MAGI1–3, and the synapse-associated protein 97 (SAP97). BAI1 also binds to MDM2, an E3 ubiquitin ligase, and prevents PSD95 ubiquitination and degradation (Fig. 3). In *Bai1*-null animals, the synaptic ultrastructure is characterized by a decreased PSD thickness associated with increased PSD95-associated MDM2 and PSD95 ubiquitination. This modification is associated with defects in hippocampal-dependent learning and memory and in synaptic plasticity. The restoration of PSD95 to physiological

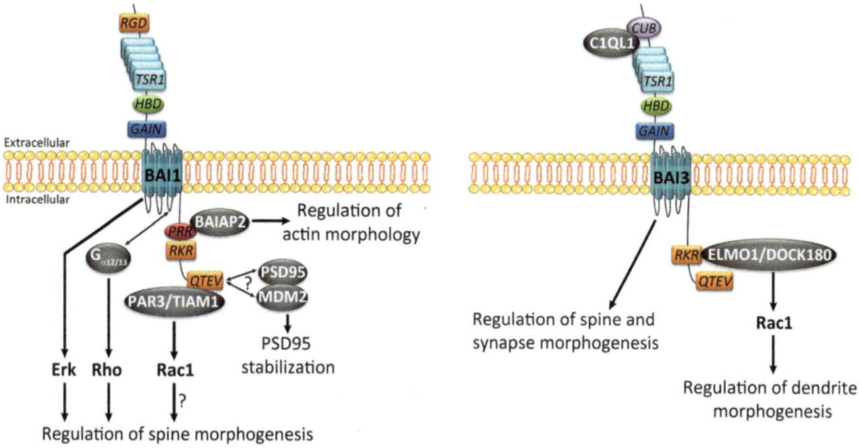

Fig. 3 Structure, partners, and functions of BAI1 and BAI3 in neurons. In the NTF, BAI proteins contain thrombospondin type 1 repeats (TSR1; five TSRs in BAI1 and four TSRs in BAI2 and BAI3), a HBD, and a GAIN domain containing the GPS site for autoproteolysis. At the N-terminal extremity, BAI3 also possesses a CUB (complement C1r/C1s, Uegf, Bmp1) domain, and BAI1 has an integrin-binding RGD motif. The N-terminal region of the BAIs is followed by the 7TM domain and a cytoplasmic domain containing an α-helical RKR motif and a terminal PDZ-binding motif (QTEV residues). BAI1 also presents a C-terminal proline-rich region (PRR). BAI1 interacts intracellularly with PAR3/TIAM1 [44], BAIAP2 [46], and $G\alpha_{12/13}$, as well as PSD95 [45] and MDM2 [47], possibly through the QTEV domain. BAI3 interacts with the complement C1Q-related molecule C1QL1 [48] extracellularly and with ELMO1/DOCK180 [41] intracellularly

levels in the hippocampus of *Bai1*-null mice restores normal synaptic plasticity, showing that plasticity defects are directly linked to the destabilization of PSD95 [47]. Knockdown experiments in cultured neurons further revealed the role of BAI1 in spinogenesis and synaptogenesis and, to a lesser extent, in synapse maintenance [45]. Rho GTPases and ERK have been previously shown to regulate the morphogenesis of dendritic spines as well as synapse formation and plasticity [38, 49]. Interestingly, BAI1 activates Rho via $G\alpha_{12/13}$ [45] and activates Rac1 by recruiting the Par3/Tiam1 complex at the PSD [50, 51]. Finally, BAI1 promotes the phosphorylation of ERK in a PDZ-motif-dependent manner. BAI1 seems thus required for the proper assembly of actin in developing spines *via* Rho GTPases and ERK signaling.

2.3.2 BAI3 Regulates Dendritogenesis and Synaptogenesis in Early Brain Development

Dendrite morphogenesis relies on the stabilization of neurites and the regulation of the actin cytoskeleton via the modulation of Rho GTPases [52, 53]. *Bai3* knockdown during early neuronal development leads to dendrite overgrowth in various types of neurons in vitro, as well as in PCs in vivo [41]. No major dendritic defects are observed in adult mice with a specific disruption of *Bai3* in PCs [54]. This discrepancy could be due to developmental timing differences in the inactivation of BAI3 in these experiments. This would suggest a requirement of BAI3 preferentially at an early stage of dendritogenesis and/or a subsequent recovery of the major defects in the adult. Dendrite morphogenesis relies on the stabilization of neurites and the regulation of the actin cytoskeleton via the modulation of Rho GTPases. BAI3 controls dendritic growth by regulating the activity of the Rho GTPase Rac1, partially through binding to the ELMO1/DOCK180 complex, a guanylate exchange factor of Rac1 (Fig. 3) [41]. Thus, BAI3 might control the rearrangement of dendritic actin cytoskeleton via Rac1, a potential role reinforced by the fact that BAI3 co-localizes with actin in growing dendrites.

Proteomic analysis of synaptic fractions revealed the localization of BAI3 at excitatory synapses in the forebrain and in the cerebellum [55, 56]. Knockdown or cell-specific knockout experiments in PCs show that BAI3 promotes the connectivity of its two types of excitatory afferents, the parallel and the climbing fibers. Indeed, *Bai3* loss of function leads to a decrease in the number and size of parallel fiber presynaptic boutons [42], as well as a decrease in the number, size, and extension territory of climbing fiber presynaptic boutons [42, 54]. During synaptogenesis, each PC is first contacted by multiple climbing fibers around embryonic day 19 in mouse. Then, only one climbing fiber "wins" and supernumerary climbing fibers are eliminated. In the PC-*Bai3*-null mice, many PCs remain multi-innervated in the adult [54]. This defect in climbing fiber elimination could be a direct effect of BAI3 in determining the "winner" climbing fiber or an indirect effect consecutive to the defective establishment of parallel fiber connections [42], since normal transmission at parallel fiber synapses is necessary for proper climbing fiber elimination.

The only extracellular partners of BAI3 reported so far are the secreted C1QL proteins from the complement C1Q-related family (Fig. 3) [48]. The four C1QLs, generated from four genes, are expressed almost exclusively in the mammalian brain, each one with a specific and partially overlapping expression pattern [42, 57]. C1QLs-BAI3 interaction was first reported to inhibit synapse formation or maintenance in cultured hippocampal neurons [48]. Two subsequent studies show that in the cerebellar cortex, C1QL1 needs to be secreted specifically at the climbing fiber to allow proper formation of climbing fiber/PC synapses [42, 54]. In addition, C1QL1 is transiently expressed in the cerebellum and promotes PC spinogenesis in a BAI3-dependent manner. Thus, BAI3 could regulate the number of parallel fiber/PC synapses early during development [42]. Both *C1ql1* and *Bai3* knockout mice exhibit defects in climbing fiber elimination and motor learning defects, highlighting the importance of the signaling pathway formed by C1QL1 and BAI3 for proper CF/PC synaptogenesis and cerebellar function [54].

3 Adhesion GPCRs in Myelinating Glia

While synapses permit direct communication between neurons, the glial-derived myelin sheath enables the rapid propagation of action potentials along axons; thus, myelinating glia are also key players in information transfer in the nervous system. Myelin is made by Schwann cells (SCs) in the PNS and oligodendrocytes (OLs) in the CNS. Myelination increases the conduction velocity of action potentials; additionally, OLs and SCs provide trophic support to aid in the proper functioning and survival of neurons [58, 59]. Relatively less is known about myelinating glia compared to neurons. Similarly, little is known about the roles of aGPCRs in myelinating glia compared to roles in synaptogenesis and dendritogenesis. Nevertheless, recent studies have established two aGPCRs, Gpr126 and Gpr56, as key regulators of both SC and OL development.

3.1 GPR126 Is Essential for Schwann Cell Myelination

Previous work has defined several steps in SC development, leading from neural crest progenitors to mature myelinating SCs [58]. During embryogenesis, SC precursors migrate with axons in the developing PNS and differentiate into immature SCs that remain associated with many axons. In a process called radial sorting, immature SCs insert cytoplasmic projections into the axon bundle and begin to envelop and separate individual axons away from the rest of the bundle according to axon diameter. Some immature SCs will remain associated with many small axons and become nonmyelinating SCs. However, immature SCs that will eventually make myelin will first become promyelinating SCs that associate in a 1:1 ratio with a single axonal segment. Upon additional cues, myelinating SCs then iteratively wrap their plasma membrane around the associated axonal segment, ultimately generating a mature myelin segment with a thickness proportional to the diameter of the axon they wrap.

The first evidence of a role for aGPCRs in myelination was the discovery that Gpr126 is required for the normal progression from a promyelinating SC to a fully mature myelinating SC in both zebrafish and mice [60, 61]. In *Gpr126* mutants, SCs express the transcription factor *Sox10*, which is required for SC specification, maintaining SC identity and proper progression of the lineage [62–64]. However, *Gpr126* mutant SCs fail to express later-stage transcription factors Oct6 (Pou3f1) and Krox20 (Egr2). Oct6 and Krox20 are required to activate the expression of many myelin-associated genes, including *myelin basic protein* (*Mbp*), which encodes a structural component of the myelin sheath. Ultrastructural analyses demonstrated that *Gpr126* mutant SCs can associate with axons in the proper 1:1 ratio, but are arrested at the promyelinating stage of development, and fail to ever elaborate a myelin sheath [60, 61].

In vitro, SCs cocultured with neurons follow a similar developmental progression as described above—they migrate and proliferate along nascent axons, perform radial sorting, and myelinate axons [65]. However, in the absence of axons, SCs fail to differentiate. This requirement for neurons can be overcome by the addition of axonal membrane fragments [66, 67] or by increasing levels of cAMP. Elevation of cAMP induces SC maturation, as illustrated by their upregulation of myelin-related gene expression and downregulation of immature SC markers [68–70]. Thus, an in vivo role for cAMP elevation in SC myelination had long been hypothesized, although the mechanism by which cAMP might be elevated has been elusive. Interestingly, myelination defects in *gpr126* mutant zebrafish can be suppressed by treating the animals with forskolin, an adenylate cyclase activator that potently elevates cAMP. Specifically, SCs of forskolin-treated *gpr126* mutant zebrafish express *oct6*, *krox20*, and *mbp* and produce ultrastructurally normal myelin [60]. Additionally, elevating cAMP levels and activating protein kinase A (PKA) restored myelination in myelinating cultures from *Gpr126*$^{-/-}$ knockout mice [71], and PKA activation also rescues myelination in zebrafish [72]. Moreover, cAMP is downregulated in the sciatic nerves of mutant mice in which *Gpr126* has been deleted specifically in SCs [71]. Finally, GPR126 was shown to directly couple to the G_s-protein and G_i-protein families [71], which are associated with modulating cAMP levels. All together, these data suggest that Gpr126 is a major, if not the sole receptor, that elevates cAMP in SCs to initiate myelin-gene expression and terminal differentiation.

3.1.1 Gpr126 Has cAMP-Independent Roles in Schwann Cell Development

Until quite recently, it was unclear if aGPCRs functioned as adhesion molecules by virtue of the NTF, as traditional GPCRs signaling through heterotrimeric G proteins by virtue of the 7TM, or if the same molecule could perform both functions. Notably, studies on Gpr126 have contributed to answering this question, as Gpr126 SC myelination studies strongly suggested that aGPCRs could behave as traditional GPCRs [71, 73]; later, the Gpr126-NTF was reported to have a CTF-independent role in heart development [74]. Together, this suggested that a given aGPCR can have domain-specific roles, at least in different cellular contexts. Interestingly, recent structure-function studies of Gpr126 have shown that an aGPCR can also have multiple independent functions in modulating the development of a single cell [73].

Preliminary analyses of *gpr126* zebrafish mutants suggested that Gpr126 functions specifically in the final maturation of SCs. This function of Gpr126 is conserved in mammals; however, both global and SC-specific *Gpr126* mutant mice display additional defects that were not observed in *gpr126* zebrafish mutants, including severe delays in radial sorting, axon degeneration, and limb contracture abnormalities [61, 71]. While species differences may certainly be at play, there are key differences between the alleles in *Gpr126* mutant mice and zebrafish. The targeted deletion in *Gpr126* null mice produces an early premature stop codon that results in the near absence of *Gpr126*-NTF mRNA, and these mice have severely impaired radial sorting and completely lack peripheral myelination as mentioned above [61, 71]. In contrast, two previously studied mutant alleles in zebrafish were likely not null alleles; the *gpr126*^{st49} allele is a point mutation resulting in a premature codon near the GPS [60], such that the NTF might be intact, while the CTF signaling domain is absent. Similarly, the *gpr126*^{st63} allele is a missense mutation that disrupts a highly conserved cysteine in the 7TM, thus the NTF and CTF both remain intact [60]. Importantly, neither *gpr126*^{st49} nor *gpr126*^{st63} displays defective radial sorting. In contrast, recently generated *gpr126*^{stl47} mutants represent a very early premature stop codon (amino acid position 96), which is more similar to the targeted deletion in *Gpr126* mutant mice [73]. Interestingly, SCs in *gpr126*^{stl47} mutants phenocopy the radial sorting defects observed in mouse mutants. These defects could be rescued by genetically "adding back" the Gpr126-NTF but could not be rescued by cAMP elevation, suggesting that Gpr126 has CTF-independent roles in SC development. These findings are consistent with earlier studies highlighting NTF- and CTF-specific functions for aGPCRs in *C. elegans* axon migration, fertility, and viability [3, 27], though future work is required to determine how the Gpr126-NTF might direct radial sorting in a cAMP-independent fashion.

3.1.2 Control of Gpr126 Signaling in Schwann Cells

Given that aGPCRs have both CTF-dependent and independent roles, an important question centers around binding partners that might modulate these functions. Notably, most aGPCRs are orphaned [1, 75], meaning their ligands and activation methods are unknown. aGPCRs are unique relative to other GPCRs in that they undergo an autoproteolysis event at the GPS motif that separates the protein into an NTF and CTF, which then remain non-covalently associated with one another at the cell membrane [2, 4]. Interestingly, a spate of recent studies have shown that several aGPCRs, including GPR126, generate a tethered agonist ligand, termed the *Stachel* sequence (German word for "stinger"), which potently activates the CTF of the receptor in vitro [76–79]. Moreover, analysis of zebrafish mutants suggested that activation of Gpr126 by the *Stachel* sequence is required for SC myelination [76], and it was recently suggested that Gpr126-interacting proteins in the extracellular matrix (ECM) might modulate availability of the *Stachel* sequence (see also [80, 81] on this topic).

The NTFs of several aGPCRs can bind to ECM molecules; Gpr126 is no exception, as the Gpr126-NTF can bind to type IV collagen and Laminin-211 [73, 82], which are both components of the SC basal lamina. In rodent SCs and GPR126-expressing heterologous cells, exogenous application of type IV collagen induced cAMP elevation, which is required for SC differentiation. Additionally, GPR126 derivatives lacking the NTF were constitutively active, suggesting that the NTF is important for inhibiting Gpr126-CTF signaling [82]. Interestingly, Laminin-211 also binds Gpr126 in a different region of the NTF than the collagen-IV binding site. Overexpression of *lama2* rescued myelination defects in hypomorphic *gpr126* ^{st63} zebrafish mutants in a cAMP-dependent manner [73]. Intriguingly, and in contrast to type IV collagen, addition of Laminin-211 to GPR126-expressing heterologous cells suppressed cAMP accumulation under static conditions. However, under dynamic culture conditions, Laminin-211 activated GPR126. The in vivo relevance of this result was underscored by the observation that a polymerization-defective mutant form of *lama2* could no longer restore myelination in *gpr126* zebrafish mutants. Taken together, these studies suggest that ECM molecules in the basal lamina may modulate the NTF and impact *Stachel* sequence signaling. Moreover, the Laminin-211/Gpr126 studies suggested that aGPCRs may be mechanically sensitive, consistent with contemporaneous work in *Drosophila melanogaster* [83]. For an in-depth discussion of the emerging concept of aGPCRs as mechanosensors, the reader is referred to [84].

Importantly, these data tie in with the model that Gpr126 has CTF-independent functions in SC development. Low levels of cAMP promote SC proliferation and facilitate an immature state, while elevated cAMP drives myelination [85]. Perhaps in early SC development, the Gpr126-NTF remains tightly associated with the CTF such that cAMP elevation is suppressed. Changes in ECM proteins in the SC basal lamina, including Laminin-211 polymerization, may modulate the NTF such that the *Stachel* sequence can bind the 7TM and induce Gpr126 signaling cascades through G_s, which elevate cAMP, activate myelin-related gene expression, and allow terminal maturation of myelinating SCs (Fig. 4).

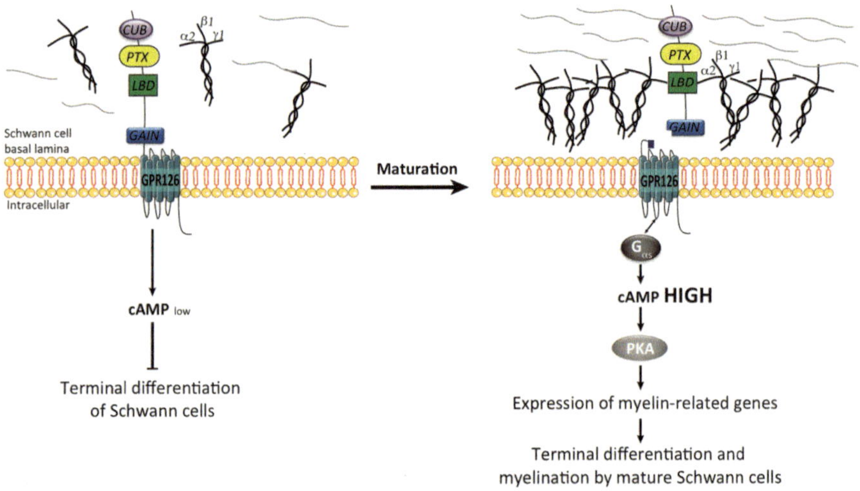

Fig. 4 Interactions with the ECM modulate Gpr126 throughout Schwann cell development. In addition to the canonical aGPCR domains (GAIN, GPS, and 7TM), GPR126 also possesses CUB and Pentraxin (PTX) domains in the NTF. Recent data showing that Gpr126-NTF interacts with Laminin-211 (α2, β1, γ1) [73] supports the addition of a novel Laminin-binding domain (LBD) in the NTF. Altogether, current data support a model in which molecules in the SC basal lamina (Laminin-211 in *black*, collagens in *gray*) interact with GPR126-NTF to stabilize the receptor in an inactive state during early development. This prevents cAMP accumulation by suppressing GPR126 G_s signaling to prevent SC differentiation. Following basal lamina maturation and Laminin-211 polymerization, an active conformation of GPR126-CTF is facilitated to induce G_s signaling and cAMP elevation, possibly through mechanical modulation of the receptor. This results in PKA activation, which promotes myelin-related gene expression and terminal SC differentiation

3.2 GPR56 Regulates Oligodendrocyte Development

OLs are the myelinating glia of the CNS that, as mentioned above, share many characteristic features with their PNS counterparts, the SCs. Like SCs, OLs generate myelin by iteratively wrapping their plasma membranes around axon segments [86], although OLs can myelinate many axonal segments, in contrast to the single axon segment myelinated by an SC. Accordingly, OLs and SCs perform similar functions using an overlapping, but not identical, set of genes [87]. Moreover, while SCs are derived from neural crest, the OL lineage is derived from neural stem cells, which ultimately give rise to all resident cells in the CNS except for microglia [88]. OL precursor cells (OPCs) migrate in waves from their germinal zones and become immature OLs as they begin extending cytoplasmic processes and associating with axons [86]. Finally, upon incompletely understood cues, immature OLs terminally differentiate into mature myelinating OLs that spiral and compact their plasma membrane around as many as 40 different axon segments [89].

Although Gpr126 is essential for SC development and myelination, it has no known function in OLs, suggesting that there might be another aGPCR that is similarly important for OL maturation and myelination. Previously published data sets indicate that *Gpr56*, an aGPCR in the same subfamily as GPR126, is highly expressed in OPCs in the developing mouse CNS [90, 91]. In humans, mutations in *GPR56* cause bilateral frontoparietal polymicrogyria (BFPP) [11], a cortical brain malformation associated with a variety of devastating neurological impairments such as epilepsy and mental retardation. Interestingly, brains of BFPP patients also exhibit reduced white matter volume by MRI and signal changes in T2-weighted images [12, 85, 92], which are indicative of myelin defects. Two complementary studies have recently defined key functions for GPR56 in OL development [93, 94].

In these studies, the authors analyzed both zebrafish and mouse *Gpr56* mutants. *gpr56*^{stl13} mutant zebrafish harbor a 6 bp deletion that removes a completely conserved tryptophan residue in the GPS motif that is essential for autoproteolysis [94]. Since previous reports show that disruption of autoproteolytic cleavage of GPR56 prevents the protein from being trafficked to the membrane and thus severely impairs GPR56 signaling [95], *stl13* is likely a strong loss-of-function allele. Although loss of Gpr56 function in *stl13* mutants did not result in abnormal OPC specification or initial OPC numbers, both *Gpr56* knockout mice and *gpr56*^{stl13} zebrafish displayed significantly fewer OL lineage cells over time [93, 94]. GPR56 is expressed in many CNS cell types including oligodendrocyte precursor cells (OPCs) and immature oligodendrocytes [93], which could impact some of the OL phenotypes observed in *Gpr56* mutants. Giera et al. also used OL-specific deletion of *Gpr56* to demonstrate that this aGPCR functions cell autonomously in OLs to regulate their development. Ultrastructural analyses show that zebrafish and mouse *Gpr56* mutants possess fewer myelinated axons as compared to controls in the spinal cord and corpus callosum, respectively. However, axons that are myelinated appear to be myelinated normally, suggesting that GPR56 regulates the number of OL lineage cells but is not required for myelination per se. Marker analyses in both species demonstrated that the reduction in OL number is not due to increased OPC cell death but instead a lack of OPC proliferation [93, 94]. Additionally, overexpression of *gpr56* in wild-type zebrafish larvae induced increased OPC proliferation, while the ventral spinal cord was precociously myelinated in *gpr56*^{stl13} zebrafish mutants at early developmental stages [94]. Together, these studies support the notion that loss of GPR56 results in OPCs prematurely exiting the cell cycle in favor of differentiation.

3.2.1 Gpr56 Regulates Oligodendrocyte Development via $G_{\alpha12/13}$ and RhoA

Previous studies established that GPR56 couples to $G_{\alpha12/13}$ and activates RhoA to control migration of neural precursor cells in vitro [96, 97]. RhoA is highly expressed in mammalian OPCs but is downregulated at later stages to facilitate terminal differentiation of OLs [98], paralleling the early function of GPR56 in these cells. In zebrafish, knockdown of the genes encoding $G_{\alpha12/13}$ proteins or overexpression of a dominant-negative RhoA phenocopied OL lineage defects in *gpr56*^{stl13} mutants, while overexpression of constitutively active RhoA rescued the defects [94]. In mouse, levels of active RhoA were significantly reduced in *Gpr56* knockout optic nerves while there were no observable changes in PKCα, the only other effector reported to be downstream of GPR56 [93]. Taken together, these experiments strongly support a model in which GPR56 couples to $G_{\alpha12/13}$ and signals through RhoA to promote OPC proliferation and inhibit terminal differentiation to ultimately regulate appropriate levels of CNS myelination (Fig. 5). In the future, it will be interesting to determine if this activation is dependent on the *Stachel* sequence as well as what ligand(s) might influence GPR56-dependent OL development.

3.3 Other aGPCRs in Myelinating Glia

Beyond GPR126 and GPR56, VIGR/GPR98 (ADGRV1) is expressed in OLs, and modulation of this aGPCR can affect levels of a key myelin protein, MAG (myelin-associated glycoprotein). In vitro, VIGR knockdown reduces, while overexpression increases, MAG expression [99]. Additionally, expression studies suggest that there may be yet undiscovered aGPCR players in glial cell development and myelination as RNAseq analyses in the mouse CNS indicates that *Gpr125* (*Adgra3*), *BAI1–3*, *Lphn1*, *Lphn3*, and *Celsr2* are all expressed (FPKM > 5) in OL lineage cells [100]. Future studies can elucidate the roles of additional aGPCRs in myelinating glia.

4 Discussion

Thus far, three groups of aGPCRs, latrophilins, Flamingo/CELSRs, and BAIs, have been implicated in morphogenesis of various neurons and/or their connectivity. Similarly, two aGPCRs from the group VIII (ADGRG) subfamily, GPR126 and

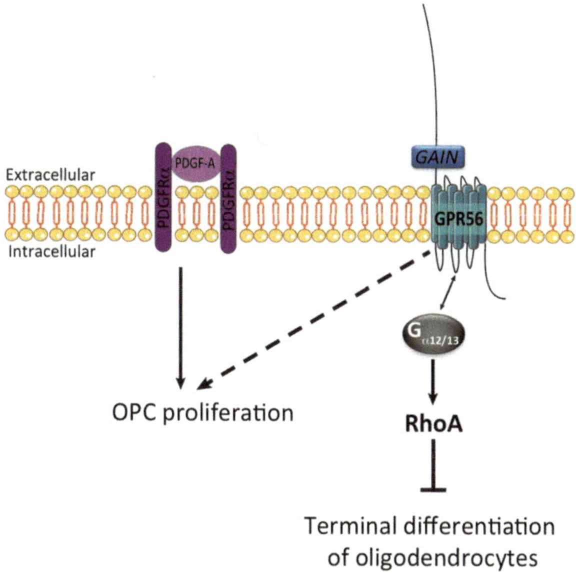

Fig. 5 Gpr56 regulates oligodendrocyte development by promoting precursor proliferation and inhibiting premature differentiation. Although GPR56 contains all of the classic aGPCR domains (GAIN, GPS, and 7TM), other protein domains in either the NTF or the intracellular tail region have not been described. Recent data from zebrafish and mouse models suggest that GPR56 couples to $G\alpha_{12/13}$ [93, 94] and activates RhoA to prevent terminal differentiation of OLs. In doing so, GPR56 indirectly promotes OPC proliferation (*dashed arrow*) by maintaining the OPC in an immature, proliferative state

GPR56, have proven to be critical regulators of myelination. These molecules are thus crucial for the establishment of a properly functioning nervous system.

The importance of understanding the molecular basis underlying dendrite and synapse formation is highlighted by the fact that any modification in neuronal microcircuits can lead to neurodevelopmental defects and brain pathologies. This is underscored by the fact that some aGPCRs regulating dendritogenesis and synaptogenesis are also associated with neurodevelopmental pathologies. In humans, deletion of *LPHN1* is associated with various symptoms including mental retardation, psychomotor and language delay, and hearing impairment [101]. SNPs in *LPHN3* increase the risk for attention deficit/hyperactivity disorder (ADHD) and substance addiction [102, 103]. In addition, human *BAI1* is present in a "hot spot" for de novo germline mutations in autism patients [104], and the BAI1-associated protein BAIAP2 is also linked to autism [105]. Finally, genetic association studies reveal that BAI proteins could contribute to behavioral defects in psychiatric disorders. Indeed, SNPs and CNVs in the *BAI3* gene have been associated with schizophrenia [106, 107], bipolar disorder [108], and addiction [109].

Similarly, the importance of understanding the molecular mechanisms that govern proper myelination is emphasized by the fact that damage to or loss of this insulation leads to debilitating symptoms in many neurological disorders. For example, in multiple sclerosis, OL myelin is attacked and destroyed by the body's immune system [110]. As noted above, BFPP patients present with white matter abnormalities; perhaps GPR56 represents a new candidate target to modulate OL differentiation in the CNS. Similarly, GPR126 may represent a viable target for patients with peripheral nerve disease as this aGPCR is required for proper myelin-gene expression in humans [111]. Furthermore, mutations in *LAMA2*, which encodes the α2 chain of Laminin-211, cause merosin-deficient congenital muscular dystrophy (MDC1A) in humans [112]. Given that GPR126 binds to Laminin-211, activation of GPR126 could be pursued in MDC1A patients. GPCRs are the most widely studied targets for the development of pharmacological drugs. Therefore, studying aGPCRs is highly pertinent to increase our understanding of these receptors in the context of human neurodevelopmental and myelin diseases and to find efficient treatments for human disorders.

Acknowledgments The authors would like to thank the following agencies for their support: the Washington University School of Medicine Institute of Clinical and Translational Sciences, NIH CTSA #UL1 TR000448 (K.R.M. and B.L.H.); the National Institutes of Health, F31NS094004 (B.L.H.), R01 NS079445 and R01 HD080601 (K.R.M.), R01 NS094164, and R56 NS085201 (X.P.); the National Multiple Sclerosis Society, RG-1501-02577 (X.P.), the Agence Nationale de la Recherche ANR-13-SAMA-0010-01 (F.S.) and ANR-10-LABX-54 MEMO LIFE (S.M.S. and F.S.); and the Centre national de la recherche scientifique (CNRS) and Institut National de la Santé et de la Recherche Médicale (INSERM), ATIP AVENIR (F.S.), and NeRF Ile de France (S.M.S.).

References

1. Hamann J, Aust G, Araç D, Engel FB, Formstone C, Fredriksson R et al (2015) International Union of Basic and Clinical Pharmacology. XCIV. Adhesion G protein-coupled receptors. Pharmacol Rev 67(2):338–367
2. Araç D, Boucard AA, Bolliger MF, Nguyen J, Soltis SM, Sudhof TC et al (2012) A novel evolutionarily conserved domain of cell-adhesion GPCRs mediates autoproteolysis. EMBO J 31(6):1364–1378
3. Prömel S, Frickenhaus M, Hughes S, Mestek L, Staunton D, Woollard A et al (2012) The GPS motif is a molecular switch for bimodal activities of adhesion class G protein-coupled receptors. Cell Rep 2(2):321–331
4. Langenhan T, Aust G, Hamann J (2013) Sticky signaling–adhesion class G protein-coupled receptors take the stage. Sci Signal 6(276):re3
5. Araç D, Sträter N, Seiradake E (2016) Understanding the structural basis of adhesion GPCR functions. In: Langenhan T, Schöneberg T (eds) Adhesion G protein-coupled receptors: molecular, physiological and pharmacological principles in health and disease. Springer, Heidelberg
6. Nieberler M, Kittel RJ, Petrenko AG, Lin H-H, Langenhan T (2016) Control of adhesion GPCR function through proteolytic processing. In: Langenhan T, Schöneberg T (eds) Adhesion G protein-coupled receptors: molecular, physiological and pharmacological principles in health and disease. Springer, Heidelberg
7. Schiöth HB, Nordstrom KJ, Fredriksson R (2010) The adhesion GPCRs; gene repertoire, phylogeny and evolution. Adv Exp Med Biol 706:1–13

8. Harty BL, Krishnan A, Sanchez NE, Schiöth HB, Monk KR (2015) Defining the gene repertoire and spatiotemporal expression profiles of adhesion G protein-coupled receptors in zebrafish. BMC Genomics 16:62
9. Krishnan A, Nijmeijer S, de Graaf C, Schiöth HB (2016) Classification, nomenclature and structural aspects of adhesion GPCRs. In: Langenhan T, Schöneberg T (eds) Adhesion G protein-coupled receptors: molecular, physiological and pharmacological principles in health and disease. Springer, Heidelberg
10. Nijmeijer S, Wolf S, Ernst OP, de Graaf C (2016) 7TM domain structure of adhesion GPCRs. In: Langenhan T, Schöneberg T (eds) Adhesion G protein-coupled receptors: molecular, physiological and pharmacological principles in health and disease. Springer, Heidelberg
11. Piao X, Hill RS, Bodell A, Chang BS, Basel-Vanagaite L, Straussberg R et al (2004) G protein-coupled receptor-dependent development of human frontal cortex. Science 303 (5666):2033–2036
12. Piao X, Chang BS, Bodell A, Woods K, Benzeev B, Topcu M et al (2005) Genotype-phenotype analysis of human frontoparietal polymicrogyria syndromes. Ann Neurol 58 (5):680–687
13. Cullen M, Elzarrad MK, Seaman S, Zudaire E, Stevens J, Yang MY et al (2011) GPR124, an orphan G protein-coupled receptor, is required for CNS-specific vascularization and establishment of the blood-brain barrier. Proc Natl Acad Sci U S A 108(14):5759–5764
14. Ichtchenko K, Bittner MA, Krasnoperov V, Little AR, Chepurny O, Holz RW et al (1999) A novel ubiquitously expressed alpha-latrotoxin receptor is a member of the CIRL family of G-protein-coupled receptors. J Biol Chem 274(9):5491–5498
15. Sugita S, Ichtchenko K, Khvotchev M, Sudhof TC (1998) alpha-Latrotoxin receptor CIRL/latrophilin 1 (CL1) defines an unusual family of ubiquitous G-protein-linked receptors. G-protein coupling not required for triggering exocytosis. J Biol Chem 273(49):32715–32724
16. Boucard AA, Maxeiner S, Sudhof TC (2014) Latrophilins function as heterophilic cell-adhesion molecules by binding to teneurins: regulation by alternative splicing. J Biol Chem 289(1):387–402
17. Kreienkamp HJ, Zitzer H, Gundelfinger ED, Richter D, Bockers TM (2000) The calcium-independent receptor for alpha-latrotoxin from human and rodent brains interacts with members of the ProSAP/SSTRIP/Shank family of multidomain proteins. J Biol Chem 275 (42):32387–32390
18. Tobaben S, Sudhof TC, Stahl B (2000) The G protein-coupled receptor CL1 interacts directly with proteins of the Shank family. J Biol Chem 275(46):36204–36210
19. Vakonakis I, Langenhan T, Prömel S, Russ A, Campbell ID (2008) Solution structure and sugar-binding mechanism of mouse latrophilin-1 RBL: a 7TM receptor-attached lectin-like domain. Structure 16(6):944–953
20. Silva JP, Lelianova VG, Ermolyuk YS, Vysokov N, Hitchen PG, Berninghausen O et al (2011) Latrophilin 1 and its endogenous ligand Lasso/teneurin-2 form a high-affinity transsynaptic receptor pair with signaling capabilities. Proc Natl Acad Sci U S A 108 (29):12113–12118
21. O'Sullivan ML, de Wit J, Savas JN, Comoletti D, Otto-Hitt S, Yates JR III et al (2012) FLRT proteins are endogenous latrophilin ligands and regulate excitatory synapse development. Neuron 73(5):903–910
22. Lelianova VG, Davletov BA, Sterling A, Rahman MA, Grishin EV, Totty NF et al (1997) Alpha-latrotoxin receptor, latrophilin, is a novel member of the secretin family of G protein-coupled receptors. J Biol Chem 272(34):21504–21508
23. O'Sullivan ML, Martini F, von Daake S, Comoletti D, Ghosh A (2014) LPHN3, a presynaptic adhesion-GPCR implicated in ADHD, regulates the strength of neocortical layer 2/3 synaptic input to layer 5. Neural Dev 9:7
24. Lange M, Norton W, Coolen M, Chaminade M, Merker S, Proft F et al (2012) The ADHD-linked gene Lphn3.1 controls locomotor activity and impulsivity in zebrafish. Mol Psychiatry 17(9):855

25. Guest M, Bull K, Walker RJ, Amliwala K, O'Connor V, Harder A et al (2007) The calcium-activated potassium channel, SLO-1, is required for the action of the novel cyclo-octadepsipeptide anthelmintic, emodepside, in Caenorhabditis elegans. Int J Parasitol 37 (14):1577–1588
26. Strutt D, Schnabel R, Fiedler F, Prömel S (2016) Adhesion GPCRs govern polarity of epithelia and cell migration. In: Langenhan T, Schöneberg T (eds) Adhesion G protein-coupled receptors: molecular, physiological and pharmacological principles in health and disease. Springer, Heidelberg
27. Steimel A, Wong L, Najarro EH, Ackley BD, Garriga G, Hutter H (2010) The Flamingo ortholog FMI-1 controls pioneer-dependent navigation of follower axons in C. elegans. Development 137(21):3663–3673
28. Bao H, Berlanga ML, Xue M, Hapip SM, Daniels RW, Mendenhall JM et al (2007) The atypical cadherin flamingo regulates synaptogenesis and helps prevent axonal and synaptic degeneration in Drosophila. Mol Cell Neurosci 34(4):662–678
29. Gao FB, Kohwi M, Brenman JE, Jan LY, Jan YN (2000) Control of dendritic field formation in Drosophila: the roles of flamingo and competition between homologous neurons. Neuron 28(1):91–101
30. Formstone CJ, Little PF (2001) The flamingo-related mouse Celsr family (Celsr1-3) genes exhibit distinct patterns of expression during embryonic development. Mech Dev 109 (1):91–94
31. Tissir F, De-Backer O, Goffinet AM, Lambert de Rouvroit C (2002) Developmental expression profiles of Celsr (Flamingo) genes in the mouse. Mech Dev 112(1–2):157–160
32. Matsubara D, Horiuchi SY, Shimono K, Usui T, Uemura T (2011) The seven-pass transmembrane cadherin Flamingo controls dendritic self-avoidance via its binding to a LIM domain protein, Espinas, in Drosophila sensory neurons. Genes Dev 25(18):1982–1996
33. Wang Y, Wang H, Li X, Li Y (2016) Epithelial microRNA-9a regulates dendrite growth through Fmi-Gq signaling in Drosophila sensory neurons. Dev Neurobiol 76(2):225–237
34. Kimura H, Usui T, Tsubouchi A, Uemura T (2006) Potential dual molecular interaction of the Drosophila 7-pass transmembrane cadherin Flamingo in dendritic morphogenesis. J Cell Sci 119(Pt 6):1118–1129
35. Shima Y, Copeland NG, Gilbert DJ, Jenkins NA, Chisaka O, Takeichi M et al (2002) Differential expression of the seven-pass transmembrane cadherin genes Celsr1-3 and distribution of the Celsr2 protein during mouse development. Dev Dyn 223(3):321–332
36. Shima Y, Kengaku M, Hirano T, Takeichi M, Uemura T (2004) Regulation of dendritic maintenance and growth by a mammalian 7-pass transmembrane cadherin. Dev Cell 7 (2):205–216
37. Shima Y, Kawaguchi SY, Kosaka K, Nakayama M, Hoshino M, Nabeshima Y et al (2007) Opposing roles in neurite growth control by two seven-pass transmembrane cadherins. Nat Neurosci 10(8):963–969
38. Tolias KF, Duman JG, Um K (2011) Control of synapse development and plasticity by Rho GTPase regulatory proteins. Prog Neurobiol 94(2):133–148
39. Koh JT, Lee ZH, Ahn KY, Kim JK, Bae CS, Kim HH et al (2001) Characterization of mouse brain-specific angiogenesis inhibitor 1 (BAI1) and phytanoyl-CoA alpha-hydroxylase-associated protein 1, a novel BAI1-binding protein. Brain Res Mol Brain Res 87(2):223–237
40. Kee HJ, Koh JT, Kim MY, Ahn KY, Kim JK, Bae CS et al (2002) Expression of brain-specific angiogenesis inhibitor 2 (BAI2) in normal and ischemic brain: involvement of BAI2 in the ischemia-induced brain angiogenesis. J Cereb Blood Flow Metab 22(9):1054–1067
41. Lanoue V, Usardi A, Sigoillot SM, Talleur M, Iyer K, Mariani J et al (2013) The adhesion-GPCR BAI3, a gene linked to psychiatric disorders, regulates dendrite morphogenesis in neurons. Mol Psychiatry 18(8):943–950
42. Sigoillot SM, Iyer K, Binda F, Gonzalez-Calvo I, Talleur M, Vodjdani G et al (2015) The secreted protein C1QL1 and its receptor BAI3 control the synaptic connectivity of excitatory inputs converging on cerebellar Purkinje cells. Cell Rep 10(5):820–832

43. Okajima D, Kudo G, Yokota H (2011) Antidepressant-like behavior in brain-specific angiogenesis inhibitor 2-deficient mice. J Physiol Sci 61(1):47–54
44. Duman JG, Tzeng CP, Tu YK, Munjal T, Schwechter B, Ho TS et al (2013) The adhesion-GPCR BAI1 regulates synaptogenesis by controlling the recruitment of the Par3/Tiam1 polarity complex to synaptic sites. J Neurosci 33(16):6964–6978
45. Stephenson JR, Paavola KJ, Schaefer SA, Kaur B, Van Meir EG, Hall RA (2013) Brain-specific angiogenesis inhibitor-1 signaling, regulation, and enrichment in the postsynaptic density. J Biol Chem 288(31):22248–22256
46. Oda K, Shiratsuchi T, Nishimori H, Inazawa J, Yoshikawa H, Taketani Y et al (1999) Identification of BAIAP2 (BAI-associated protein 2), a novel human homologue of hamster IRSp53, whose SH3 domain interacts with the cytoplasmic domain of BAI1. Cytogenet Cell Genet 84(1–2):75–82
47. Zhu D, Li C, Swanson AM, Villalba RM, Guo J, Zhang Z et al (2015) BAI1 regulates spatial learning and synaptic plasticity in the hippocampus. J Clin Invest 125(4):1497–1508
48. Bolliger MF, Martinelli DC, Sudhof TC (2011) The cell-adhesion G protein-coupled receptor BAI3 is a high-affinity receptor for C1q-like proteins. Proc Natl Acad Sci U S A 108 (6):2534–2539
49. Goldin M, Segal M (2003) Protein kinase C and ERK involvement in dendritic spine plasticity in cultured rodent hippocampal neurons. Eur J Neurosci 17(12):2529–2539
50. Tolias KF, Bikoff JB, Burette A, Paradis S, Harrar D, Tavazoie S et al (2005) The Rac1-GEF Tiam1 couples the NMDA receptor to the activity-dependent development of dendritic arbors and spines. Neuron 45(4):525–538
51. Zhang H, Macara IG (2006) The polarity protein PAR-3 and TIAM1 cooperate in dendritic spine morphogenesis. Nat Cell Biol 8(3):227–237
52. Van Aelst L, Cline HT (2004) Rho GTPases and activity-dependent dendrite development. Curr Opin Neurobiol 14(3):297–304
53. Cerri C, Fabbri A, Vannini E, Spolidoro M, Costa M, Maffei L et al (2011) Activation of Rho GTPases triggers structural remodeling and functional plasticity in the adult rat visual cortex. J Neurosci 31(42):15163–15172
54. Kakegawa W, Mitakidis N, Miura E, Abe M, Matsuda K, Takeo YH et al (2015) Anterograde C1ql1 signaling is required in order to determine and maintain a single-winner climbing fiber in the mouse cerebellum. Neuron 85(2):316–329
55. Collins MO, Husi H, Yu L, Brandon JM, Anderson CN, Blackstock WP et al (2006) Molecular characterization and comparison of the components and multiprotein complexes in the postsynaptic proteome. J Neurochem 97(Suppl 1):16–23
56. Selimi F, Cristea IM, Heller E, Chait BT, Heintz N (2009) Proteomic studies of a single CNS synapse type: the parallel fiber/purkinje cell synapse. PLoS Biol 7(4), e83
57. Iijima T, Miura E, Watanabe M, Yuzaki M (2010) Distinct expression of C1q-like family mRNAs in mouse brain and biochemical characterization of their encoded proteins. Eur J Neurosci 31(9):1606–1615
58. Jessen KR, Mirsky R (2005) The origin and development of glial cells in peripheral nerves. Nat Rev Neurosci 6(9):671–682
59. Nave KA (2010) Myelination and support of axonal integrity by glia. Nature 468 (7321):244–252
60. Monk KR, Naylor SG, Glenn TD, Mercurio S, Perlin JR, Dominguez C et al (2009) A G protein-coupled receptor is essential for Schwann cells to initiate myelination. Science 325 (5946):1402–1405
61. Monk KR, Oshima K, Jors S, Heller S, Talbot WS (2011) Gpr126 is essential for peripheral nerve development and myelination in mammals. Development 138(13):2673–2680
62. Kuhlbrodt K, Herbarth B, Sock E, Hermans-Borgmeyer I, Wegner M (1998) Sox10, a novel transcriptional modulator in glial cells. J Neurosci 18(1):237–250

63. Britsch S, Goerich DE, Riethmacher D, Peirano RI, Rossner M, Nave KA et al (2001) The transcription factor Sox10 is a key regulator of peripheral glial development. Genes Dev 15 (1):66–78
64. Finzsch M, Schreiner S, Kichko T, Reeh P, Tamm ER, Bosl MR et al (2010) Sox10 is required for Schwann cell identity and progression beyond the immature Schwann cell stage. J Cell Biol 189(4):701–712
65. Wood PM (1976) Separation of functional Schwann cells and neurons from normal peripheral nerve tissue. Brain Res 115(3):361–375
66. Salzer JL, Williams AK, Glaser L, Bunge RP (1980) Studies of Schwann cell proliferation. II. Characterization of the stimulation and specificity of the response to a neurite membrane fraction. J Cell Biol 84(3):753–766
67. Sobue G, Brown MJ, Kim SU, Pleasure D (1984) Axolemma is a mitogen for human Schwann cells. Ann Neurol 15(5):449–452
68. Mokuno K, Sobue G, Reddy UR, Wurzer J, Kreider B, Hotta H et al (1988) Regulation of Schwann cell nerve growth factor receptor by cyclic adenosine 3',5'-monophosphate. J Neurosci Res 21(2-4):465–472
69. Monuki ES, Weinmaster G, Kuhn R, Lemke G (1989) SCIP: a glial POU domain gene regulated by cyclic AMP. Neuron 3(6):783–793
70. Scherer SS, Wang DY, Kuhn R, Lemke G, Wrabetz L, Kamholz J (1994) Axons regulate Schwann cell expression of the POU transcription factor SCIP. J Neurosci 14(4):1930–1942
71. Mogha A, Benesh AE, Patra C, Engel FB, Schoneberg T, Liebscher I et al (2013) Gpr126 functions in Schwann cells to control differentiation and myelination via G-protein activation. J Neurosci 33(46):17976–17985
72. Glenn TD, Talbot WS (2013) Analysis of Gpr126 function defines distinct mechanisms controlling the initiation and maturation of myelin. Development 140(15):3167–3175
73. Petersen SC, Luo R, Liebscher I, Giera S, Jeong SJ, Mogha A et al (2015) The adhesion GPCR GPR126 has distinct, domain-dependent functions in Schwann cell development mediated by interaction with laminin-211. Neuron 85(4):755–769
74. Patra C, van Amerongen MJ, Ghosh S, Ricciardi F, Sajjad A, Novoyatleva T et al (2013) Organ-specific function of adhesion G protein-coupled receptor GPR126 is domain-dependent. Proc Natl Acad Sci U S A 110(42):16898–16903
75. Yona S, Lin HH, Siu WO, Gordon S, Stacey M (2008) Adhesion-GPCRs: emerging roles for novel receptors. Trends Biochem Sci 33(10):491–500
76. Liebscher I, Schon J, Petersen SC, Fischer L, Auerbach N, Demberg LM et al (2014) A tethered agonist within the ectodomain activates the adhesion G protein-coupled receptors GPR126 and GPR133. Cell Rep 9(6):2018–2026
77. Stoveken HM, Hajduczok AG, Xu L, Tall GG (2015) Adhesion G protein-coupled receptors are activated by exposure of a cryptic tethered agonist. Proc Natl Acad Sci U S A 112 (19):6194–6199
78. Demberg LM, Rothemund S, Schoneberg T, Liebscher I (2015) Identification of the tethered peptide agonist of the adhesion G protein-coupled receptor GPR64/ADGRG2. Biochem Biophys Res Commun 464(3):743–747
79. Wilde C, Fischer L, Lede V, Kirchberger J, Rothemund S, Schoneberg T et al (2016) The constitutive activity of the adhesion GPCR GPR114/ADGRG5 is mediated by its tethered agonist. FASEB J 30(2):666–673
80. Liebscher I, Schöneberg T (2016) Tethered agonism: a common activation mechanism of adhesion GPCRs. In: Langenhan T, Schöneberg T (eds) Adhesion G protein-coupled receptors: molecular, physiological and pharmacological principles in health and disease. Springer, Heidelberg
81. Kishore A, Hall RA (2016) Versatile signaling activity of adhesion GPCRs. In: Langenhan T, Schöneberg T (eds) Adhesion G protein-coupled receptors: molecular, physiological and pharmacological principles in health and disease. Springer, Heidelberg

82. Paavola KJ, Sidik H, Zuchero JB, Eckart M, Talbot WS (2014) Type IV collagen is an activating ligand for the adhesion G protein-coupled receptor GPR126. Sci Signal 7(338): ra76
83. Scholz N, Gehring J, Guan C, Ljaschenko D, Fischer R, Lakshmanan V et al (2015) The adhesion GPCR latrophilin/CIRL shapes mechanosensation. Cell Rep 11(6):866–874
84. Scholz N, Monk KR, Kittel RJ, Langenhan T (2016) Adhesion GPCRs as a putative class of metabotropic mechanosensors. In: Langenhan T, Schöneberg T (eds) Adhesion G protein-coupled receptors: molecular, physiological and pharmacological principles in health and disease. Springer, Heidelberg
85. Arthur-Farraj P, Wanek K, Hantke J, Davis CM, Jayakar A, Parkinson DB et al (2011) Mouse Schwann cells need both NRG1 and cyclic AMP to myelinate. Glia 59(5):720–733
86. Emery B (2010) Regulation of oligodendrocyte differentiation and myelination. Science 330 (6005):779–782
87. Aggarwal S, Yurlova L, Simons M (2011) Central nervous system myelin: structure, synthesis and assembly. Trends Cell Biol 21(10):585–593
88. Campbell K, Gotz M (2002) Radial glia: multi-purpose cells for vertebrate brain development. Trends Neurosci 25(5):235–238
89. Simons M, Snaidero N, Aggarwal S (2012) Cell polarity in myelinating glia: from membrane flow to diffusion barriers. Biochim Biophys Acta 1821(8):1146–1153
90. Lein ES, Hawrylycz MJ, Ao N, Ayres M, Bensinger A, Bernard A et al (2007) Genome-wide atlas of gene expression in the adult mouse brain. Nature 445(7124):168–176
91. Cahoy JD, Emery B, Kaushal A, Foo LC, Zamanian JL, Christopherson KS et al (2008) A transcriptome database for astrocytes, neurons, and oligodendrocytes: a new resource for understanding brain development and function. J Neurosci 28(1):264–278
92. Piao X, Basel-Vanagaite L, Straussberg R, Grant PE, Pugh EW, Doheny K et al (2002) An autosomal recessive form of bilateral frontoparietal polymicrogyria maps to chromosome 16q12.2-21. Am J Hum Genet 70(4):1028–1033
93. Giera S, Deng Y, Luo R, Ackerman SD, Mogha A, Monk KR et al (2015) The adhesion G protein-coupled receptor GPR56 is a cell-autonomous regulator of oligodendrocyte development. Nat Commun 6:6121
94. Ackerman SD, Garcia C, Piao X, Gutmann DH, Monk KR (2015) The adhesion GPCR Gpr56 regulates oligodendrocyte development via interactions with Galpha12/13 and RhoA. Nat Commun 6:6122
95. Jin Z, Tietjen I, Bu L, Liu-Yesucevitz L, Gaur SK, Walsh CA et al (2007) Disease-associated mutations affect GPR56 protein trafficking and cell surface expression. Hum Mol Genet 16 (16):1972–1985
96. Iguchi T, Sakata K, Yoshizaki K, Tago K, Mizuno N, Itoh H (2008) Orphan G protein-coupled receptor GPR56 regulates neural progenitor cell migration via a G alpha 12/13 and Rho pathway. J Biol Chem 283(21):14469–14478
97. Luo R, Jeong SJ, Jin Z, Strokes N, Li S, Piao X (2011) G protein-coupled receptor 56 and collagen III, a receptor-ligand pair, regulates cortical development and lamination. Proc Natl Acad Sci U S A 108(31):12925–12930
98. Liang X, Draghi NA, Resh MD (2004) Signaling from integrins to Fyn to Rho family GTPases regulates morphologic differentiation of oligodendrocytes. J Neurosci 24 (32):7140–7149
99. Shin D, Lin ST, Fu YH, Ptacek LJ (2013) Very large G protein-coupled receptor 1 regulates myelin-associated glycoprotein via Galphas/Galphaq-mediated protein kinases A/C. Proc Natl Acad Sci U S A 110(47):19101–19106
100. Zhang Y, Chen K, Sloan SA, Bennett ML, Scholze AR, O'Keeffe S et al (2014) An RNA-sequencing transcriptome and splicing database of glia, neurons, and vascular cells of the cerebral cortex. J Neurosci 34(36):11929–11947

101. Bonaglia MC, Marelli S, Novara F, Commodaro S, Borgatti R, Minardo G et al (2010) Genotype-phenotype relationship in three cases with overlapping 19p13.12 microdeletions. Eur J Hum Genet 18(12):1302–1309
102. Arcos-Burgos M, Jain M, Acosta MT, Shively S, Stanescu H, Wallis D et al (2010) A common variant of the latrophilin 3 gene, LPHN3, confers susceptibility to ADHD and predicts effectiveness of stimulant medication. Mol Psychiatry 15(11):1053–1066
103. Domené S, Stanescu H, Wallis D, Tinloy B, Pineda DE, Kleta R et al (2011) Screening of human LPHN3 for variants with a potential impact on ADHD susceptibility. Am J Med Genet B Neuropsychiatr Genet 156B(1):11–18
104. Michaelson JJ, Shi Y, Gujral M, Zheng H, Malhotra D, Jin X et al (2012) Whole-genome sequencing in autism identifies hot spots for de novo germline mutation. Cell 151 (7):1431–1442
105. Toma C, Hervas A, Balmana N, Vilella E, Aguilera F, Cusco I et al (2011) Association study of six candidate genes asymmetrically expressed in the two cerebral hemispheres suggests the involvement of BAIAP2 in autism. J Psychiatr Res 45(2):280–282
106. DeRosse P, Lencz T, Burdick KE, Siris SG, Kane JM, Malhotra AK (2008) The genetics of symptom-based phenotypes: toward a molecular classification of schizophrenia. Schizophr Bull 34(6):1047–1053
107. Liao HM, Chao YL, Huang AL, Cheng MC, Chen YJ, Lee KF et al (2012) Identification and characterization of three inherited genomic copy number variations associated with familial schizophrenia. Schizophr Res 139(1-3):229–236
108. McCarthy MJ, Nievergelt CM, Kelsoe JR, Welsh DK (2012) A survey of genomic studies supports association of circadian clock genes with bipolar disorder spectrum illnesses and lithium response. PLoS One 7(2), e32091
109. Liu QR, Drgon T, Johnson C, Walther D, Hess J, Uhl GR (2006) Addiction molecular genetics: 639,401 SNP whole genome association identifies many "cell adhesion" genes. Am J Med Genet B Neuropsychiatr Genet 141B(8):918–925
110. Lubetzki C, Stankoff B (2014) Demyelination in multiple sclerosis. Handb Clin Neurol 122:89–99
111. Ravenscroft G, Nolent F, Rajagopalan S, Meireles AM, Paavola KJ, Gaillard D et al (2015) Mutations of GPR126 are responsible for severe arthrogryposis multiplex congenita. Am J Hum Genet 96(6):955–961
112. Shorer Z, Philpot J, Muntoni F, Sewry C, Dubowitz V (1995) Demyelinating peripheral neuropathy in merosin-deficient congenital muscular dystrophy. J Child Neurol 10 (6):472–475

Control of Skeletal Muscle Cell Growth and Size Through Adhesion GPCRs

James P. White

Graphical Abstract

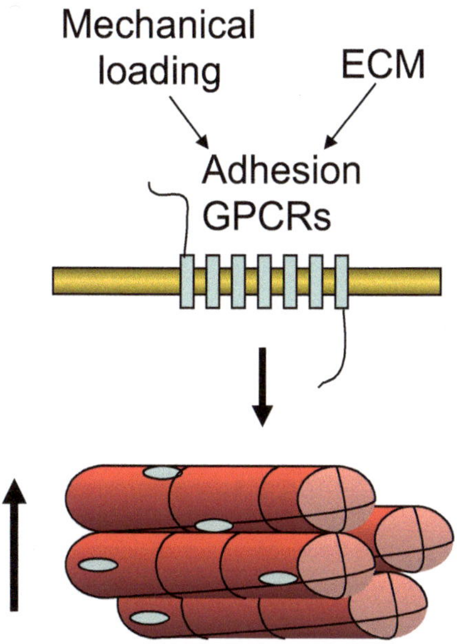

J.P. White (✉)
Duke University Medical Center, 300 North Duke Street, Durham, NC 27701, USA
e-mail: james.white@duke.edu

T. Langenhan, T. Schöneberg (eds.), *Adhesion G Protein-coupled Receptors*,
Handbook of Experimental Pharmacology 234, DOI 10.1007/978-3-319-41523-9_13

Contents

Abstract

Skeletal muscle homeostasis is regulated by a constant influx of chemicals and exposure to mechanical stimuli. A number of key signaling pathways that translate these stimuli into changes in muscle physiology have been established. The GPCR family known as adhesion GPCRs (aGPCRs) has largely elusive roles in skeletal muscle biology; however, their unique capacity to activate adhesion and G protein signaling pathways makes them an attractive point of investigation. The skeletal muscle myofiber contains a highly organized cyto-architecture to ensure contractile function. This requires intricate interactions with components of the extracellular matrix (ECM) surrounding each fiber. aGPCRs possess extended N-termini known to interact with ECM proteins and complexes suggesting a compatible role in skeletal muscle biology. Furthermore, recent work demonstrated the involvement of certain aGPCRs in whole muscle hypertrophy and differentiation of muscle progenitor cells. Signaling pathways downstream of aGPCRs are still incompletely understood; however, initial findings show involvement of the Gα12/13 subunit signaling to the pro-anabolic Akt/mTOR pathway. Together, this chapter will review the emerging role of aGPCRs in skeletal muscle biology and putative mechanism (s) employed to regulate skeletal muscle growth.

Keywords

Skeletal muscle hypertrophy • Mechanical overload • mTOR • Adhesion GPCRs • Myogenesis

1 Introduction

Skeletal muscle is a highly plastic tissue capable of responding to both chemical and mechanical stimuli. The balance of chemical stimuli derives from factors that promote muscle growth [e.g., growth hormones, anabolic steroids, insulin-like growth factor 1 (IGF-1)] or atrophy [e.g., myostatin, interleukin 6 (IL-6), tumor necrosis factor α (TNF-α)] [1]. These factors along with their downstream signaling pathways have been heavily researched over the last several decades. In addition to these chemical stimuli, mechanical loading/unloading constitutes a potent regulator of muscle growth and atrophy [2]. However, unlike the chemical affecters of muscle mass, the load-sensitive mechanism(s) that regulate muscle size remain ill-defined. Despite the growing body of research investigating the mechanisms of muscle growth, the role of G protein-coupled receptors (GPCRs) and specifically aGPCRs in this process is currently unclear. Intriguingly, aGPCRs are structurally similar to integrin receptors yet functionally signal through G protein pathways allowing for potentially novel signaling mode, which could integrate with known anabolic pathways in muscle. Further investigation of the biology of aGPCRs in skeletal muscle is certainly warranted.

1.1 Molecular Regulators of Muscle Mass

Under physiological conditions, skeletal muscle regulates mass by increasing/decreasing myofibrillar protein accumulation. This can be achieved through expansion or depletion of myofibrillar proteins with or without adding or subtracting myonuclei. The addition of myonuclei is a process referred to as myogenesis. Muscle stem cells, called satellite cells, are located around the periphery of the myofiber and remain quiescent until the local environment cues for satellite cell activation and fusion into the adjacent myofiber [3]. The additional myonuclei can increase the capacity to transcribe mRNA and facilitate accretion of more myofibrillar protein. The process of myogenesis is particularly crucial in conditions of muscle injury when protein synthesis is not sufficient for myofiber repair (REF). The IGF-1/Akt/mTOR pathway (Fig. 1) is a well-established signaling cascade of muscle protein balance (reviewed in [4, 5]). The IGF-1 signaling pathway is well established and known to regulate skeletal muscle mass by controlling protein synthesis and degradation as well as apoptotic pathways [4, 5]. Binding of IGF-1/insulin activates the receptor tyrosine kinase IGF-1 receptor and recruits insulin receptor substrate (IRS). Subsequently, this leads to the activation of phosphatidylinositol 3′-kinase (PI3K) and potentially serine-threonine kinase Akt (PKB) through a phosphorylation of a serine residue at position 473 [6–8]. Interestingly, during skeletal muscle hypertrophy, Akt activity is increased when examined in vivo [9] and in cultured myotubes [10]. In addition, a genetically altered, constitutively active Akt was able to induce muscle hypertrophy without growth factor stimulation [9, 11]. A known downstream target of Akt is the mammalian target of rapamycin (mTOR). mTOR signals downstream to enhance translational capacity through p70S6K and 4E-BP1 to promote muscle protein accretion. In addition to activating anabolic signaling, the IGF-1 pathway can also suppress atrophic signaling through inhibition of FOXO, a key transcriptional regulator of atrophy-related gene expression [12]. Together, the Akt/mTOR pathway drives muscle hypertrophy by increasing protein synthesis and suppressing protein breakdown.

In addition to growth factor stimulation, skeletal muscle is very sensitive to mechanical load. Overload-induced muscle hypertrophy is evident with continuous stretch [13], mechanical overload [14, 15], and intermittent lengthening contractions [16, 17]. In terms of load-sensitive signaling, mTOR plays a central

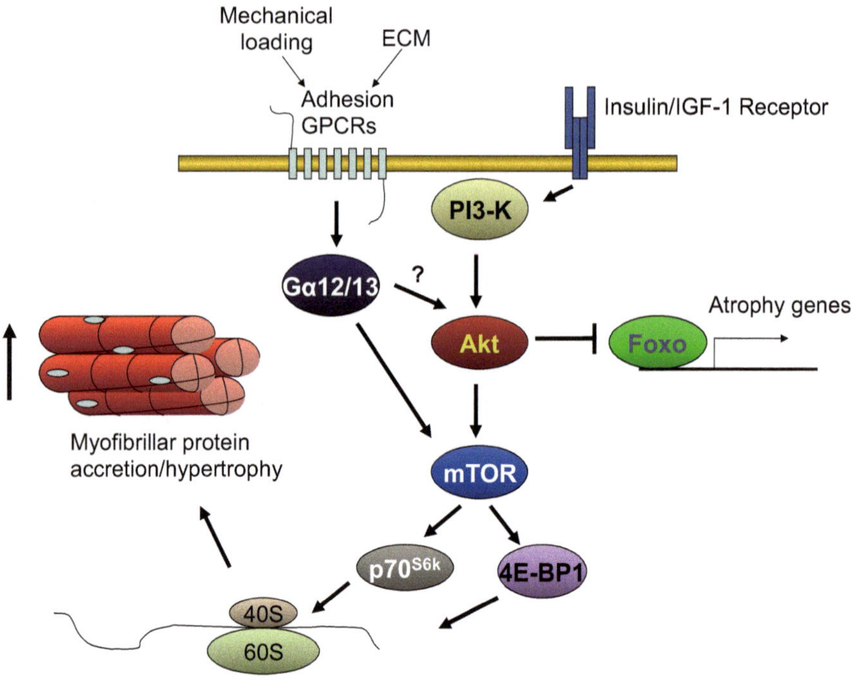

Fig. 1 Anabolic signaling pathways in skeletal muscle. The canonical insulin/IGF-1 pathway signals through the PI-3k/Akt/mTOR pathway to induce ribosomal translation and increased myofiber protein accretion. Activation of the adhesion GPCR is initiated through binding to ECM components and/or mechanical stress. Interaction of adhesion GPCR signaling and Akt/mTOR signaling pathways is mediated by the Gα12/13 subunit and downstream activation of mTOR

role in regulating mechanical overload-induced muscle adaptations [18]. Overload-induced activation of mTOR and its target p70S6K is independent of PI3K or Akt [19, 20], which suggests dispensability of growth factors for overload-induced hypertrophy. Mechanical overload in both mice [21] and humans [22] increases mTOR phosphorylation and its downstream target p70S6K. Integrin signaling has been implied as a mechanism to sense mechanical load and integrate downstream signaling events [23]. Currently, no evidence directly links canonical integrin signaling to skeletal muscle mTOR activation. However, in nonmuscle cells, integrin signaling activates Ras and its major effectors phosphoinositide-3 kinase (PI3K) and raf [24] suggesting a potential integrin-mTOR link in muscle.

In terms of physiological hypertrophy, overexpression of α7β1 integrin in skeletal muscle enhances the hypertrophic effect of downhill running exercise (an eccentric contraction-induced hypertrophy model) [25]. However, no signaling mechanism was implied to explain these observations. aGPCRs have integrin-like signaling capacity and could conceivably act as a load sensor in the muscle. The extended N-terminal tail can signal to ECM proteins such as collagen I/IV [14, 26] similar to integrins. Furthermore some aGPCRs signal through Rho [26, 27], another well-established downstream effector of integrin activation. GPCRs can signal through the Akt pathway in 3T3 fibroblast [28]; however, this connection has not been described in skeletal muscle. The role of aGPCRs in both canonical Akt/mTOR signaling and mechanical loading-induced hypertrophy is unclear and could act as a focal point to connect several known anabolic pathways.

2 Adhesion GPCRs and Skeletal Muscle Biology

2.1 Muscle Mass Regulation by Adhesion GPCRs

aGPCRs form one of five GPCR families and have been shown to facilitate cell–cell and cell–matrix interaction [29]. One such aGPCR, GPR56, has been heavily investigated in the brain as mutations in *GPR56* result in a condition known as bilateral frontoparietal polymicrogyria (BFPP), an autosomal recessively inherited developmental disorder of the brain [30] (see also [31]). In terms of skeletal muscle, *GPR56* was recently discovered to be a transcriptional target of the PGC-1α isoform 4 (PGC-1α4) known to promote muscle hypertrophy and to accumulate during resistance-type exercise [32]. Furthermore, GPR56 is necessary for PGC-1α4-induced muscle hypertrophy and regulates mechanical loading-induced hypertrophy [14]. GPR56 can independently cause muscle hypertrophy as overexpression of GPR56 resulted in myotube hypertrophy in vitro. In addition, mRNA expression of GPR56 and its ligand collagen III were increased in muscle during various mouse models of loading-based muscle hypertrophy and human resistance exercise. Interestingly, *GPR56* loss of function mice did not display an overt muscle phenotype until challenged with an overload stimulus. Under overload conditions, the *GPR56* KO mice did not hypertrophy to the same extent as wild types [14]. This suggests that GPR56 may function as a "load-sensitive" driver of muscle growth, which is not active during sedentary conditions. In terms of myogenesis, GPR56 is highly expressed during the initial days of in vitro myoblast differentiation [33]. Knockdown of *GPR56* attenuates myotube formation, expression of myogenic genes, and nuclear factor of activated T cell (NFAT) (Fig. 2). Furthermore, *GPR56* null mice were able to regenerate from muscle injury similar to wild-type mice [33].

The aGPCR brain-specific angiogenesis inhibitor (BAI3) has also been shown to have a role in myogenesis [34]. Knockdown of BAI3 in C2C12 myotubes results in a severe suppression of myotube formation. The function of BAI3 in myotube formation depends on binding to the ELMO1/DOCK1 complex, a downstream component of integrin signaling. The third and most recently identified adhesion GPCR associated with skeletal muscle biology is CD97. CD97 localizes to the muscle fiber sarcolemma and the sarcoplasmic reticulum. In mouse, deletion of CD97 was shown to alter the morphology of the sarcoplasmic reticulum, while the skeletal muscle function was unaltered in vivo [35]. The signaling mechanism for CD97 is currently unknown in skeletal muscle. Together, the role of aGPCRs in

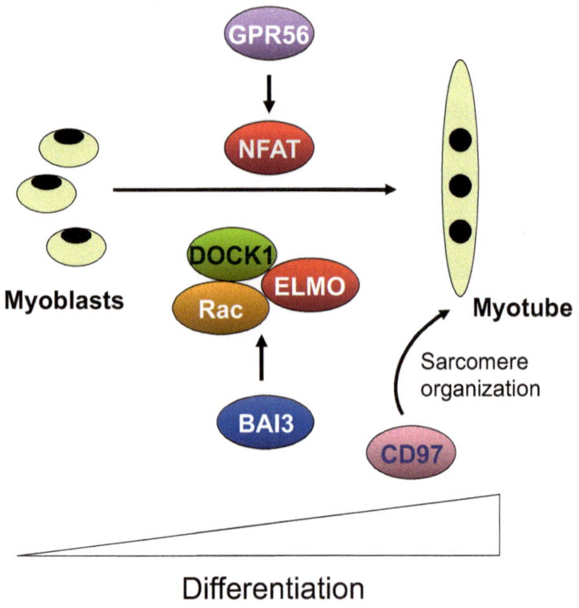

Fig. 2 Regulation of myogenesis by adhesion GPCRs. The differentiation process of myoblasts into mature myotubes is a highly complex process in which certain mechanisms remain unclear. aGPCRs are known to regulate myogenesis through different pathways. GPR56 regulates myotube formation through expression of NFAT and coordination of myogenic regulatory factors. BAI3 binds to the Rac/DOCK1/ELMO complex to promote differentiation, while CD97 has a role in sarcomere organization during myotube formation

skeletal muscle is slowly expanding and interestingly can be identified throughout various locations within the myofiber. In addition, aGPCRs effect various functions in muscle ranging from anabolic signaling (Fig. 1) to myoblast differentiation (Fig. 2).

2.2 Adhesion GPCR Signaling in Skeletal Muscle

The convergence between aGPCR signaling and established regulatory pathways in muscle is currently unknown and could elucidate novel points of regulation. Outside the aGPCR family, there is evidence for Gα signaling, specifically through $G\alpha_{i2,}$ to promote skeletal muscle hypertrophy and muscle stem cell (satellite cell) myogenic capacity [36, 37]. The anabolic effect of $G\alpha_{i2}$ is dependent on PKC to inhibit GSK-3β, a negative regulator of protein synthesis, and to activate p70S6k, a positive regulator of translation initiation [37]. The functional role of GPR56 in skeletal muscle is dependent on Gα12/13 activity [14]. Interestingly, the Gα13/Rho pathway has been associated with muscle hypertrophy in pressure overloaded cardiac muscle [38], supporting the concept of the Gα12/13 pathway being an integrin-like "mechano"-sensitive anabolic pathway (see also [39]). The Rho/SRF pathway is sensitive to mechanical loading in muscle and associated with muscle hypertrophy [40, 41]. In addition, the load-sensitive Rho pathway has been shown to regulate satellite cell proliferation [42] and differentiation [43]. GPR56 overexpression in myotubes was associated with an increase in mTOR activation, while inhibition of the Gα12/13 subunit was sufficient to block GPR56-induced mTOR activation (summarized in Fig. 1) [14]. Together, GPR56 activation may promote muscle anabolism through multiple pathways. The direct mechanism for Gα12/13 to activate mTOR is unclear and may require additional signaling arms specific to muscle, as Gα12/13 activation in 293 cells inhibited Akt signaling [44] suggesting this pathway may not be so straightforward. Elucidating downstream signaling through other aGPCRs in skeletal muscle deserves further investigation.

2.3 Endocrine Signaling from ECM to Adhesion GPRCs

An intriguing concept of aGPCRs especially in the case of GPR56 is the idea of trans-activation through ECM fragments originated from other muscle groups. As previously mentioned, collagen III is a known ligand of GPR56, which activates Rho signaling through the Gα12/13 pathway [26]. The N-terminal pro-peptide of collagen type III (P-III-NP) has been suggested as a biomarker of muscle anabolism with testosterone and growth hormone treatment [45]. There are two possible hypotheses for this observation: (1) the correlation of muscle growth could simply be a by-product of collagen synthesis used for muscle hypertrophy, or (2) it could be a novel "feed-forward" endocrine mechanism to promote muscle anabolism. This paracrine/endocrine mechanism could work in parallel with secretion of typical growth factors such as IGF-1 [46] and other myokines including IL-6 [47], IL-10 [48], irisin [49], and meteorin-like [50].

3 Conclusions

The integrin/adhesion-like properties of aGPCRs make for an easy transition into the biology of skeletal muscle. Integrin signaling is an established "load-sensitive" pathway associated with muscle anabolism; however, direct mechanisms are currently unclear. aGPCRs may represent a novel functional unit to ensure adequate and reliable translation of mechanical stimuli into anabolic signaling events. The second role for aGPCRs is in muscle precursor cell function including regeneration after injury. The investigation of GPR56, BAI3, and CD97 function in muscle is the start of an understanding for the role of these novel receptors. Interestingly, despite a small literature base, aGPCRs play a role in both whole muscle in vivo and myoblasts in vitro suggesting their role in regulating myofiber growth and myogenic progenitor function.

In addition to promising functional capacity in skeletal muscle, in terms of pharmacology, GPCRs are often druggable targets due to their expression on the cell membrane. Currently, no pharmacological agents are available to treat muscle disease. Therefore, investigating the role of adhesion, and other GPCR family members, in skeletal muscle is warranted and could lead to new pharmacol targets to improve muscle anabolism.

References

1. Egerman MA, Glass DJ (2014) Signaling pathways controlling skeletal muscle mass. Crit Rev Biochem Mol Biol 49:59–68
2. Jacobs BL, Goodman CA, Hornberger TA (2014) The mechanical activation of mTOR signaling: an emerging role for late endosome/lysosomal targeting. J Muscle Res Cell Motil 35:11–21
3. Mauro A (1961) Satellite cell of skeletal muscle fibers. J Biophys Biochem Cytol 9:493–495
4. Glass DJ (2010) PI3 kinase regulation of skeletal muscle hypertrophy and atrophy. Curr Top Microbiol Immunol 346:267–278
5. Frost RA, Lang CH (2007) Protein kinase B/Akt: a nexus of growth factor and cytokine signaling in determining muscle mass. J Appl Physiol 103:378–387
6. Moelling K, Schad K, Bosse M, Zimmermann S, Schweneker M (2002) Regulation of Raf-Akt Cross-talk. J Biol Chem 277:31099–31106
7. Alessi DR, James SR, Downes CP, Holmes AB, Gaffney PR et al (1997) Characterization of a 3-phosphoinositide-dependent protein kinase which phosphorylates and activates protein kinase Balpha. Curr Biol 7:261–269
8. Andjelkovic M, Alessi DR, Meier R, Fernandez A, Lamb NJ et al (1997) Role of translocation in the activation and function of protein kinase B. J Biol Chem 272:31515–31524
9. Bodine SC, Stitt TN, Gonzalez M, Kline WO, Stover GL et al (2001) Akt/mTOR pathway is a crucial regulator of skeletal muscle hypertrophy and can prevent muscle atrophy in vivo. Nat Cell Biol 3:1014–1019
10. Rommel C, Bodine SC, Clarke BA, Rossman R, Nunez L et al (2001) Mediation of IGF-1-induced skeletal myotube hypertrophy by PI(3)K/Akt/mTOR and PI(3)K/Akt/GSK3 pathways. Nat Cell Biol 3:1009–1013
11. Lai KM, Gonzalez M, Poueymirou WT, Kline WO, Na E et al (2004) Conditional activation of akt in adult skeletal muscle induces rapid hypertrophy. Mol Cell Biol 24:9295–9304
12. Stitt TN, Drujan D, Clarke BA, Panaro F, Timofeyva Y et al (2004) The IGF-1/PI3K/Akt pathway prevents expression of muscle atrophy-induced ubiquitin ligases by inhibiting FOXO transcription factors. Mol Cell 14:395–403
13. Carson JA, Yan Z, Booth FW, Coleman ME, Schwartz RJ et al (1995) Regulation of skeletal alpha-actin promoter in young chickens during hypertrophy caused by stretch overload. Am J Physiol 268:C918–C924
14. White JP, Wrann CD, Rao RR, Nair SK, Jedrychowski MP et al (2014) G protein-coupled receptor 56 regulates mechanical overload-induced muscle hypertrophy. Proc Natl Acad Sci U S A 111:15756–15761
15. White JP, Reecy JM, Washington TA, Sato S, Le ME et al (2009) Overload-induced skeletal muscle extracellular matrix remodelling and myofibre growth in mice lacking IL-6. Acta Physiol (Oxf) 197:321–332
16. Farthing JP, Chilibeck PD (2003) The effects of eccentric and concentric training at different velocities on muscle hypertrophy. Eur J Appl Physiol 89:578–586
17. Tsumiyama W, Oki S, Takamiya N, Umei N, Shimizu ME et al (2014) Induction of muscle hypertrophy in rats through low intensity eccentric contraction. J Phys Ther Sci 26:1623–1625
18. Hornberger TA (2011) Mechanotransduction and the regulation of mTORC1 signaling in skeletal muscle. Int J Biochem Cell Biol 43:1267–1276
19. Hornberger TA, Stuppard R, Conley KE, Fedele MJ, Fiorotto ML et al (2004) Mechanical stimuli regulate rapamycin-sensitive signalling by a phosphoinositide 3-kinase-, protein kinase B- and growth factor-independent mechanism. Biochem J 380:795–804
20. Miyazaki M, McCarthy JJ, Fedele MJ, Esser KA (2011) Early activation of mTORC1 signalling in response to mechanical overload is independent of phosphoinositide 3-kinase/Akt signalling. J Physiol 589:1831–1846
21. Goodman CA, Frey JW, Mabrey DM, Jacobs BL, Lincoln HC et al (2011) The role of skeletal muscle mTOR in the regulation of mechanical load-induced growth. J Physiol 589:5485–5501

22. Drummond MJ, Fry CS, Glynn EL, Dreyer HC, Dhanani S et al (2009) Rapamycin administration in humans blocks the contraction-induced increase in skeletal muscle protein synthesis. J Physiol 587:1535–1546
23. Carson JA, Wei L (2000) Integrin signaling's potential for mediating gene expression in hypertrophying skeletal muscle. J Appl Physiol 88:337–343
24. Frisch SM, Francis H (1994) Disruption of epithelial cell-matrix interactions induces apoptosis. J Cell Biol 124:619–626
25. Zou K, Meador BM, Johnson B, Huntsman HD, Mahmassani Z et al (2011) The alpha(7)beta (1)-integrin increases muscle hypertrophy following multiple bouts of eccentric exercise. J Appl Physiol 111:1134–1141
26. Luo R, Jeong SJ, Jin Z, Strokes N, Li S et al (2011) G protein-coupled receptor 56 and collagen III, a receptor-ligand pair, regulates cortical development and lamination. Proc Natl Acad Sci U S A 108:12925–12930
27. Ackerman SD, Garcia C, Piao X, Gutmann DH, Monk KR (2015) The adhesion GPCR Gpr56 regulates oligodendrocyte development via interactions with Galpha12/13 and RhoA. Nat Commun 6:6122
28. Murga C, Fukuhara S, Gutkind JS (2000) A novel role for phosphatidylinositol 3-kinase beta in signaling from G protein-coupled receptors to Akt. J Biol Chem 275:12069–12073
29. Langenhan T, Aust G, Hamann J (2013) Sticky signaling–adhesion class G protein-coupled receptors take the stage. Sci Signal 6:re3
30. Piao X, Hill RS, Bodell A, Chang BS, Basel-Vanagaite L et al (2004) G protein-coupled receptor-dependent development of human frontal cortex. Science 303:2033–2036
31. Sigoillot SM, Monk KR, Piao X, Selimi F, Harty BL (2016) Adhesion G protein-coupled receptors in the nervous system: from synapse and dendrite morphogenesis to myelination. In: Langenhan T, Schöneberg T (eds) Adhesion G protein-coupled receptors: molecular, physiological and pharmacological principles in health and disease. Springer, Heidelberg
32. Ruas JL, White JP, Rao RR, Kleiner S, Brannan KT et al (2012) A PGC-1alpha isoform induced by resistance training regulates skeletal muscle hypertrophy. Cell 151:1319–1331
33. Wu MP, Doyle JR, Barry B, Beauvais A, Rozkalne A et al (2013) G-protein coupled receptor 56 promotes myoblast fusion through serum response factor- and nuclear factor of activated T-cell-mediated signalling but is not essential for muscle development in vivo. FEBS J 280: 6097–6113
34. Hamoud N, Tran V, Croteau LP, Kania A, Cote JF (2014) G-protein coupled receptor BAI3 promotes myoblast fusion in vertebrates. Proc Natl Acad Sci U S A 111:3745–3750
35. Zyryanova T, Schneider R, Adams V, Sittig D, Kerner C et al (2014) Skeletal muscle expression of the adhesion-GPCR CD97: CD97 deletion induces an abnormal structure of the sarcoplasmatic reticulum but does not impair skeletal muscle function. PLoS One 9, e100513
36. Minetti GC, Feige JN, Bombard F, Heier A, Morvan F et al (2014) Galphai2 signaling is required for skeletal muscle growth, regeneration, and satellite cell proliferation and differentiation. Mol Cell Biol 34:619–630
37. Minetti GC, Feige JN, Rosenstiel A, Bombard F, Meier V et al (2011) Galphai2 signaling promotes skeletal muscle hypertrophy, myoblast differentiation, and muscle regeneration. Sci Signal 4:ra80
38. Takefuji M, Wirth A, Lukasova M, Takefuji S, Boettger T et al (2012) G(13)-mediated signaling pathway is required for pressure overload-induced cardiac remodeling and heart failure. Circulation 126:1972–1982
39. Scholz N, Monk KR, Kittel RJ, Langenhan T (2016) Adhesion GPCRs as a putative class of metabotropic mechanosensors. In: Langenhan T, Schöneberg T (eds) Adhesion G protein-coupled receptors: molecular, physiological and pharmacological principles in health and disease. Springer, Heidelberg

40. McClung JM, Lee WJ, Thompson RW, Lowe LL, Carson JA (2003) RhoA induction by functional overload and nandrolone decanoate administration in rat skeletal muscle. Pflugers Arch 447:345–355
41. Gordon SE, Fluck M, Booth FW (2001) Selected contribution: skeletal muscle focal adhesion kinase, paxillin, and serum response factor are loading dependent. J Appl Physiol 90: 1174–1183, discussion 1165
42. Castellani L, Salvati E, Alema S, Falcone G (2006) Fine regulation of RhoA and Rock is required for skeletal muscle differentiation. J Biol Chem 281:15249–15257
43. Shin J, McFarland DC, Velleman SG (2013) Migration of turkey muscle satellite cells is enhanced by the syndecan-4 cytoplasmic domain through the activation of RhoA. Mol Cell Biochem 375:115–130
44. Wu EH, Tam BH, Wong YH (2006) Constitutively active alpha subunits of G(q/11) and G (12/13) families inhibit activation of the pro-survival Akt signaling cascade. FEBS J 273: 2388–2398
45. Bhasin S, He EJ, Kawakubo M, Schroeder ET, Yarasheski K et al (2009) N-terminal propeptide of type III procollagen as a biomarker of anabolic response to recombinant human GH and testosterone. J Clin Endocrinol Metab 94:4224–4233
46. Adams GR, McCue SA (1998) Localized infusion of IGF-I results in skeletal muscle hypertrophy in rats. J Appl Physiol 84:1716–1722
47. Pal M, Febbraio MA, Whitham M (2014) From cytokine to myokine: the emerging role of interleukin-6 in metabolic regulation. Immunol Cell Biol 92:331–339
48. Petersen AM, Pedersen BK (2005) The anti-inflammatory effect of exercise. J Appl Physiol 98:1154–1162
49. Bostrom P, Wu J, Jedrychowski MP, Korde A, Ye L et al (2012) A PGC1-alpha-dependent myokine that drives brown-fat-like development of white fat and thermogenesis. Nature 481: 463–468
50. Rao RR, Long JZ, White JP, Svensson KJ, Lou J et al (2014) Meteorin-like is a hormone that regulates immune-adipose interactions to increase beige fat thermogenesis. Cell 157: 1279–1291

Adhesion GPCR Function in Pulmonary Development and Disease

Marie-Gabrielle Ludwig, Klaus Seuwen, and James P. Bridges

Graphical Abstract

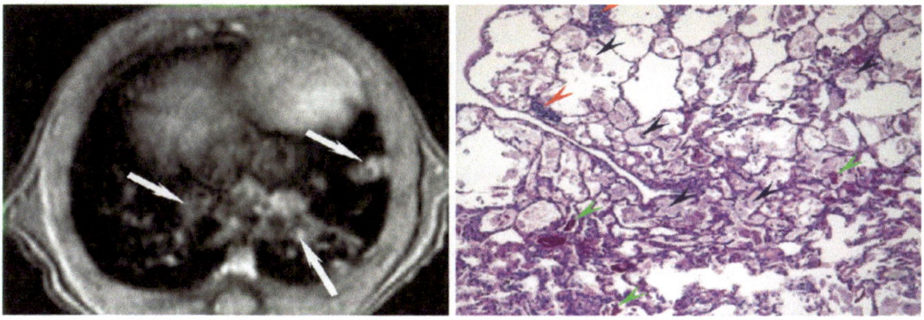

Contents

M.-G. Ludwig • K. Seuwen
Novartis Institutes for Biomedical Research, Basel 4056, Switzerland

J.P. Bridges, PhD (✉)
Department of Pediatrics, Section of Pulmonary Biology, Perinatal Institute, Cincinnati Children's Hospital Medical Center, Cincinnati, OH 45229, USA

Division of Pulmonary Biology, Perinatal Institute, Cincinnati Children's Hospital Medical Center, 3333 Burnet Avenue, ML7029, Cincinnati, OH 45229, USA
e-mail: James.Bridges@cchmc.org

T. Langenhan, T. Schöneberg (eds.), *Adhesion G Protein-coupled Receptors*,
Handbook of Experimental Pharmacology 234, DOI 10.1007/978-3-319-41523-9_14

Abstract

Classic G-protein-coupled receptors (GPCRs) control multiple aspects of pulmonary physiology as demonstrated by loss-of-function experiments in mice and pharmacologic targeting of GPCRs for treatment of several pulmonary diseases. Emerging data demonstrate critical roles for members of the adhesion GPCR (aGPCR) family in pulmonary development, homeostasis, and disease. Although this field is still in its infancy, this chapter will review all available data regarding aGPCRs in pulmonary biology, with a particular focus on the aGPCR for which the most substantial data to date exist: Adgrf5.

Keywords

Lung development • Lung homeostasis • Lung disease • COPD • Pulmonary surfactant • Alveolar epithelium

1 Introduction: Overview of Lung Development

The lung is comprised of approximately 40 distinct cell types that are derived from the endoderm (epithelial cells) or the mesoderm (fibroblasts, vascular and airway smooth muscle cells, pericytes, endothelial cells, and mesothelial cells). A complex interplay of multiple signaling pathways between epithelial and mesenchymal cells is essential for branching morphogenesis and maturation of the lung (reviewed in [1]). Anatomically speaking, the lung can be broadly parsed into two compartments: the proximal and distal lung. Epithelial cell types within the proximal lung include basal cells, club cells, ciliated cells, mucous cells, and neuroendocrine cells, while the distal lung epithelium consists of two epithelial cell types, type I and type II epithelial cells. Structural and functional maturation of the developing distal lung is required for the transition of a fluid-filled fetal lung to one that is equipped for air breathing at birth. Key processes during this period of development include: (1) thinning of the mesenchyme within the presumptive alveolar septae, culminating in a tight juxtaposition of the capillary vascular bed with the distal epithelium to promote efficient gas exchange at birth, (2) differentiation of alveolar type 1 (AT1) and type 2 epithelial cells (AT2), and (3) maturation of the pulmonary surfactant system within AT2 epithelial cells. Defects in any one of these processes contribute to pulmonary dysfunction at birth or in early perinatal life.

At birth, the lung must rapidly transition from a fluid-filled organ to one that is compatible with air breathing. This multifaceted process is achieved through increased respiratory rates, closing of the ductus arteriosus resulting in increased blood flow through the pulmonary vasculature, adsorption of fluid from the airways and airspaces via active trans-epithelial fluid transport and lymphatic drainage, and increased secretion of pulmonary surfactant from AT2 cells. Surfactant is a mixture of 90 % lipids/10 % protein by weight and is unequivocally required for air breathing as evidenced by the fact that premature infants are susceptible to respiratory distress at birth due to insufficient surfactant levels in the distal airspaces. Following birth, the lung continues to grow and mature as nascent alveoli are formed in the distal lung, a process that continues throughout childhood and adolescence in humans [2]. Recent studies have provided insight into the multiple stem/progenitor cell populations that exist in the fetal lung to promote branching morphogenesis and differentiation and in the adult lung to mediate homeostasis and repair following injury (reviewed in [3]). The potential role of aGPCR family members in these developmental, homeostatic, and reparative processes in the lung is just beginning to be explored.

2 Non-adhesion GPCRs in Lung Homeostasis and Disease

Several classic GPCRs, including members of the adrenergic, muscarinic, protease-activated receptor (PAR), endothelin, and purinergic families, regulate various aspects of pulmonary physiology including airway and vascular tone, mucous production, and modulation of the immune system (reviewed in [4–6]). The clinical importance of the adrenergic and muscarinic GPCR families in pulmonary disease is exemplified by the fact that agonists and antagonists for this family of receptors comprise the mainstay therapy for chronic lung diseases including chronic obstructive pulmonary disease (COPD) and asthma [4, 7].

Recent data implicate bitter taste GPCRs, specifically those in the TAS2R family, as regulators of airway smooth muscle tone in mice and humans [8, 9]. Bitter taste GPCRs are expressed on solitary chemosensory epithelial cells lining the digestive and respiratory tract, ciliated cells in the upper airways, airway smooth muscle cells, and on subsets of immune cells [10]. Stimulation of TAS2Rs with agonists including saccharin, chloroquine, or denatonium elicited increased intracellular calcium in a Gβγ-, PLCβ-, and IP3-receptor-dependent manner that resulted in relaxation of isolated smooth muscle cells and decreased airway contraction in a mouse model of allergen-induced asthma [8]. This area of research represents an interesting and unexpected role for bitter taste GPCRs in airway physiology, opening up potential new avenues for pharmacologic intervention in pulmonary disease.

3 Adhesion GPCR Expression and Function in the Developing Lung and in Adult AT2 Cells

Review of the published literature for all identified aGPCRs revealed that three receptors in this GPCR subfamily are expressed in the developing lung including Adgrf5, Adgra2, and Adgrc1 [11–13]. To determine the repertoire of aGPCR expression in distal pulmonary epithelial cells, we performed a GPCR screen on isolated murine AT2 cells using a microfluidic card-based approach. As shown in Fig. 1a, several aGPCRs were detected in wild type (WT) AT2 cells, in addition to Adgfr5, including Adgrg1, Adgre5, Adgrc1, Adgrl2, Adgrl4, Adgra3, Adgre1, Adgrg3, Adgrd1, Adgra2, Adgrl3, and Adgrc3. Of the aGPCRs that were detected at a level consistent with potential functionality (Adgrf5, Adgrg1, Adgre5, and Adgrc1), two have reported lung phenotypes following gene deletion: Adgrf5 and Adgrc1 [11, 14]. We performed qPCR analysis on separate samples of murine WT AT2 cells (1) to confirm the top five expressing aGPCRs identified in the microfluidic screen data, (2) to analyze an aGPCR that was absent from the chip (Adgrg6), and (3) to determine the relative expression levels of the G alpha protein family members Gnaq, Gna11, Gna12, Gna13, and Gna15. As shown in Fig. 1b, the top five highly expressed aGPCRs identified in the microfluidic screen were confirmed by conventional qPCR analysis. Of the Gna proteins, Gna11 is the most abundant isoform, followed by Gna13, Gna12, and Gnaq. Gna15 mRNA levels are barely detectable in AT2 cells (Fig. 1c), suggesting little if any functionality of this Gna family member in AT2 cell function. Below is a summary of the current published data regarding the role of aGPCRs in pulmonary fetal and perinatal development.

3.1 Adgrf5 (Gpr116)

Adgrf5 mRNA is developmentally regulated in the murine lung, reaching peak expression in newborn animals with maintenance of high expression levels throughout adulthood [11]. The sharp increase in Adgrf5 expression in the late gestation lung is temporally consistent with maturation of the surfactant system in AT2 epithelial cells. Regarding cell-specific expression, Adgrf5 mRNA is readily detected in pulmonary endothelial cells and alveolar epithelial cells. Expression of Adgrf5 protein in AT2 cells has been demonstrated by multiple groups [11, 15, 16], while Adgrf5 protein expression in pulmonary endothelial cells remains controversial; Adgrf5 protein has been detected in areas consistent with endothelial cells in mouse and rat lung sections using immunohistochemical staining approaches [15, 16], while Adgrf5 protein was undetectable in isolated CD31+ primary mouse lung endothelial cells [11]. Interestingly, a recent report demonstrated a role for Adgrf5 in regulating blood-brain barrier permeability and retinal regrowth following oxygen-induced injury [17]. A definitive role for Adgrf5 in pulmonary endothelial cells has yet to be determined.

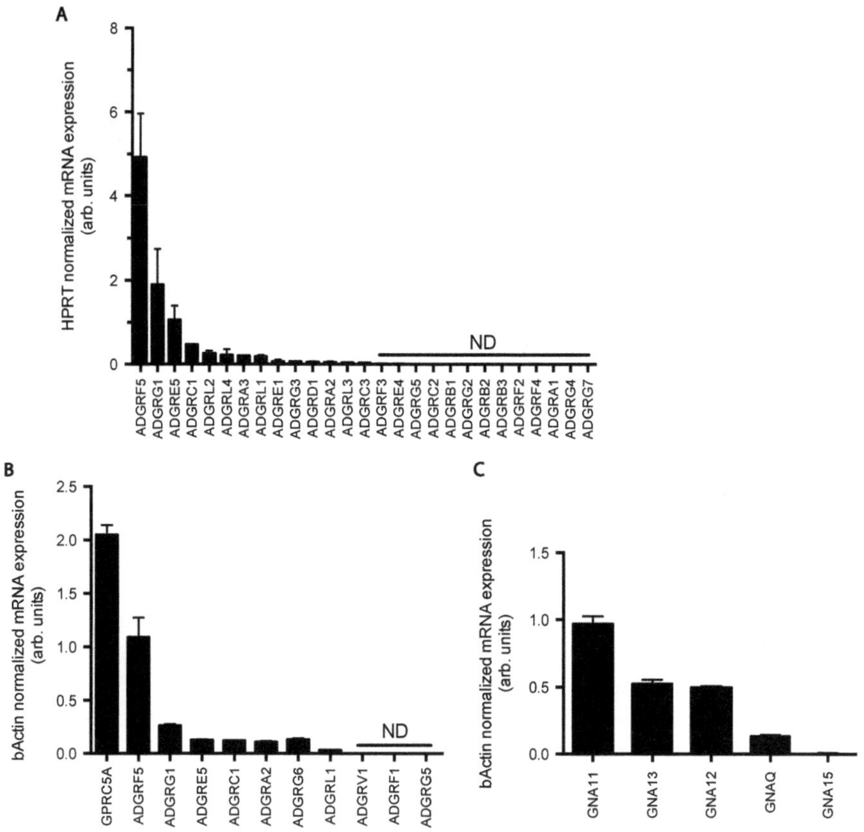

Fig. 1 aGPCR and G-protein alpha mRNA expression in alveolar type 2 (AT2) epithelial cells. (**a**) aGPCR expression in primary wild type, murine AT2 cells isolated from 4-week-old animals as determined by qPCR-based microfluidic card platform (TaqMan® Array Mouse GPCR Panel, Thermo Fisher Scientific). Samples were analyzed in triplicate and data were normalized to HPRT expression. *ND* not detected. (**b**) Taqman-based qPCR validation of aGPCR expression in murine AT2 cells. Samples were analyzed in triplicate and data were normalized to β-actin expression. (**c**) Expression of selected G-protein alpha subunit mRNAs in murine AT2 cells by TaqMan®-based qPCR method. Samples were analyzed in triplicate and data were normalized to β-actin expression

Despite the developmental increase in Adgrf5 mRNA expression in whole lung tissue and enrichment of Adgrf5 in AT2 cells, fetal lung development is normal in the absence of Adgrf5 and Adgrf5$^{-/-}$; mice are born at expected Mendelian ratios. While Adgrf5$^{-/-}$ mice do not display any overt respiratory phenotype at birth or in the early postnatal period, the pulmonary phenotype in juvenile and adult mice is profound and is discussed in detail in the following section.

3.2 Adgrg6 (Gpr126)

In addition to expression in the heart and peripheral nervous system, Adgrg6 mRNA is detected at high levels in rat and mouse lungs [12, 13]. In the developing lung, Adgrg6 mRNA expression appears to be enriched pericytes and smooth muscle cells (mural cells) at embryonic day 16.5 (E16.5) (unpublished data, https://research.cchmc.org/pbge/lunggens/). Germline deletion of Adgrg6 in mice results in uniform embryonic lethality associated with cardiac defects in one line [18] and partial embryonic lethality in a separate line [19]. Recent reports have identified laminin-211 as an Adgrg6 ligand in the developing peripheral nervous system of mice [20]. While Adgrg6 is known to be important for peripheral nervous system and heart development, the cell-specific expression pattern and potential function of this receptor in pulmonary tissues has yet to be explored.

3.3 Adgrc1 (Celsr1)

Adgrc1 has been implicated in planar cell polarity and is abundantly expressed in the embryonic murine lung. While Adgrc1$^{-/-}$ mice are viable and devoid of gross respiratory deficits, embryonic Adgrc1$^{-/-}$ mice have hypoplastic lungs characterized by condensed mesenchyme and decreased distal airspaces [14]. Alveolar type II cell maturation, as demonstrated by expression of the differentiation marker surfactant protein C, appeared to be normal although a thorough assessment of AT2 epithelial cell maturation was not performed in this particular study.

4 Adhesion GPCR Function in Lung Cancer and Disease

According to the American Cancer Society, lung cancer is the second most common type of cancer in both men and women and is the leading cause of cancer death in both sexes. To date, six aGPCRs have been associated with human lung cancer, either in primary tissues or in transformed cell lines. Adgrg2 (Gpr64) expression was found to be upregulated in human lung carcinomas and in transformed lung cancer cell lines; inhibition of Adgrg2 led to decreased cell growth and metastasis, associated with reduced cellular matrix protein degradation and invasiveness in vitro [21]. Similarly, Adgrf1 (Gpr110) has been identified as an oncogene in a mouse mutagenesis study and was found to be overexpressed in pulmonary adenocarcinoma specimens in two separate studies [22, 23]. A high frequency of somatic mutations in Adgrg5 (Gpr114) have been associated with air pollution-related lung cancer in a cohort of patients in China [24]. Expression of Adgrb1 (Bai1) mRNA was detected in resected pulmonary adenocarcinoma specimens and expression levels correlated with high vascular density in comparison to Adgrb1-negative carcinomas [25]. Adgrb3 (Bai3), in combination with CDX2 and VIL1 expression, serves as a diagnostic lung cancer biomarker, differentiating highly aggressive small cell lung carcinomas (SCLC) from large-cell neurocrine carcinomas (LCNEC)

[26]. In addition to its role in alveolar homeostasis, Adgrf5 has also been associated with lung cancer as it was found to be overexpressed in adenocarcinomas [27]. The precise role of the aforementioned aGPCRs in the initiation and/or progression of lung cancer remains unknown and requires further mechanistic studies.

In addition to the association of certain aGPCRs with lung tumorigenesis, three aGPCR family members, Adgrg6, Adgrg1, and Adgrf5, have been associated with pulmonary disease in humans and/or have been experimentally proven to be regulators of pulmonary homeostasis in animal models. Below is a summary of these aGPCRs in relation to perinatal and adult lung disease.

4.1 Adgrg6 (Gpr126)

As stated previously, Adgrg6 mRNA is highly expressed in embryonic and perinatal mouse lung tissue with enrichment in pericytes and smooth muscle cells at E16.5. Interestingly, two independent studies have reported single nucleotide polymorphisms (SNPs) in the Adgrg6 locus associated with severe chronic obstructive pulmonary disease (COPD) in humans [28, 29]. While a specific lung phenotype has not been reported in Adgrg6$^{-/-}$ mice, based on human genetic data, it is interesting to speculate that Adgrg6 may play an important role in adult pulmonary function and/or disease.

4.2 Adgrg1 (Gpr56)

Our aGPCR expression screen determined that Adgrg1 mRNA is highly expressed in murine AT2 cells (Fig. 1a, b). Loss-of-function studies demonstrated that Adgrg1-null mice are viable and display central nervous system defects that closely recapitulate those observed in human patients with mutations in the Adgrg1 locus [30]. While pulmonary defects in Adgrg1$^{-/-}$ mice have not been reported, decreased expression of Adgrg1 has been associated with extracellular matrix production and migration in primary fibroblasts isolated from patients with idiopathic pulmonary fibrosis (IPF), a devastating lung disease with a mean survival rate of approximately 5 years [31]. Furthermore, methylation of the Adgrg1 promoter was found in the lung adenocarcinoma epithelial cell line A549 treated with the chemotherapeutic cisplatin [32], suggesting that modulation of Adgrg1 levels may contribute to fibroblast migration/invasion in the context of lung fibrosis, and/or tumorigenesis when repressed in pulmonary epithelial cells.

4.3 Adgrf5 (Gpr116)

To ascertain the function of Adgrf5, transgenic mice were generated in which exon17 of Adgrf5, encoding the seven transmembrane domains of the receptor, was deleted from the germline (Adgrf5$^{\Delta exon17}$). Homozygous loss of Adgrf5 in this model resulted in a fully penetrant, rapid, and progressive accumulation of pulmonary surfactant in the distal airspaces and lung tissue [11] (Fig. 2). This phenotype is accompanied by massive inflammatory cell infiltrates, including lipid-laden alveolar macrophages, neutrophils, and B-cells, as evidenced by histopathological analysis and the predominance of inflammatory gene induction in microarray analysis comparing WT and Adgrf5$^{-/-}$ AT2 cells from our previous study [11]. In Adgrf5$^{\Delta exon17}$ mice, increases in the predominant surfactant phospholipid species, dipalmitoylphosphatidylcholine (SatPC), were first detected as early as 2 days of age. Augmented alveolar and tissue surfactant phospholipid pools correlated with age in Adgrf5$^{\Delta exon17}$ mice, associated with lipid-laden ("foamy") alveolar macrophages, increased inflammatory infiltrates and alveolar simplification (emphysematous-like structural alterations), ultimately resulting in compromised gas exchange and respiratory distress in aged mice. The primary pulmonary pathological features in Adgrf5$^{\Delta exon17}$ mice were remarkably consistent with three other mouse models of Adgrf5 deficiency in which distinct exons were targeted [16, 17, 33].

Analysis of surfactant lipid and protein composition in Adgrf5$^{-/-}$ mice revealed commonalities as well as differences between the distinct models. For example, all of the major phospholipid species present in alveolar surfactant, namely, phosphatidylcholine (PC), phosphatidylglycerol (PG), phosphatidylserine (PS),

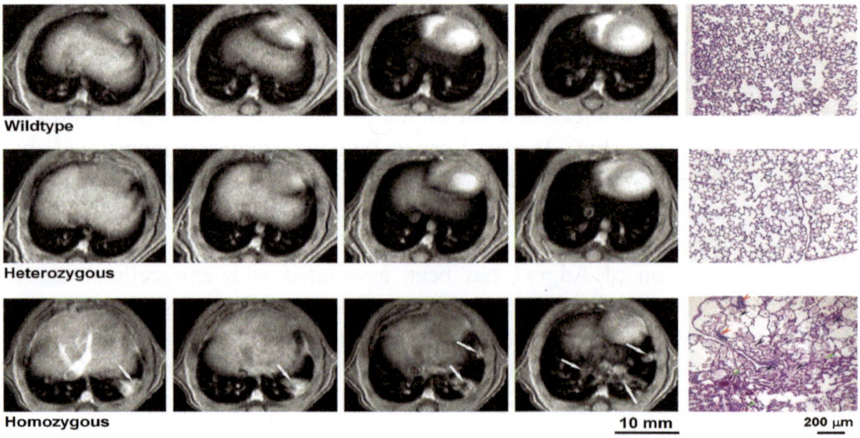

Fig. 2 Pulmonary imaging and pathology of adult Adgrf5$^{-/-}$ mice. Analysis of Adgrf5$^{-/-}$ mice at 3 months of age by magnetic resonance imaging (MRI) and histology. (*Left*) Axial MRI images from three representative mice. Acquisitions were performed at 4.7 T on spontaneously breathing, isoflurane-anesthetized animals without administration of contrast material. A single-slice image (0.75 mm thickness) was acquired in 74 seconds. Fluid signals (*white arrows*) were seen only in homozygous Adgrf5$^{-/-}$ mice. (*Right*) Corresponding PAS-stained histological slices. Lung anatomy was normal in wild-type and heterozygous animals. In contrast, eosinophilic amorphous material in the alveoli (*green arrowheads*), phagocytic macrophages (*blue arrowheads*), and infiltration foci (*red arrowheads*) were detected in the lungs of homozygous Adgrf5$^{-/-}$ knockout mice

phosphatidylinositol (PI), and phosphatidylethanolamine (PE), were increased in both Adgrf5$^{-/-}$ mouse lines for which they were reported, with modest increases noted in PS, PI, and PE ratios in one line [16] and lyso-PC in another [11] between knockout and control animals. Interestingly, a mass spectrometry-based lipidomic analysis showed alterations in the saturation status of individual PC and lyso-PC species in Adgrf5$^{-/-}$ surfactant, with a marked shift from the saturated and monosaturated species normally found in wild type surfactant to polyunsaturated species [16]. Large increases in cholesterol neutral lipid were also detected in Adgrf5$^{-/-}$ lavage fluid of adult mice [16]. It remains to be determined if alterations in the saturation status of surfactant phospholipids or increased cholesterol levels in Adgrf5$^{-/-}$ mice impact surfactant function.

In contrast to the lipidosis phenotype, surfactant protein levels differed across the Adgrf5$^{-/-}$ models. In two of the knockout lines, surfactant protein levels were increased proportionate to phospholipid levels, while in two other lines, surfactant proteins appeared discordant with phospholipid levels [11, 16]. The reason for the reported discrepancy is unknown but may be related to differences in sample isolation, preparation, and/or normalization procedures among individual labs. Irrespective of surfactant protein levels, mRNA levels for the four major surfactant proteins were unchanged in Adgrf5$^{-/-}$ mice [11, 16, 34]. In addition to the phospholipidosis phenotype, Ariestanti et al. have reported that the emphysematous-like alterations in Adgrf5$^{-/-}$ mice were associated with increased oxidative stress, phospholipid oxidation and increased expression of activated MMP2 and MMP9, two metalloproteinases associated with macrophage activation, and lung remodeling [33]. The temporal order of the pathological sequelae downstream of alveolar surfactant accumulation in the lungs of Adgrf5$^{-/-}$ mice greater than 1 month of age has yet to be determined.

Epithelial-specific deletion of Adgrf5 using Adgrf5$^{\text{flox/flox}}$ mice mated to mice carrying *Sftpc*-rtTA and *TetO7*-Cre alleles, which specifically targets AT2 cells [16], or to mice with an *Shh*Cre allele (our unpublished data), which uniformly targets all epithelial lineages in the lung, results in a phenotype nearly identical to that observed in Adgrf5$^{-/-}$ mice. Surprisingly, given the high level of Adgrf5 mRNA expression in endothelial cells, deletion of Adgrf5 in this lineage had no effect on pulmonary or alveolar homeostasis at the histological level [16]; likewise, alveolar surfactant phospholipid levels were normal in 4-week-old mice in which Adgrf5 was deleted in endothelial cells using the *Tie2*Cre driver (our unpublished data). Furthermore, bone marrow transplant of WT donor cells into Adgrf5$^{-/-}$ hosts failed to rescue the lung phenotype [16]. Taken together, these data demonstrate that cell-specific expression of Adgrf5 in alveolar epithelial cells is critical for alveolar homeostasis.

Pulmonary surfactant levels within the lumen of the alveolus are tightly controlled through a balance of three interconnected mechanisms: (1) surfactant synthesis and secretion by the AT2 cell; (2) surfactant uptake, recycling, and secretion by the AT2 cell; and (3) surfactant uptake and catabolism by alveolar macrophages. Perturbations in any one of these three mechanisms can result in either surfactant overload or surfactant deficiency in the distal lung. The surfactant phospholipidosis phenotype in Adgrf5$^{-/-}$ mice is phenotypically similar to that observed in mice homozygous deficient for Sftpd [35, 36], Csf [37], Csf2rb [38], or Slc34a2 [39]; however, the severity and mechanistic basis of the phenotype varies significantly between the models with Adgrf5$^{-/-}$ and Slc34a2$^{-/-}$ being the most severe. Patients with mutations in Csf2rb in Csf2ra, or in those that autoantibodies against Csf2, present with a surfactant overload pathology that closely resembles the disease seen in correspondingly targeted mice, a rare disease known as pulmonary alveolar proteinosis (PAP) (for review, see [40]). Studies in genetically modified mice, and in primary cells from patients with PAP, have shown that surfactant accumulation related to defects in the Csf2 signaling pathway are due to alveolar macrophage dysfunction resulting in impaired surfactant uptake/catabolism [41, 42]. Conversely, in vivo surfactant catabolism studies in Sftpd$^{-/-}$ mice have demonstrated that the primary defect lies in decreased surfactant uptake by alveolar epithelial cells, owed to altered ultrastructure of surfactant in the airway [43, 44].

The endogenous ligand and intracellular pathways by which Adgrf5 signals in AT2 cells to control alveolar homeostasis have yet to be identified. In the transformed human breast cancer cell line MDA-MB-231, endogenous Adgfr5 regulates cell motility, F-actin dynamics, and mammary tumor metastases in a Gq-p63RhoGEF-RhoA-dependent manner [45]. Whether these intracellular pathways are operative in AT2 cells downstream of Adgrf5 to modulate surfactant secretion and/or uptake remains unknown. Hirose et al. have postulated that Sftpd may be a putative ligand for Adgrf5 based on co-immunoprecipitation of Sftpd with Adgrf5 in a heterologous overexpression system [34]. The fact that Sftpd$^{-/-}$ mice and Adgrf5$^{-/-}$ mice display a similar pulmonary phenotype, albeit at differing degrees of severity, and the abundance of Sftpd protein in the alveolar lumen supports this hypothesis. However, transgenic mice that overexpress Sftpd in the distal lung do not have decreased alveolar surfactant pools nor do they exhibit signs of respiratory distress [46], as would be predicted if Sftpd activated Adgfr5 signaling based on Adgrf5 loss-of-function data. In addition, it remains to be determined if purified Sftpd protein is sufficient to activate G-protein signaling downstream of Adgrf5 in transfected cells or in primary AT2 cells.

5 Transcriptional Responses in Adgrf5$^{-/-}$ Lung Tissue

We undertook an additional microarray analysis of Adgrf5$^{-/-}$ lung tissue from 3.5-month-old mice versus WT controls in order to determine the transcriptional responses in Adgrf5$^{-/-}$ lungs at the whole lung level (Table 1). In addition to induction of inflammatory pathways consistent with previously published data [33], several additional physiological responses were identified including increased expression of genes involved in lipid synthesis and vesicle secretion (LIPA, LRP2, LCN2, FABP5, PLIN2, LPL, FADS2, SCD1, LPCAT, members of the SLC27A and SCAMP families, and the V-ATPase ATP6V), as well as increased expression of CSF2, CSF2RA, two genes known to be essential for macrophage function, and surfactant uptake/catabolism in both mice and humans [37, 47, 48].

Table 1 Genes differentially expressed in adult Adgrf5$^{-/-}$ lung tissue

Gene symbol	Adgrf5 KO/WT log fold change	Description	No. probe sets[a]
Aqp11	−1.3	Aquaporin 11	1
Aqp3	2.2	Aquaporin 3	2
Aqp9	2.4	Aquaporin 9	2
Atp1a1	0.5	ATPase, Na+/K+ transporting, alpha 1 polypeptide	1
Atp1a2	−0.1	ATPase, Na+/K+ transporting, alpha 2 polypeptide	4
Atp1a3	3.5	ATPase, Na+/K+ transporting, alpha 3 polypeptide	1
Atp1a4	−0.7	ATPase, Na+/K+ transporting, alpha 4 polypeptide	1
Atp1b1	0.1	ATPase, Na+/K+ transporting, beta 1 polypeptide	1
Atp1b2	0.4	ATPase, Na+/K+ transporting, beta 2 polypeptide	1
Atp1b3	0.2	ATPase, Na+/K+ transporting, beta 3 polypeptide	1
Atp2a2	−0.8	ATPase, Ca++ transporting, cardiac muscle, slow twitch 2	2
Atp2b1	−0.2	ATPase, Ca++ transporting, plasma membrane 1	1
Atp2b4	−1.3	ATPase, Ca++ transporting, plasma membrane 4	1
Atp2c1	−0.6	ATPase, Ca++-sequestering	4
Atp6ap1	0.2	ATPase, H+ transporting, lysosomal accessory protein 1	1
Atp6ap2	0.2	ATPase, H+ transporting, lysosomal accessory protein 2	2
Atp6v0a1	0.6	ATPase, H+ transporting, lysosomal V0 subunit A1	3
Atp6v0b	0.4	ATPase, H+ transporting, lysosomal V0 subunit B	3
Atp6v0c	0.6	ATPase, H+ transporting, lysosomal V0 subunit C	3
Atp6v0d1	0.6	ATPase, H+ transporting, lysosomal V0 subunit D1	1
Atp6v0d2	1.6	ATPase, H+ transporting, lysosomal V0 subunit D2	2
Atp6v0e2	0.3	ATPase, H+ transporting, lysosomal V0 subunit E2	1
Atp6v1a	0.4	ATPase, H+ transporting, lysosomal V1 subunit A	2
Atp6v1b2	0.8	ATPase, H+ transporting, lysosomal V1 subunit B2	3
Atp6v1c1	0.7	ATPase, H+ transporting, lysosomal V1 subunit C1	3
Atp6v1c2	0.4	ATPase, H+ transporting, lysosomal V1 subunit C2	1
Atp6v1e1	0.4	ATPase, H+ transporting, lysosomal V1 subunit E1	3
Atp6v1f	0.5	ATPase, H+ transporting, lysosomal V1 subunit F	1
Atp6v1g1	0.3	ATPase, H+ transporting, lysosomal V1 subunit G1	1
Cacna1a	0.4	Calcium channel, voltage-dependent, P/Q type, alpha 1A subunit	1
Cacna1c	−0.3	Calcium channel, voltage-dependent, L type, alpha 1C subunit	1
Cacna1d	−0.5	Calcium channel, voltage-dependent, L type, alpha 1D subunit	1
Cacna2d1	−0.7	Calcium channel, voltage-dependent, alpha 2/delta subunit 1	5
Cacna2d4	−0.8	Calcium channel, voltage-dependent, alpha 2/delta subunit 4	1

(continued)

Table 1 (continued)

Gene symbol	Adgrf5 KO/WT log fold change	Description	No. probe sets[a]
Cacnb2	−0.3	Calcium channel, voltage-dependent, beta 2 subunit	1
Cacnb3	0.7	Calcium channel, voltage-dependent, beta 3 subunit	1
Cacnb4	−1.4	Calcium channel, voltage-dependent, beta 4 subunit	2
Car10	0.6	Carbonic anhydrase 10	1
Car13	1.2	Carbonic anhydrase 13	3
Car14	0.5	Carbonic anhydrase 14	1
Car2	0.2	Carbonic anhydrase 2	1
Car4	−0.6	Carbonic anhydrase 4	2
Car5b	0.4	Carbonic anhydrase 5b, mitochondrial	1
Car7	0.2	Carbonic anhydrase 7	1
Car8	0.3	Carbonic anhydrase 8	3
Cftr	−0.8	Cystic fibrosis transmembrane conductance regulator homologue	1
Clca1	1.2	Chloride channel calcium activated 1	3
Clca3	3.8	Chloride channel calcium activated 3	2
Clca5	1.6	Chloride channel calcium activated 5	1
Clcc1	−0.8	Chloride channel CLIC-like 1	1
Clcn5	0.6	Chloride channel 5	3
Clcn7	0.6	Chloride channel 7	1
Clic1	0.3	Chloride intracellular channel 1	1
Clic4	−0.1	Chloride intracellular channel 4 (mitochondrial)	2
Clic5	−0.3	Chloride intracellular channel 5	4
Clic6	0.3	Chloride intracellular channel 6	2
Ece1	−1.2	Endothelin converting enzyme 1	1
Ece2	0.9	Endothelin converting enzyme 2	1
Edn1	0.5	Endothelin 1	1
Ednra	−1.0	Endothelin receptor type A	3
Ednrb	−1.6	Endothelin receptor type B	3
Fxyd2	1.0	FXYD domain-containing ion transport regulator 2	3
Fxyd3	0.3	FXYD domain-containing ion transport regulator 3	1
Fxyd4	2.5	FXYD domain-containing ion transport regulator 4	1
Fxyd5	0.6	FXYD domain-containing ion transport regulator 5	1
Fxyd6	1.0	FXYD domain-containing ion transport regulator 6	1
Hvcn1	2.1	Hydrogen voltage-gated channel 1	1
Kcnab1	0.4	Potassium voltage-gated channel, shaker-related subfamily, beta member 1	1
Kcnab2	1.7	Potassium voltage-gated channel, shaker-related subfamily, beta member 2	1
Kcnd3	−0.5	Potassium voltage-gated channel, Shal-related family, member 3	1
Kcne2	−0.5	Potassium voltage-gated channel, Isk-related subfamily, gene 2	1

(continued)

Table 1 (continued)

Gene symbol	Adgrf5 KO/WT log fold change	Description	No. probe sets[a]
Kcnf1	−1.4	Potassium voltage-gated channel, subfamily F, member 1	1
Kcnh1	0.4	Potassium voltage-gated channel, subfamily H (eag-related), member 1	1
Kcnip4	−0.9	Kv channel interacting protein 4	3
Kcnj13	−0.6	Potassium inwardly-rectifying channel, subfamily J, member 13	1
Kcnj15	−1.3	Potassium inwardly-rectifying channel, subfamily J, member 15	1
Kcnj5	-0.6	Potassium inwardly-rectifying channel, subfamily J, member 5	1
Kcnj8	0.4	Potassium inwardly-rectifying channel, subfamily J, member 8	1
Kcnk1	0.2	Potassium channel, subfamily K, member 1	1
Kcnk13	0.3	Potassium channel, subfamily K, member 13	1
Kcnk2	−1.0	Potassium channel, subfamily K, member 2	1
Kcnmb2	−0.5	Potassium large conductance calcium-activated channel, subfamily M, beta member 2	2
Kcnn4	3.1	Potassium intermediate/small conductance calcium-activated channel, subfamily N, member 4	2
Kcnq1	0.4	Potassium voltage-gated channel, subfamily Q, member 1	1
Kcnq4	−1.3	Potassium voltage-gated channel, subfamily Q, member 4	1
Kcnq5	−1.6	Potassium voltage-gated channel, subfamily Q, member 5	2
Kcnt1	−2.8	Potassium channel, subfamily T, member 1	1
Kctd1	0.4	Potassium channel tetramerization domain containing 1	1
Kctd11	0.3	Potassium channel tetramerization domain containing 11	1
Kctd17	0.6	Potassium channel tetramerization domain containing 17	1
Kctd18	−0.7	Potassium channel tetramerization domain containing 18	2
Kctd20	0.3	Potassium channel tetramerization domain containing 20	1
Kctd3	−0.7	Potassium channel tetramerization domain containing 3	1
Kctd5	0.2	Potassium channel tetramerization domain containing 5	2
Kctd6	−0.1	Potassium channel tetramerization domain containing 6	1
Kctd8	−1.4	Potassium channel tetramerization domain containing 8	1

(continued)

Table 1 (continued)

Gene symbol	Adgrf5 KO/WT log fold change	Description	No. probe sets[a]
Kctd9	−0.2	Potassium channel tetramerization domain containing 9	1
Nppb	2.4	Natriuretic peptide precursor type B	1
Nppc	1.3	Natriuretic peptide precursor type C	1
Npr2	−0.4	Natriuretic peptide receptor 2	1
Npr3	−1.1	Natriuretic peptide receptor 3	3
Ptgds	0.8	Prostaglandin D2 synthase (brain)	1
Ptger2	0.2	Prostaglandin E receptor 2 (subtype EP2)	1
Ptger3	1.0	Prostaglandin E receptor 3 (subtype EP3)	2
Ptger4	0.8	Prostaglandin E receptor 4 (subtype EP4)	2
Ptges	1.1	Prostaglandin E synthase	3
Ptges3	−0.2	Prostaglandin E synthase 3 (cytosolic)	2
Ptgfr	−0.7	Prostaglandin F receptor	3
Ptgir	1.9	Prostaglandin I receptor (IP)	1
Ptgr1	0.3	Prostaglandin reductase 1	1
Ptgs1	1.1	Prostaglandin-endoperoxide synthase 1	2
Ptgs2	0.3	Prostaglandin-endoperoxide synthase 2	1
Sclt1	−2.0	Sodium channel and clathrin linker 1	2
Scn1b	0.6	Sodium channel, voltage-gated, type I, beta	1
Scn2b	0.9	Sodium channel, voltage-gated, type II, beta	1
Scn3a	−1.0	Sodium channel, voltage-gated, type III, alpha	1
Scn3b	−1.6	Sodium channel, voltage-gated, type III, beta	2
Scn7a	−1.4	Sodium channel, voltage-gated, type VII, alpha	3
Scn8a	−3.2	Sodium channel, voltage-gated, type VIII, alpha	1
Scnm1	0.3	Sodium channel modifier 1	1
Slc12a2	−0.5	Solute carrier family 12, member 2	3
Slc12a4	0.4	Solute carrier family 12, member 4	1
Slc12a5	−0.8	Solute carrier family 12, member 5	1
Slc12a6	−0.5	Solute carrier family 12, member 6	3
Slc12a9	0.4	Solute carrier family 12 (potassium/chloride transporters), member 9	1
Slc19a1	0.3	Solute carrier family 19 (sodium/hydrogen exchanger), member 1	1
Slc19a3	−0.8	Solute carrier family 19 (sodium/hydrogen exchanger), member 3	1
Slc26a11	0.4	Solute carrier family 26, member 11	1
Slc26a3	0.5	Solute carrier family 26, member 3	2
Slc26a4	3.8	Solute carrier family 26, member 4	1
Slc27a1	0.8	Solute carrier family 27 (fatty acid transporter), member 1	2
Slc27a2	0.7	Solute carrier family 27 (fatty acid transporter), member 2	1

(continued)

Table 1 (continued)

Gene symbol	Adgrf5 KO/WT log fold change	Description	No. probe sets[a]
Slc4a1	1.9	Solute carrier family 4 (anion exchanger), member 1	2
Slc4a10	−2.1	Solute carrier family 4, sodium bicarbonate cotransporter-like, member 10	1
Slc4a11	0.4	Solute carrier family 4, sodium bicarbonate transporter-like, member 11	1
Slc4a4	−0.4	Solute carrier family 4 (anion exchanger), member 4	3
Slc4a5	−0.5	Solute carrier family 4, sodium bicarbonate cotransporter, member 5	1
Slc9a1	0.6	Solute carrier family 9 (sodium/hydrogen exchanger), member 1	1

[a]The number of probe sets on the genechip for each gene

Alveolar epithelial cells are critical regulators of fluid balance between the surface epithelium and the surfactant layer present at the air liquid interface. In addition to the lipid transcriptional response, several electrolyte channels and solute transporters were differentially expressed in Adgrf5$^{-/-}$ lung tissue, suggesting adaptation of the lung to maintain fluid homeostasis in the absence of Adgrf5. The major electrolytes involved in maintaining homeostasis of the alveolar fluid layer include Na^+, Cl^-, and K^+. Levels of these ions within the lining fluid are primarily regulated by the following channels and transporters: ENAC, CNG, and Na^+/K^+ ATPase for Na^+; CFTR, CLCs, NKKC and KCC transporters, and $HCO3^-/Cl^-$ exchangers for Cl^-; and voltage-dependent, Ca^{2+}-activated, 2-pore domain, and inward-rectifying channels for K^+.

Management of Na^+ flux is the principal regulator of fluid resorption in the distal lung, in conjunction with aquaporin-mediated water transport. Na^+ trans-epithelial transporters (mainly ENAC, CNG, and Na^+/K^+ ATPase), as well as the major aquaporins, were not modulated at the transcriptional level in Adgrf5$^{-/-}$ mice, which correlates with seemingly normal alveolar fluid clearance in these animals following birth [49, 50].

Although K^+ channels are known to play an important role in ion and liquid transport, only KCNQ5 and KCNK2 were decreased significantly, suggesting a minor role in the Adgrf5$^{-/-}$ phenotype. Similarly, classical Cl^- channels known to maintain electrolyte balance in the alveoli (CFTR, CLCs, NKKC and KCC transporters, $HCO3^-/Cl^-$ exchangers) were not strongly modulated in the lungs of Adgrf5$^{-/-}$ mice. In contrasts, the Ca^{2+}-activated Cl^- channel CLCA3, virtually undetectable in the WT lung, was significantly upregulated in Adgrf5$^{-/-}$ lung tissue, suggesting it may influence currents by modulating the activity, rather than the expression, of Ca^{2+}-activated Cl^- channels present in AT2 pneumocytes [51].

Upregulation the $HCO3^-/Cl^-$ exchangers SLC26A4 and SLC4A1 in Adgrf5$^{-/-}$ lungs may play an important role in fluid as well as acid-base homeostasis in the distal lung [52, 53]. The observed higher expression of HVCN1 further supports the hypothesis of an alteration in the pH of the alveolar lining fluid, possibly toward a more alkaline state, and the need for adaptations [54]. The increase in regulatory subunits of V-ATPases may be part of these coordinated adjustments for maintenance of extra- and intracellular pH, leading to acidification of lamellar bodies and potentially participating in the increased synthesis, secretion, and/or recycling of surfactant in Adgrf5$^{-/-}$ lungs [55–57].

With regard to Ca^{2+} management, most voltage-dependent Ca^{2+} channels were slightly suppressed in Adgrf5$^{-/-}$ lungs (CACNA/D). However, the regulatory subunit of voltage-dependent calcium channels, CACNB3, was upregulated, while the Ca^{2+}-sequestering ATPase ATP2C1 was downregulated [58], which may result in increased Ca^{2+} entry and cytosolic concentration in AT2 cells, thereby augmenting Ca^{2+} signaling and Ca^{2+}-dependent intracellular processes including membrane and vesicle trafficking [56, 59]. It remains to be determined if membrane and vesicular trafficking are altered in Adgrf5$^{-/-}$ AT2 cells and if these processes contribute to the phospholipidosis phenotype observed in Adgrf5$^{-/-}$ mice.

In summary, these data highlight important transcriptional responses induced by Adgrf5 deficiency, including the modulation of several ion transporters and channels, possibly functioning to promote proper alveolar fluid levels and pH to maintain tissue homeostasis in the context of Adgrf5 loss. Dissecting the molecular mechanisms responsible for these transcriptional responses in Adgrf5 knockout mice requires further study to assess the functionality of channels through modulation with agonists or antagonists and/or through assessment of alveolar fluid clearance capacity in vitro and in vivo. Early responses following Adgrf5 loss may involve sensing changes in oxygenation levels and/or modulation of fluid acid-base balance, linked directly or indirectly to increased surfactant accumulation and associated sequelae observed in the lungs of Adgrf5$^{-/-}$ mice.

6 Conclusions

The field of aGPCR research is still in its infancy, particularly with respect to pulmonary physiology and disease. To date, a handful of aGPCRs have been associated with human lung disease and numerous questions remain as to the precise role of these particular receptors, and others within this family, in the field of pulmonary biology. The aGPCR for which the most lung data exists is Adgrf5, primarily owed to the surprising robust pulmonary phenotype in loss-of-function mouse models. As additional reagents for aGPCRs are generated, including genetic mouse models, and ligands are identified, the contribution of other aGPCRs to various aspects of pulmonary development, function, and disease are likely to be revealed.

Acknowledgments J.P.B. acknowledges support from the American Heart Association (AHA) 13SDG17090028 and the National Institutes of Health (NIH) HL131634. The authors thank Nicolau Beckmann at Novartis for contributing data presented in Fig. 2.

References

1. Morrisey EE, Cardoso WV, Lane RH, Rabinovitch M, Abman SH et al (2013) Molecular determinants of lung development. Ann Am Thorac Soc 10:S12–S16
2. Narayanan M, Owers-Bradley J, Beardsmore CS, Mada M, Ball I et al (2012) Alveolarization continues during childhood and adolescence: new evidence from helium-3 magnetic resonance. Am J Respir Crit Care Med 185:186–191
3. Hogan BL, Barkauskas CE, Chapman HA, Epstein JA, Jain R et al (2014) Repair and regeneration of the respiratory system: complexity, plasticity, and mechanisms of lung stem cell function. Cell Stem Cell 15:123–138
4. Pera T, Penn RB (2014) Crosstalk between beta-2-adrenoceptor and muscarinic acetylcholine receptors in the airway. Curr Opin Pharmacol 16:72–81
5. Peters T, Henry PJ (2009) Protease-activated receptors and prostaglandins in inflammatory lung disease. Br J Pharmacol 158:1017–1033
6. Alberelli MA, De Candia E (2014) Functional role of protease activated receptors in vascular biology. Vascul Pharmacol 62:72–81
7. Wasilewski NV, Lougheed MD, Fisher JT (2014) Changing face of beta2-adrenergic and muscarinic receptor therapies in asthma. Curr Opin Pharmacol 16:148–156
8. Deshpande DA, Wang WC, McIlmoyle EL, Robinett KS, Schillinger RM et al (2010) Bitter taste receptors on airway smooth muscle bronchodilate by localized calcium signaling and reverse obstruction. Nat Med 16:1299–1304
9. Robinett KS, Koziol-White CJ, Akoluk A, An SS, Panettieri RAJ et al (2014) Bitter taste receptor function in asthmatic and nonasthmatic human airway smooth muscle cells. Am J Respir Cell Mol Biol 50:678–683
10. Kinnamon SC (2012) Taste receptor signalling – from tongues to lungs. Acta Physiol (Oxf) 204:158–168
11. Bridges JP, Ludwig MG, Mueller M, Kinzel B, Sato A et al (2013) Orphan G protein-coupled receptor GPR116 regulates pulmonary surfactant pool size. Am J Respir Cell Mol Biol 49:348–357
12. Moriguchi T, Haraguchi K, Ueda N, Okada M, Furuya T et al (2004) DREG, a developmentally regulated G protein-coupled receptor containing two conserved proteolytic cleavage sites. Genes Cells 9:549–560
13. Haitina T, Olsson F, Stephansson O, Alsio J, Roman E et al (2008) Expression profile of the entire family of Adhesion G protein-coupled receptors in mouse and rat. BMC Neurosci 9:43
14. Yates LL, Schnatwinkel C, Murdoch JN, Bogani D, Formstone CJ et al (2010) The PCP genes Celsr1 and Vangl2 are required for normal lung branching morphogenesis. Hum Mol Genet 19:2251–2267
15. Abe J, Suzuki H, Notoya M, Yamamoto T, Hirose S (1999) Ig-hepta, a novel member of the G protein-coupled hepta-helical receptor (GPCR) family that has immunoglobulin-like repeats in a long N-terminal extracellular domain and defines a new subfamily of GPCRs. J Biol Chem 274:19957–19964
16. Yang MY, Hilton MB, Seaman S, Haines DC, Nagashima K et al (2013) Essential regulation of lung surfactant homeostasis by the orphan G protein-coupled receptor GPR116. Cell Rep 3:1457–1464
17. Niaudet C, Hofmann JJ, Mae MA, Jung B, Gaengel K et al (2015) Gpr116 receptor regulates distinctive functions in pneumocytes and vascular endothelium. PLoS One 10:e0137949
18. Waller-Evans H, Promel S, Langenhan T, Dixon J, Zahn D et al (2010) The orphan adhesion-GPCR GPR126 is required for embryonic development in the mouse. PLoS One 5:e14047
19. Monk KR, Oshima K, Jors S, Heller S, Talbot WS (2011) Gpr126 is essential for peripheral nerve development and myelination in mammals. Development 138:2673–2680
20. Petersen SC, Luo R, Liebscher I, Giera S, Jeong SJ et al (2015) The adhesion GPCR GPR126 has distinct, domain-dependent functions in Schwann cell development mediated by interaction with laminin-211. Neuron 85:755–769

21. Richter GH, Fasan A, Hauer K, Grunewald TG, Berns C et al (2013) G-Protein coupled receptor 64 promotes invasiveness and metastasis in Ewing sarcomas through PGF and MMP1. J Pathol 230:70–81
22. Lum AM, Wang BB, Beck-Engeser GB, Li L, Channa N et al (2010) Orphan receptor GPR110, an oncogene overexpressed in lung and prostate cancer. BMC Cancer 10:40
23. Hasan AN, Ahmad MW, Madar IH, Grace BL, Hasan TN (2015) An in silico analytical study of lung cancer and smokers datasets from gene expression omnibus (GEO) for prediction of differentially expressed genes. Bioinformation 11:229–235
24. Yu XJ, Yang MJ, Zhou B, Wang GZ, Huang YC et al (2015) Characterization of somatic mutations in air pollution-related lung cancer. EBioMed 2:583–590
25. Hatanaka H, Oshika Y, Abe Y, Yoshida Y, Hashimoto T et al (2000) Vascularization is decreased in pulmonary adenocarcinoma expressing brain-specific angiogenesis inhibitor 1 (BAI1). Int J Mol Med 5:181–183
26. Bari MF, Brown H, Nicholson AG, Kerr KM, Gosney JR et al (2014) BAI3, CDX2 and VIL1: a panel of three antibodies to distinguish small cell from large cell neuroendocrine lung carcinomas. Histopathology 64:547–556
27. Davidson B, Stavnes HT, Risberg B, Nesland JM, Wohlschlaeger J et al (2012) Gene expression signatures differentiate adenocarcinoma of lung and breast origin in effusions. Hum Pathol 43:684–694
28. Hancock DB, Eijgelsheim M, Wilk JB, Gharib SA, Loehr LR et al (2010) Meta-analyses of genome-wide association studies identify multiple loci associated with pulmonary function. Nat Genet 42:45–52
29. Soler Artigas M, Wain LV, Miller S, Kheirallah AK, Huffman JE et al (2015) Sixteen new lung function signals identified through 1000 Genomes Project reference panel imputation. Nat Commun 6:8658
30. Li S, Jin Z, Koirala S, Bu L, Xu L et al (2008) GPR56 regulates pial basement membrane integrity and cortical lamination. J Neurosci 28:5817–5826
31. Yang J, Wang Z, Leng D, Dai H, Wang J et al (2014) G protein-coupled receptor 56 regulates matrix production and motility of lung fibroblasts. Exp Biol Med (Maywood) 239:686–696
32. Guo R, Wu G, Li H, Qian P, Han J et al (2013) Promoter methylation profiles between human lung adenocarcinoma multidrug resistant A549/cisplatin (A549/DDP) cells and its progenitor A549 cells. Biol Pharm Bull 36:1310–1316
33. Ariestanti DM, Ando H, Hirose S, Nakamura N (2015) Targeted disruption of Ig-Hepta/Gpr116 causes emphysema-like symptoms that are associated with alveolar macrophage activation. J Biol Chem 290:11032–11040
34. Fukuzawa T, Ishida J, Kato A, Ichinose T, Ariestanti DM et al (2013) Lung surfactant levels are regulated by Ig-Hepta/GPR116 by monitoring surfactant protein D. PLoS One 8:e69451
35. Korfhagen TR, Sheftelyevich V, Burhans MS, Bruno MD, Ross GF et al (1998) Surfactant protein-D regulates surfactant phospholipid homeostasis in vivo. J Biol Chem 273:28438–28443
36. Botas C, Poulain F, Akiyama J, Brown C, Allen L et al (1998) Altered surfactant homeostasis and alveolar type II cell morphology in mice lacking surfactant protein D. Proc Natl Acad Sci U S A 95:11869–11874
37. Dranoff G, Crawford AD, Sadelain M, Ream B, Rashid A et al (1994) Involvement of granulocyte-macrophage colony-stimulating factor in pulmonary homeostasis. Science 264:713–716
38. Nishinakamura R, Nakayama N, Hirabayashi Y, Inoue T, Aud D et al (1995) Mice deficient for the IL-3/GM-CSF/IL-5 beta c receptor exhibit lung pathology and impaired immune response, while beta IL3 receptor-deficient mice are normal. Immunity 2:211–222
39. Saito A, Nikolaidis NM, Amlal H, Uehara Y, Gardner JC et al (2015) Modeling pulmonary alveolar microlithiasis by epithelial deletion of the Npt2b sodium phosphate cotransporter reveals putative biomarkers and strategies for treatment. Sci Transl Med 7:313ra181

40. Carey B, Trapnell BC (2010) The molecular basis of pulmonary alveolar proteinosis. Clin Immunol 135:223–235
41. Ikegami M, Ueda T, Hull W, Whitsett JA, Mulligan RC et al (1996) Surfactant metabolism in transgenic mice after granulocyte macrophage-colony stimulating factor ablation. Am J Physiol Lung Cell Mol Physiol 14:L650–L658
42. Yoshida M, Ikegami M, Reed JA, Chroneos ZC, Whitsett JA (2001) GM-CSF regulates protein and lipid catabolism by alveolar macrophages. Am J Physiol 280:L379–L386
43. Ikegami M, Whitsett JA, Jobe A, Ross G, Fisher J et al (2000) Surfactant metabolism in SP-D gene-targeted mice. Am J Physiol Lung Cell Mol Physiol 279:L468–L476
44. Ikegami M, Na CL, Korfhagen TR, Whitsett JA (2005) Surfactant protein D influences surfactant ultrastructure and uptake by alveolar type II cells. Am J Physiol Lung Cell Mol Physiol 288:L552–L561
45. Tang X, Jin R, Qu G, Wang X, Li Z et al (2013) GPR116, an adhesion G-protein-coupled receptor, promotes breast cancer metastasis via the Galphaq-p63RhoGEF-Rho GTPase pathway. Cancer Res 73:6206–6218
46. Fisher JH, Sheftelyevich V, Ho YS, Fligiel S, McCormack FX et al (2000) Pulmonary-specific expression of SP-D corrects pulmonary lipid accumulation in SP-D gene-targeted mice. Am J Physiol Lung Cell Mol Physiol 278:L365–L373
47. Kitamura T, Tanaka N, Watanabe J, Uchida Kanegasaki S et al (1999) Idiopathic pulmonary alveolar proteinosis as an autoimmune disease with neutralizing antibody against granulocyte/macrophage colony-stimulating factor. J Exp Med 190:875–880
48. Suzuki T, Sakagami T, Rubin BK, Nogee LM, Wood RE et al (2008) Familial pulmonary alveolar proteinosis caused by mutations in CSF2RA. J Exp Med 205:2703–2710
49. Matalon S, Bartoszewski R, Collawn JF (2015) Role of epithelial sodium channels in the regulation of lung fluid homeostasis. Am J Physiol Lung Cell Mol Physiol 309:L1229–L1238
50. Berthiaume Y, Matthay MA (2007) Alveolar edema fluid clearance and acute lung injury. Respir Physiol Neurobiol 159:350–359
51. Ebbinghaus M, Gajda M, Holtzman MJ, Schulz S, Schaible HG (2014) Does chloride channel accessory 3 have a role in arthritis pain? A study on murine antigen-induced arthritis. Neurosci Lett 576:40–44
52. Nakagami Y, Favoreto SJ, Zhen G, Park SW, Nguyenvu LT et al (2008) The epithelial anion transporter pendrin is induced by allergy and rhinovirus infection, regulates airway surface liquid, and increases airway reactivity and inflammation in an asthma model. J Immunol 181:2203–2210
53. Romero MF, Chen AP, Parker MD, Boron WF (2013) The SLC4 family of bicarbonate (HCO (3)(-)) transporters. Mol Aspects Med 34:159–182
54. Capasso M, DeCoursey TE, Dyer MJ (2011) pH regulation and beyond: unanticipated functions for the voltage-gated proton channel, HVCN1. Trends Cell Biol 21:20–28
55. Feingold KR (2012) Lamellar bodies: the key to cutaneous barrier function. J Invest Dermatol 132:1951–1953
56. Chintagari NR, Mishra A, Su L, Wang Y, Ayalew S et al (2010) Vacuolar ATPase regulates surfactant secretion in rat alveolar type II cells by modulating lamellar body calcium. PLoS One 5:e9228
57. Chander A, Sen N, Wadsworth S, Spitzer AR (2000) Coordinate packaging of newly synthesized phosphatidylcholine and phosphatidylglycerol in lamellar bodies in alveolar type II cells. Lipids 35:35–43
58. Behne MJ, Tu CL, Aronchik I, Epstein E, Bench G et al (2003) Human keratinocyte ATP2C1 localizes to the Golgi and controls Golgi Ca2+ stores. J Invest Dermatol 121:688–694
59. Micaroni M (2012) Calcium around the Golgi apparatus: implications for intracellular membrane trafficking. Adv Exp Med Biol 740:439–460

Adhesion GPCRs as Modulators of Immune Cell Function

Jörg Hamann, Cheng-Chih Hsiao, Chang Sup Lee,
Kodi S. Ravichandran, and Hsi-Hsien Lin

Graphical Abstract

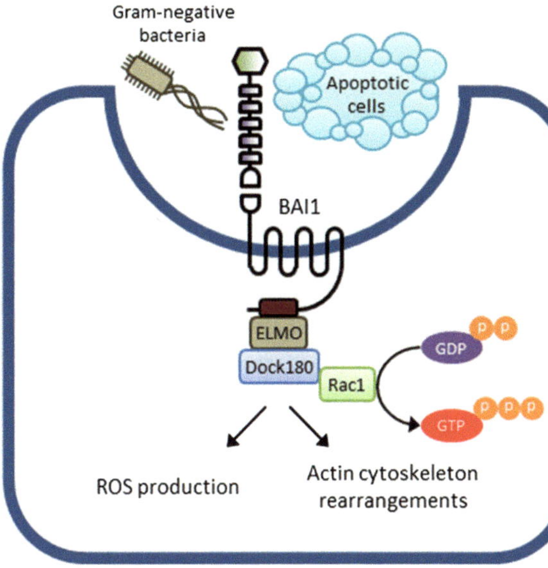

J. Hamann (✉) • C.-C. Hsiao
Department of Experimental Immunology, K0-144, Academic Medical Center, University of Amsterdam, Meibergdreef 9, Amsterdam 1105AZ, The Netherlands
e-mail: j.hamann@amc.uva.nl

C.S. Lee • K.S. Ravichandran
Department of Microbiology, Immunology, and Cancer Biology, University of Virginia, Jordan Hall 7315, Charlottesville, VA 22908, USA

H.-H. Lin (✉)
Department of Microbiology and Immunology, Chang Gung University, 259 Wen-Hwa 1st Road, Kwei-Shan, Tao-Yuan, Taiwan

Chang Gung Immunology Consortium and Department of Anatomic Pathology, Chang Gung Memorial Hospital-Linkou, Tao-Yuan 333, Taiwan
e-mail: hhlin@mail.cgu.edu.tw

T. Langenhan, T. Schöneberg (eds.), *Adhesion G Protein-coupled Receptors*,
Handbook of Experimental Pharmacology 234, DOI 10.1007/978-3-319-41523-9_15

Contents

Abstract

Immune cells express several adhesion G protein-coupled receptors (aGPCRs), including the ADGRE subfamily members EMR1 (F4/80, ADGRE1), EMR2 (ADGRE2), EMR3 (ADGRE3), EMR4 (FIRE, ADGRE4), and CD97 (ADGRE5), the ADGRB subfamily member BAI1 (ADGRB1), and the ADGRG subfamily members GPR56 (ADGRG1), GPR97 (Pb99, ADGRG3), and GPR114 (ADGRG5). Expression of these molecules in hematopoietic stem and progenitor cells, monocytes/macrophages (Mφs), dendritic cells, granulocytes, and lymphocytes depends on lineage diversification and maturation, making them suitable markers for individual leukocyte subsets (e.g., F4/80 on mouse Mφs). Recent studies revealed intriguing activities of aGPCRs in tolerance induction (EMR1), granulopoiesis (CD97), engulfment of apoptotic cells and bacteria (BAI1), hematopoietic stem cell formation (GPR56), and control of cytotoxicity (GPR56). Here, we review these findings and discuss their biological and translational implications.

Keywords

Adhesion GPCRs • Immunity • Macrophages • Granulocytes • T cells • Phagocytosis

Abbreviations

7TM	Seven transmembrane
Ab	Antibody
ABCA	ATP-binding cassette transporter
ACAID	Anterior chamber-associated immune deviation
Ag	Antigen
AML	Acute myeloid leukemia

BAI	Brain-specific angiogenesis inhibitor
BDCA	Blood dendritic cell antigen
CTF	C-terminal fragment
DC	Dendritic cell
EGF	Epidermal growth factor
ELMO	Engulfment and cell motility
EMR	EGF-like module-containing mucin-like hormone receptor-like
GAIN	GPCR autoproteolysis inducing
GEF	Guanine nucleotide exchange factor
GPCR	G protein-coupled receptor
GPS	GPCR proteolysis site
HDL	High-density lipoprotein
HRM	Hormone receptor motif
HSPC	Hematopoietic stem and progenitor cell
IFN	Interferon
IL	Interleukin
LDL	Low-density lipoprotein
LPS	Lipopolysaccharide
mAb	Monoclonal antibody
Mφ	Macrophages
NK	Natural killer
NTF	N-terminal fragment
PtdSer	Phosphatidylserine
ROS	Reactive oxygen species
SIRS	Systemic inflammatory response syndrome
TGF	Tumor growth factor
TNF	Tumor necrosis factor
Treg	Regulatory T
TSR	Thrombospondin type 1 repeat

1 Adhesion GPCRs in the Immune System

Though regulating numerous cellular processes, G protein-coupled receptors (GPCRs) in the immune system are mostly known for their ability to steer cell migration and homing in response to chemokine gradients [1]. However, leukocytes express many more GPCRs, including members of the adhesion family. In fact, the first aGPCRs that have been described—F4/80 and CD97—are molecules expressed on mouse macrophages (Mφs) and human activated T cells, respectively [2, 3]. Since then, ADGRE, ADGRB, and ADGRG subfamily members on hematopoietic cells have been identified and extensively studied, both in vitro and in vivo. Here, we review current ideas about the role of these molecules in relation to immune cell development and function and discuss the biological and translational implications of these findings.

2 EGF-TM7/ADGRE Subfamily

The five members of the ADGRE subfamily, also known as EGF-TM7 receptors, are encoded by two gene clusters on chromosome 19p13.3 (EMR1 and EMR4) and chromosome 19p13.1 (EMR2, EMR3, and CD97) [4–6]. EGF-TM7 receptors possess N-terminal epidermal growth factor (EGF)-like domains encoded by a single exon. Alternative splicing of these exons in EMR1, EMR2, and CD97 gives rise to isoforms with a different EGF-domain composition. Expression of EMR1–4 is restricted to myeloid cells. In contrast, CD97 is found on virtually all hematopoietic cells as well as on many non-hematopoietic cell types (Fig. 1).

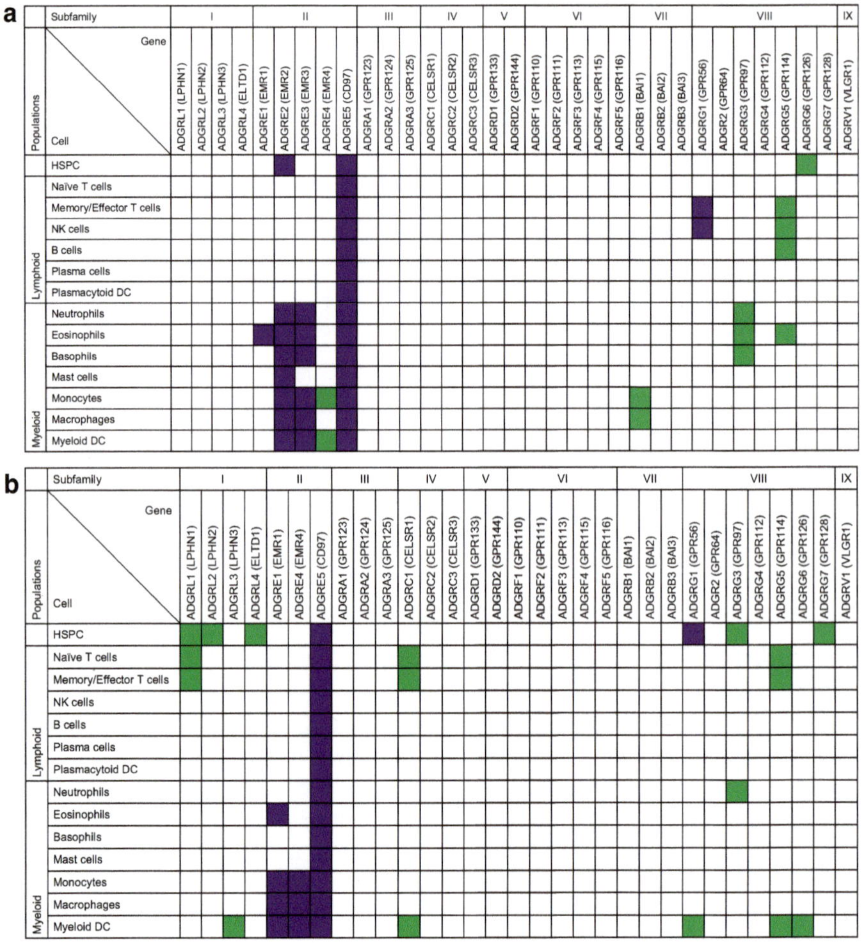

Fig. 1 Expression of adhesion GPCRs in immune cells. Summary of available published data on the expression of aGPCRs in hematopoietic cells in (**a**) human and (**b**) mice, obtained by transcriptional profiling (*green squares*) and protein analysis techniques (*purple squares*). For additional information and references, see the descriptions of individual aGPCRs at www.guidetopharmacology.org/

2.1 EMR1 (F4/80, ADGRE1)

The F4/80 antigen (Ag) was initially described as a novel cell surface marker of mouse Mφs by Jon Austyn and Siamon Gordon in 1981, following the development of the F4/80 monoclonal antibody (mAb) in rats against isolated thioglycollate-elicited mouse peritoneal Mφs [2]. Subsequent experiments established the limited expression pattern of F4/80 in myeloid cells, including monocytes, most tissue Mφs, and some dendritic cells (DCs) [7, 8]. Eosinophils also express the F4/80 Ag, whose expression is enhanced after parasitic infection [9]. No F4/80 reactivity was detected in other cells of the myeloid lineage, such as neutrophilic or basophilic granulocytes and osteoclasts, in cells of the lymphoid lineage, and in non-hematopoietic cells. Due to its robust reactivity and specificity in restricted Mφ subsets, the F4/80 Ag has been extensively used as a tissue Mφ marker in mice ever since [10–12]. The F4/80 Ag is detected as early as embryonic day 8/9 in Mφs in the aorta–gonad–mesonephros region and yolk sac, in fetal livers, and in splenic red pulp and bone marrow before birth [13]. In the adult mouse, F4/80 is weak on circulating monocytes, but is highly expressed on sinusoidal Mφs in the liver (Kupffer cells), spleen, and adrenal glands and on almost all tissue extravascular Mφs. F4/80 has also been used to identify tumor-associated Mφs [14].

F4/80 was identified as the murine ortholog of human EMR1 through genomic and phylogenetic analyses [4–6]. F4/80 and EMR1 possess seven and six N-terminal EGF-like domains, respectively [15–17]. They are the only members of the ADGRE subfamily that are not proteolytically processed due to the presence of an atypical GPCR proteolysis site (GPS) sequence. Indeed, both F4/80 and EMR1 are expressed as single-chain polypeptide receptor molecules. Notably, EMR1 is not expressed in Mφs, but instead is a specific marker of human eosinophilic granulocytes [18, 19]. Therefore, it is believed that F4/80 and EMR1 are structural orthologs that possess different cell-type- and species-specific expression patterns and functions.

Several approaches, including Ab treatment and gene targeting, have been used to investigate the physiological function(s) of F4/80 ex vivo and in vivo. Using the F4/80 mAb in an in vitro model of facultative intracellular pathogen infection, Warschkau et al. demonstrated a role for F4/80 in interferon (IFN)-γ production by severe combined immunodeficiency (SCID) mouse splenic cells in response to heat-killed *Listeria monocytogenes* [20]. IFN-γ is critical for the control of bacterial growth during *L. monocytogenes* infection in vivo. It was shown that heat-killed *L. monocytogenes*-stimulated Mφs secreted tumor necrosis factor (TNF) and interleukin (IL)-12, which activated natural killer (NK) cells to produce IFN-γ. It was concluded that F4/80 is critical in the Mφ–NK cell interaction during *L. monocytogenes* infection, because the F4/80 mAb treatment downregulated TNF, IL-12, and IFN-γ production [20]. The F4/80 molecule was believed to interact with a potential cellular ligand on NK cell, which delivered a stimulatory signal for cytokine release.

Two strains of F4/80 knockout mice were generated using different gene targeting approaches; one involved the production of F4/80–Cre knock-in mice, where the first coding exon of F4/80 was replaced with the cDNA of the Cre recombinase, while the other used a conventional method by deleting the first coding exon [21, 22]. Interestingly, the F4/80-deficient animals were healthy and phenotypically normal. No obvious defect in the maturation and development of tissue Mϕs was noted in the mutant mice either. This indicated that F4/80 is not required for the differentiation and development of cells of the monocyte–Mφ lineage. In contrast to what was found in vitro, the F4/80-deficient animals show normal responses to infectious challenge with *L. monocytogenes*, suggesting a minimum role for F4/80 in the immune responses against *L. monocytogenes* infection in vivo [22].

In another study, the F4/80-deficient mice were employed to confirm the role of F4/80 in the induction of peripheral immune tolerance [21]. Indeed, the F4/80-deficient mice did not develop anterior chamber-associated immune deviation (ACAID) because the animals failed to generate functional Ag-specific efferent $CD8^+$ regulatory T (Treg) cells. Similarly, it was shown that efferent Treg cells were only produced when the APCs are $F4/80^+$ in an in vitro ACAID model using both the F4/80 mAb and F4/80-deficient APC. Furthermore, the F4/80-deficient animals were found unable to produce an efficient immune tolerance response in a low-dose oral tolerance model. Importantly, reconstitution of the F4/80-deficient animals with the $F4/80^+$ APCs restored their ability to induce ACAID and low-dose oral tolerance [21]. Therefore, the F4/80 molecule is required for the generation of Ag-specific efferent Treg cells [23]. The molecular mechanism mediated by the F4/80 receptor to induce peripheral immune tolerance remains elusive, but a putative cellular ligand of F4/80 on interacting immune cells is a likely candidate.

2.2 EMR2 (ADGRE2)

EMR2 seems to be a chimera of CD97 and another EGF-TM7 receptor, EMR3, as a result of a relatively recent evolution, comprising gene conversion and duplication [24]. Indeed, EMR2 shares ~97 % amino acid sequence identity in the EGF-like domains with CD97 and ~90 % identity in the seven-transmembrane (7TM) region with EMR3, respectively [6, 24–26]. This suggests that EMR2 might interact with similar cellular ligand(s) as CD97 while sharing with EMR3 comparable signaling activities. The EMR2 gene is lost in rodents, suggesting a specific need for EMR2 in the immune reactions of human [24].

Unlike F4/80 and human EMR1 in Mφs and eosinophils, EMR2 is more widely expressed in myeloid cells, including monocytes, Mφs, DCs, and granulocytes, suggesting a role for EMR2 in these innate immune cells [27]. Resting or activated lymphocytes do not express any detectable EMR2. $CD16^+$ blood monocytes, Mφs, and $BDCA\text{-}3^+$ myeloid DCs express the highest levels of EMR2. Interestingly, EMR2 expression is upregulated during the differentiation and maturation of Mφs in vitro, but is downregulated during DC maturation [28]. Lipopolysaccharide (LPS) treatment upregulated EMR2 expression in monocytes and Mφs in an IL-10-dependent manner. In tissues, EMR2 expression was detected in certain Mφ subpopulations of the lung, skin, spleen, tonsil, and placenta, but not in those of the liver and kidney. In inflamed tissues, EMR2 staining was found in subpopulations of Mφs and neutrophils [28]. Recently, high levels of EMR2 were detected in foamy Mϕs in atherosclerotic vessels and in Gaucher cells in the spleen. On the contrary, multiple sclerosis brain foam cells express little if any EMR2 [29].

The full-length EMR2 protein contains a total of five EGF-like motifs. Alternative splicing results in four isoforms: EMR2(EGF1,2), EMR2(EGF1,2,5), EMR2 (EGF1,2,3,5), and EMR2(EGF1–5) [25]. EMR2(EGF1–5) shares with CD97 (EGF1–5) a glycosaminoglycan ligand, chondroitin sulfate B (dermatan sulfate), while EMR2(EGF1,2,5) binds to CD55 with a tenfold weaker affinity than does CD97(EGF1,2,5) [30, 31]. These findings suggest possible overlapping, but also specific functions for EMR2 and CD97.

Studies using the mAb 2A1, which binds downstream of the EGF domains, have demonstrated a role for EMR2 in potentiating the activation and recruitment of human neutrophils [32]. Specifically, ligation of EMR2 by 2A1 enhanced neutrophil adherence and migration under both static and shear stress conditions. In addition, ligation and activation of EMR2 was shown to work synergistically with a number of pro-inflammatory mediators in potentiating neutrophil respiratory burst and degranulation [32]. Patients suffering from noninfectious systemic inflammatory response syndrome (SIRS) and sepsis often die through multiple organ failure, caused by increased numbers and hyperactivation of neutrophils. A more recent study revealed higher EMR2 expression in neutrophils of SIRS patients and a positive correlation between the percentage of $EMR2^+$ neutrophils and the occurrence of organ failure. In fact, EMR2 in neutrophil was considered a potential biomarker for SIRS [33]. Huang et al. have investigated the potential role of EMR2 in Mφ activation. They demonstrated that cross-linking of EMR2 in Mφs by 2A1 induced the translocation and colocalization of N-terminal fragment (NTF) and the C-terminal fragment (CTF) within lipid rafts. This in turn led to receptor signaling and production of inflammatory cytokines, such as IL-8 and TNF [34]. Therefore, EMR2 seems to act as an activation receptor in myeloid cells, including neutrophils and Mφs.

Notably, a search for genetic variants causing the autoinflammatory disorder vibratory urticaria recently identified a missense substitution 26 amino acids upstream from the GPS in EMR2 [35] (see also [36, 37]). This C492Y variant destabilized the association between the NTF and the CTF, such that shear stress that is incurred during vibration disrupted their noncovalent attachment. The remaining CTF elicited greater degranulation from transfected LAD2 mast cells, which is consistent with numerous observations indicating enhanced signaling activity of the CTF of aGPCRs upon removal of the NTF (Fig. 2a) (see also [38, 39]).

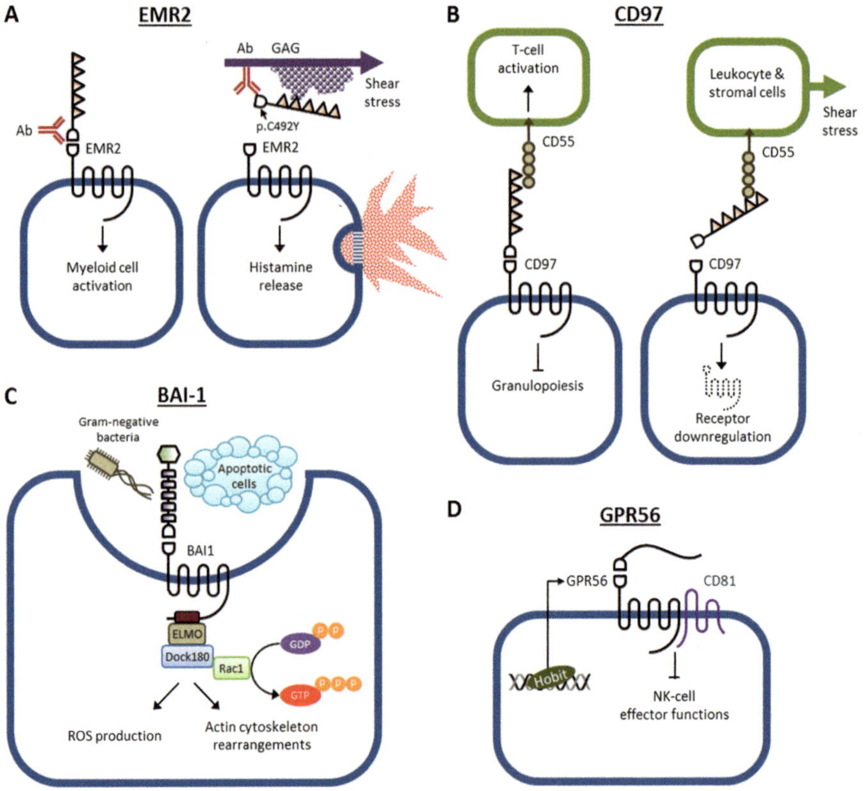

Fig. 2 Functions of adhesion GPCRs in immune cells. (**a**) *Left*: A mAb binding at the NTF of EMR2 underneath the EGF-domain region potentiates the activation and recruitment of human neutrophils and Mφ. *Right*: A mutation near the GPS of EMR2 facilitates sheer stress-induced dissociation of the NTF from the CTF upon ligation by antibody or glycosaminoglycan (GAG). The remaining CTF enhances vibration-induced degranulation of mast cells in patients suffering from vibratory urticaria. (**b**) *Left*: CD97 triggers different activities. Mice lacking either CD97 or CD55 display a comparable phenotype with about twofold more circulating granulocytes. Augmented numbers of Gr-1-positive cells in cell cycle in the bone marrow indicate a higher granulopoietic activity. In addition, the NTF of CD97 can activate T cells by engaging CD55. *Right*: In the circulation, CD97 expression is constantly downregulated by contact with CD55 on blood and stromal cells, possible to restrict CD97–CD55-mediated cell adhesion to tissue sites. (**c**) BAI1 on Mφ binds PtdSer on apoptotic cells and LPS on gram-negative bacteria via the TSRs in its extracellular domain. BAI1 interacts with the N-terminal domain of ELMO via an alpha-helical domain in its intracellular tail. ELMO recruits Dock180 and promotes guanine nucleotide exchange factor activity. The GEF activity of the ELMO–Dock180 complex promotes Rac1 association with GTP and thereby mediates actin cytoskeletal reorganization during engulfment and ROS production. (**d**) GPR56 on NK cells forms a complex with the tetraspanin CD81 to inhibit cytotoxicity. Expression of GPR56 is induced by the transcription factor Hobit and terminated upon physiological activation through shedding of the NTF and transcriptional downregulation

2.3 EMR3 (ADGRE3)

EMR3 was first identified in a DNA database, based on its high similarity with EMR2. cDNA cloning revealed that EMR3, due to concerted evolution of the encoding genes, shares ~90 % identity in the 7TM region with EMR2 [26]. EMR3 possesses two N-terminal EGF-like domains that may recognize a ligand on human Mφs and activated neutrophils [26]. Flow cytometric analysis revealed that all types of myeloid cells express EMR3 [40]. In peripheral blood, the highest expression of EMR3 was found on polymorphonuclear granulocytes. Moreover, mature $CD16^{+}$ monocytes express high levels of EMR3, while $CD16^{-}$ monocytes and myeloid DCs are $EMR3^{dim/low}$. Lymphocytes and plasmacytoid DCs lack EMR3. Interestingly, $CD34^{+}CD33^{-}/CD38^{-}$ committed hematopoietic stem cells and $CD34^{+}CD33^{+}/CD38^{+}$ progenitors in bone marrow do not express EMR3, which is only upregulated during late granulopoiesis. Accordingly, EMR3 can be used to monitor granulocyte maturation [41]. In rodents, the EMR3 gene has been lost together with gene for EMR2 [24].

2.4 EMR4 (FIRE, ADGRE4)

Like EMR1, EMR4 was first identified in mouse, where it was called F4/80-like receptor (FIRE) [42]. FIRE has two N-terminal EGF-like domains and is expressed exclusively in myeloid cells, including $CD8^{-}$ DCs, but not by $CD8^{+}$ DCs, monocytes, or Mφs [42, 43]. Immunohistochemical analysis revealed that $FIRE^{+}$ cells reside predominantly in the red pulp and in the marginal zone of the spleen; yet, very few $FIRE^{+}$ cells are seen in the white pulp. FIRE is downregulated during DC and Mφ activation, although the latter is not conclusive [43]; also the distribution of EMR4 on tissue Mφ warrants further investigation. A ligand for FIRE, present on the B lymphoma cell line A20, remains to be identified [43].

Human EMR4 is not expressed at the cell surface. Due to a one-nucleotide deletion in exon 8, translation of EMR4 would result in a truncated 232-amino acid protein, lacking the entire 7TM region [44]. Whether human cells express a soluble secreted fragment of EMR4 is not clear; transcripts are present in $BDCA\text{-}1^{+}$ DCs and monocytes, but protein expression remains to be demonstrated [45]. Of note, the deletion disabling EMR4 surface expression is not found in nonhuman primates, including chimpanzees, suggesting that the gene became nonfunctional only after human speciation, about 5 million years ago [44]. Thus, EMR4 accounts for a genetic difference between humans and primates related to immunity.

2.5 CD97 (ADGRE5)

CD97 is a prototypical member of the EGF-TM7 subfamily of aGPCRs [46]. Evolutionary conserved, the gene encoding CD97 is found in all vertebrate genomes analyzed so far. Moreover, the structure and cellular distribution of CD97 are similar in human and mouse, indicating a conserved biological role. Initially, CD97 has been described as an activation-dependent Ag on lymphocytes [3]. Later, it became clear that hematopoietic stem and progenitor cells (HSPCs) and all leukocytes in human and mouse express CD97^{+}, with highest levels found in myeloid cells [47, 48]. Moreover, many normal and malignant non-hematopoietic cell types express CD97 [49, 50].

Like EMR2, CD97 possesses up to five N-terminal EGF-like domains. Alternative splicing gives rise to three isoforms in human, namely CD97(EGF1,2,5), CD97 (EGF1,2,3,5), and CD97(EGF1–5) [51]. CD97 isoforms expressed in the mouse are CD97(EGF1,2,4), CD97(EGF1,2,3,4), and CD97(EGF1,2,X,3,4), with X indicating a 45-amino acids sequence that does not correspond to any known protein module [52, 53]. EGF-like domains 1 and 2 interact with CD55 (decay-accelerating factor), a membrane-based regulator of complement activation [54, 55], also deposited on collagen fibers [56]. EGF-like domain 4, which is present only in the largest isoform, binds chondroitin sulfate B, a glycosaminoglycan that is found abundantly on cell surfaces and extracellular matrix [31, 57]. An RGD (Arg-Gly-Asp) motif 3′ to the EGF-domain region of human CD97 is bound by the integrin $\alpha5\beta1$ and, possibly, $\alpha v\beta3$ [58]. Finally, through the GPCR autoproteolysis-inducing (GAIN) domain, CD97 interacts with the stromal and T-cell Ag Thy-1 (CD90) [59]. Thus, CD97 interacts with at least four matricellular ligands.

Based on its molecular structure and ligand interactions, a role for CD97 in leukocyte adhesion and/or trafficking has been proposed [54]. To investigate this assumption, mAbs directed against individual EGF domains were generated, interfering with either CD55 or chondroitin sulfate B binding. The functional consequences of targeting CD97 and blocking its ligand interactions were studied using several mouse inflammation models, each representing different aspects of innate and adaptive immunity. CD97 mAbs inhibited the accumulation of neutrophils at sites of inflammation, thereby affecting antibacterial host defense, inflammatory disorders, and stem cell mobilization from bone marrow [60–62]. Unexpectedly, they did so independent of the ligand interaction they interfered with. Moreover, the tested mAbs had no impact on Ag-specific (adaptive) responses, such as delayed-type hypersensitivity or experimental autoimmune encephalomyelitis [63]. Comparison of the consequences of Ab treatment and gene targeting implied that CD97 mAbs actively inhibit the innate response, presumably at the level of granulocyte and/or Mφ recruitment to sites of inflammation. Detailed investigation showed that CD97 mAbs deplete polymorphonuclear granulocytes in bone marrow and blood, which involved Fc receptors, but not complement activation, and was associated with an increase in serum levels of TNF and other pro-inflammatory cytokines [64]. Of note, depletion of granulocytes by CD97 mAbs requires acute inflammation, suggesting a mechanism of conditional, Ab-mediated granulocytopenia.

Two independently developed CD97-deficient mice showed no overt phenotype, except for a mild granulocytosis that increased under inflammatory conditions [50, 65]. A comparable phenotype with about twofold more circulating granulocytes was found in mice deficient for the CD97 ligand CD55 [66]. Augmented numbers of Gr-1-positive cells in cell cycle in the bone marrow indicated a higher granulopoietic activity in mice lacking either CD97 or CD55 (Fig. 2b). Concomitant with the increase in blood granulocyte numbers, CD55 knockout mice, challenged with the respiratory pathogen *Streptococcus pneumoniae*, developed less bacteremia and died later after infection [66], while CD97 knockout mice displayed an improved immune response toward acute infection with *L. monocytogenes* [65]. Moreover, possibly related to the interaction of CD55 with CD97, amelioration of collagen-induced arthritis and a trend toward less severe K/BxN serum transfer arthritis was found in mice that lack CD97 or CD55 [67].

Cellular and molecular assays have proved the interaction between CD97 and its various ligands in vitro. For example, soluble CD97 can cross-link CD55 on T cells in vitro, thereby increasing $CD4^+$ T-cell proliferation, activation, and IL-10 and granulocyte–macrophage colony-stimulating factor (GM-CSF) secretion [68]. Yet, it has been difficult to prove ligand interactions of CD97 in vivo. Conclusive evidence was obtained when leukocytes from CD55-deficient mice were shown to express significantly increased levels of cell surface CD97 that normalized after transfer into wild-type mice because of contact with CD55 on both leukocytes and stromal cells [69]. Downregulation of both CD97 subunits occurred within minutes after first contact with CD55 in vivo and was strictly dependent on shear stress in the circulation. Of note, de novo ligation of CD97 did not activate signaling molecules constitutively engaged by CD97 in cancer cells, suggesting that CD55 downregulates CD97 surface expression on circulating leukocytes by a process that requires physical forces, but may not induce receptor signaling (Fig. 2b).

3 BAI/ADGRB Subfamily

The members of the ADGRB subfamily, commonly referred to as brain-specific angiogenesis inhibitors (BAIs), possess a large NTF that contains, next to the GAIN domain, a single hormone receptor motif (HRM) and 4–5 thrombospondin type 1 repeat (TSR) domains [70–72]. Moreover, BAIs contain a long C-terminal tail, downstream of the 7TM region, which comprises the PDZ-binding motif QTEV (Gln-Thr-Glu-Val) for intracellular signal transduction. BAIs were initially known for their ability to inhibit angiogenesis and tumor formation. Later studies unraveled key roles in the clearance of apoptotic cells and bacteria and in the regulation of immune responses. Finally, BAIs have been involved in synaptogenesis and dendritic spine formation. Below, we focus on the roles of BAI family proteins in cell clearance and immune regulation.

3.1 BAI1 (ADGRB1)

BAI1 (human chromosome location at 8q24) was first cloned in 1997 as a target of the tumor suppressor gene p53 [73]; yet, it is not clear whether the expression of BAI1 indeed is dependent on p53 [74]. Abundantly found in the brain on neurons, astrocytes, and microglia, BAI1 is also expressed in other tissues (bone marrow, spleen, testis, and colon) and in distinct cell types (Mφs, skeletal muscle myoblasts) [70, 72] (Fig. 1). The TSR region of BAI1 comprises five repeats, which can bind phosphatidylserine (PtdSer) and LPS [75, 76]. Moreover, an N-terminal RGD motif can interact with the integrin $\alpha5\beta1$ that regulates extracellular matrix attachment and motility [77]. Autoproteolysis of BAI1 at the GPS or behind the first TSR generates soluble ectodomains with antiangiogenic properties, known as vasculostatin 120 (kDa) and vasculostatin 40 (kDa) [78–80].

Park et al. identified BAI1 as an engulfment receptor for apoptotic cells that can both, bind the corpses and signal intracellularly in the phagocyte to facilitate their internalization [75]. BAI1 was found to function as a phagocytic receptor both, in Mφs and nonprofessional phagocytes, such as fibroblasts and epithelial cells. In embryonic zebra fish brains, BAI1 mediates engulfment of dying neurons by microglia [81]. BAI1 directly binds PtdSer, which is exposed on the outer leaflet of the plasma membrane of cells undergoing apoptosis, via the TSR regions, as determined by a combination of biochemical approaches [75]. Upon PtdSer binding, BAI1 transmits signals via the downstream intracellular engulfment molecule ELMO1 (engulfment and cell motility 1). The specific interaction occurs via an α-helical motif within the cytoplasmic tail of BAI1 and the N-terminal region of ELMO1 [75]. In fact, all three ELMO family members (ELMO1, ELMO2, and ELMO3) can bind to the cytoplasmic tail of BAI1. ELMO1 acts as a guanine nucleotide exchange factor for Rac and activates Rac, which promotes actin cytoskeletal rearrangements, necessary to mediate the phagocytosis of apoptotic cells (Fig. 2c). When phagocytes engulf apoptotic cells, they release anti-inflammatory mediators, such as tumor growth factor (TGF)β, IL-10, and prostaglandin E2, and these, in turn, reduce the levels of pro-inflammatory TNF in the local tissue. BAI1-mediated apoptotic cell clearance has been implicated in anti-inflammatory response.

Upon stimulation with apoptotic cells, BAI1 signaling in Mφs leads to upregulation of ABCA1 (ATP-binding cassette transporter 1) [82]. ABCA1 is a large transmembrane transporter that mediates cholesterol efflux from Mφs and is a key player for high-density lipoprotein (HDL) biogenesis. Genetic loss of BAI1 prevented ABCA1 upregulation in response to apoptotic cells and reduced serum levels of total cholesterol, HDL, low-density lipoprotein (LDL), and triglycerides [82]. Furthermore, transgenic mice that overexpressed BAI1 had an improved HDL/LDL ratio [82]. HDL has potent anti-inflammatory functions. Therefore, BAI1 in Mφs can regulate lipid homeostasis and anti-inflammatory responses via mediating apoptotic cell clearance.

A more recent study revealed a novel role for BAI1 in colonic epithelial cells [83]. Lee et al. showed that healthy epithelial cells mediate the engulfment of their apoptotic neighbors that arise due to acute injury, and this BAI1-mediated apoptotic cell removal beneficially influenced the level of tissue inflammation. Genetic deficiency of BAI1 caused increased numbers of uncleared apoptotic cells, greater inflammatory colon regions, and higher inflammatory cytokine levels (IL-1α and TNF) in the dextran sulfate sodium-induced colitis model. Conversely, transgenic overexpression of BAI1 in colonic epithelial cells resulted in a reduction of uncleared apoptotic cells, smaller inflammatory colon regions, and lower inflammatory cytokine levels. Furthermore, transgenic mice expressing a mutant version of BAI1 that cannot signal failed to protect mice suggesting that BAI1 signaling (BAI1–ELMO) is required for the beneficial effects in experimental colitis.

In addition to its role as a PtdSer receptor, BAI1 mediates engulfment of gram-negative bacteria [76]. The TSR region of BAI1 directly interacts with the LPS inner core of gram-negative bacteria. BAI1-mediated engulfment of bacteria by Mφs through this interaction induced production of the pro-inflammatory cytokine TNF [76]. Furthermore, a recent study revealed that BAI1 mediates reactive oxygen species (ROS) production and bacterial clearance in vivo [84]. BAI1-deficient mice showed impaired bacterial clearance and were more susceptible to peritoneal infection. These studies suggest that BAI1 can act as a pattern recognition receptor that mediates bacterial clearance through microbicidal activity and local inflammation (Fig. 2c). An interesting question that arises from these observations is how the signaling downstream of BAI1 remains anti-inflammatory during apoptotic cell engulfment while being pro-inflammatory during bacterial engulfment; whether this is mediated through other receptors that interact with BAI1 during apoptotic cell recognition or a different set of intracellular signaling molecules that might be attracted to the tail of BAI1 during these recognition events remains to be determined.

Besides its immune-related activities, BAI1 possesses other unique functions. As a PtdSer receptor, BAI1 induces myoblast fusion into multinucleated myofibers in response to apoptotic or PtdSer-exposing myoblasts [85]. Moreover, BAI1 regulates spinogenesis and synaptogenesis [86]. Finally, in a study of BAI1-deficient mice, BAI1 was found to regulate spatial learning and memory as well as synaptic plasticity [87].

3.2 BAI2 (ADGRB2) and BAI3 (ADGRB3)

BAI2 (human chromosome location at 1p35) and BAI3 (human chromosome location at 6q12) were identified as homologs of BAI1 [88]. Although expression patterns are not well characterized, BAI2 and BAI3 are found in many tissues, including the brain, heart, skeletal muscle, intestine, and thymus [70, 72]. Both receptors possess four TSR motifs, and although immune-related functions have not been reported yet, BAI3 possesses the motif necessary for binding ELMO proteins that allows BAI1 to engulf apoptotic cells [89]. Moreover, similar to BAI1, BAI3 has been implicated in skeletal myoblast fusion and muscle fiber formation [90].

4 ADGRG Subfamily

The ADGRG subfamily member GPR56 is one of the best-studied aGPCRs, well known for its causal involvement in neuronal and cortical development [91]. The gene encoding GPR56 clusters with the genes for GPR97 and GPR114 on chromosome 16q21, suggesting a common evolutionary origin. In contrast to GPR56, which is expressed in many different cell types, very little is known about GPR97 and GPR114, which both seem to have a rather restricted cellular distribution [92]. All three genes are expressed in leukocytes (Fig. 1).

4.1 GPR56 (ADGRG1)

GPR56 has been strongly linked with development. HSPCs in the mouse embryo and adult bone marrow abundantly express GPR56, but levels substantially decrease as cells differentiate. Recent studies have explored the role of GPR56 in HSPCs. Solaimani-Kartalaei et al. found a critical role of GPR56 in the formation of hematopoietic clusters during endothelial to hematopoietic cell transition [93]. In contrast, GPR56 deficiency does not impair HSPC maintenance or function during steady-state or myeloablative stress-induced hematopoiesis, and GPR56-deficient cells respond normally to physiological and pharmacological mobilization signals, despite the reported role of this aGPCR as a regulator of cell adhesion and migration in neuronal cells [94]. Thus, GPR56 expression is required for generating the first hematopoietic stem cells, but largely dispensable for steady-state and regenerative hematopoiesis.

Flow cytometric analysis with newly generated mAbs detected the presence of GPR56 in human cytotoxic NK and T lymphocytes, including $CD8^+$, $CD4^+$, and $\gamma\delta$ T cells [95, 96]. Primary infection with cytomegalovirus, which generates a vast population of $CD8^+$ T cells with an effector phenotype, induced a strong increase in GPR56 expression in virus-specific $CD8^+$ T cells that remained detectable during latency. In NK-92 cells, ectopic expression of GPR56 inhibits spontaneous and SDF-1-stimulated cell migration [96]. Investigation of NK cells from polymicrogyria patients with a null mutation in the GPR56 gene and NK-92 cells overexpressing GPR56 revealed that GPR56 suppresses the production of inflammatory cytokines and cytolytic proteins, degranulation, and target cell killing [97]. GPR56 pursues this activity by associating with the tetraspanin CD81. Expression of GPR56 was triggered by Hobit, a homolog of the transcription factor Blimp-1, and declined upon cell activation. Thus, GPR56 controls natural cytotoxicity displayed by NK cells in order to protect the body against harmful viruses and neoplasms (Fig. 2d). Circulating T and NK cells in the mouse do not express GPR56, which may be due to the different way they acquire cytotoxic capacity.

4.2 GPR97 (PB99, ADGRG3) and GPR114 (ADGRG5)

A microarray study of mature human leukocyte populations detected GPR97 transcripts in neutrophils and eosinophils [96]. Moreover, GPR97 gene expression that was found in murine pre-B cells and thymocytes, but not in mature B and T cells, suggested a role in early lymphoid development. Investigation of independently raised knockout mice revealed no necessity of GPR97 for B- or T-cell development [98], but implied an essential role in follicular versus marginal zone B-lymphocyte fate decision [99]. In spite of its expression in eosinophilic granulocytes, GPR97 was found to be dispensable for inflammation in ovalbumin-induced asthmatic mice [100]. GPR114 is another aGPCR transcribed in lymphoid and myeloid cells [96]; yet, as for GPR97, expression needs to be confirmed at the protein level.

5 Biological and Translational Implications

aGPCRs have been implicated in the development and function of HSPCs, monocytes, Mφs, DCs, granulocytes, lymphocytes, and, thus, most lineages of hematopoietic cells. Nevertheless, it has not been clear for quite some time whether these noncanonical GPCRs regulate cellular homeostasis and activation or whether they also control specific immune functions. Studies on the role of BAI1 in Mφs and GPR56 in cytotoxic lymphocytes imply that aGPCRs indeed regulate innate, and possibly also adaptive, effector functions. Thereby, these studies paved the way for further exploring the biological role of these receptors using novel genetic models and pharmacological tools. As a prime example, BAI1 facilitates clearance of apoptotic cells by Mφs and intestinal epithelial cells and modulates the inflammatory status within tissue.

EGF-TM7 subfamily aGPCRs figure prominently in the biology of polymorphonuclear granulocytes, an aspect that has been overlooked in the beginning, due to the prominent presence of the EMR1 ortholog F4/80 on Mφs in the mouse and the upregulation of CD97 during lymphocyte activation. Legrand et al. recently showed that an afucosylated mAb directed against EMR1 efficiently eradicates eosinophils through Ab-dependent cell-mediated cytotoxicity, resulting in long-term in vivo depletion of eosinophils in monkeys and, thus, providing a treatment option for eosinophilic disorders [19]. Similarly, mAbs to mouse CD97 eliminate neutrophils (and possible also eosinophils) in an Fc receptor-dependent manner, thereby ameliorating various inflammatory conditions [64]. These studies imply that biologicals targeting aGPCRs may be useful in treating immune cell-related diseases.

Association of aGPCRs with hematopoietic malignancies has not been investigated systematically so far, despite their role in solid tumors (see [101]). A recent study reported the association of CD97 expression with internal tandem duplications within the juxtamembrane region of the FMS-like tyrosine kinase receptor FLT3 (FLT3-ITD) in patients with acute myeloid leukemia (AML), which is associated with an aggressive clinical phenotype [102]. CD97 knockdown resulted in reduced cell adhesion and trans-well migration in vitro. Moreover, GPR56 expression identifies primary human AML cells with high repopulating potential in vivo [103]. Elevated expression of GPR56 was found in AML cells with high ecotropic viral integration site-1 expression, a refractory type of the disease with a poor prognosis. Knockdown of GPR56 expression decreased the cellular adhesion ability through inactivation of RhoA signaling, resulting in a reduction of cellular growth rates and enhanced apoptosis [104].

Despite an increasing number of studies that have linked aGPCRs with immune function(s), main uncertainties persist. For example, though many of the protein folds found in the NTF of aGPCRs are involved in cell–cell interactions, evidence is scarce that aGPCRs contribute to the communication between immune cells. Further, it is not clear to what extent aGPCR autoproteolysis and signaling facilitate immune functions. Moreover, functional redundancy between members of the EGF-TM7 subfamily and the GPR56/GPR97/GPR114 cluster seems possible and warrants further investigation. Related to this, the aGPCR signature of immune cells need to be unraveled comprehensively. Exploring these issues eventually will improve our understanding of the biology of aGPCRs and, thereby, may widen the canon of GPCR activities in the immune system.

Acknowledgments We thank the members of our laboratories for generating a large part of the data discussed in this chapter. This work was supported by grants to J.H. from the Deutsche Forschungsgemeinschaft (Research Unit 2149) and the Thyssen Foundation (2015-00387), to K.S.R. from the National Institutes of Health, USA (GM064709, HD074981, and MH096484), and to H.H.L. from the Ministry of Science and Technology, Taiwan (MOST-104-2320-B-182-035-MY3) and the Chang Gung Memorial Hospital (CMRPD1C0633, CMRPD1D0072-3, and CMRPD1D0392).

Author Contributions J.H., C.C.H., H.H.L., C.S.L., and K.S.R. wrote the manuscript.

Competing Financial Interests The authors declare no competing financial interests.

References

1. Rot A, von Andrian UH (2004) Chemokines in innate and adaptive host defense: basic chemokinese grammar for immune cells. Annu Rev Immunol 22:891–928
2. Austyn JM, Gordon S (1981) F4/80, a monoclonal antibody directed specifically against the mouse macrophage. Eur J Immunol 11:805–815
3. Eichler W, Aust G, Hamann D (1994) Characterization of an early activation-dependent antigen on lymphocytes defined by the monoclonal antibody BL-Ac(F2). Scand J Immunol 39:111–115
4. McKnight AJ, Gordon S (1996) EGF-TM7: a novel subfamily of seven-transmembrane-region leukocyte cell-surface molecules. Immunol Today 17:283–287
5. McKnight AJ, Gordon S (1998) The EGF-TM7 family: unusual structures at the leukocyte surface. J Leukoc Biol 63:271–280
6. Kwakkenbos MJ, Kop EN, Stacey M, Matmati M, Gordon S, Lin H-H et al (2004) The EGF-TM7 family: a postgenomic view. Immunogenetics 55:655–666
7. Lin H-H, Stacey M, Stein-Streilein J, Gordon S (2010) F4/80: the macrophage-specific adhesion-GPCR and its role in immunoregulation. Adv Exp Med Biol 706:149–156

8. Gordon S, Hamann J, Lin H-H, Stacey M (2011) F4/80 and the related adhesion-GPCRs. Eur J Immunol 41:2472–2476
9. McGarry MP, Stewart CC (1991) Murine eosinophil granulocytes bind the murine macrophage-monocyte specific monoclonal antibody F4/80. J Leukoc Biol 50:471–478
10. Taylor PR, Martinez-Pomares L, Stacey M, Lin HH, Brown GD, Gordon S (2005) Macrophage receptors and immune recognition. Annu Rev Immunol 23:901–944
11. Hume DA, Gordon S (1983) Mononuclear phagocyte system of the mouse defined by immunohistochemical localization of antigen F4/80. Identification of resident macrophages in renal medullary and cortical interstitium and the juxtaglomerular complex. J Exp Med 157:1704–1709
12. Hume DA, Robinson AP, MacPherson GG, Gordon S (1983) The mononuclear phagocyte system of the mouse defined by immunohistochemical localization of antigen F4/80. Relationship between macrophages, Langerhans cells, reticular cells, and dendritic cells in lymphoid and hematopoietic organs. J Exp Med 158:1522–1536
13. Morris L, Graham CF, Gordon S (1991) Macrophages in haemopoietic and other tissues of the developing mouse detected by the monoclonal antibody F4/80. Development 112:517–526
14. Qian B-Z, Pollard JW (2010) Macrophage diversity enhances tumor progression and metastasis. Cell 141:39–51
15. Baud V, Chissoe SL, Viegas-Péquignot E, Diriong S, N'Guyen VC, Roe BA et al (1995) EMR1, an unusual member in the family of hormone receptors with seven transmembrane segments. Genomics 26:334–344
16. Lin HH, Stubbs LJ, Mucenski ML (1997) Identification and characterization of a seven transmembrane hormone receptor using differential display. Genomics 41:301–308
17. McKnight AJ, Macfarlane AJ, Dri P, Turley L, Willis AC, Gordon S (1996) Molecular cloning of F4/80, a murine macrophage-restricted cell surface glycoprotein with homology to the G-protein-linked transmembrane 7 hormone receptor family. J Biol Chem 271:486–489
18. Hamann J, Koning N, Pouwels W, Ulfman LH, van Eijk M, Stacey M et al (2007) EMR1, the human homolog of F4/80, is an eosinophil-specific receptor. Eur J Immunol 37:2797–2802
19. Legrand F, Tomasevic N, Simakova O, Lee C-CR, Wang Z, Raffeld M et al (2014) The eosinophil surface receptor epidermal growth factor-like module containing mucin-like hormone receptor 1 (EMR1): a novel therapeutic target for eosinophilic disorders. J Allergy Clin Immunol 133:1439–1447 1447.e1–8
20. Warschkau H, Kiderlen AF (1999) A monoclonal antibody directed against the murine macrophage surface molecule F4/80 modulates natural immune response to Listeria monocytogenes. J Immunol 163:3409–3416
21. Lin H-H, Faunce DE, Stacey M, Terajewicz A, Nakamura T, Zhang-Hoover J et al (2005) The macrophage F4/80 receptor is required for the induction of antigen-specific efferent regulatory T cells in peripheral tolerance. J Exp Med 201:1615–1625
22. Schaller E, Macfarlane AJ, Rupec RA, Gordon S, McKnight AJ, Pfeffer K (2002) Inactivation of the F4/80 glycoprotein in the mouse germ line. Mol Cell Biol 22:8035–8043
23. van den Berg TK, Kraal G (2005) A function for the macrophage F4/80 molecule in tolerance induction. Trends Immunol 26:506–509
24. Kwakkenbos MJ, Matmati M, Madsen O, Pouwels W, Wang Y, Bontrop RE et al (2006) An unusual mode of concerted evolution of the EGF-TM7 receptor chimera EMR2. FASEB J 20:2582–2584
25. Lin HH, Stacey M, Hamann J, Gordon S, McKnight AJ (2000) Human EMR2, a novel EGF-TM7 molecule on chromosome 19p13.1, is closely related to CD97. Genomics 67:188–200
26. Stacey M, Lin HH, Hilyard KL, Gordon S, McKnight AJ (2001) Human epidermal growth factor (EGF) module-containing mucin-like hormone receptor 3 is a new member of the EGF-TM7 family that recognizes a ligand on human macrophages and activated neutrophils. J Biol Chem 276:18863–18870

27. Kwakkenbos MJ, Chang G-W, Lin H-H, Pouwels W, de Jong EC, van Lier RAW et al (2002) The human EGF-TM7 family member EMR2 is a heterodimeric receptor expressed on myeloid cells. J Leukoc Biol 71:854–862
28. Chang G-W, Davies JQ, Stacey M, Yona S, Bowdish DME, Hamann J et al (2007) CD312, the human adhesion-GPCR EMR2, is differentially expressed during differentiation, maturation, and activation of myeloid cells. Biochem Biophys Res Commun 353:133–138
29. van Eijk M, Aust G, Brouwer MSM, van Meurs M, Voerman JSA, Dijke IE et al (2010) Differential expression of the EGF-TM7 family members CD97 and EMR2 in lipid-laden macrophages in atherosclerosis, multiple sclerosis and Gaucher disease. Immunol Lett 129:64–71
30. Lin HH, Stacey M, Saxby C, Knott V, Chaudhry Y, Evans D et al (2001) Molecular analysis of the epidermal growth factor-like short consensus repeat domain-mediated protein-protein interactions: dissection of the CD97-CD55 complex. J Biol Chem 276:24160–24169
31. Stacey M, Chang G-W, Davies JQ, Kwakkenbos MJ, Sanderson RD, Hamann J et al (2003) The epidermal growth factor-like domains of the human EMR2 receptor mediate cell attachment through chondroitin sulfate glycosaminoglycans. Blood 102:2916–2924
32. Yona S, Lin H-H, Dri P, Davies JQ, Hayhoe RPG, Lewis SM et al (2008) Ligation of the adhesion-GPCR EMR2 regulates human neutrophil function. FASEB J 22:741–751
33. Lewis SM, Treacher DF, Edgeworth J, Mahalingam G, Brown CS, Mare TA et al (2015) Expression of CD11c and EMR2 on neutrophils: potential diagnostic biomarkers for sepsis and systemic inflammation. Clin Exp Immunol 182:184–194
34. Huang Y-S, Chiang N-Y, Hu C-H, Hsiao C-C, Cheng K-F, Tsai W-P et al (2012) Activation of myeloid cell-specific adhesion class G protein-coupled receptor EMR2 via ligation-induced translocation and interaction of receptor subunits in lipid raft microdomains. Mol Cell Biol 32:1408–1420
35. Boyden SE, Desai A, Cruse G, Young ML, Bolan HC, Scott LM et al (2016) Vibratory urticaria associated with a missense variant in ADGRE2. N Engl J Med 374:656–663
36. Nieberler M, Kittel RJ, Petrenko AG, Lin H-H, Langenhan T (2016) Control of adhesion GPCR function through proteolytic processing. In: Langenhan T, Schöneberg T (eds) Adhesion G protein-coupled receptors: molecular, physiological and pharmacological principles in health and disease. Springer, Heidelberg
37. Scholz N, Monk KR, Kittel RJ, Langenhan T (2016) Adhesion GPCRs as a putative class of metabotropic mechanosensors. In: Langenhan T, Schöneberg T (eds) Adhesion G protein-coupled receptors: molecular, physiological and pharmacological principles in health and disease. Springer, Heidelberg
38. Liebscher I, Schöneberg T (2016) Tethered agonism: a common activation mechanism of adhesion GPCRs. In: Langenhan T, Schöneberg T (eds) Adhesion G protein-coupled receptors: molecular, physiological and pharmacological principles in health and disease. Springer, Heidelberg
39. Kishore A, Hall RA (2016) Versatile signaling activity of adhesion GPCRs. In: Langenhan T, Schöneberg T (eds) Adhesion G protein-coupled receptors: molecular, physiological and pharmacological principles in health and disease. Springer, Heidelberg
40. Matmati M, Pouwels W, van Bruggen R, Jansen M, Hoek RM, Verhoeven AJ et al (2007) The human EGF-TM7 receptor EMR3 is a marker for mature granulocytes. J Leukoc Biol 81:440–448
41. Drewniak A, van Raam BJ, Geissler J, Tool ATJ, Mook ORF, van den Berg TK et al (2009) Changes in gene expression of granulocytes during in vivo granulocyte colony-stimulating factor/dexamethasone mobilization for transfusion purposes. Blood 113:5979–5998
42. Caminschi I, Lucas KM, O'Keeffe MA, Hochrein H, Laâbi Y, Köntgen F et al (2001) Molecular cloning of F4/80-like-receptor, a seven-span membrane protein expressed differentially by dendritic cell and monocyte-macrophage subpopulations. J Immunol 167:3570–3576

43. Stacey M, Chang G-W, Sanos SL, Chittenden LR, Stubbs L, Gordon S et al (2002) EMR4, a novel epidermal growth factor (EGF)-TM7 molecule up-regulated in activated mouse macrophages, binds to a putative cellular ligand on B lymphoma cell line A20. J Biol Chem 277:29283–29293
44. Hamann J, Kwakkenbos MJ, de Jong EC, Heus H, Olsen AS, van Lier RAW (2003) Inactivation of the EGF-TM7 receptor EMR4 after the Pan-Homo divergence. Eur J Immunol 33:1365–1371
45. Caminschi I, Vandenabeele S, Sofi M, McKnight AJ, Ward N, Brodnicki TC et al (2006) Gene structure and transcript analysis of the human and mouse EGF-TM7 molecule, FIRE. DNA Seq 17:8–14
46. Hamann J, Eichler W, Hamann D, Kerstens HM, Poddighe PJ, Hoovers JM et al (1995) Expression cloning and chromosomal mapping of the leukocyte activation antigen CD97, a new seven-span transmembrane molecule of the secretion receptor superfamily with an unusual extracellular domain. J Immunol 155:1942–1950
47. van Pel M, Hagoort H, Hamann J, Fibbe WE (2008) CD97 is differentially expressed on murine hematopoietic stem-and progenitor-cells. Haematologica 93:1137–1144
48. Kop EN, Matmati M, Pouwels W, Leclercq G, Tak PP, Hamann J (2009) Differential expression of CD97 on human lymphocyte subsets and limited effect of CD97 antibodies on allogeneic T-cell stimulation. Immunol Lett 123:160–168
49. Jaspars LH, Vos W, Aust G, van Lier RA, Hamann J (2001) Tissue distribution of the human CD97 EGF-TM7 receptor. Tissue Antigens 57:325–331
50. Veninga H, Becker S, Hoek RM, Wobus M, Wandel E, van der Kaa J et al (2008) Analysis of CD97 expression and manipulation: antibody treatment but not gene targeting curtails granulocyte migration. J Immunol 181:6574–6583
51. Gray JX, Haino M, Roth MJ, Maguire JE, Jensen PN, Yarme A et al (1996) CD97 is a processed, seven-transmembrane, heterodimeric receptor associated with inflammation. J Immunol 157:5438–5447
52. Qian YM, Haino M, Kelly K, Song WC (1999) Structural characterization of mouse CD97 and study of its specific interaction with the murine decay-accelerating factor (DAF, CD55). Immunology 98:303–311
53. Hamann J, van Zeventer C, Bijl A, Molenaar C, Tesselaar K, van Lier RA (2000) Molecular cloning and characterization of mouse CD97. Int Immunol 12:439–448
54. Hamann J, Vogel B, van Schijndel GM, van Lier RA (1996) The seven-span transmembrane receptor CD97 has a cellular ligand (CD55, DAF). J Exp Med 184:1185–1189
55. Hamann J, Stortelers C, Kiss-Toth E, Vogel B, Eichler W, van Lier RA (1998) Characterization of the CD55 (DAF)-binding site on the seven-span transmembrane receptor CD97. Eur J Immunol 28:1701–1707
56. Karpus ON, Kiener HP, Niederreiter B, Yilmaz-Elis AS, van der Kaa J, Ramaglia V et al (2015) CD55 deposited on synovial collagen fibers protects from immune complex-mediated arthritis. Arthritis Res Ther 17:6
57. Kwakkenbos MJ, Pouwels W, Matmati M, Stacey M, Lin H-H, Gordon S et al (2005) Expression of the largest CD97 and EMR2 isoforms on leukocytes facilitates a specific interaction with chondroitin sulfate on B cells. J Leukoc Biol 77:112–119
58. Wang T, Ward Y, Tian L, Lake R, Guedez L, Stetler-Stevenson WG et al (2005) CD97, an adhesion receptor on inflammatory cells, stimulates angiogenesis through binding integrin counterreceptors on endothelial cells. Blood 105:2836–2844
59. Wandel E, Saalbach A, Sittig D, Gebhardt C, Aust G (2012) Thy-1 (CD90) is an interacting partner for CD97 on activated endothelial cells. J Immunol 188:1442–1450
60. Leemans JC, te Velde AA, Florquin S, Bennink RJ, de Bruin K, van Lier RAW et al (2004) The epidermal growth factor-seven transmembrane (EGF-TM7) receptor CD97 is required for neutrophil migration and host defense. J Immunol 172:1125–1131

61. Kop EN, Adriaansen J, Smeets TJM, Vervoordeldonk MJ, van Lier RAW, Hamann J et al (2006) CD97 neutralisation increases resistance to collagen-induced arthritis in mice. Arthritis Res Ther 8:R155
62. de Groot DM, Vogel G, Dulos J, Teeuwen L, Stebbins K, Hamann J et al (2009) Therapeutic antibody targeting of CD97 in experimental arthritis: the role of antigen expression, shedding, and internalization on the pharmacokinetics of anti-CD97 monoclonal antibody 1B2. J Immunol 183:4127–4134
63. Hamann J, Veninga H, de Groot DM, Visser L, Hofstra CL, Tak PP et al (2010) CD97 in leukocyte trafficking. Adv Exp Med Biol 706:128–137
64. Veninga H, de Groot DM, McCloskey N, Owens BM, Dessing MC, Verbeek JS et al (2011) CD97 antibody depletes granulocytes in mice under conditions of acute inflammation via a Fc receptor-dependent mechanism. J Leukoc Biol 89:413–421
65. Wang T, Tian L, Haino M, Gao J-L, Lake R, Ward Y et al (2007) Improved antibacterial host defense and altered peripheral granulocyte homeostasis in mice lacking the adhesion class G protein receptor CD97. Infect Immun 75:1144–1153
66. Veninga H, Hoek RM, de Vos AF, de Bruin AM, An F-Q, van der Poll T et al (2011) A novel role for CD55 in granulocyte homeostasis and anti-bacterial host defense. PLoS One 6: e24431
67. Hoek RM, de Launay D, Kop EN, Yilmaz-Elis AS, Lin F, Reedquist KA et al (2010) Deletion of either CD55 or CD97 ameliorates arthritis in mouse models. Arthritis Rheum 62:1036–1042
68. Capasso M, Durrant LG, Stacey M, Gordon S, Ramage J, Spendlove I (2006) Costimulation via CD55 on human CD4+ T cells mediated by CD97. J Immunol 177:1070–1077
69. Karpus ON, Veninga H, Hoek RM, Flierman D, van Buul JD, Vandenakker CC et al (2013) Shear stress-dependent downregulation of the adhesion-G protein-coupled receptor CD97 on circulating leukocytes upon contact with its ligand CD55. J Immunol 190:3740–3748
70. Cork SM, Van Meir EG (2011) Emerging roles for the BAI1 protein family in the regulation of phagocytosis, synaptogenesis, neurovasculature, and tumor development. J Mol Med 89:743–752
71. Park D, Ravichandran KS (2010) Emerging roles of brain-specific angiogenesis inhibitor 1. Adv Exp Med Biol 706:167–178
72. Stephenson JR, Purcell RH, Hall RA (2014) The BAI subfamily of adhesion GPCRs: synaptic regulation and beyond. Trends Pharmacol Sci 35:208–215
73. Nishimori H, Shiratsuchi T, Urano T, Kimura Y, Kiyono K, Tatsumi K et al (1997) A novel brain-specific p53-target gene, BAI1, containing thrombospondin type 1 repeats inhibits experimental angiogenesis. Oncogene 15:2145–2150
74. Kaur B, Brat DJ, Calkins CC, Van Meir EG (2003) Brain angiogenesis inhibitor 1 is differentially expressed in normal brain and glioblastoma independently of p53 expression. Am J Pathol 162:19–27
75. Park D, Tosello-Trampont A-C, Elliott MR, Lu M, Haney LB, Ma Z et al (2007) BAI1 is an engulfment receptor for apoptotic cells upstream of the ELMO/Dock180/Rac module. Nature 450:430–434
76. Das S, Owen KA, Ly KT, Park D, Black SG, Wilson JM et al (2011) Brain angiogenesis inhibitor 1 (BAI1) is a pattern recognition receptor that mediates macrophage binding and engulfment of Gram-negative bacteria. Proc Natl Acad Sci U S A 108:2136–2141
77. Koh JT, Kook H, Kee HJ, Seo Y-W, Jeong BC, Lee JH et al (2004) Extracellular fragment of brain-specific angiogenesis inhibitor 1 suppresses endothelial cell proliferation by blocking alphavbeta5 integrin. Exp Cell Res 294:172–184
78. Kaur B, Brat DJ, Devi NS, Van Meir EG (2005) Vasculostatin, a proteolytic fragment of brain angiogenesis inhibitor 1, is an antiangiogenic and antitumorigenic factor. Oncogene 24:3632–3642

79. Kaur B, Cork SM, Sandberg EM, Devi NS, Zhang Z, Klenotic PA et al (2009) Vasculostatin inhibits intracranial glioma growth and negatively regulates in vivo angiogenesis through a CD36-dependent mechanism. Cancer Res 69:1212–1220
80. Cork SM, Kaur B, Devi NS, Cooper L, Saltz JH, Sandberg EM et al (2012) A proprotein convertase/MMP-14 proteolytic cascade releases a novel 40 kDa vasculostatin from tumor suppressor BAI1. Oncogene 31:5144–5152
81. Mazaheri F, Breus O, Durdu S, Haas P, Wittbrodt J, Gilmour D et al (2014) Distinct roles for BAI1 and TIM-4 in the engulfment of dying neurons by microglia. Nat Commun 5:4046
82. Fond AM, Lee CS, Schulman IG, Kiss RS, Ravichandran KS (2015) Apoptotic cells trigger a membrane-initiated pathway to increase ABCA1. J Clin Invest 125:2748–2758
83. Lee CS, Penberthy KK, Wheeler KM, Juncadella IJ, Vandenabeele P, Lysiak JJ et al (2016) Boosting apoptotic cell clearance by colonic epithelial cells attenuates inflammation in vivo. Immunity 44(4):807–820
84. Billings EA, Lee CS, Owen KA, D'Souza RS, Ravichandran KS, Casanova JE (2016) The adhesion GPCR BAI1 mediates macrophage ROS production and microbicidal activity against Gram-negative bacteria. Sci Signal 9:ra14
85. Hochreiter-Hufford AE, Lee CS, Kinchen JM, Sokolowski JD, Arandjelovic S, Call JA et al (2013) Phosphatidylserine receptor BAI1 and apoptotic cells as new promoters of myoblast fusion. Nature 497:263–267
86. Duman JG, Tzeng CP, Tu Y-K, Munjal T, Schwechter B, Ho TS-Y et al (2013) The adhesion-GPCR BAI1 regulates synaptogenesis by controlling the recruitment of the Par3/Tiam1 polarity complex to synaptic sites. J Neurosci 33:6964–6978
87. Zhu D, Li C, Swanson AM, Villalba RM, Guo J, Zhang Z et al (2015) BAI1 regulates spatial learning and synaptic plasticity in the hippocampus. J Clin Invest 125:1497–1508
88. Shiratsuchi T, Nishimori H, Ichise H, Nakamura Y, Tokino T (1997) Cloning and characterization of BAI2 and BAI3, novel genes homologous to brain-specific angiogenesis inhibitor 1 (BAI1). Cytogenet Cell Genet 79:103–108
89. Lanoue V, Usardi A, Sigoillot SM, Talleur M, Iyer K, Mariani J et al (2013) The adhesion-GPCR BAI3, a gene linked to psychiatric disorders, regulates dendrite morphogenesis in neurons. Mol Psychiatry 18:943–950
90. Hamoud N, Tran V, Croteau L-P, Kania A, Côté J-F (2014) G-protein coupled receptor BAI3 promotes myoblast fusion in vertebrates. Proc Natl Acad Sci U S A 111:3745–3750
91. Piao X, Hill RS, Bodell A, Chang BS, Basel-Vanagaite L, Straussberg R et al (2004) G protein-coupled receptor-dependent development of human frontal cortex. Science 303:2033–2036
92. Hamann J, Aust G, Araç D, Engel FB, Formstone C, Fredriksson R et al (2015) International Union of Basic and Clinical Pharmacology. XCIV. Adhesion G protein-coupled receptors. Pharmacol Rev 67:338–367
93. Solaimani Kartalaei P, Yamada-Inagawa T, Vink CS, de Pater E, van der Linden R, Marks-Bluth J et al (2015) Whole-transcriptome analysis of endothelial to hematopoietic stem cell transition reveals a requirement for Gpr56 in HSC generation. J Exp Med 212:93–106
94. Rao TN, Marks-Bluth J, Sullivan J, Gupta MK, Chandrakanthan V, Fitch SR et al (2015) High-level Gpr56 expression is dispensable for the maintenance and function of hematopoietic stem and progenitor cells in mice. Stem Cell Res 14:307–322
95. Chiesa Della M, Falco M, Parolini S, Bellora F, Petretto A, Romeo E et al (2010) GPR56 as a novel marker identifying the CD56dull CD16+ NK cell subset both in blood stream and in inflamed peripheral tissues. Int Immunol 22:91–100
96. Peng Y-M, van de Garde MDB, Cheng K-F, Baars PA, Remmerswaal EBM, van Lier RAW et al (2011) Specific expression of GPR56 by human cytotoxic lymphocytes. J Leukoc Biol 90:735–740
97. Chang G-W, Hsiao C-C, Peng Y-M, Vieira Braga FA, Kragten NAM, Remmerswaal EBM et al (2016) The adhesion G protein-coupled receptor GPR56/ADGRG1 is an inhibitory receptor on human NK cells. Cell Rep 15:1757–1770

98. Sleckman BP, Khan WN, Xu W, Bassing CH, Malynn BA, Copeland NG et al (2000) Cloning and functional characterization of the early-lymphocyte-specific Pb99 gene. Mol Cell Biol 20:4405–4410
99. Wang J-J, Zhang L-L, Zhang H-X, Shen C-L, Lu S-Y, Kuang Y et al (2013) Gpr97 is essential for the follicular versus marginal zone B-lymphocyte fate decision. Cell Death Dis 4:e853
100. Shi J-P, Li X-N, Zhang X-Y, Du B, Jiang W-Z, Liu M-Y et al (2015) Gpr97 is dispensable for inflammation in OVA-induced asthmatic mice. PLoS One 10:e0131461
101. Aust G, Zhu D, Van Meir EG, Xu L (2016) Adhesion GPCRs in tumorigenesis. In: Langenhan T, Schöneberg T (eds) Adhesion G protein-coupled receptors: molecular, physiological and pharmacological principles in health and disease. Springer, Heidelberg
102. Wobus M, Bornhäuser M, Jacobi A, Kräter M, Otto O, Ortlepp C et al (2015) Association of the EGF-TM7 receptor CD97 expression with FLT3-ITD in acute myeloid leukemia. Oncotarget 6:38804–38815
103. Pabst C, Bergeron A, Lavallée V-P, Yeh J, Gendron P, Norddahl GL et al (2016) GPR56 identifies primary human acute myeloid leukemia cells with high repopulating potential in vivo. Blood 127:2018–2027
104. Saito Y, Kaneda K, Suekane A, Ichihara E, Nakahata S, Yamakawa N et al (2013) Maintenance of the hematopoietic stem cell pool in bone marrow niches by EVI1-regulated GPR56. Leukemia 27:1637–1649

Heart Development, Angiogenesis, and Blood-Brain Barrier Function Is Modulated by Adhesion GPCRs

Gentian Musa, Felix B. Engel, and Colin Niaudet

Graphical Abstract

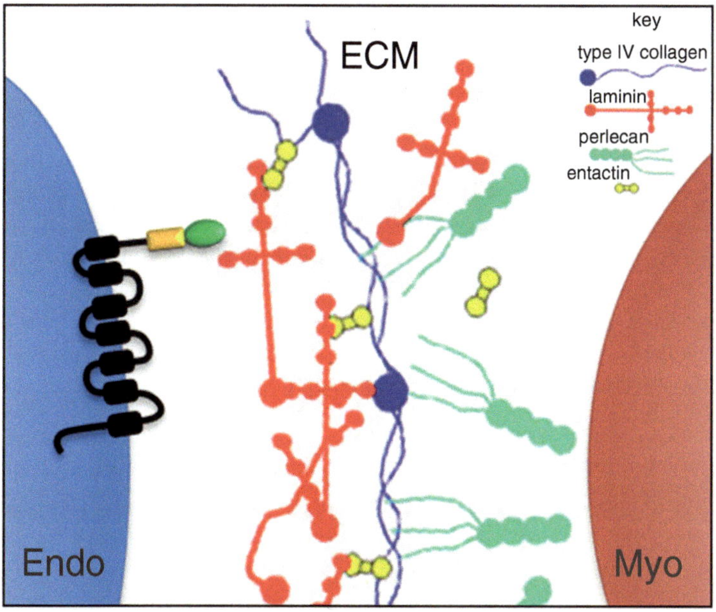

G. Musa • F.B. Engel (✉)
Experimental Renal and Cardiovascular Research, Department of Nephropathology, Institute of Pathology, Friedrich-Alexander-Universität Erlangen-Nürnberg, Schwabachanlage 12, Erlangen 91054, Germany
e-mail: felix.engel@uk-erlangen.de

C. Niaudet (✉)
Department of Immunology, Genetics and Pathology, Rudbeck Laboratory, Uppsala University, Dag Hammarskjölds väg 20, Uppsala 751 85, Sweden
e-mail: colin.niaudet@gmail.com

T. Langenhan, T. Schöneberg (eds.), *Adhesion G Protein-coupled Receptors*,
Handbook of Experimental Pharmacology 234, DOI 10.1007/978-3-319-41523-9_16

Contents

Abstract

The cardiovascular system in adult organisms forms a network of interconnected endothelial cells, supported by mural cells and displaying a high degree of hierarchy: arteries emerging from the heart ramify into arterioles and then capillaries, which return to the venous systems through venules and veins. The cardiovascular system allows blood circulation, which in turn is essential for hemostasis through gas diffusion, nutrient distribution, and cell trafficking. In this chapter, we have summarized the current knowledge on how adhesion GPCRs (aGPCRs) impact heart development, followed by their role in modulating vascular angiogenesis.

Keywords

Heart development • Trabeculation • Blood-brain-barrier • Cardiovascular • Angiogenesis • Endothelium • Adhesion GPCRs

1 Introduction

The cardiovascular system in adult organisms forms a network of interconnected endothelial cells, supported by mural cells and displaying a high degree of hierarchy: arteries emerging from the heart ramify into arterioles and then capillaries, which return to the venous systems through venules and veins. The cardiovascular system allows blood circulation, which in turn is essential for hemostasis through gas diffusion, nutrient distribution, and cell trafficking. In this chapter, we have summarized the current knowledge on how adhesion GPCRs (aGPCRs) impact heart development, followed by their role in modulating vascular angiogenesis.

2 Adhesion GPCRs in Cardiac Development

2.1 Overview of Cardiac Development

The mammalian heart is the first organ to develop in the embryo. It evolves from a simple tubular structure, the linear heart tube, into a highly complex four-chambered organ (Fig. 1). The cells destined to form the linear heart are symmetrically arrayed in two groups: the primary (PHF) and the secondary (SHF) heart field (Fig. 1a) which together form the precardiac mesoderm or cardiac crescent (Fig. 1b). Later, the cells of the PHF initiate to differentiate into the endocardial and myocardial lineages and fuse in the embryonal midline forming the primitive linear heart tube (Fig. 1c) [2]. At this stage, the heart is formed by only two cell layers, the myocardium and endocardium, separated by an extracellular matrix (ECM) called cardiac jelly. Subsequently, the heart tube loops to the right, and

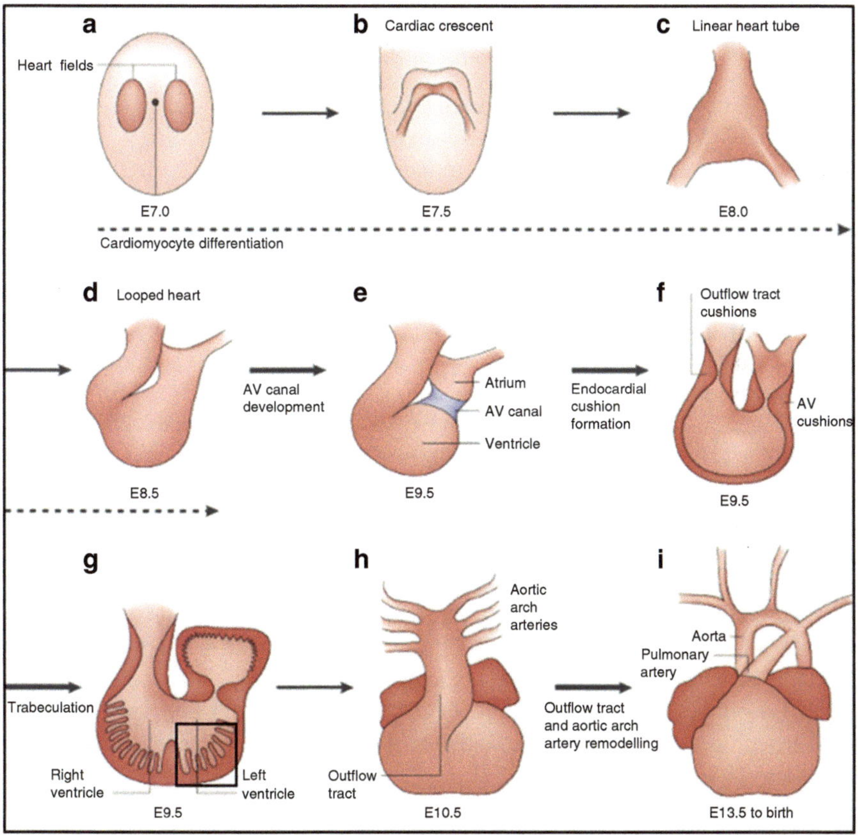

Fig. 1 Overview of cardiac development. Representation of the key events in the development of the murine heart starting from embryonic day 7 (E7) to embryonic day 13.5 (E13.5). Trabeculation appears at E9.5 at the basis of the ventricular walls. Figure adapted from [1]

cells of the SHF are added culminating in the formation of the embryonal heart (Fig. 1d, e) with different distinguishable cardiac regions: the inflow tract (IFT), embryonal atrium, atrioventricular canal (AVC), embryonal ventricle, and outflow tract (OFT). Each of these regions has a differential pattern of gene expression as well as different functional characteristics. Later on in development, the AVC and OFT present endocardial cushions on their inner surface, which are focal expansions of the cardiac jelly (Fig. 1f). The endocardial cells that line the AVC and OFT cushions undergo epithelial to mesenchymal transition (EMT) to populate the cushions with mesenchymal cells initiating the transformation into the AV and OFT valves. Signals exchanged between specialized myocardium and cushion endocardium are required for endocardial EMT [3]. Subsequently, trabeculae appear during the ventricular chamber development and maturation (Figs. 1g and 2), which consist of highly organized cardiomyocyte muscular protrusions of the ventricular wall, lined by a single layer of endocardial cells [2, 4].

2.2 Regulation of Cardiac Development by Adhesion GPCRs

Currently, six aGPCRs have been detected in the endocardium and/or myocardium of the heart. Among these are three members of the subfamily I. ADGRL1 (latrophilin-1) has been detected via Northern blot analysis during rat heart development with an intense expression at postnatal day 3. Yet, the cellular origin of ADGRL1 expression as well as its function remains unknown [5]. ADGRL2 (latrophilin-2) has been shown to be expressed in the AVC and to be required for EMT [6]. In situ hybridizations revealed that ADGRL2 is expressed in the mesoderm leaving the primitive streak in chicken. At later stages ADGRL2 is expressed in the endocardium as well as the myocardium during cushion formation and subsequently in the mesenchymal cells populating the AVC cushion. Knockdown experiments in explant cultures demonstrated that ADGRL2 mediates cell-cell separation and is required for the population of the cardiac cushion with mesenchymal cells. ADGRL4 (ELTD1) has been detected by Northern blot, RT-PCR, and in situ hybridization analyses in rat hearts. Northern blot analyses of isolated cardiomyocytes from E16 to 12-week-old rats and fetal vs. human heart tissue revealed that the expression levels of ADGRL4 increase with age, which was verified by RT-PCR for rats. In situ hybridization of 2-week-old rats demonstrated that ADGRL4 is expressed in cardiomyocytes and vascular smooth muscle cells [7]. Patients undergoing heart transplant due to heart failure show a decrease in ADGRL4 protein expression. Moreover mice with ADGRL4 knockout exhibit a more severe induced hypertrophy, suggesting ADGRL4 as a target in the treatment of hypertrophic cardiomyopathies [8].

Of the subfamilies II and VI, only the members ADGRE5 (CD97) and ADGRF5 (Gpr116) have been detected in the adult mouse heart. Studies utilizing a reporter mouse as well as antibody staining revealed expression of ADGRE5 in cardiomyocytes [9], while the expression of ADGRF5 was detected by RT-PCR utilizing RNA from whole hearts [10]. The function of both adhesion GPCRs in the heart is unknown.

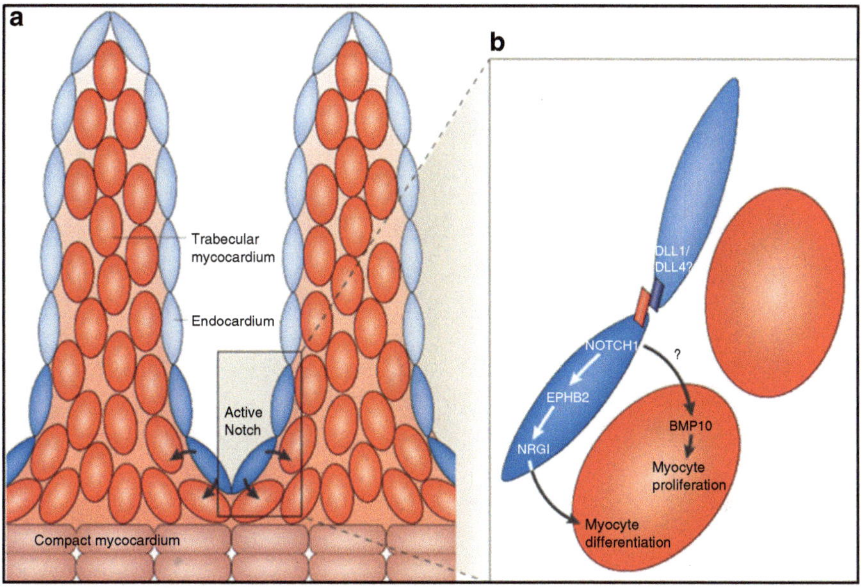

Fig. 2 Ventricular trabeculation. (**a**) The myocardial cells of the trabeculae (*red oval* shaped) are outlined by the endocardial cells (*blue oval* shaped) arranged in a single-cell layer. (**b**) The main pathways involved in heart trabeculation. Figure adapted from [1]

Finally, it has been demonstrated that the subfamily VIII member ADGRG6 (Gpr126) is required for trabeculation during zebrafish and mouse heart development [11–13]. Communication between myocardium and endocardium is essential for the proper formation of the trabeculae even though the genetic regulatory network behind this is not well known. Among key molecules with a well-established role in trabeculation are NRG1/ErbB2/ErbB4 [14–16], EphrinB2/EphB4 [17, 18], and BMP10 [19] (Fig. 2). Lately also the Notch signaling network has been shown to regulate trabeculation via the expression of EphrinB2, NRG1, and BMP10 [20]. Notch1 activity is restricted to the trabecular endocardium and has been shown to be an upstream regulator of ADGRG6. The analysis of the Notch1-ADGRG6 relationship has confirmed ADGRG6 among the key regulatory elements of trabeculation [21].

Interestingly, pathways required for trabeculation, such as NRG1/ErbB [22] and Notch [23], are also involved in another developmental process that is known to be dependent on ADGRG6, namely, Schwann cell differentiation and myelination of peripheral nerves [24] (see also [25] on this topic). Moreover, the ECM, which is known to be required for trabeculation [26–28], plays an important role in the function of ADGRG6. For example, it has been shown that the ECM proteins laminin and type IV collagen, which are known to be crucial for the regulation of Schwann cell function [29], are in vitro activating ligands of ADGRG6 [30, 31]. These data suggest that the interaction of ADGRG6 with the ECM might also underlie trabeculation (Fig. 3).

Collectively, several members of the aGPCR family are expressed in the heart in cardiomyocytes and/or endocardial cells, and at least two members are required for

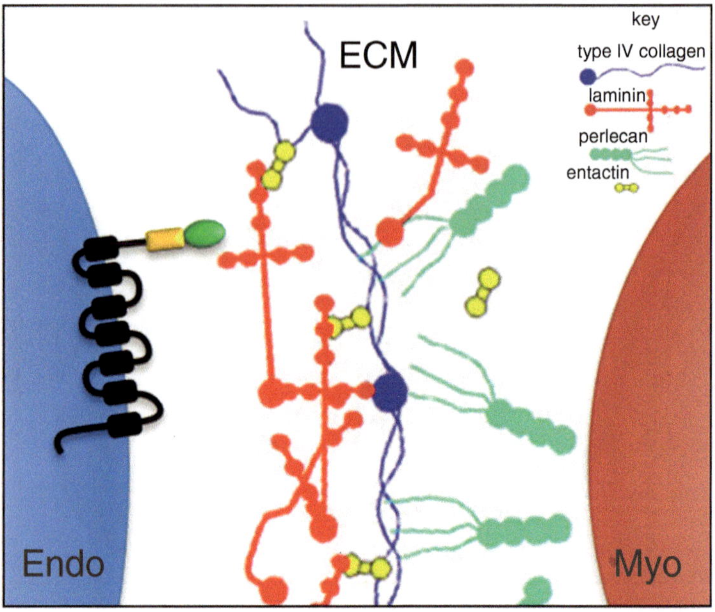

Fig. 3 Extracellular matrix and trabeculation. Between the endocardial (Endo) and the myocardial cells (Myo) lies the extracellular matrix (ECM) mainly composed of type IV collagen and laminin. The figure suggests a putative interaction between ADGRG6 expressed in the endocardial cells and the ECM

proper early heart development. However, the downstream targets of these aGPCRs as well as the cellular functions controlled by them remain elusive.

3 Adhesion GPCRs in Angiogenesis

3.1 Adhesion GPCRs and Endothelium Heterogeneity

Endothelial cells line the inner side of the entire blood vessel tree. Several layers of specialization maintain a high diversity in between the different vascular beds. First is the aforementioned hierarchy of vessels, from large arterial vessels to capillaries, and back to the venous system. Overimposed on this global architecture, vascular beds acquire local properties in the vicinity of each organ or depending on hemodynamic conditions. Hence, a mix of transmitted differentiation and microenvironment-induced changes leads to a very high heterogeneity in between each vascular bed, increased again by the salt and pepper pattern of expression of certain genes. Among the most conspicuous differences lies the nature of the junctions connecting endothelial cells to each other and the presence of openings like fenestrations. As a consequence, the number of specific, pan-vascular genes is rather limited, and as explained thereafter, no aGPCR seems to be expressed at a high level on the whole endothelium [32, 33].

3.2 Adhesion GPCRs Expression in Vascular Beds

Analysis of microarray data revealed that 58 genes are broadly and specifically enriched in microvascular endothelium [34]. Correlating with the known endothelial markers (VEGFR2, PECAM), two aGPCRs appeared highly expressed in vascular endothelium: ADGRL4 and ADGRF5 are, respectively, on the third and the fifth position of this ranking, suggesting they might be broadly expressed as pan-vascular genes. However, this study was not based on a profiling of vascular beds in all organs.

A more detailed analysis came from the profiling of organ-specific vascular beds, established by Shahin Rafii's team [35]. For the majority of aGPCRs, the expression over the different vascular beds is homogenous. However, 13 of them show a peak of expression in a discrete number of organ vasculature, summarized in the following chart (analyzed by Dr. Liqun He):

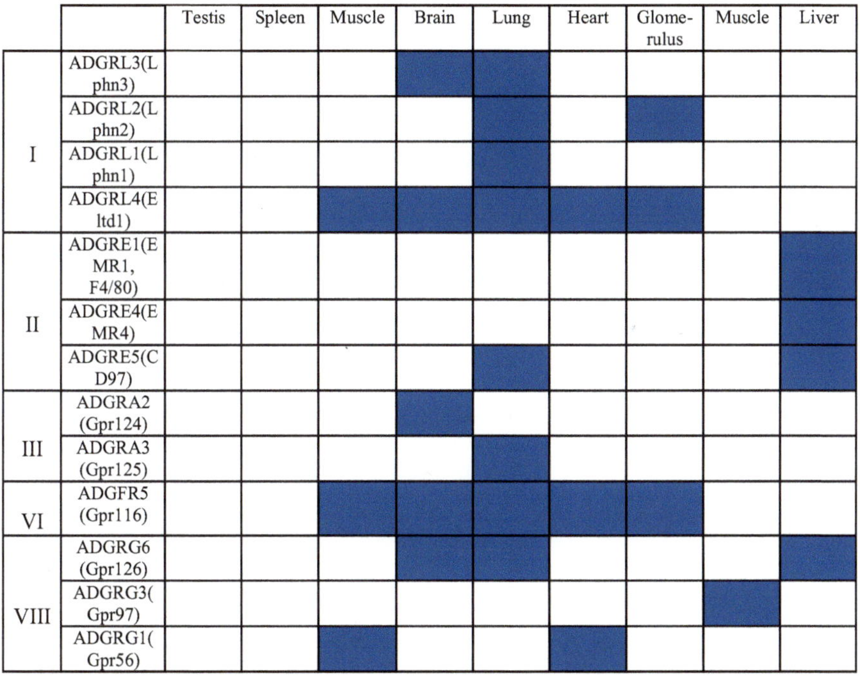

Each peak of expression might indicate that the aGPCRs intervene in the acquisition or maintenance of the organ-specific phenotype of vascular beds, like the lung (ADGRE5, ADGRA3, ADGRL1), the liver [ADGRE1 (EMR1, F4/80) and ADGRE4 (EMR4)], the brain [ADGRA2 (Gpr124)], or the bone marrow [ADGRG3 (Gpr97)]. Those results also suggest that ADGRL4 as well as ADGRF5 to some lesser extent might be somewhat less expressed in the vasculature of liver, bone marrow, spleen, and testis, rather than fully pan-vascular.

The recent progresses in single-cell profiling are progressively adding up a complementary parameter: are some aGPCRs expressed *exclusively* in the endothelial cells of a given organ? This requires establishing the transcriptome profiles of each cell type from a given organ.

The brain cortex of adult mice has been to date the subject of two intensive single-cell profiling, from Ben Barres and Sten Linnarsson laboratories [36, 37]. They revealed that 13 aGPCRs are exclusive to a specific cell type of the cortex, among which, two exclusively and abundantly expressed in the cortical endothelium (ADGRF5 and ADGRL4), in line with profiling of endothelial cells from the retina in p8 Tie2-Cre/GFP mice [38]. Besides, a group of aGPCR is abundant in the endothelium but co-expressed at lower levels by additional cell types: ADGRL2 (astrocytes, microglia, neuron), ADGRE5 (microglia, oligodendrocytes precursor cells, astrocytes), ADGRA2 (oligodendrocytes precursor cells), ADGRG6 (astrocytes, neurons), and ADGRG3 (neuron) [36].

Profiling of other organs is still less advanced, even if significant progress has been done in mapping the lung endothelium proteome [39].

Taken together, those expression profiles suggest a group of 13 genes with potential impact on endothelial functions:

- ADGRL4 and ADGRF5 possess the higher vascular expression, especially in the brain, lung, heart, and kidney endothelium.
- ADGRA2, ADGRG6 in the brain.
- ADGRL3 (Lphn3), ADGRL2, ADGRL1, ADGRG3, ADGRG1 (Gpr56), ADGRE5, ADGRE1, ADGRE4, and ADGRA3 have peaks of expression in a restricted subset of organ-specific vascular beds.

Extensive transcriptomic analysis of lymphatic vessels is still missing, yet the available data are promising when it comes to expression of aGPCRs. For instance, comparison in between freshly sorted lymphatic endothelial cells and blood vascular endothelial cells from mouse intestines showed that the lymphatic vessels expressed more ADGRG3, ADGRA2, and ADGRC1 (Celsr1) than their vascular counterpart and on the contrary seem to repress ADGRL2 and ADGRG1 [40].

As described below, to this list should be added aGPCRs with a steady expression, regulating key ubiquitous functions such as control of planar cell polarity, which might kick in on discrete spatio-temporal window in vascular development (see ADGRC1 below), as well as aGPCRs whose expression is induced very transiently, under normal or challenging conditions (see Sect. 3.4). Such identification requires in-depth profiling of organ-specific vascular beds, covering various developmental time points. For instance, very little is known about differential expression of aGPCRs upon arterial-venous differentiation, even if profiling studies suggest that arteries express more ADGRE5 and ADGRG1 but less ADGRA2, ADGRD1(Gpr133), and ADGRG6 than their venous counterpart in the umbilical cord [41, 42].

3.3 Adhesion GPCRs Knockouts in Physiological Vasculature

3.3.1 Formation of the Initial Vascular Network

Genes regulating critical, nonredundant events in vascular development lead to temporal specific lethality when knocked down in mice models. They pinpoint the sequence of morphologic events that lead to a fully functional vascular network. Two successive processes shape the vascular network. First, a primary vascular plexus is formed by the fusion of endothelial precursors, the angioblasts, at different sites of vascularization, among which the yolk sac, at E7.5. Mice lacking VEGFR2, a receptor for the key angiogenic factor VEGF, die around E8.5/E9.5 when those angioblasts fail to form [43]. In the brain, the angioblast migrates to create the perineural vascular plexus that covers the surface of the brain (E7.5–E8.5) [44].

Later, starting at about E9, capillaries will emerge from those primitive vessels through proliferation and migration of endothelial cells or splitting of preexisting vessels. Together those processes are known as angiogenesis. In the brain still, intracerebral vessels sprout from this network and dive gradually deeper into the neuroectodermal tissue toward the subventricular zone at E9.5. Animals in which this cerebral sprouting angiogenesis is impaired are severely affected, such as the (VEGF flox/flox; nestin Cre) mouse strain which dies at about E13.5 [45].

3.3.2 Maturation of the Vascular Network

Newly formed vessels mature and acquire location-specific differentiation. In certain organs, the vascular maturation might largely expand over the postnatal period. In the brain, vessels establish the blood-brain barrier (BBB) by tightening of the inter-endothelial junctions, downregulation of a number of transporters, and expression of specific markers such as Glut-1. The mouse BBB becomes functional at embryonic day 15.5 (E15.5) [46]. However, angiogenesis and parallel maturation persist at the postnatal stage [47], and therefore late lethality might occur such as in the Claudin-5-deficient mice [48]. Another example is the dorsal aorta, which forms very early on E8.5. Alteration in its maturation can lead to ectopic branching and glomeruloid vascular malformations, such as in the $S1pr1^{-/-}$ embryos (E11.5–E12.5), leading to rapid demise and synchronous death of the embryos between E13.5 and E14 [49].

3.3.3 Early Lethality Associated with Vascular Deficiency in Adhesion GPCRs Knockouts

However, in mice, only 3 genes out of the 13 aGPCRs defined above are associated with early lethality. In the case of ADGRA2, developed below, central nervous system hemorrhages result in lethality starting at E15.5 and preventing the survival of any mutant beyond E17.5 [50]. ADGRL2 mutant mice might also die according to a very similar temporal window (European Mouse Mutant Archive, unpublished results). Finally, ADGRG6 knockout mice are not viable, but the mutants die over a period ranging from late embryonic stages to early postnatal days (p8–p16, [24]). In the case of ADGRL2 and ADGRG6, a potential vascular phenotype has not been described yet, and ADGRG6 knockout is affected by a very severe demyelination phenotype.

The rest of the genes do not affect viability of the mice upon genetic inactivation. In particular, ADGRL4 and ADGRF5, the broadest expressed aGPCRs in endothelial cells, are perfectly viable [51] as well as ADGRL1 [52], ADGRL3 [53], ADGRG3 [54], ADGRG1 [55], ADGRE1 [56], and ADGRA3 [57], while the viability of the ADGRE4 knockout mouse is still unknown.

3.3.4 Adhesion GPCRs Regulating Blood-Brain Barrier

ADGRA2 (GPR124) is specifically expressed in the vessels of the brain and spinal cord, as well as within the arteries located outside the CNS. It is also expressed in the embryonic epithelium of the lung and esophagus and becomes exclusively vascular in adult tissues, with endothelial expression in CNS including the brain and retina and pericytic expression in the brain and peripheric organs. Global deletion of ADGRA2 or its endothelial-specific deletion in mouse results in embryonic lethality at E15.5. Vascular fronts fail to invade the neuroepithelium in the forebrain and spinal cord of the knockout animals at E11, forming abnormal tufts and profound CNS hemorrhages. Interestingly, this defect in migration and patterning is also accompanied by a defect in maturation of the vessels. Plasmalemma vesicle-associated protein, normally downregulated upon BBB induction, is maintained in ADGRA2-deficient vessels, whereas Glut-1, a marker of mature cerebral blood vessel, is absent. The absence of maturation of the blood vessels was functionally demonstrated as large deposits of extravascular fibrin surrounding glomeruloid tufts, and more importantly, biotin injected in living embryos was detected in both intra- and extravascular regions in the forebrain and ventral spinal cord, evidencing vascular leakage and hence a dysfunctional BBB. Overexpression of ADGRA2 in endothelial cells in culture enhanced their barrier properties, but interestingly overexpression in mouse leads to a progressive brain-specific vascular phenotype characterized by areas of hypervascularity. These lesions resemble venous angiomas, are restricted to the cortex, and frequently accompanied by calcifications [50, 58, 59].

Mechanistic investigations revealed how ADGRA2 functions as a WNT7A-/WNT7B-specific costimulator of β-catenin signaling in brain endothelium, both in mouse and zebrafish. ADGRA2$^{-/-}$ CNS angiogenesis defects can be rescued by artificially activating canonical Wnt signaling in the endothelium. ADGRA2 activity seems critical only in the tip cells leafing the sprouts, as the restoration of single wild-type tip cells in mosaic Wnt/β-catenin-deficient perineural vessels is sufficient to initiate the formation of CNS vessels. ADGRA2 could therefore ensure a high signal-to-noise ratio for WNT7a and WNT7b induced, in a context where signal strength of the Wnt pathway is close to the minimal threshold required for normal development and function [60–62].

As indicated before, Adgrf5 is expressed at high levels in various vascular beds and especially in the CNS vasculature. Genetic ablation in the mouse showed an anatomically normal and largely functional vascular network, but a disrupted blood-brain barrier. At 1.5 months of age, cerebral vascular leakage could be evidenced by tracer accumulation in the brain parenchyma. This leakage is maintained over age and phenocopied by an endothelial-specific deletion of ADGRF5, suggesting a cell-autonomous defect in late maturation or maintenance of the cerebral endothelium [51].

3.3.5 Regulation of Valve Formation by Adhesion GPCRs

The ADGRC1 case also shows that important aspects of the vascular role of aGPCRs might still be unknown. ADGRC1 is a critical determinant of planar cell polarity. Its complete knockout is partially lethal and leads to various defects, but its specific knockout in lymphatic vessels leads to abnormal organization of the valves in mesenteric vessels, at E18.5. This suggests that aGPCRs mediating general functions like planar cell polarity might control vascular phenotypes, if deleted in a tissue-specific fashion [63].

3.4 Adhesion GPCRs in Challenged Vasculature

In this last part more elusive functions regulated by aGPCRs in the endothelium but that were not evidenced during development will be reviewed: a pathological context is necessary to reveal this function or an in vitro model necessary to assess a fine mechanistic aspect, and in both cases it remains to be verified if those functions are regulated by the studied aGPCR during vascular development, as cultured cells might lack important cues derived from the in vivo microenvironment.

At least five aGPCRs have been subjected to mechanistic studies in the endothelium:

ADGRA2, whose role has been extensively studied in vivo, is also demonstrating peculiar properties in vitro: its expression is induced upon capillary morphogenesis in cultured venous endothelial cells, after which its N-terminal part is cleaved, through thrombin activity, and then mediates cell survival by bridging glycosaminoglycans of the ECM to integrin $\alpha_V\beta_3$ [64–66].

In a parallel fashion, the extracellular domain of ADGRG3 can be used as a substrate for cultured venous endothelial cells, via an interaction with integrin $\alpha5\beta1$ and $\alpha v\beta3$. It can also act as a chemoattractant when used in migration assays [67].

The extracellular fragment of ADGRB1 (BAI1) on the contrary acts as an inhibitor for the migration of dermal microvascular cells [68].

In the case of ADGRL4, silencing in venous endothelial cells had a slight effect on cell adhesion and no effect on endothelial migration in 2D but reduced quite dramatically the migration in a 3D beads assay. This inhibition might be paralleled with results in zebrafish, where the ADGRL4 morphants displayed blocking of intersegmental vessels, never reaching the dorsal position observed in wild type [69].

In lymphatic endothelial cells, silencing of ADGRG3 leads to dichotomous effects on the migration, with an enhanced collective cell migration and cell adhesion and reduced single-cell migration. In this system, ADGRG3 tilts the

Cdc42/RhoA balance, leading to F-actin redistribution and focal adhesion rearrangements [40].

The role of aGPCRs in tumor endothelium has been tackled mostly through orthotropic models (see also [70]).

ADGRE5 is induced in the human metastatic breast cancer cell line MDA-MB-231 [71], as well as elevated in various types of carcinomas. 3T3-derived tumor cells expressing no ADGRE5 or permanently transfected with ADGRE5 isoforms were injected into mice and used to demonstrate a greater vessel density in the tumors derived from ADGRE5-expressing cells, suggesting that ADGRE5 expression on the surface of tumor cells might stimulate angiogenesis [67]. On the opposite, ADGRA2 silencing in human endothelial cells inhibited angiogenic vessel formation and tumor growth in mouse xenograft [72].

ADGRB1 has been studied extensively in relation to tumor. While present in normal tissues, its expression is reduced or lost in cerebral or colorectal tumors [73, 74]. Its expression is controlled by p53, and it gets cleaved to release its 120 kDa extracellular domain, named vasculostatin, upon action of a proprotein convertase/MMP-14 cascade (Kaur et al. 2005 [75]). ADGRB1 or vasculostatin expression limits tumor growth and vascularization after human pancreatic adenocarcinoma cell line injection, murine renal cell carcinoma, gliomas, and in matrigel plug assay [76–78].

As mentioned before, ADGRL4 is a broad marker of the endothelium. Its involvement in the angiogenic response of tumors has also been solidly demonstrated: first, it was identified as part of a vascular cluster of genes expressed in vivo in human tumor samples derived from different cancer types (neck squamous cell carcinomas, breast cancers, and clear cell renal cell carcinomas). This cluster was first validated by studying modulation of the signature genes in response to antiangiogenic treatments in preclinical models. Second, detection of ADGRL4 by immunohistochemistry of tissue microarrays containing primary human tumor samples (head and neck, renal, colorectal, and ovarian cancer samples) confirmed an increased expression of the aGPCR in tumor vessels. Finally, the role of ADGRL4 in tumor was studied by injecting siRNA incorporated into chitosan nanoparticles into an orthotropic model of ovarian carcinoma and a colorectal cancer model. This resulted in a significant decrease in tumor weight, associated with a reduced metastatic dissemination and a steep increase in survival [69].

It would be interesting to see if some of these results hold true in mouse models harboring oncogenic lesions, in which the cooptation of the endogenous microenvironment reflects the patient situation more closely.

Finally, a number of challenging situations, often related to pathological context, might unravel unexpected roles for aGPCRs, following the example of ADGRF5, which upon genetic inactivation completely reverses the pathological neovascularization in the oxygen-induced retinopathy model [51]. In this perspective, a number of expression data suggest modulation of aGPCRs in the endothelium upon challenge:

- Shear stress: ADGRA2 and ADGRG6 expression is elevated upon shear stress [79, 80].
- Ischemia: ADGRB2 (BAI2) was discovered in the brain upon ischemic challenge [81].
- Hypoxia: ADGRE5, ADGRC2 (Celsr2), ADGRG5 (Gpr114), and ADGRG1 are induced, while ADGRD1, ADGRB3 (BAI3), and ADGRG7 (Gpr128) are repressed [82].
- Inflammation/cytokine exposure: ADGRL4, ADGRE2 (EMR2), ADGRF4 (Gpr115), ADGRB1, and ADGRG5 are induced, while ADGRL1, ADGRL2, ADGRE5, and ADGRD2 (Gpr144) are downregulated [83–85].
- Cirrhosis: ADGRA3 and ADGRC1 expression is induced, while ADGRL1, ADGFR5, and ADGRL4 are repressed [86].

4 Conclusions

Cardiac and vascular functions of aGPCRs are still largely unknown. Deletion of two aGPCRs, ADGRG6 and ADGRA2, has resulted in a severe cardiac and brain vascular phenotype, respectively. While the observed phenotypes of aGPCRs knockout mice demonstrate the importance of aGPCRs in the cardiovascular field, the ligands or mechanisms activating aGPCRs as well as their downstream signaling pathways during heart development, angiogenesis and in blood-brain barrier function remain elusive.

Acknowledgments F.B.E. acknowledges support from the Deutsche Forschungsgemeinschaft (FOR2149, Project 7, EN 453/10-1). C.N.'s work in the Betsholtz lab was supported by grants from the European Research Council (ERC-AdG #294556 BBBARRIER), the European Union (ITN-2012-317250-VESSEL), the Swedish Cancer Foundation and the Swedish Research Council, the Knut and Alice Wallenberg Foundation, and the Leducq Foundation through the Sphingonet transatlantic network. Bioinformatics analyses were kindly communicated by Dr. Liqun He.

References

1. High FA, Epstein JA (2008) The multifaceted role of Notch in cardiac development and disease. Nat Rev Genet 9:49–61
2. Pérez-Pomares JM, González-Rosa JM, Muñoz-Chápuli R (2009) Building the vertebrate heart – an evolutionary approach to cardiac development. Int J Dev Biol 53:1427–1443
3. von Gise A, Pu WT (2012) Endocardial and epicardial epithelial to mesenchymal transitions in heart development and disease. Circ Res 110:1628–1645
4. Samsa LA, Yang B, Liu J (2013) Embryonic cardiac chamber maturation: trabeculation, conduction, and cardiomyocyte proliferation. Am J Med Genet C Semin Med Genet 163C:157–168
5. Ferrand N, Pessah M, Frayon S, Marais J, Garel JM (1999) Olfactory receptors, Golf alpha and adenylyl cyclase mRNA expressions in the rat heart during ontogenic development. J Mol Cell Cardiol 31:1137–1142

6. Doyle SE, Scholz MJ, Greer KA, Hubbard AD, Darnell DK, Antin PB et al (2006) Latrophilin-2 is a novel component of the epithelial-mesenchymal transition within the atrioventricular canal of the embryonic chicken heart. Dev Dyn 235:3213–3221
7. Nechiporuk T, Urness LD, Keating MT (2001) ETL, a novel seven-transmembrane receptor that is developmentally regulated in the heart. ETL is a member of the secretin family and belongs to the epidermal growth factor-seven-transmembrane subfamily. J Biol Chem 276:4150–4157
8. Xiao J, Jiang H, Zhang R, Fan G, Zhang Y, Jiang D et al (2012) Augmented cardiac hypertrophy in response to pressure overload in mice lacking ELTD1. PLoS One 7:e35779
9. Veninga H, Becker S, Hoek RM, Wobus M, Wandel E, van der Kaa J et al (2008) Analysis of CD97 expression and manipulation: antibody treatment but not gene targeting curtails granulocyte migration. J Immunol 181:6574–6583
10. Prömel S, Waller-Evans H, Dixon J, Zahn D, Colledge WH, Doran J et al (2012) Characterization and functional study of a cluster of four highly conserved orphan adhesion-GPCR in mouse. Dev Dyn 241:1591–1602
11. Waller-Evans H, Prömel S, Langenhan T, Dixon J, Zahn D, Colledge WH et al (2010) The orphan adhesion-GPCR GPR126 is required for embryonic development in the mouse. PLoS One 5:e14047
12. Patra C, van Amerongen MJ, Ghosh S, Ricciardi F, Sajjad A, Novoyatleva T et al (2013) Organ-specific function of adhesion G protein-coupled receptor GPR126 is domain-dependent. Proc Natl Acad Sci U S A 110:16898–16903
13. Patra C, Monk KR, Engel FB (2014) The multiple signaling modalities of adhesion G protein-coupled receptor GPR126 in development. Recept Clin Investig 1:79
14. Gassmann M, Casagranda F, Orioli D, Simon H, Lai C, Klein R et al (1995) Aberrant neural and cardiac development in mice lacking the ErbB4 neuregulin receptor. Nature 378:390–394
15. Lee KF, Simon H, Chen H, Bates B, Hung MC, Hauser C (1995) Requirement for neuregulin receptor erbB2 in neural and cardiac development. Nature 378:394–398
16. Meyer D, Birchmeier C (1995) Multiple essential functions of neuregulin in development. Nature 378:386–390
17. Wang HU, Chen ZF, Anderson DJ (1998) Molecular distinction and angiogenic interaction between embryonic arteries and veins revealed by ephrin-B2 and its receptor Eph-B4. Cell 93:741–753
18. Gerety SS, Wang HU, Chen ZF, Anderson DJ (1999) Symmetrical mutant phenotypes of the receptor EphB4 and its specific transmembrane ligand ephrin-B2 in cardiovascular development. Mol Cell 4:403–414
19. Chen H, Shi S, Acosta L, Li W, Lu J, Bao S et al (2004) BMP10 is essential for maintaining cardiac growth during murine cardiogenesis. Development 131:2219–2231
20. Grego-Bessa J, Luna-Zurita L, del Monte G, Bolós V, Melgar P, Arandilla A et al (2007) Notch signaling is essential for ventricular chamber development. Dev Cell 12:415–429
21. D'Amato G, Luxán G, Del Monte-Nieto G, Martínez-Poveda B, Torroja C, Walter W et al (2016) Sequential Notch activation regulates ventricular chamber development. Nat Cell Biol 18:7–20
22. Newbern J, Birchmeier C (2010) Nrg1/ErbB signaling networks in Schwann cell development and myelination. Semin Cell Dev Biol 21:922–928
23. Woodhoo A, Alonso MBD, Droggiti A, Turmaine M, D'Antonio M, Parkinson DB et al (2009) Notch controls embryonic Schwann cell differentiation, postnatal myelination and adult plasticity. Nat Neurosci 12:839–847. doi:10.1038/nn.2323
24. Monk KR, Oshima K, Jörs S, Heller S, Talbot WS (2011) Gpr126 is essential for peripheral nerve development and myelination in mammals. Development 138:2673–2680
25. Sigoillot SM, Monk KR, Piao X, Selimi F, Harty BL (2016) Adhesion G protein-coupled receptors in the nervous system: from synapse and dendrite morphogenesis to myelination. In: Langenhan T, Schöneberg T (eds) Adhesion G protein-coupled receptors: molecular, physiological and pharmacological principles in health and disease. Springer, Heidelberg

26. Camenisch TD, Spicer AP, Brehm-Gibson T, Biesterfeldt J, Augustine ML, Calabro A et al (2000) Disruption of hyaluronan synthase-2 abrogates normal cardiac morphogenesis and hyaluronan-mediated transformation of epithelium to mesenchyme. J Clin Invest 106:349–360
27. Stankunas K, Hang CT, Tsun Z-Y, Chen H, Lee NV, Wu JI et al (2008) Endocardial Brg1 represses ADAMTS1 to maintain the microenvironment for myocardial morphogenesis. Dev Cell 14:298–311
28. Patra C, Diehl F, Ferrazzi F, van Amerongen MJ, Novoyatleva T, Schaefer L et al (2011) Nephronectin regulates atrioventricular canal differentiation via Bmp4-Has2 signaling in zebrafish. Development 138:4499–4509
29. Chernousov MA, Yu W-M, Chen Z-L, Carey DJ, Strickland S (2008) Regulation of Schwann cell function by the extracellular matrix. Glia 56:1498–1507
30. Paavola KJ, Sidik H, Zuchero JB, Eckart M, Talbot WS (2014) Type IV collagen is an activating ligand for the adhesion G protein-coupled receptor GPR126. Sci Signal 7:ra76
31. Petersen SC, Luo R, Liebscher I, Giera S, Jeong S-J, Mogha A et al (2015) The adhesion GPCR GPR126 has distinct, domain-dependent functions in Schwann cell development mediated by interaction with laminin-211. Neuron 85:755–769
32. Aird WC (2007) Phenotypic heterogeneity of the endothelium: I. Structure, function, and mechanisms. Circ Res 100:158–173. doi:10.1161/01.RES.0000255691.76142.4a
33. Aird WC (2012) Endothelial cell heterogeneity. Cold Spring Harb Perspect Med 2:a006429
34. Wallgard E, Larsson E, He L, Hellström M, Armulik A, Nisancioglu MH et al (2008) Identification of a core set of 58 gene transcripts with broad and specific expression in the microvasculature. Arterioscler Thromb Vasc Biol 28:1469–1476
35. Nolan DJ, Ginsberg M, Israely E, Palikuqi B, Poulos MG, James D et al (2013) Molecular signatures of tissue-specific microvascular endothelial cell heterogeneity in organ maintenance and regeneration. Dev Cell 26:204–219
36. Zhang Y, Chen K, Sloan SA, Bennett ML, Scholze AR, O'Keeffe S et al (2014) An RNA-sequencing transcriptome and splicing database of glia, neurons, and vascular cells of the cerebral cortex. J Neurosci 34:11929–11947
37. Zeisel A, Muñoz-Manchado AB, Codeluppi S, Lönnerberg P, La Manno G, Juréus A et al (2015) Brain structure. Cell types in the mouse cortex and hippocampus revealed by single-cell RNA-seq. Science 347:1138–1142
38. Fukushima Y, Okada M, Kataoka H, Hirashima M, Yoshida Y, Mann F et al (2011) Sema3E-PlexinD1 signaling selectively suppresses disoriented angiogenesis in ischemic retinopathy in mice. J Clin Invest 121:1974–1985
39. Durr E, Yu J, Krasinska KM, Carver LA, Yates JR, Testa JE et al (2004) Direct proteomic mapping of the lung microvascular endothelial cell surface in vivo and in cell culture. Nat Biotechnol 22:985–992
40. Valtcheva N, Primorac A, Jurisic G, Hollmén M, Detmar M (2013) The orphan adhesion G protein-coupled receptor GPR97 regulates migration of lymphatic endothelial cells via the small GTPases RhoA and Cdc42. J Biol Chem 288:35736–35748
41. Liu NM, Yokota T, Maekawa S, Lü P, Zheng Y-W, Tei I et al (2013) Transcription profiles of endothelial cells in the rat ductus arteriosus during a perinatal period. PLoS One 8:e73685
42. Aranguren XL, Agirre X, Beerens M, Coppiello G, Uriz M, Vandersmissen I et al (2013) Unraveling a novel transcription factor code determining the human arterial-specific endothelial cell signature. Blood 122:3982–3992
43. Shalaby F, Rossant J, Yamaguchi TP, Gertsenstein M, Wu XF, Breitman ML et al (1995) Failure of blood-island formation and vasculogenesis in Flk-1-deficient mice. Nature 376:62–66
44. Wittko-Schneider IM, Schneider FT, Plate KH (2014) Cerebral angiogenesis during development: who is conducting the orchestra? Methods Mol Biol 1135:3–20

45. Raab S, Beck H, Gaumann A, Yüce A, Gerber H-P, Plate K et al (2004) Impaired brain angiogenesis and neuronal apoptosis induced by conditional homozygous inactivation of vascular endothelial growth factor. Thromb Haemost 91:595–605
46. Ben-Zvi A, Lacoste B, Kur E, Andreone BJ, Mayshar Y, Yan H et al (2014) Mfsd2a is critical for the formation and function of the blood-brain barrier. Nature 509:507–511
47. Wälchli T, Mateos JM, Weinman O, Babic D, Regli L, Hoerstrup SP et al (2015) Quantitative assessment of angiogenesis, perfused blood vessels and endothelial tip cells in the postnatal mouse brain. Nat Protoc 10:53–74
48. Nitta T, Hata M, Gotoh S, Seo Y, Sasaki H, Hashimoto N et al (2003) Size-selective loosening of the blood-brain barrier in claudin-5-deficient mice. J Cell Biol 161:653–660
49. Gaengel K, Niaudet C, Hagikura K, Laviña B, Siemsen BL, Muhl L et al (2012) The sphingosine-1-phosphate receptor S1PR1 restricts sprouting angiogenesis by regulating the interplay between VE-cadherin and VEGFR2. Dev Cell 23:587–599
50. Kuhnert F, Mancuso MR, Shamloo A, Wang H-T, Choksi V, Florek M et al (2010) Essential regulation of CNS angiogenesis by the orphan G protein-coupled receptor GPR124. Science 330:985–989
51. Niaudet C, Hofmann JJ, Mäe MA, Jung B, Gaengel K, Vanlandewijck M et al (2015) Gpr116 receptor regulates distinctive functions in pneumocytes and vascular endothelium. PLoS One 10:e0137949
52. Tobaben S, Südhof TC, Stahl B (2002) Genetic analysis of alpha-latrotoxin receptors reveals functional interdependence of CIRL/latrophilin 1 and neurexin 1 alpha. J Biol Chem 277:6359–6365
53. Wallis D, Hill DS, Mendez IA, Abbott LC, Finnell RH, Wellman PJ et al (2012) Initial characterization of mice null for Lphn3, a gene implicated in ADHD and addiction. Brain Res 1463:85–92
54. Wang J-J, Zhang L-L, Zhang H, Shen C-L, Lu SY, Kuang Y et al (2013) Gpr97 is essential for the follicular versus marginal zone B-lymphocyte fate decision. Cell Death Dis 4, e853
55. Li S, Jin Z, Koirala S, Bu L, Xu L, Hynes RO et al (2008) GPR56 regulates pial basement membrane integrity and cortical lamination. J Neurosci 28:5817–5826
56. Lin H-H, Faunce DE, Stacey M, Terajewicz A, Nakamura T, Zhang-Hoover J et al (2005) The macrophage F4/80 receptor is required for the induction of antigen-specific efferent regulatory T cells in peripheral tolerance. J Exp Med 201:1615–1625
57. Seandel M, James D, Shmelkov SV, Falciatori I, Kim J, Chavala S et al (2007) Generation of functional multipotent adult stem cells from GPR125+ germline progenitors. Nature 449:346–350
58. Anderson KD, Pan L, Yang X, Hughes VC, Walls JR, Dominguez MG et al (2011) Angiogenic sprouting into neural tissue requires Gpr124, an orphan G protein-coupled receptor. Proc Natl Acad Sci U S A 108:2807–2812
59. Cullen M, Elzarrad MK, Seaman S, Zudaire E, Stevens J, Yang MY et al (2011) GPR124, an orphan G protein-coupled receptor, is required for CNS-specific vascularization and establishment of the blood-brain barrier. Proc Natl Acad Sci U S A 108:5759–5764
60. Zhou Y, Nathans J (2014) Gpr124 controls CNS angiogenesis and blood-brain barrier integrity by promoting ligand-specific canonical wnt signaling. Dev Cell 31:248–256
61. Vanhollebeke B, Stone OA, Bostaille N, Cho C, Zhou Y, Maquet E, et al (2015) Tip cell-specific requirement for an atypical Gpr124- and Reck-dependent Wnt/β-catenin pathway during brain angiogenesis. eLife 4
62. Posokhova E, Shukla A, Seaman S, Volate S, Hilton MB, Wu B et al (2015) GPR124 functions as a WNT7-specific coactivator of canonical β-catenin signaling. Cell Rep 10:123–130
63. Tatin F, Taddei A, Weston A, Fuchs E, Devenport D, Tissir F et al (2013) Planar cell polarity protein Celsr1 regulates endothelial adherens junctions and directed cell rearrangements during valve morphogenesis. Dev Cell 26:31–44

64. Vallon M, Essler M (2006) Proteolytically processed soluble tumor endothelial marker (TEM) 5 mediates endothelial cell survival during angiogenesis by linking integrin alpha(v)beta3 to glycosaminoglycans. J Biol Chem 281:34179–34188
65. Vallon M, Rohde F, Janssen K-P, Essler M (2010) Tumor endothelial marker 5 expression in endothelial cells during capillary morphogenesis is induced by the small GTPase Rac and mediates contact inhibition of cell proliferation. Exp Cell Res 316:412–421
66. Vallon M, Aubele P, Janssen K-P, Essler M (2012) Thrombin-induced shedding of tumour endothelial marker 5 and exposure of its RGD motif are regulated by cell-surface protein disulfide-isomerase. Biochem J 441:937–944
67. Wang T, Ward Y, Tian L, Lake R, Guedez L, Stetler-Stevenson WG et al (2005) CD97, an adhesion receptor on inflammatory cells, stimulates angiogenesis through binding integrin counterreceptors on endothelial cells. Blood 105:2836–2844
68. Kaur B, Brat DJ, Devi NS, Van Meir EG (2005) Vasculostatin, a proteolytic fragment of brain angiogenesis inhibitor 1, is an antiangiogenic and antitumorigenic factor. Oncogene 24:3632–3642
69. Masiero M, Simões FC, Han HD, Snell C, Peterkin T, Bridges E et al (2013) A core human primary tumor angiogenesis signature identifies the endothelial orphan receptor ELTD1 as a key regulator of angiogenesis. Cancer Cell 24:229–241
70. Aust G, Zhu D, Van Meir EG, Xu L (2016) Adhesion GPCRs in tumorigenesis. In: Langenhan T, Schöneberg T (eds) Adhesion G protein-coupled receptors: molecular, physiological and pharmacological principles in health and disease. Springer, Heidelberg
71. Lu YY, Sweredoski MJ, Huss D, Lansford R, Hess S, Tirrell DA (2014) Prometastatic GPCR CD97 is a direct target of tumor suppressor microRNA-126. ACS Chem Biol 9:334–338
72. Wang Y, Cho S-G, Wu X, Siwko S, Liu M (2014) G-protein coupled receptor 124 (GPR124) in endothelial cells regulates vascular endothelial growth factor (VEGF)-induced tumor angiogenesis. Curr Mol Med 14:543–554
73. Fukushima Y, Oshika Y, Tsuchida T, Tokunaga T, Hatanaka H, Kijima H et al (1998) Brain-specific angiogenesis inhibitor 1 expression is inversely correlated with vascularity and distant metastasis of colorectal cancer. Int J Oncol 13:967–970
74. Nam D-H, Park K, Suh YL, Kim J-H (2004) Expression of VEGF and brain specific angiogenesis inhibitor-1 in glioblastoma: prognostic significance. Oncol Rep 11:863–869
75. Cork SM, Kaur B, Devi NS, Cooper L, Saltz JH, Sandberg EM et al (2012) A proprotein convertase/MMP-14 proteolytic cascade releases a novel 40 kDa vasculostatin from tumor suppressor BAI1. Oncogene 31:5144–5152
76. Duda DG, Sunamura M, Lozonschi L, Yokoyama T, Yatsuoka T, Motoi F et al (2002) Overexpression of the p53-inducible brain-specific angiogenesis inhibitor 1 suppresses efficiently tumour angiogenesis. Br J Cancer 86:490–496
77. Kudo S, Konda R, Obara W, Kudo D, Tani K, Nakamura Y et al (2007) Inhibition of tumor growth through suppression of angiogenesis by brain-specific angiogenesis inhibitor 1 gene transfer in murine renal cell carcinoma. Oncol Rep 18:785–791
78. Kaur B, Cork SM, Sandberg EM, Devi NS, Zhang Z, Klenotic PA et al (2009) Vasculostatin inhibits intracranial glioma growth and negatively regulates in vivo angiogenesis through a CD36-dependent mechanism. Cancer Res 69:1212–1220
79. Andersson M, Karlsson L, Svensson P-A, Ulfhammer E, Ekman M, Jernås M et al (2005) Differential global gene expression response patterns of human endothelium exposed to shear stress and intraluminal pressure. J Vasc Res 42:441–452
80. White SJ, Hayes EM, Lehoux S, Jeremy JY, Horrevoets AJG, Newby AC (2011) Characterization of the differential response of endothelial cells exposed to normal and elevated laminar shear stress. J Cell Physiol 226:2841–2848
81. Kee HJ, Koh JT, Kim M-Y, Ahn KY, Kim JK, Bae CS et al (2002) Expression of brain-specific angiogenesis inhibitor 2 (BAI2) in normal and ischemic brain: involvement of BAI2 in the ischemia-induced brain angiogenesis. J Cereb Blood Flow Metab 22:1054–1067

82. Costello CM, Howell K, Cahill E, McBryan J, Konigshoff M, Eickelberg O et al (2008) Lung-selective gene responses to alveolar hypoxia: potential role for the bone morphogenetic antagonist gremlin in pulmonary hypertension. Am J Physiol Lung Cell Mol Physiol 295: L272–L284
83. Mayer H, Bilban M, Kurtev V, Gruber F, Wagner O, Binder BR et al (2004) Deciphering regulatory patterns of inflammatory gene expression from interleukin-1-stimulated human endothelial cells. Arterioscler Thromb Vasc Biol 24:1192–1198
84. Viemann D, Goebeler M, Schmid S, Nordhues U, Klimmek K, Sorg C et al (2006) TNF induces distinct gene expression programs in microvascular and macrovascular human endothelial cells. J Leukoc Biol 80:174–185
85. Rajashekhar G, Grow M, Willuweit A, Patterson CE, Clauss M (2007) Divergent and convergent effects on gene expression and function in acute versus chronic endothelial activation. Physiol Genomics 31:104–113
86. Tugues S, Morales-Ruiz M, Fernandez-Varo G, Ros J, Arteta D, Muñoz-Luque J et al (2005) Microarray analysis of endothelial differentially expressed genes in liver of cirrhotic rats. Gastroenterology 129:1686–1695

Adhesion GPCRs in Tumorigenesis

Gabriela Aust, Dan Zhu, Erwin G. Van Meir, and Lei Xu

Graphical Abstract

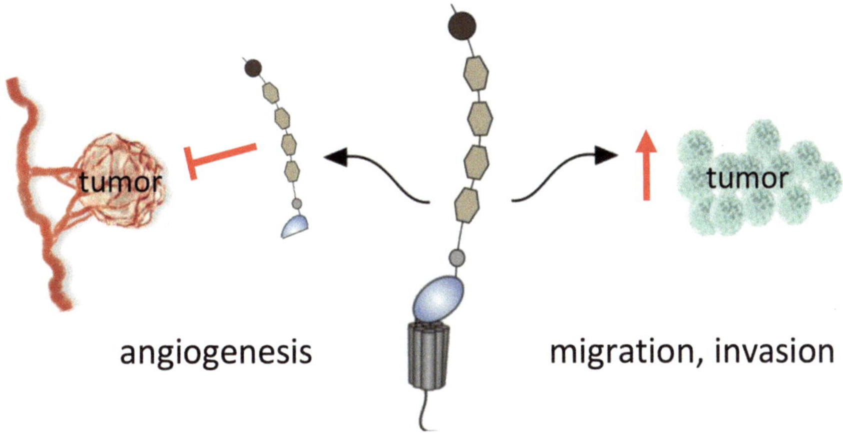

G. Aust (✉)
Department of Surgery, Research Laboratories, University of Leipzig, Liebigstraße 19, Leipzig 04103, Germany
e-mail: gabriela.aust@medizin.uni-leipzig.de

D. Zhu • E.G. Van Meir
Department of Neurosurgery and Hematology & Medical Oncology, School of Medicine and Winship Cancer Institute, Emory University, Atlanta, GA 30322, USA
e-mail: dzhu2@emory.edu; evanmei@emory.edu

L. Xu
Department of Biomedical Genetics, University of Rochester Medical Center, Rochester, NY 14642, USA
e-mail: lei_xu@urmc.rochester.edu

T. Langenhan, T. Schöneberg (eds.), *Adhesion G Protein-coupled Receptors*, Handbook of Experimental Pharmacology 234, DOI 10.1007/978-3-319-41523-9_17

Contents

Abstract

Alterations in the homeostasis of several adhesion GPCRs (aGPCRs) have been observed in cancer. The main cellular functions regulated by aGPCRs are cell adhesion, migration, polarity, and guidance, which are all highly relevant to tumor cell biology. Expression of aGPCRs can be induced, increased, decreased, or silenced in the tumor or in stromal cells of the tumor microenvironment, including fibroblasts and endothelial and/or immune cells. For example, ADGRE5 (CD97) and ADGRG1 (GPR56) show increased expression in many cancers, and initial functional studies suggest that both are relevant for tumor cell migration and invasion. aGPCRs can also impact the regulation of angiogenesis by releasing soluble fragments following the cleavage of their extracellular domain (ECD) at the conserved GPCR-proteolytic site (GPS) or other more distal cleavage sites as typical for the ADGRB (BAI) family. Interrogation of in silico cancer databases suggests alterations in other aGPCR members and provides the impetus for further exploration of their potential role in cancer. Integration of knowledge on the expression, regulation, and function of aGPCRs in tumorigenesis is currently spurring the first preclinical studies to examine the potential of aGPCR or the related pathways as therapeutic targets.

Keywords

Tumor cell migration • Tumor invasion • Metastasis • Tumor angiogenesis • Tumor therapy

Abbreviations

CCLE	Cancer Cell Line Encyclopedia
CTF	C-terminal fragment
ECD	Extracellular domain
ECM	Extracellular matrix
GAIN	GPCR autoproteolysis inducing
GPS	GPCR-proteolytic site
NTF	N-terminal fragment
TG2	Tissue transglutaminase
TSR	Thrombospondin repeat

1 Introduction

Cancer arises through the sequential accumulation of driver gene mutations which ultimately confer a growth advantage upon the cells in which they have occurred, leading to tumor formation through a clonal selection process. An important cancer hallmark is the ability of cancer cells to acquire invasive and metastatic properties which is associated with progression to systemic disease and reduced survival. Invasion and metastasis require tumor cell detachment and migration, extravasation into the circulation, and subsequent seeding at distant sites accompanied by altered extracellular matrix (ECM) turnover and initiation of tumor neo-angiogenesis. Each of these steps involves direct tumor cell-cell and tumor cell-microenvironment interactions, including stromal elements such as tumor-associated fibroblasts, endothelial cells, and tumor-infiltrating immune cells as well as the ECM. aGPCRs regulate cell adhesion, migration, polarity, and guidance; cellular functions that are also required for tumorigenesis. Characteristic for the aGPCRs is the large ECD with multiple adhesive folds, which is coupled to a seven-span transmembrane (7TM) domain and an intracellular domain (ICD) (see also [1–4]). Many aGPCRs are cleaved at the GPS in the juxtamembrane region which is part of and regulated by a GPCR autoproteolysis-inducing (GAIN) domain and results in an N- (NTF) and a C-terminal fragment (CTF), which remain associated at the cell surface (discussed in depth in [3, 5]). The various possible signaling scenarios resulting from this bipartite structure have been reviewed recently [6, 7] and in [8–10].

2 Software for the In Silico Interrogation of Cancer-Related Databases

A number of online resources provide user-friendly software for the rapid interrogation of large-scale cancer genomics data sets, and this in silico data mining provides useful starting information on aGPCR expression or mutation in a variety of cancer samples and cell lines.

1. The Cancer Cell Line Encyclopedia (CCLE) (www.broadinstitute.org/ccle) contains detailed genetic characterization for more than 1000 human cancer cell lines [11].
2. The cBioPortal for Cancer Genomics (www.cbioportal.org) provides analysis tools to query, analyze, and visualize large-scale cancer genomics data sets, including The Cancer Genome Atlas (TCGA) [12, 13].
3. The Human Protein Atlas (HPA) database (www.proteinatlas.org) contains high-resolution images showing the spatial distribution of proteins in dozens of normal and malignant tissues [14]. Note: the specificity of the used paraffin-suitable antibodies is not verified in depth.
4. The Catalogue of Somatic Mutations in Cancer (COSMIC) (cancer.sanger.ac.uk/cosmic) is designed to store and display somatic mutation information.
5. The SurvExpress is a biomarker validation tool and database for cancer gene expression with survival analysis [15].

3 Identification of Adhesion GPCRs in Cancer

In the last two decades, it has become apparent that changes in aGPCRs occur in cancer, either at the gene or protein expression level. Studies addressing the relevance of these changes to the tumorigenesis process are currently limited. CD97 was the first aGPCR identified to be correlated to cancer [16]. While it is not expressed in normal thyrocytes and well-differentiated papillary and in part follicular thyroid cancers, the protein was found to be elevated in anaplastic carcinomas. In the following years other aGPCRs such as GPR56 were identified to be cancer related [17]. Non-biased screens such as mRNA microarrays, genome-wide association (GWA) studies, or proteome analyses confirmed previously identified cancer-related changes in aGPCRs [18], showed alterations in new tumor entities [19], or identified new aGPCRs involved in cancer [20]. Recently, a whole-genome sequencing (WGS) approach in 441 human breast, lung, ovarian, and prostate cancers identified the aGPCRs *ADGRB3 (BAI3)* and *ADGRL3 (LPHN3)* among 77 significantly mutated genes [21]. Further studies are warranted to determine whether these mutations are functionally significant to cancer development or progression or simply passenger mutations.

4 Alterations in Adhesion GPCR Expression in Tumors Compared to Normal Tissue

aGPCRs may be induced, increased, decreased, or silenced in tumor compared to the corresponding normal cells. A quick way to determine whether a particular aGPCR may be present in tumor cells is to interrogate the CCLE [11]. This database shows high levels of mRNA for ADGRA3 (GPR125), ADGRB2 (BAI2), ADGRC2 (CELSR2) and ADGRC3 (CELSR3), CD97, GPR56, ADGRG6 (GPR126), ADGRL1 (LPHN1), and ADGRL2 (LPHN2) genes in hundreds of cell lines derived from nearly all tumor entities (note: the normal corresponding cells are not included in the database). The mRNA of other aGPCRs is absent in these cell lines, including many members of the ADGRF and ADGRG families. Whether expression in cultured cell lines reflects expression in tumor cells in vivo needs to be addressed in each individual aGPCR and tumor type.

There is also evidence for changes in posttranslational modification of aGPCRs associated with cancer, which could have functional consequences. This was first shown for CD97. In normal smooth and skeletal muscle cells, the non-modified CD97 protein core is present, whereas in their malignant counterparts, i.e., in leiomyosarcoma and rhabdomyosarcoma cells, CD97 is found N-glycosylated [66, 67]. It is well known that glycosylation patterns are modified in cancer [88, 89]; therefore, it is important to determine whether such changes are relevant to the tumorigenic process for each individual aGPCR. N-glycosylation of CD97 is mapped to the adhesive EGF-like domains and is needed for the binding of the interaction partner CD55 [67], which is consistent with CD97 and CD55 often being co-expressed in cancer [53, 54, 90–93].

Regulation of aGPCR expression in stromal cells, such as tumor-associated fibroblasts and endothelial or tumor-infiltrating immune cells, as shown for ADGRA2 (GPR124) and all BAIs, may also support or suppress tumor progression. For example, BAI1 has been shown to play a role as an engulfment receptor on macrophages that can recognize apoptotic cells through binding of membrane-exposed phosphatidylserine groups to BAI1's extracellular thrombospondin repeats (TSRs) [94]. Variation in BAI1 expression on macrophages or possibly tumor cells may change the dynamics of dying tumor cell clearance from tumors, related inflammation, and cancer metabolism.

5 Soluble Adhesion GPCR and Cleavage Fragments in Tumorigenesis

As a direct consequence of the autocatalytic cleavage of aGPCRs at the GPS, as well as through (additional) independent proteolysis events within their ECD, the NTF or smaller fragments may be released from the receptor and appear in body fluids. The existence of circulating soluble NTFs is reported for GPR124 [33], BAI1 [69, 85], BAI2 [95], CD97 [96], ADGRF5 (GPR116) [97], GPR126 [98], and LPHN1 [99].

Theoretically, soluble NTFs could engage with interaction partners over large distances far away from the site at which they were released and thereby modulate heterotypic tumor-stromal interactions that may be relevant to tumorigenesis. This has been shown experimentally for the NTFs of the various isoforms of CD97, which differ in the number of EGF-like domains in their ECDs. Soluble CD97 stimulates tumor angiogenesis in vitro and in vivo through binding of integrins and chondroitin sulfate, a constituent of the ECM, in an isoform-specific manner [100]. Soluble CD97 is found at sites of inflammation [96, 101, 102], most likely released from CD97-positive immune cells, but not in body fluids of patients suffering from CD97-positive tumors [48]. These data suggest either that unknown cell-specific mechanisms control the release of the NTF or that an additional independent proteolysis event, restricted to immune cells, is necessary for this release.

In contrast to the pro-tumorigenic activity of CD97 NTFs, the NTFs of BAI1 have anti-angiogenic and anti-tumorigenic activities [69, 70] as further discussed below.

6 Biological Functions of Adhesion GPCRs in Cancer

6.1 Proliferation, Apoptosis, and Cell Cycle Regulation

Little is known about the role of aGPCRs in tumor growth, either through alteration in tumor cell division or cell death mechanisms including apoptosis. A few aGPCRs have been shown to promote cell proliferation and/or prevent apoptosis, although their role in signaling is not clarified. Downregulation of ADGRL4 (ELTD1) by microRNA-139-5p leads to an inhibition of glioblastoma cell proliferation in vitro [103]. CD97 protects fibrosarcoma HT1080 and cervical HeLa cancer cell lines from serum starvation- and staurosporine-induced intrinsic apoptosis [104].

Preliminary studies examining the effect of GPR56 on in vitro tumor cell proliferation and apoptosis have yielded contradictory results and may reflect the different cancer types and studies performed. One study reported that the knock-down of GPR56 expression induces a transient activation of apoptosis in several cancer cell lines, suggesting that GPR56 may be important for cell survival and have pro-tumorigenic functions [32]. This pro-apoptotic response could be overcome, as this research group was able to generate stable cell lines with GPR56 knockdown. These clones show a reduction in anchorage-independent growth in vitro, and some also have decreased tumor growth in vivo [32]. Similar observations were reported in the EVI-1^{high} leukemia cell line, in which GPR56 knockdown induced apoptosis although its effects on leukemia progression in vivo were not explored [22]. This pro-growth function of GPR56 contrasts with observations by another group in metastatic melanoma (A375P and MC-1) cell lines. This study suggests that GPR56 is necessary for the subcutaneous growth of these cells in mice, but does not affect their in vitro proliferation [60, 105]. The mechanisms linking GPR56 to in vivo tumor growth are found to depend on regulation of proper ECM deposition by the melanoma cells [60]. A larger panel of cell lines as well as a combination of in vitro and in vivo studies will be needed to fully understand the function of GPR56 during cancer progression.

6.2 Cell Adhesion and Interaction with ECM

Some aGPCRs have been implicated in regulating tumor cell adhesion, but potential mechanisms of these effects have not been investigated in depth. Knocking down *Gpr125* in murine myeloid sarcoma lines that harbor $KRAS^{G12C}$ and MLL/AF10(OM-LZ) oncogenes leads to reduced myeloid sarcoma burden in mice, and a decrease in cell adhesion in culture, possibly contributing to their reduced tumorigenicity [106].

ECM proteins are large peptides that assemble into highly ordered structures modulating various aspects of cell behavior, including cell adhesion, migration, survival, and proliferation, which are key factors during tumorigenesis [107]. GPR56 NTF was discovered to bind tissue transglutaminase (TG2) [105] and collagen III [108]. It was reported that GPR56 internalizes TG2 from the surface of melanoma cells, resulting in defects in ECM deposition in melanomas [60]. This defect may contribute to the inhibitory function of GPR56 on melanoma growth and metastasis. Knocking down GPR56 inhibits the adhesion of $EVI\text{-}1^{high}$ leukemia cells to ECM [22], and granule cells from the rostral cerebellum of $Gpr56^{-/-}$ mice are defective in adhering to fibronectin and laminin [109]. The adhesion effect of GPR56 could be blocked by its NTF: purified recombinant GPR56 NTF inhibits adhesion of glioma cells to fibronectin [82]. Taken together, the above studies point to a promoting role of GPR56 in cell-ECM adhesion, but whether this is a shared mechanism among different tumor cells and whether it contributes to the function of GPR56 in cancer progression remain to be investigated.

Several other aGPCRs such as ADGRE2 (EMR2) and GPR126 have also been shown to interact with the ECM [110–112], but the impact of these interactions on cancer has not been explored.

6.3 Cell Migration and Invasion

Among all aGPCRs, CD97 has been the most studied in the context of cell migration and invasion (see also [113] on this topic). Its association with tumor invasion is reported in several studies. It is upregulated at the invasive tumor front [42, 48, 91], and its high expression correlates with lymph node invasion [48, 54], advanced tumor stages [48, 53, 54, 91, 93], and patient survival [53, 77, 93]. The positive effects of CD97 on tumor invasion are verified in experimental models both in vitro and in vivo [49, 58]. Consistent with its function in promoting migration and invasion, CD97 is reported to enhance trans-well cell migration toward fetal calf serum in 15 different colorectal cancer cell lines [48] and toward fetal calf serum and lysophosphatidic acid (LPA) in AML cells (MV4-1 cell line) [29]. Consistently, directed cell migration and invasion stimulated by fetal calf serum are decreased following CD97 knockdown in a prostate cancer cell line (DU145) [58]. Conflicting results have been published on the effect of CD97 in a fibrosarcoma cell line (HT1080) on in vitro migration and in vivo formation of tumors and metastases: in one study CD97 overexpression increases migration in vitro and induces earlier tumor growth [49], whereas another study found the opposite [114].

Other aGPCRs have also been reported to regulate cancer cell migration. GPR56 inhibits cell migration toward stromal cell-derived factor 1 in leukemia cell lines [22], and GPR116 promotes cell migration and invasion toward serum in a breast cancer cell line (MDA-MB-231) through the Gαq-p63RhoGEF-RhoA/Rac1 pathway [65].

6.4 Angiogenesis

Several aGPCRs are found to regulate tumor angiogenesis, the process through which tumors elicit vessel formation that is critical for their sustained growth (see also [115] on this topic). The most studied one is BAI1, which inhibits angiogenesis in glioblastomas, the most malignant primary intracranial brain tumor which is incurable [69, 70, 116, 117] as well as other cancer types [36, 37, 118, 119]. The ECD of BAIs contains TSRs, and some TSRs are known to negatively regulate angiogenesis [120, 121]. Cleavage of the BAI1 ECD leads to the release of two fragments, dubbed vasculostatin-40 (Vstat40) and vasculostatin-120 (Vstat120) due to their molecular size (Fig. 1), and both were shown to suppress blood vessel formation in a variety of in vitro and in vivo assays [69, 70, 116]. When overexpressed, GPR56 can also inhibit tumor angiogenesis in melanoma, probably by blocking the secretion of vascular endothelial growth factor (VEGF) [61].

A number of aGPCRs also stimulate neovascularization in tumors. GPR124 was described first as a tumor endothelial marker (TEM5) [125]. ELTD1 is identified as part of a core angiogenic signature composed of genes whose expression jointly correlates with that of several well-recognized angiogenesis and/or endothelial cells "seed" genes, in more than 1000 human primary tumors, and is subsequently found to be essential for angiogenesis and tumor growth [34]. The NTF of CD97 promotes angiogenesis via binding to integrin $\alpha5\beta1$ and $\alpha v\beta3$ and elicits a chemotactic response in endothelial cells leading to their recruitment to the tumor [100].

7 Families of Adhesion GPCR in Tumorigenesis

Table 1 gives an overview on published studies of aGPCRs in human tumors.

7.1 ADGRA

GPR124 was originally identified as an endothelial marker that is upregulated during tumor angiogenesis [126]. A soluble GPR124 fragment, probably the NTF, is shed from *GPR124*-transfected endothelial cells. Further proteolytic processing creates a protein subunit that mediates endothelial survival and subsequent tumor angiogenesis via interactions with glycosaminoglycans and the integrin αvβ3 [33]. Only cell lines derived from Ewing sarcoma, chondrosarcoma, and osteosarcoma strongly express GPR124 [11]. microRNA miR-138-5p, found to be decreased in gefitinib-resistant non-small cell lung cancer (NSCLC) cell lines (PC9GR, H1975), directly targets GPR124. Thus, downregulation of GPR124 is discussed as a therapeutic approach to overcome NSCLC gefitinib resistance [127]. Knockdown of *Gpr125* reduces tumor cell aggregation and diminishes myeloid sarcoma formation induced by transplantation of immortalized acute myeloid leukemia [106].

7.2 ADGRB

The BAI1–3 proteins are predominantly expressed in the brain [116, 128]. The human *ADGRB1 (BAI1)* gene is located on chromosome 8q24, and BAI1 is the most studied member in this aGPCR subfamily. BAI1 contains several well-defined protein modules in the N-terminus such as an integrin-binding Arg-Gly-Asp (RGD) motif followed by five TSRs, a hormone binding domain and a GAIN domain with a GPS (Fig. 1) [85]. BAI1 can be cleaved at the GPS releasing a soluble, 120 kDa anti-angiogenic NTF called vasculostatin (Vstat120), which was reported to suppress the proliferation of endothelial cells by blocking αvβ5 integrin [129], reduce the migration of cultured microvascular endothelial cells in a CD36-dependent manner [70], and inhibit angiogenesis and glioma growth in vivo [69, 71, 72, 116, 130]. Subsequently, a novel more distal proteolytic processing event was identified in BAI1, and it generates a more abundant cleaved NTF containing only the RGD and the first TSR (Vstat40). This fragment was also able to inhibit angiogenesis in vitro and in vivo [71]. At the other end of the protein, the C-terminus of BAI1 contains an intracellular proline-rich region (PRR) and a terminal PDZ-binding domain (Fig. 1), which associates with a number of intracellular signaling and scaffolding proteins [131, 132], but the role of these signaling events in cancer has not been studied to date. BAI1 also plays an important role in myogenesis, synaptic plasticity, and phagocytosis [124], but it is unclear whether any of these functions can also intersect with cancer.

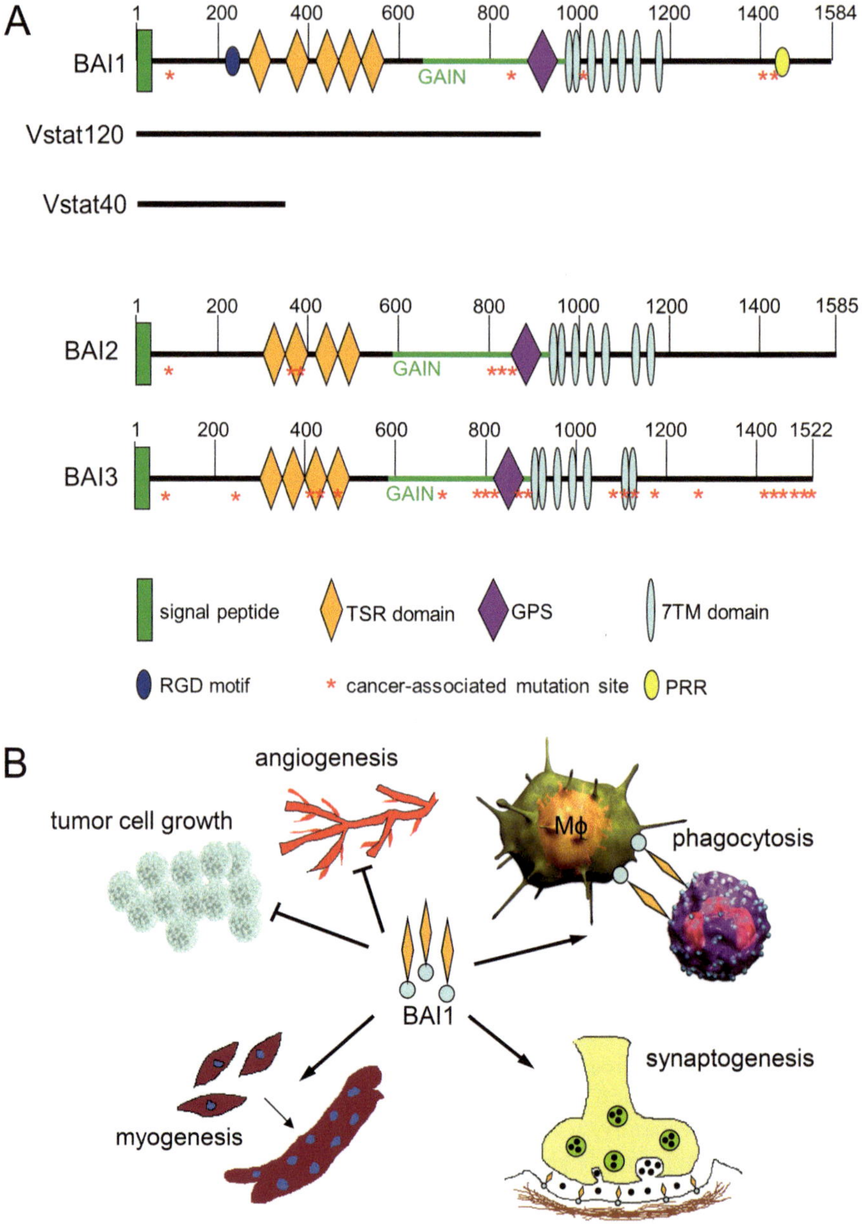

Fig. 1 Structure and functions of BAIs (ADGRBs). (**a**) Schematic of the three BAI proteins, representing the major known structural and functional features as well as known proteolysis events that generate either Vstat120 [69] or Vstat40 [71], and cancer-associated somatic mutations [21, 122]. Abbreviations: *TSR* thrombospondin type 1 repeat, *GAIN* GPCR autoproteolysis-inducing domain [123], *GPS* GPCR-proteolytic site, *7TM* seven-transmembrane region, *RGD* Arg-Gly-Asp integrin-binding motif, *PRR* proline-rich region. (**b**) Multiple functions of BAI1:

BAI1 mRNA levels are consistently downregulated in primary glioma specimens and cell lines [72, 73], and in brain metastases from lung adenocarcinoma [37]. Pulmonary adenocarcinomas, gastric, and colorectal cancers also show reduced BAI1 expression compared with normal tissue [36, 37, 44, 46]. A recent study also suggests that BAI1 expression is significantly reduced in breast cancer and correlates with poorer patient survival [62].

The human *BAI2* and *BAI3* genes were discovered as a result of their sequence identity with BAI1 [128] and are localized on chromosomes 1p35 and 6q12, respectively. The amino acid sequence is highly conserved in all three members; however, BAI1 contains five TSRs, while BAI2 or BAI3 each contains only four. In addition, the extracellular RGD motif and the intracellular proline-rich region (PRR), a domain known to interact with Src homology 3 (SH3) and WW domain-containing proteins [133], are unique to BAI1. Unlike BAI1, the expression of BAI2 or BAI3 is not silenced in glioma cells [72]. Moreover, recent studies determined higher BAI2 expression in tumor metastasis and advanced stages [134, 135].

Importantly, recent discoveries have shown that the BAI genes are silenced and/or undergo somatic mutations in several cancers (Fig. 1), including the lung, breast, ovarian, and brain [21, 72, 122], suggesting that eliminating their function might be required for tumor formation. Whether the identified mutations alter BAI1 function and may contribute to tumor formation has not been investigated to date. The function of BAI2 and BAI3 in tumorigenesis needs to be further investigated.

7.3 ADGRC

Although the role of ADGRCs (CELSRs) as key components in planar cell polarity (PCP) was clarified elegantly in knockout mice, our knowledge on their expression and function in human malignancies is rather limited. Unbiased screening approaches for cancer-related proteins often scored *CELSRs* as interesting hits, but most of them await further validation as to their role in cancer. *CELSR1* is upregulated in B cells of patients with chronic lymphocytic leukemia (CLL) [25]; *CELSR1* and *CELSR3* are found to be present in gastrointestinal and brain tumors, respectively [47]; and *Celsr1* is overexpressed in Graffi murine leukemia virus (MuLV)-induced hematologic malignancies with a lymphoid subtype [24]. Gene expression profiling of mantle cell lymphoma shows that *CELSR1* is downregulated in the non-nodal form compared to other lymphomas [27], and *CELSR1* expression is increased in pure ductal breast carcinomas in situ (DCIS) compared to such

Fig. 1 (continued) BAI1 inhibits tumor cell growth in part by inhibiting angiogenesis. Moreover, its TSRs interact with exposed phosphatidylserine at the outer leaflet on apoptotic cells and elicit the engulfment in macrophages (Mϕ). BAI1 also promotes myogenesis. Deficiency of BAI1 promotes PSD-95 degradation at the synapses and induces enhanced long-term potentiation. Figure adapted from [124]

Table 1 Expression of adhesion GPCRs in tumors

			ADGRA2	ADGRB1	ADGRB2	ADGRB3	ADGRC1	ADGRC3	ADGRE2	ADGRE3	ADGRE5	ADGRF5	ADGRG1	ADGRG2	ADGRG4	ADGRL1	ADGRL4
System	Tissue	Tumor	GPR124	BAI1	BAI2	BAI3	CELSR1	CELSR3	EMR2	EMR3	CD97	GPR116	GPR56	GPR64	GPR112	LPHN1	ELTD1
Hematopoietic		Leukemia											[22, 23]				
		B-cell leukemia					[24]										
		Chronic lymphocytic leukemia (CLL)					[25]										
		Acute lymphatic leukemia (ALL)									[19, 26]						
		Mantle cell lymphoma (MCL)					[27]										
		Acute myeloid leukemia (AML)									[28], [29]		[30]				
		Other leukemias											[31]				
Cardiovascular	Heart												[32]				
	Endothelial cells		[33]														[34, 35]
Respiratory	Lung	Non-small cell lung cancer (NSCLC)		[36, 37]	[21]	[38]						[20]	[32, 39]			[40]	
		Small cell lung cancer (SCLC)				[41]											
Digestive	Esophagus										[42]		[43]				
	Stomach	Gastric cancer		[44]							[42]		[45]				
	Large bowel	Colorectal		[46]			[47]				[48, 49]		[32]				
	Neuroendocrine	Ileal carcinoid													[50]		
	Liver	Hepatocellular carcinoma					[51]						[52]				
	Gall bladder, bile duct	Gall bladder carcinoma									[53]						
	Pancreas	Pancreatic cancer									[42, 54]		[32, 55]				
Urinary	Kidney	Wilms' tumor (pediatric)															
		Renal cancer											[32, 56]				
Endocrine	Thyroid	Thyroid cancer									[16, 57]						

Reproductive	Prostate	Prostate cancer									[58]						
	Ovary	Ovarian carcinoma			[21]	[21]							[32]				
	Uterus	Cervical											[32]				
Skin	Skin	Skin tumors														[59]	
		Melanoma											[32, 60, 61]				
	Mammary	Breast cancer, invasive ductal		[62]	[21]		[63]		[64]			[65]	[32]				
Musculoskeletal	Skeletal muscle	Rhabdomyosarcoma									[66]						
	Smooth muscle	Leiomyosarcoma									[67]						
	Bone related	Ewing sarcoma												[68]			
Brain/nervous system	Neurons, glia	Brain tumors		[37]				[47]					[32]				
		Glioblastoma multiforme (GBM)		[69–73]					[74, 75]	[76]	[18, 21, 77–81]		[82, 83]				[84]
		Malignant glioma		[85, 86]	[72]	[72]							[82]				
		Astrocytoma											[82]				
		Medulloblastoma												[87]			

Most studies were performed at the protein level. Only aGPCRs with published data are included. The direction of regulation is not indicated because of the partly complex investigated scenarios tumors with invasive components [63]. Moreover, the gene locus of *CELSR1* is hypermethylated in hepatocellular carcinoma (HCC) compared to control tissue [51].

7.4 ADGRE

ADGRE1–3 (EMR1–3) expression is restricted to leukocytes. Thus, almost only cell lines derived from hematologic malignancies express them [11]. EMR2 is not or rarely expressed in gastric, pancreatic, esophageal, and colorectal carcinomas [42, 136]. Here, only a subpopulation of tumor-infiltrating macrophages is strongly EMR2 positive. Aberrant expression of EMR2 protein is reported in breast carcinomas. It is associated with poor patient survival [64]. *EMR2* is found upregulated in glioblastoma, in particular in the mesenchymal subtype [74], and is associated with poor overall survival and an invasive phenotype [75].

EMR3 is also found to be increased in glioblastoma and associated with poor survival [76]. In colon cancer EMR3 is restricted to tumor-infiltrating immune cells [137].

CD97 is the only ADGRE family member whose expression is not restricted to immune cells. In human, its expression varies in cells of epithelial and mesenchymal origin from negative, as in keratinocytes and thyrocytes, to low in enterocytes, and high in pneumocytes and leukocytes. Cancer cell lines are almost all moderately or strongly *CD97* positive, suggesting that, compared to normal tissues, CD97 is frequently induced and/or increased in the corresponding malignancies. Low *CD97* levels are found only in cell lines derived from neuroblastomas. Small cell lung cancer (SCLC) cell lines are nearly all CD97 negative, while non-small cell lung cancer (NSCLC) cell lines are strongly CD97 positive, a fact that could be helpful in discriminant analysis on lung cancers.

In thyroid cancer, CD97 is induced and expression levels correlate with malignant progression [16, 57], i.e., only few papillary and follicular thyroid carcinomas are CD97 positive, whereas most anaplastic cancers strongly express CD97. In normal human intestinal epithelial cells, CD97 resides at low levels in lateral cell contacts, whereas in colorectal cancer, the molecule is increased. In part, this upregulation parallels its new cellular location within the cytoplasm [48, 49]. Overexpression of CD97 in scattered tumor cells is observed at the tumor invasion front [48]. The presence of these CD97-positive scattered tumor cells correlates with higher tumor stage and higher lymphatic vessel infiltration, both prognostic factors in colorectal cancer. CD97 is also upregulated in gastric cancer [42]. Further supporting an oncogenic role, in mouse models of colorectal and gastric cancer, CD97 supports local tumor growth and promoted metastatic spread [49, 138].

In gall bladder carcinoma, CD97 expression is an independent risk factor for overall survival [53]. CD97 is highly expressed in tumor cells of poorly differentiated pancreatic ductal adenocarcinoma and in tumor-infiltrating leukocytes in chronic pancreatitis samples, but not on normal pancreatic epithelial cells [139]. In a study of

37 pancreatic cancer patients, CD97 correlates with aggressiveness and was associated with prognosis [54]. In esophageal cancer CD97 is among the 19 genes with promoter hypomethylation and upregulation [140].

CD97 mediates invasion in prostate cancer cells, at least in part, by associating with lysophosphatidic acid receptor 1 (LPAR1), leading to enhanced LPA-dependent RHO and extracellular signal-regulated kinase (ERK) activation. Consistent with its role in invasion, depletion of CD97 in prostate cancer cells (PC3 cell line) results in decreased bone metastasis without affecting subcutaneous tumor growth. CD97 and LPAR1 are significantly co-expressed in clinical prostate cancer specimens.

CD97 is highly expressed in glioblastoma (WHO grade IV) and has prognostic significance in two independent cohorts of 187 and 539 glioblastoma patients, respectively [18, 77]. CD97 is also identified as a potential biomarker for glioma-initiating cells through an in vivo phage display screen [78], suggesting it represents a potential new therapeutic target [77]. In contrast, CD97 expression is minimal in WHO grade II and III astrocytomas [79]. Experimental data suggest that CD97 regulates tumor cell invasion in glioblastomas. Decreasing CD97 by siRNA reduces migration and invasion, but not proliferation in two glioblastoma cell lines [77]. Mass spectroscopy-based proteomic analysis showed that CD97 is enriched in membrane fractions of invadopodia, actin-rich protrusions, of invasive glioblastoma cells [18]. The mechanism underlying CD97-mediated promotion of invasion of glioblastoma cells is under investigation. Suppression of Wilms' tumor gene product WT1 by siRNA led to a decrease in the invasiveness of glioblastoma cell lines paralleled by a suppression of *CD97* RNA [80].

Further evidence for a role of CD97 in tumorigenesis comes from screening studies demonstrating that it is a direct target of the tumor suppressor microRNA-126 in the breast cancer cell line MDA-MB-231 [141].

Additionally, evaluation of CD97 expression in leukemia is highly informative. Gene expression studies and characterization of the leukemia cell surface proteome identified CD97 as a marker for minimal residual disease in acute lymphoblastic leukemia (ALL) [19]. CD97 expression also accounts for the most informative differences between normal and malignant cells in ALL [26]. CD97 has further been identified as a leukemic stem cell marker in acute myeloid leukemia (AML) [28]. In AML, expression levels of CD97 are associated with internal tandem duplications within the juxtamembrane region of the FMS-like tyrosine kinase receptor FLT3 (LFT3-ITD) [29].

Whether the function of CD97 in tumorigenesis could be modulated by other interaction partners such as LPAR1 is unknown. CD55 [142], chondroitin sulfate B [110], $\alpha_5\beta_1$ and $\alpha_v\beta_3$ integrins [100], and CD90 [143] bind distinct sites within the NTF of CD97 at immune cells; nothing is known whether these interaction partners also bind CD97 in tumors. A few articles have been published on co-expression of CD55 and CD97 in solid tumors but not on the resulting functional consequence [53, 54, 91–93].

7.5 ADGRF

GPR116, highly enriched in fetal and adult lung [144–146], is part of a gene expression signature that differentiates adenocarcinoma of lung and breast origin in effusions [20]. Knockdown of *GPR116* suppresses migration and invasion, whereas ectopic expression enhances invasion of the breast cancer cell line MDA-MB-231 [65]. GPR116 promotes constitutively breast cancer metastasis via the Gαq-p63RhoGEF-Rho GTPase pathway [65]. Its expression is significantly correlated with breast tumor progression, recurrence, and poor prognosis.

7.6 ADGRG

Among the ADGRG members, GPR56 is the most studied in the context of cancer progression. It was reported in 1999 to be downregulated in highly metastatic melanoma cell lines compared with poorly metastatic lines [17]. Its inverse correlation with metastatic potential was evaluated in more depth later in an experimental metastasis model which shows the downregulation of GPR56 in several highly metastatic melanoma derivatives compared with their poorly metastatic parental line (A375P). Re-expression of GPR56 inhibits the growth and metastasis of melanoma cells [105], supporting a tumor suppressor role for GPR56 in melanoma. Mechanistic studies in one of the metastatic derivatives (MC-1) revealed that GPR56 overexpression inhibits the activation of PKCα, resulting in the suppression of VEGF secretion from melanoma cells, leading to angiogenesis inhibition [61]. In contrast to the full-length receptor, expression of the CTF of GPR56 or deletion of the serine threonine proline (STP)-rich segment (ΔSTP-GPR56) leads to enhanced activation of PKCα, VEGF secretion, and angiogenesis. These opposing functions of GPR56 and its CTF indicate that GPR56 might exist in different activation states, although the signaling mechanisms of GPR56 in cancer have not been elucidated. The NTF of GPR56 binds to tissue transglutaminase (TG2) [105] and collagen III [108]. The interaction between GPR56 and collagen III in cancer progression is not reported. TG2 is a cross-linking enzyme in the ECM and thought to play pleiotropic roles in cancer progression [147]. The interaction of GPR56 with TG2 in melanoma growth was evaluated recently in a xenograft model using the immunodeficient $Tg2^{-/-}$ mice [60]. TG2-knockdown melanomas growing in these mice are depleted of TG2 in both cancer cells and stroma and found to be much smaller than the control tumors growing in $Tg2^{+/+}$ mice, arguing that TG2 promotes melanoma growth. This tumor-promoting function is abolished by GPR56 overexpression, probably via receptor-mediated TG2 internalization from the cell surface [60]. Whether GPR56 functions similarly in metastasis and in other cancer types remains to be determined.

In contrast to the situation in melanoma, GPR56 is often found upregulated in other cancer types relative to the corresponding normal tissues [32, 43, 148, 149], so perhaps its function in cancer is cell of origin and/or stage specific. Knocking down of *GPR56* induces transient apoptosis and reduces the in vitro growth of several cancer cell lines derived from colon (HCT116), melanoma (M14), and cervix (HeLa). Notwithstanding this pro-apoptotic response, the authors were able to derive stable GPR56 knockdown in all the cancer cell lines they tested, including those from the cervix (HeLa), colon (HCT116), melanoma (A2058 and M14), ovary (OVCAR3 and OVCAR8), prostate (PC3), pancreas (AsPC1), and lung (NCI-H460). A reduction in anchorage-independent growth in vitro was observed in all the knockdown lines, and decreased tumor growth in vivo was observed in the knockdown of A2058, PC3, and HCT116 cell lines [32]. This study suggests that GPR56 plays a tumor-promoting role in certain cancer cells. Similar pro-proliferation effects of GPR56 are observed in EVI1^{high} leukemia cell lines AML1 and HNT34, in which knocking down of *GPR56* enriches cells at the subG1 phase of the cell cycle and induces apoptosis [22].

The discrepancies among the above studies need to be followed up in future investigations. It is possible that GPR56 has both pro- and anti-growth functions during cancer progression, for example, via distinct binding partners. In fact, in a recent study [60], knocking down of both GPR56 and TG2 leads to a much more severe reduction in subcutaneous growth of MC-1 cells than *TG2*-knockdown alone, suggesting that GPR56 may have tumor-promoting functions in MC-1 cells in the absence of TG2. This effect of GPR56 was not observed in the WM115 melanoma cell line, pointing again to the complexity of GPR56 function in cancer.

Very recently, GPR56 was identified as a novel and stable marker for leukemic subpopulations with high repopulating capacity, a key feature attributed to leukemia stem cells, for the majority of acute myeloid leukemia (AML) samples [30]. High GPR56 expression was significantly associated with high-risk genetic subgroups and poor outcome.

The second member of the ADGRG group, *GPR64*, was found upregulated in Ewing sarcomas (ES) compared with normal tissues and other sarcomas [68]. *Gpr64* knockdown in Ewing sarcoma lines led to a reduction in tumor growth and metastasis. This tumor-promoting function of Gpr64 was mediated by the induction of placental growth factor (PGF) and matrix metalloproteinase 1 (MMP1) [68]. The mRNA of GPR64 was detected at high levels in cell lines from prostate cancer, non-small cell lung cancer (NSCLC), and melanomas and at moderate to low levels in cell lines from brain, ovary, breast, and colon cancers [68]. It was also identified as a marker for a subgroup of medulloblastomas characterized by overactive WNT signaling [87]. *GPR64* silencing in the highly motile cancer cell lines Hs578T and MDA-MB-231 resulted in a reduction of cell adhesion and migration [150].

ADGRG4 (GPR112) was identified as a marker for neuroendocrine carcinoma cells [151] and was predicted to be one of the candidate genes for targeted therapy of ileal carcinoids [50].

None of the other ADGRG members (ADGRG3, GPR97; ADGRG5, GPR114; GPR126; ADGRG7, GPR128) have been reported to affect tumorigenesis or be dysregulated in cancer samples.

7.7 ADGRL

The CCLE predicts high expression of LPHN1 and LPHN2 in many tumors [11]. A gene identification study in breast tumors led to *Latrophilin-1* [59]. Indeed, a number of breast tumor cell lines apparently overexpress the gene [152]. An invasion-associated four-gene signature including *Latrophilin-1* obtained from the NCI-60 cell line panel has significant prediction in non-small cell lung cancer (NSCLC) [40]. A genome-wide screen between a recurrent muscle-invasive cisplatin-resistant urothelial bladder carcinoma and its adjacent non-tumor tissue found cancer-associated alternative splicing with differential exon usage for *LPHN2* [153]. Linkage studies have revealed numerous *Latrophilin-3* loss-of-function mutations in breast, lung, ovarian, and prostate cancers [21].

ELTD1 is a regulator of physiological and tumor angiogenesis in vitro and in vivo [34]. ELTD1 showed higher expression on endothelium in peritumoral vessels compared to vessels of matched normal tissues [34]. Silencing of *ELTD1* in human ovarian and colorectal cancer xenografts implanted in mice inhibited tumor growth [34]. The transcriptome of the microvasculature associated with glioblastoma identified *ETLD1* as 1 of 95 upregulated genes [35]. Indeed, ELTD1 can be used as a vascular biomarker in glioblastoma [84]. It displays higher expression in high-grade compared with low-grade gliomas. miR-139-5p suppresses glioma cell proliferation by targeting *ELTD1* and regulating the cell cycle [103].

7.8 ADGRV

To identify molecular biomarkers associated with low-grade glioma-associated epileptic seizures, one of the initial symptoms of this tumor, RNA sequence data were collected [154]. A lower expression level of *ADGRV1 (VLGR1)* was found in patients with epileptic seizures compared to seizure-free patients and was confirmed by quantitative RT-PCR.

8 Tumor Therapy and Adhesion GPCRs

Accumulating evidence of the direct involvement of aGPCRs in tumor pathogenesis provides a solid foundation for their further study as potential therapeutic targets. Some aGPCRs are silenced with tumor formation and appear to have tumor suppressor activity (BAI1), while others are overexpressed in cancer and may act as oncogenic factors (CD97, GPR116, ELTD1).

The observation that BAI1 is epigenetically silenced in malignant glioma [72, 155], along with the recent findings of multiple somatic point mutations in the BAI1–3s in several cancers [21, 122], suggests that tumorigenesis may select for BAI1 silencing or inactivation. Importantly, exogenous restoration of BAI1 expression reduces growth and vascularization of tumors derived from gliomas and pancreatic and renal cell carcinomas [36, 119, 156, 157]. Taken together, these studies suggest that BAI1 fulfills the criteria for a *bona fide* tumor suppressor, and there is significant potential for BAI1 and its extracellular fragments as therapies for the treatment of human cancers. *BAI1* is silenced in glioblastoma multiforme due to methylation of a CpG island in the gene regulatory region, which leads to binding of methyl-CpG-binding domain protein 2 (MBD2) and transition to a suppressive chromatin conformation [72]. Treatment of glioma cells with 5-aza-2-′-deoxycytidine (5-Aza-dC) or knockdown of *MBD2* by shRNA resulted in reactivation of *BAI1* expression and restoration of BAI1 functional activity in that it conferred potent anti-angiogenic activity to glioma cell conditioned media in vitro and in vivo [72]. These findings have therapeutic implications since inhibiting MBD2 could offer a strategy to reactivate BAI1 expression and suppress tumor growth. Sequence-specific antisense inhibitors of MBD2 have been shown to inhibit both anchorage-independent growth of human cancer cell lines in vitro and the growth of human tumor xenografts in vivo [158, 159]. At present, epigenetic approaches in cancer therapy have focused primarily on inhibitors of the DNA methyltransferases and histone modifiers (e.g., HDACs). Thus, targeting MBD2 to reactivate BAI1 may represent a novel promising cancer therapeutic intervention.

In contrast to BAI1, some aGPCRs are overexpressed in tumors. Expression of CD97 was found in human thyroid carcinomas, but not in the corresponding normal thyrocytes [16, 57]. In human thyroid cancer cell lines, CD97 depletion reduced Rho-GTP and decreased lysophosphatidic acid (LPA)-stimulated invasion [57]. Similarly, GPR116 protein expression correlates with clinical progression stages in breast cancer and peaks in tumors with distant metastases [20, 65]. Knockdown of GPR116 in highly metastatic breast cancer cells (MDA-MB-231) suppressed cell migration, invasion, and metastasis in vivo [65]. These studies suggest that modulation of aGPCR expression in tumors may be effective therapeutics. A recent study identified ELTD1 as a novel angiogenesis regulator [34, 35, 84]. ELTD1 is upregulated in tumor endothelial cells and is a good prognostic marker. Moreover, targeting *Eltd1* blocks tumor angiogenesis and substantially inhibits tumor growth in vivo [34]. The function of GPR56 in different cancer types was found to differ, with both pro- and anti-tumorigenic properties, so the therapeutic implication of targeting GPR56 in cancer in general awaits careful evaluation. ADGRG4 (GPR112) was reported to serve as a marker for neuroendocrine carcinoma cells [151] and proposed a therapeutic target for treating ileal carcinoids [50].

In summary, alterations in the expression pattern of different aGPCRs have been documented in several cancers, suggesting that they may potentially serve as biomarkers of disease or therapeutic targets. Undoubtedly, the interest in the therapeutic targeting of the aGPCR family will continue to grow in the upcoming years, but first requires a more comprehensive characterization of their fundamental biological function.

Acknowledgments G.A. was supported by grants of the German Research Foundation (AU 132/7-3; FOR2149 Project 8 AU 132/8-1), L.X. by the National Institute of General Medicine Sciences (NIGMS; grant R01GM098591), D.Z. and E.G.V.M. in part by the US National Cancer Institute (grants CA086335 and NS096236), and the Southeastern Brain Tumor, Cure Childhood Cancer, and St. Baldrick's Foundations.

References

1. Krishnan A, Nijmeijer S, de Graaf C, Schiöth HB (2016) Classification, nomenclature and structural aspects of adhesion GPCRs. In: Langenhan T, Schöneberg T (eds) Adhesion G protein-coupled receptors: molecular, physiological and pharmacological principles in health and disease. Springer, Heidelberg
2. Kovacs P, Schöneberg T (2016) The relevance of genomic signatures at adhesion GPCR loci in humans. In: Langenhan T, Schöneberg T (eds) Adhesion G protein-coupled receptors: molecular, physiological and pharmacological principles in health and disease. Springer, Heidelberg
3. Araç D, Sträter N, Seiradake E (2016) Understanding the structural basis of adhesion GPCR functions. In: Langenhan T, Schöneberg T (eds) Adhesion G protein-coupled receptors: molecular, physiological and pharmacological principles in health and disease. Springer, Heidelberg
4. Nijmeijer S, Wolf S, Ernst OP, de Graaf C (2016) 7TM domain structure of adhesion GPCRs. In: Langenhan T, Schöneberg T (eds) Adhesion G protein-coupled receptors: molecular, physiological and pharmacological principles in health and disease. Springer, Heidelberg
5. Nieberler M, Kittel RJ, Petrenko AG, Lin H-H, Langenhan T (2016) Control of adhesion GPCR function through proteolytic processing. In: Langenhan T, Schöneberg T (eds) Adhesion G protein-coupled receptors: molecular, physiological and pharmacological principles in health and disease. Springer, Heidelberg
6. Langenhan T, Aust G, Hamann J (2013) Sticky signaling – Adhesion class G protein-coupled receptors take the stage. Sci Signal 6(276):re3. doi:10.1126/scisignal.2003825
7. Liebscher I, Schon J, Petersen SC, Fischer L, Auerbach N, Demberg LM et al (2014) A tethered agonist within the ectodomain activates the adhesion G protein-coupled receptors GPR126 and GPR133. Cell Rep 9:2018–2026
8. Liebscher I, Schöneberg T (2016) Tethered agonism: a common activation mechanism of adhesion GPCRs. In: Langenhan T, Schöneberg T (eds) Adhesion G protein-coupled receptors: molecular, physiological and pharmacological principles in health and disease. Springer, Heidelberg
9. Kishore A, Hall RA (2016) Versatile signaling activity of adhesion GPCRs. In: Langenhan T, Schöneberg T (eds) Adhesion G protein-coupled receptors: molecular, physiological and pharmacological principles in health and disease. Springer, Heidelberg
10. Scholz N, Monk KR, Kittel RJ, Langenhan T (2016) Adhesion GPCRs as a putative class of metabotropic mechanosensors. In: Langenhan T, Schöneberg T (eds) Adhesion G protein-coupled receptors: molecular, physiological and pharmacological principles in health and disease. Springer, Heidelberg
11. Barretina J, Caponigro G, Stransky N, Venkatesan K, Margolin AA, Kim S et al (2012) The Cancer Cell Line Encyclopedia enables predictive modelling of anticancer drug sensitivity. Nature 483:603–607
12. Cerami E, Gao J, Dogrusoz U, Gross BE, Sumer SO, Aksoy BA et al (2012) The cBio cancer genomics portal: an open platform for exploring multidimensional cancer genomics data. Cancer Discov 2:401–404
13. Gao J, Aksoy BA, Dogrusoz U, Dresdner G, Gross B, Sumer SO et al (2013) Integrative analysis of complex cancer genomics and clinical profiles using the cBioPortal. Sci Signal 6 (269):pl1. doi:10.1126/scisignal.2004088

14. Uhlen M, Fagerberg L, Hallstrom BM, Lindskog C, Oksvold P, Mardinoglu A et al (2015) Proteomics. Tissue-based map of the human proteome. Science 347(6220):1260419. doi:10.1126/science.1260419
15. Aguirre-Gamboa R, Gomez-Rueda H, Martinez-Ledesma E, Martinez-Torteya A, Chacolla-Huaringa R, Rodriguez-Barrientos A et al (2013) SurvExpress: an online biomarker validation tool and database for cancer gene expression data using survival analysis. PLoS One 8 (9), e74250. doi:10.1371/journal.pone.0074250
16. Aust G, Eichler W, Laue S, Lehmann I, Heldin N-E, Lotz O et al (1997) CD97: a dedifferentiation marker in human thyroid carcinomas. Cancer Res 57:1798–1806
17. Zendman AJ, Cornelissen IM, Weidle UH, Ruiter DJ, van Muijen GN (1999) TM7XN1, a novel human EGF-TM7-like cDNA, detected with mRNA differential display using human melanoma cell lines with different metastatic potential. FEBS Lett 446:292–298
18. Mallawaaratchy DM, Buckland ME, McDonald KL, Li CC, Ly L, Sykes EK et al (2015) Membrane proteome analysis of glioblastoma cell invasion. J Neuropathol Exp Neurol 74:425–441
19. Coustan-Smith E, Song G, Clark C, Key L, Liu P, Mehrpooya M et al (2011) New markers for minimal residual disease detection in acute lymphoblastic leukemia. Blood 117:6267–6276
20. Davidson B, Stavnes HT, Risberg B, Nesland JM, Wohlschlaeger J, Yang Y et al (2012) Gene expression signatures differentiate adenocarcinoma of lung and breast origin in effusions. Hum Pathol 43:684–694
21. Kan Z, Jaiswal BS, Stinson J, Janakiraman V, Bhatt D, Stern HM et al (2010) Diverse somatic mutation patterns and pathway alterations in human cancers. Nature 466:869–873
22. Saito Y, Kaneda K, Suekane A, Ichihara E, Nakahata S, Yamakawa N et al (2013) Maintenance of the hematopoietic stem cell pool in bone marrow niches by EVI1-regulated GPR56. Leukemia 27:1637–1649
23. Silveira VS, Scrideli CA, Moreno DA, Yunes JA, Queiroz RG, Toledo SC et al (2013) Gene expression pattern contributing to prognostic factors in childhood acute lymphoblastic leukemia. Leuk Lymphoma 54:310–314
24. Charfi C, Edouard E, Rassart E (2014) Identification of GPM6A and GPM6B as potential new human lymphoid leukemia-associated oncogenes. Cell Oncol 37:179–191
25. Kaucka M, Plevova K, Pavlova S, Janovska P, Mishra A, Verner J et al (2013) The planar cell polarity pathway drives pathogenesis of chronic lymphocytic leukemia by the regulation of B-lymphocyte migration. Cancer Res 73:1491–1501
26. Mirkowska P, Hofmann A, Sedek L, Slamova L, Mejstrikova E, Szczepanski T et al (2013) Leukemia surfaceome analysis reveals new disease-associated features. Blood 121:e149–e159
27. Del Giudice I, Messina M, Chiaretti S, Santangelo S, Tavolaro S, De Propris MS et al (2012) Behind the scenes of non-nodal MCL: downmodulation of genes involved in actin cytoskeleton organization, cell projection, cell adhesion, tumour invasion, TP53 pathway and mutated status of immunoglobulin heavy chain genes. Br J Haematol 156:601–611
28. Bonardi F, Fusetti F, Deelen P, van Gosliga D, Vellenga E, Schuringa JJ (2013) A proteomics and transcriptomics approach to identify leukemic stem cell (LSC) markers. Mol Cell Proteomics 12:626–637
29. Wobus M, Bornhauser M, Jacobi A, Krater M, Otto O, Ortlepp C et al (2015) Association of the EGF-TM7 receptor CD97 expression with FLT3-ITD in acute myeloid leukemia. Oncotarget 6:38804–38815
30. Pabst C, Bergeron A, Lavallee VP, Yeh J, Gendron P, Norddahl GL et al (2016) GPR56 identifies primary human acute myeloid leukemia cells with high repopulating potential in vivo. Blood 27:2018–27
31. Hricik T, Federici G, Zeuner A, Alimena G, Tafuri A, Tirelli V et al (2013) Transcriptomic and phospho-proteomic analyzes of erythroblasts expanded in vitro from normal donors and from patients with polycythemia vera. Am J Hematol 88:723–729

32. Ke N, Sundaram R, Liu G, Chionis J, Fan W, Rogers C et al (2007) Orphan G protein-coupled receptor GPR56 plays a role in cell transformation and tumorigenesis involving the cell adhesion pathway. Mol Cancer Ther 6:1840–1850
33. Vallon M, Essler M (2006) Proteolytically processed soluble tumor endothelial marker (TEM) 5 mediates endothelial cell survival during angiogenesis by linking integrin alpha (v)beta3 to glycosaminoglycans. J Biol Chem 281:34179–34188
34. Masiero M, Simoes FC, Han HD, Snell C, Peterkin T, Bridges E et al (2013) A core human primary tumor angiogenesis signature identifies the endothelial orphan receptor ELTD1 as a key regulator of angiogenesis. Cancer Cell 24:229–241
35. Dieterich LC, Mellberg S, Langenkamp E, Zhang L, Zieba A, Salomaki H et al (2012) Transcriptional profiling of human glioblastoma vessels indicates a key role of VEGF-A and TGFbeta2 in vascular abnormalization. J Pathol 228:378–390
36. Fukushima Y, Oshika Y, Tsuchida T, Tokunaga T, Hatanaka H, Kijima H et al (1998) Brain-specific angiogenesis inhibitor 1 expression is inversely correlated with vascularity and distant metastasis of colorectal cancer. Int J Oncol 13:967–970
37. Hatanaka H, Oshika Y, Abe Y, Yoshida Y, Hashimoto T, Handa A et al (2000) Vascularization is decreased in pulmonary adenocarcinoma expressing brain-specific angiogenesis inhibitor 1 (BAI1). Int J Mol Med 5:181–183
38. Hao C, Wang L, Peng S, Cao M, Li H, Hu J et al (2015) Gene mutations in primary tumors and corresponding patient-derived xenografts derived from non-small cell lung cancer. Cancer Lett 357:179–185
39. Guo R, Wu G, Li H, Qian P, Han J, Pan F et al (2013) Promoter methylation profiles between human lung adenocarcinoma multidrug resistant A549/cisplatin (A549/DDP) cells and its progenitor A549 cells. Biol Pharm Bull 36:1310–1316
40. Hsu YC, Yuan S, Chen HY, Yu SL, Liu CH, Hsu PY et al (2009) A four-gene signature from NCI-60 cell line for survival prediction in non-small cell lung cancer. Clin Cancer Res 15:7309–7315
41. Bari MF, Brown H, Nicholson AG, Kerr KM, Gosney JR, Wallace WA et al (2014) BAI3, CDX2 and VIL1: a panel of three antibodies to distinguish small cell from large cell neuroendocrine lung carcinomas. Histopathology 64:547–556
42. Steinert M, Wobus M, Schütz A, Aust G (2002) CD97, but not its closely related EGF-TM7 family member EMR2, is expressed on gastric pancreatic and esophageal carcinomas. Am J Clin Pathol 118(5):699–707
43. Kausar T, Sharma R, Hasan MR, Tripathi SC, Saraya A, Chattopadhyay TK et al (2011) Clinical significance of GPR56, transglutaminase 2, and NF-kappaB in esophageal squamous cell carcinoma. Cancer Invest 29:42–48
44. Miyamoto N, Yamamoto H, Taniguchi H, Miyamoto C, Oki M, Adachi Y et al (2007) Differential expression of angiogenesis-related genes in human gastric cancers with and those without high-frequency microsatellite instability. Cancer Lett 254:42–53
45. Miao R, Guo X, Zhi Q, Shi Y, Li L, Mao X et al (2013) VEZT, a novel putative tumor suppressor, suppresses the growth and tumorigenicity of gastric cancer. PLoS One 8:e74409
46. Yoshida Y, Oshika Y, Fukushima Y, Tokunaga T, Hatanaka H, Kijima H et al (1999) Expression of angiostatic factors in colorectal cancer. Int J Oncol 15:1221–1225
47. Katoh M, Katoh M (2007) Comparative integromics on non-canonical WNT or planar cell polarity signaling molecules: transcriptional mechanism of PTK7 in colorectal cancer and that of SEMA6A in undifferentiated ES cells. Int J Mol Med 20:405–409
48. Steinert M, Wobus M, Boltze C, Schütz A, Wahlbuhl M, Hamann J et al (2002) Expression and regulation of CD97 in colorectal carcinoma cell lines and tumor tissues. Am J Pathol 161:1657–1667
49. Galle J, Sittig D, Hanisch I, Wobus M, Wandel E, Loeffler M et al (2006) Individual cell-based models of tumor–environment interactions. Multiple effects of CD97 on tumor invasion. Am J Pathol 169:1802–1811
50. Nilsson O (2013) Profiling of ileal carcinoids. Neuroendocrinology 97:7–18

51. Ammerpohl O, Pratschke J, Schafmayer C, Haake A, Faber W, von Kampen O et al (2012) Distinct DNA methylation patterns in cirrhotic liver and hepatocellular carcinoma. Int J Cancer 130:1319–1328
52. Tomimaru Y, Koga H, Yano H, de la Monte S, Wands JR, Kim M (2013) Upregulation of T-cell factor-4 isoform-responsive target genes in hepatocellular carcinoma. Liver Int 33:1100–1112
53. Wu J, Lei L, Wang S, Gu D, Zhang J (2012) Immunohistochemical expression and prognostic value of CD97 and its ligand CD55 in primary gallbladder carcinoma. J Biomed Biotechnol 2012:587672. doi:10.1155/2012/587672
54. He Z, Wu H, Jiao Y, Zheng J (2015) Expression and prognostic value of CD97 and its ligand CD55 in pancreatic cancer. Oncol Lett 9:793–797
55. Huang Y, Fan J, Yang J, Zhu GZ (2008) Characterization of GPR56 protein and its suppressed expression in human pancreatic cancer cells. Mol Cell Biochem 308:133–139
56. Maina EN, Morris MR, Zatyka M, Raval RR, Banks RE, Richards FM et al (2005) Identification of novel VHL target genes and relationship to hypoxic response pathways. Oncogene 24:4549–4558
57. Ward Y, Lake R, Martin PL, Killian K, Salerno P, Wang T et al (2013) CD97 amplifies LPA receptor signaling and promotes thyroid cancer progression in a mouse model. Oncogene 32:2726–2738
58. Ward Y, Lake R, Yin JJ, Heger CD, Raffeld M, Goldsmith PK et al (2011) LPA receptor heterodimerizes with CD97 to amplify LPA-initiated RHO-dependent signaling and invasion in prostate cancer cells. Cancer Res 71:7301–7311
59. White GR, Varley JM, Heighway J (1998) Isolation and characterization of a human homologue of the latrophilin gene from a region of 1p31.1 implicated in breast cancer. Oncogene 17:3513–3519
60. Yang L, Friedland S, Corson N, Xu L (2014) GPR56 inhibits melanoma growth by internalizing and degrading its ligand TG2. Cancer Res 74:1–10
61. Yang L, Chen G, Mohanty S, Scott G, Fazal F, Rahman A et al (2011) GPR56 regulates VEGF production and angiogenesis during melanoma progression. Cancer Res 71:5558–5568
62. Meisen WH, Dubin S, Sizemore ST, Mathsyaraja H, Thies K, Lehman NL et al (2015) Changes in BAI1 and nestin expression are prognostic indicators for survival and metastases in breast cancer and provide opportunities for dual targeted therapies. Mol Cancer Ther 14:307–314
63. Liao S, Desouki MM, Gaile DP, Shepherd L, Nowak NJ, Conroy J et al (2012) Differential copy number aberrations in novel candidate genes associated with progression from in situ to invasive ductal carcinoma of the breast. Genes Chromosomes Cancer 51:1067–1078
64. Davies JQ, Lin HH, Stacey M, Yona S, Chang GW, Gordon S et al (2011) Leukocyte adhesion-GPCR EMR2 is aberrantly expressed in human breast carcinomas and is associated with patient survival. Oncol Rep 25:619–627
65. Tang X, Jin R, Qu G, Wang X, Li Z, Yuan Z et al (2013) GPR116, an adhesion G-protein-coupled receptor, promotes breast cancer metastasis via the Galphaq-p63RhoGEF-Rho GTPase pathway. Cancer Res 73:6206–6218
66. Zyryanova T, Schneider R, Adams V, Sittig D, Kerner C, Gebhardt C et al (2014) Skeletal muscle expression of the adhesion-GPCR CD97: CD97 deletion induces an abnormal structure of the sarcoplasmatic reticulum but does not impair skeletal muscle function. PLoS One 9, e100513
67. Aust G, Wandel E, Boltze C, Sittig D, Schütz A, Horn LC et al (2006) Diversity of CD97 in smooth muscle cells (SMCs). Cell Tissue Res 323:1–9
68. Richter GH, Fasan A, Hauer K, Grunewald TG, Berns C, Rossler S et al (2013) G-Protein coupled receptor 64 promotes invasiveness and metastasis in Ewing sarcomas through PGF and MMP1. J Pathol 230:70–81

69. Kaur B, Brat DJ, Devi NS, Van Meir EG (2005) Vasculostatin, a proteolytic fragment of brain angiogenesis inhibitor 1, is an antiangiogenic and antitumorigenic factor. Oncogene 24:3632–3642
70. Kaur B, Cork SM, Sandberg EM, Devi NS, Zhang Z, Klenotic PA et al (2009) Vasculostatin inhibits intracranial glioma growth and negatively regulates in vivo angiogenesis through a CD36-dependent mechanism. Cancer Res 69:1212–1220
71. Cork SM, Kaur B, Devi NS, Cooper L, Saltz JH, Sandberg EM et al (2012) A proprotein convertase/MMP-14 proteolytic cascade releases a novel 40 kDa vasculostatin from tumor suppressor BAI1. Oncogene 31:5144–5152
72. Zhu D, Hunter SB, Vertino PM, Van Meir EG (2011) Overexpression of MBD2 in glioblastoma maintains epigenetic silencing and inhibits the antiangiogenic function of the tumor suppressor gene BAI1. Cancer Res 71:5859–5870
73. Kaur B, Brat DJ, Calkins CC, Van Meir EG (2003) Brain angiogenesis inhibitor 1 is differentially expressed in normal brain and glioblastoma independently of p53 expression. Am J Pathol 162:19–27
74. Ivan ME, Safaee M, Oh T, Clark AJ, Sun MZ, Kim J et al (2015) Epidermal growth factor-like module containing mucin-like hormone receptor 2 expression in gliomas. J Neurooncol 121:53–61
75. Rutkowski MJ, Sughrue ME, Kane AJ, Kim JM, Bloch O, Parsa AT (2011) Epidermal growth factor module-containing mucin-like receptor 2 is a newly identified adhesion G protein-coupled receptor associated with poor overall survival and an invasive phenotype in glioblastoma. J Neurooncol 105:165–171
76. Kane AJ, Sughrue ME, Rutkowski MJ, Phillips JJ, Parsa AT (2010) EMR-3: a potential mediator of invasive phenotypic variation in glioblastoma and novel therapeutic target. Neuroreport 21:1018–1022
77. Safaee M, Clark AJ, Oh MC, Ivan ME, Bloch O, Kaur G et al (2013) Overexpression of CD97 confers an invasive phenotype in glioblastoma cells and is associated with decreased survival of glioblastoma patients. PLoS One 8(4), e62765. doi:10.1371/journal.pone.0062765
78. Liu JK, Lubelski D, Schonberg DL, Wu Q, Hale JS, Flavahan WA et al (2014) Phage display discovery of novel molecular targets in glioblastoma-initiating cells. Cell Death Differ 21:1325–1339
79. Safaee M, Fakurnejad S, Bloch O, Clark AJ, Ivan ME, Sun MZ et al (2015) Proportional upregulation of CD97 isoforms in glioblastoma and glioblastoma-derived brain tumor initiating cells. PLoS One 10, e0111532
80. Chidambaram A, Fillmore HL, Van Meter TE, Dumur CI, Broaddus WC (2012) Novel report of expression and function of CD97 in malignant gliomas: correlation with Wilms tumor 1 expression and glioma cell invasiveness. J Neurosurg 116:843–853
81. Somasundaram A, Ardanowski N, Opalak CF, Fillmore HL, Chidambaram A, Broaddus WC (2014) Wilms tumor 1 gene, CD97, and the emerging biogenetic profile of glioblastoma. Neurosurg Focus 37:E14
82. Shashidhar S, Lorente G, Nagavarapu U, Nelson A, Kuo J, Cummins J et al (2005) GPR56 is a GPCR that is overexpressed in gliomas and functions in tumor cell adhesion. Oncogene 24:1673–1682
83. Taatjes DJ (2010) The human Mediator complex: a versatile, genome-wide regulator of transcription. Trends Biochem Sci 35:315–322
84. Towner RA, Jensen RL, Colman H, Vaillant B, Smith N, Casteel R et al (2013) ELTD1, a potential new biomarker for gliomas. Neurosurgery 72:77–90
85. Cork SM, Van Meir EG (2011) Emerging roles for the BAI1 protein family in the regulation of phagocytosis, synaptogenesis, neurovasculature, and tumor development. J Mol Med (Berl) 89:743–752
86. Fujii K, Kurozumi K, Ichikawa T, Onishi M, Shimazu Y, Ishida J et al (2013) The integrin inhibitor cilengitide enhances the anti-glioma efficacy of vasculostatin-expressing oncolytic virus. Cancer Gene Ther 20:437–444

87. Whittier KL, Boese EA, Gibson-Corley KN, Kirby PA, Darbro BW, Qian Q et al (2013) G-protein coupled receptor expression patterns delineate medulloblastoma subgroups. Acta Neuropathol Commun 1:66
88. Boligan KF, Mesa C, Fernandez LE, von Gunten S (2015) Cancer intelligence acquired (CIA): tumor glycosylation and sialylation codes dismantling antitumor defense. Cell Mol Life Sci 72:1231–1248
89. Pinho SS, Reis CA (2015) Glycosylation in cancer: mechanisms and clinical implications. Nat Rev Cancer 15:540–555
90. Durrant LG, Chapman MA, Buckley DJ, Spendlove I, Robins RA, Armitage NC (2003) Enhanced expression of the complement regulatory protein CD55 predicts a poor prognosis in colorectal cancer patients. Cancer Immunol Immunother 52:638–642
91. Liu Y, Chen L, Peng S, Chen Z, Gimm O, Finke R et al (2005) The expression of CD97EGF and its ligand CD55 on marginal epithelium is related to higher stage and depth of tumor invasion of gastric carcinomas. Oncol Rep 14:1413–1420
92. Loberg RD, Wojno KJ, Day LL, Pienta KJ (2005) Analysis of membrane-bound complement regulatory proteins in prostate cancer. Urology 66:1321–1326
93. Han SL, Xu C, Wu XL, Li JL, Liu Z, Zeng QQ (2010) The impact of expressions of CD97 and its ligand CD55 at the invasion front on prognosis of rectal adenocarcinoma. Int J Colorectal Dis 25:695–702
94. Park D, Tosello-Trampont AC, Elliott MR, Lu M, Haney LB, Ma Z et al (2007) BAI1 is an engulfment receptor for apoptotic cells upstream of the ELMO/Dock180/Rac module. Nature 450:430–434
95. Okajima D, Kudo G, Yokota H (2010) Brain-specific angiogenesis inhibitor 2 (BAI2) may be activated by proteolytic processing. J Recept Signal Transduct Res 30:143–153
96. Gray JX, Haino M, Roth MJ, Maguire JE, Jensen PN, Yarme A et al (1996) CD97 is a processed, seven-transmembrane, heterodimeric receptor associated with inflammation. J Immunol 157:5438–5447
97. Abe J, Fukuzawa T, Hirose S (2002) Cleavage of Ig-Hepta at a "SEA" module and at a conserved G protein-coupled receptor proteolytic site. J Biol Chem 277:23391–23398
98. Moriguchi T, Haraguchi K, Ueda N, Okada M, Furuya T, Akiyama T (2004) DREG, a developmentally regulated G protein-coupled receptor containing two conserved proteolytic cleavage sites. Genes Cells 9:549–560
99. Krasnoperov V, Deyev IE, Serova OV, Xu C, Lu Y, Buryanovsky L et al (2009) Dissociation of the subunits of the calcium-independent receptor of alpha-latrotoxin as a result of two-step proteolysis. Biochemistry 48:3230–3238
100. Wang T, Ward Y, Tian L, Lake R, Guedez L, Stetler-Stevenson WG et al (2004) CD97, an adhesion receptor on inflammatory cells, stimulates angiogenesis through binding integrin counter receptors on endothelial cells. Blood 105:2836–2844
101. Hamann J, Wishaupt JO, van Lier RA, Smeets TJ, Breedveld FC, Tak PP (1999) Expression of the activation antigen CD97 and its ligand CD55 in rheumatoid synovial tissue. Arthritis Rheum 42:650–658
102. de Groot DM, Vogel G, Dulos J, Teeuwen L, Stebbins K, Hamann J et al (2009) Therapeutic antibody targeting of CD97 in experimental arthritis: the role of antigen expression, shedding, and internalization on the pharmacokinetics of anti-CD97 monoclonal antibody 1B2. J Immunol 183:4127–4134
103. Dai S, Wang X, Li X, Cao Y (2015) MicroRNA-139-5p acts as a tumor suppressor by targeting ELTD1 and regulating cell cycle in glioblastoma multiforme. Biochem Biophys Res Commun 467:204–210
104. Hsiao CC, Keysselt K, Chen HY, Sittig D, Hamann J, Lin HH et al (2015) The Adhesion GPCR CD97 inhibits apoptosis. Int J Biochem Cell Biol 65:197–208
105. Xu L, Begum S, Hearn JD, Hynes RO (2006) GPR56, an atypical G protein-coupled receptor, binds tissue transglutaminase, TG2, and inhibits melanoma tumor growth and metastasis. Proc Natl Acad Sci U S A 103:9023–9028

106. Fu JF, Yen TH, Chen Y, Huang YJ, Hsu CL, Liang DC et al (2013) Involvement of Gpr125 in the myeloid sarcoma formation induced by cooperating MLL/AF10(OM-LZ) and oncogenic KRAS in a mouse bone marrow transplantation model. Int J Cancer 133:1792–1802
107. Mouw JK, Ou G, Weaver VM (2014) Extracellular matrix assembly: a multiscale deconstruction. Nat Rev Mol Cell Biol 15:771–785
108. Luo R, Jin Z, Deng Y, Strokes N, Piao X (2012) Disease-associated mutations prevent GPR56-collagen III interaction. PLoS One 1, e29818. doi:10.1371/journal.pone.0029818
109. Koirala S, Jin Z, Piao X, Corfas G (2009) GPR56-regulated granule cell adhesion is essential for rostral cerebellar development. J Neurosci 29:7439–7449
110. Stacey M, Chang GW, Davies JQ, Kwakkenbos MJ, Sanderson RD, Hamann J et al (2003) The epidermal growth factor-like domains of the human EMR2 receptor mediate cell attachment through chondroitin sulphate glycosaminoglycans. Blood 102:2916–2924
111. Paavola KJ, Sidik H, Zuchero JB, Eckart M, Talbot WS (2014) Type IV collagen is an activating ligand for the adhesion G protein-coupled receptor GPR126. Sci Signal 7(338): ra76. doi:10.1126/scisignal.2005347
112. Petersen SC, Luo R, Liebscher I, Giera S, Jeong SJ, Mogha A et al (2015) The adhesion GPCR GPR126 has distinct, domain-dependent functions in Schwann cell development mediated by interaction with laminin-211. Neuron 85:755–769
113. Strutt D, Schnabel R, Fiedler F, Prömel S (2016) Adhesion GPCRs govern polarity of epithelia and cell migration. In: Langenhan T, Schöneberg T (eds) Adhesion G protein-coupled receptors: molecular, physiological and pharmacological principles in health and disease. Springer, Heidelberg
114. Hsiao CC, Wang WC, Kuo WL, Chen HY, Chen TC, Hamann J et al (2014) CD97 inhibits cell migration in human fibrosarcoma cell by modulating TIMP-2/MT1-MMP/MMP-2 activity. FEBS J 281:4878–4891
115. Musa G, Engel FB, Niaudet C (2016) Heart development, angiogenesis and blood-brain barrier function is modulated by adhesion GPCRs. In: Langenhan T, Schöneberg T (eds) Adhesion G protein-coupled receptors: molecular, physiological and pharmacological principles in health and disease. Springer, Heidelberg
116. Nishimori H, Shiratsuchi T, Urano T, Kimura Y, Kiyono K, Tatsumi K et al (1997) A novel brain-specific p53-target gene, BAI1, containing thrombospondin type 1 repeats inhibits experimental angiogenesis. Oncogene 15:2145–2150
117. Kang X, Xiao X, Harata M, Bai Y, Nakazaki Y, Soda Y et al (2006) Antiangiogenic activity of BAI1 in vivo: implications for gene therapy of human glioblastomas. Cancer Gene Ther 13:385–392
118. Lee JH, Koh JT, Shin BA, Ahn KY, Roh JH, Kim YJ et al (2001) Comparative study of angiostatic and anti-invasive gene expressions as prognostic factors in gastric cancer. Int J Oncol 18:355–361
119. Kudo S, Konda R, Obara W, Kudo D, Tani K, Nakamura Y et al (2007) Inhibition of tumor growth through suppression of angiogenesis by brain-specific angiogenesis inhibitor 1 gene transfer in murine renal cell carcinoma. Oncol Rep 18:785–791
120. de Fraipont F, Nicholson AC, Feige JJ, Van Meir EG (2001) Thrombospondins and tumor angiogenesis. Trends Mol Med 7:401–407
121. Nicholson AC, Malik SB, Logsdon JM Jr, Van Meir EG (2005) Functional evolution of ADAMTS genes: evidence from analyses of phylogeny and gene organization. BMC Evol Biol 5:11
122. Pleasance ED, Cheetham RK, Stephens PJ, McBride DJ, Humphray SJ, Greenman CD et al (2010) A comprehensive catalogue of somatic mutations from a human cancer genome. Nature 463:191–196
123. Arac D, Boucard AA, Bolliger MF, Nguyen J, Soltis SM, Sudhof TC et al (2012) A novel evolutionarily conserved domain of cell-adhesion GPCRs mediates autoproteolysis. EMBO J 31:1364–1378
124. Zhu D, Van Meir EG (2016) BAI1: from cancer to neurological disease. Oncotarget

125. St Croix B, Rago C, Velculescu V, Traverso G, Romans KE, Montgomery E et al (2000) Genes expressed in human tumor endothelium. Science 289:1197–1202
126. Carson-Walter EB, Watkins DN, Nanda A, Vogelstein B, Kinzler KW, St CB (2001) Cell surface tumor endothelial markers are conserved in mice and humans. Cancer Res 61:6649–6655
127. Gao Y, Fan X, Li W, Ping W, Deng Y, Fu X (2014) miR-138-5p reverses gefitinib resistance in non-small cell lung cancer cells via negatively regulating G protein-coupled receptor 124. Biochem Biophys Res Commun 446:179–186
128. Shiratsuchi T, Nishimori H, Ichise H, Nakamura Y, Tokino T (1997) Cloning and characterization of BAI2 and BAI3, novel genes homologous to brain-specific angiogenesis inhibitor 1 (BAI1). Cytogenet Cell Genet 79:103–108
129. Koh JT, Kook H, Kee HJ, Seo YW, Jeong BC, Lee JH et al (2004) Extracellular fragment of brain-specific angiogenesis inhibitor 1 suppresses endothelial cell proliferation by blocking alphavbeta5 integrin. Exp Cell Res 294:172–184
130. Klenotic PA, Huang P, Palomo J, Kaur B, Van Meir EG, Vogelbaum MA et al (2010) Histidine-rich glycoprotein modulates the anti-angiogenic effects of vasculostatin. Am J Pathol 176:2039–2050
131. Shiratsuchi T, Futamura M, Oda K, Nishimori H, Nakamura Y, Tokino T (1998) Cloning and characterization of BAI-associated protein 1: a PDZ domain-containing protein that interacts with BAI1. Biochem Biophys Res Commun 247:597–604
132. Stephenson JR, Paavola KJ, Schaefer SA, Kaur B, Van Meir EG, Hall RA (2013) Brain-specific angiogenesis inhibitor-1 signaling, regulation, and enrichment in the postsynaptic density. J Biol Chem 288:22248–22256
133. Oda K, Shiratsuchi T, Nishimori H, Inazawa J, Yoshikawa H, Taketani Y et al (1999) Identification of BAIAP2 (BAI-associated protein 2), a novel human homologue of hamster IRSp53, whose SH3 domain interacts with the cytoplasmic domain of BAI1. Cytogenet Cell Genet 84:75–82
134. Kim JC, Kim SY, Roh SA, Cho DH, Kim DD, Kim JH et al (2008) Gene expression profiling: canonical molecular changes and clinicopathological features in sporadic colorectal cancers. World J Gastroenterol 14:6662–6672
135. Wang Q (2015) Identification of biomarkers for metastatic osteosarcoma based on DNA microarray data. Neoplasma 62:365–371
136. Aust G, Hamann J, Schilling N, Wobus M (2003) Detection of alternatively spliced EMR2 mRNAs in colorectal tumor cell lines but rare expression of the molecule in the corresponding adenocarcinomas. Virchows Arch 443:32–37
137. Grone J, Lenze D, Jurinovic V, Hummel M, Seidel H, Leder G et al (2011) Molecular profiles and clinical outcome of stage UICC II colon cancer patients. Int J Colorectal Dis 26:847–858
138. Liu D, Trojanowicz B, Ye L, Li C, Zhang L, Li X et al (2012) The invasion and metastasis promotion role of CD97 small isoform in gastric carcinoma. PLoS One 7, e39989
139. Boltze C, Schneider-Stock R, Aust G, Mawrin C, Dralle H, Roessner A et al (2002) CD97, CD95 and Fas-L clearly discriminate between chronic pancreatitis and pancreatic ductal adenocarcinoma in perioperative evaluation of cryocut sections. Pathol Int 52:83–88
140. Singh V, Singh LC, Vasudevan M, Chattopadhyay I, Borthakar BB, Rai AK et al (2015) Esophageal cancer epigenomics and integrome analysis of genome-wide methylation and expression in high risk northeast Indian population. OMICS 19:688–699
141. Lu YY, Sweredoski MJ, Huss D, Lansford R, Hess S, Tirrell DA (2013) Prometastatic GPCR CD97 is a direct target of tumor suppressor microRNA-126. ACS Chem Biol 9:334–338
142. Hamann J, Vogel B, van Schijndel GM, van Lier RA (1996) The seven-span transmembrane receptor CD97 has a cellular ligand (CD55, DAF). J Exp Med 184:1185–1189
143. Wandel E, Saalbach A, Sittig D, Gebhardt C, Aust G (2012) Thy-1 (CD90) is an interaction partner for CD97 on activated endothelial cells. J Immunol 188:1442–1450

144. Fukuzawa T, Ishida J, Kato A, Ichinose T, Ariestanti DM, Takahashi T et al (2013) Lung surfactant levels are regulated by Ig-Hepta/GPR116 by monitoring surfactant protein D. PLoS One 8, e69451
145. Yang MY, Hilton MB, Seaman S, Haines DC, Nagashima K, Burks CM et al (2013) Essential regulation of lung surfactant homeostasis by the orphan G protein-coupled receptor GPR116. Cell Rep 3:1457–64
146. Bridges JP, Ludwig MG, Mueller M, Kinzel B, Sato A, Xu Y et al (2013) Orphan G protein-coupled receptor GPR116 regulates pulmonary surfactant pool size. Am J Respir Cell Mol Biol 49:348–57
147. Yang L, Xu L (2012) GPR56 in cancer progression: current status and future perspective. Future Oncol 8:431–440
148. Mori A, Arii S, Furutani M, Hanaki K, Takeda Y, Moriga T et al (1999) Vascular endothelial growth factor-induced tumor angiogenesis and tumorigenicity in relation to metastasis in a HT1080 human fibrosarcoma cell model. Int J Cancer 80:738–743
149. Duda DG, Sunamura M, Lozonschi L, Yokoyama T, Yatsuoka T, Motoi F et al (2002) Overexpression of the p53-inducible brain-specific angiogenesis inhibitor 1 suppresses efficiently tumour angiogenesis. Br J Cancer 86:490–496
150. Peeters MC, Fokkelman M, Boogaard B, Egerod KL, van de Water B, Ijzerman AP et al (2015) The adhesion G protein-coupled receptor G2 (ADGRG2/GPR64) constitutively activates SRE and NFkappaB and is involved in cell adhesion and migration. Cell Signal 27:2579–2588
151. Leja J, Essaghir A, Essand M, Wester K, Oberg K, Totterman TH et al (2009) Novel markers for enterochromaffin cells and gastrointestinal neuroendocrine carcinomas. Mod Pathol 22:261–272
152. White GR, Varley JM, Heighway J (2000) Genomic structure and expression profile of LPHH1, a 7TM gene variably expressed in breast cancer cell lines. Biochim Biophys Acta 1491:75–92
153. Zhang S, Liu Y, Liu Z, Zhang C, Cao H, Ye Y et al (2014) Transcriptome profiling of a multiple recurrent muscle-invasive urothelial carcinoma of the bladder by deep sequencing. PLoS One 9, e91466
154. Wang Y, Fan X, Zhang W, Zhang C, Wang J, Jiang T et al (2015) Deficiency of very large G-protein-coupled receptor-1 is a risk factor of tumor-related epilepsy: a whole transcriptome sequencing analysis. J Neurooncol 121:609–616
155. Mosmann TR, Li L, Sad S (1997) Functions of CD8 T-cell subsets secreting different cytokine patterns. Semin Immunol 9:87–92
156. Nam DH, Park K, Suh YL, Kim JH (2004) Expression of VEGF and brain specific angiogenesis inhibitor-1 in glioblastoma: prognostic significance. Oncol Rep 11:863–869
157. Yoon KC, Ahn KY, Lee JH, Chun BJ, Park SW, Seo MS et al (2005) Lipid-mediated delivery of brain-specific angiogenesis inhibitor 1 gene reduces corneal neovascularization in an in vivo rabbit model. Gene Ther 12:617–624
158. Slack A, Bovenzi V, Bigey P, Ivanov MA, Ramchandani S, Bhattacharya S et al (2002) Antisense MBD2 gene therapy inhibits tumorigenesis. J Gene Med 4:381–389
159. Campbell PM, Bovenzi V, Szyf M (2004) Methylated DNA-binding protein 2 antisense inhibitors suppress tumourigenesis of human cancer cell lines in vitro and in vivo. Carcinogenesis 25:499–507

Erratum to: 7TM Domain Structure of Adhesion GPCRs

Saskia Nijmeijer, Steffen Wolf, Oliver P. Ernst, and Chris de Graaf

T. Langenhan, T. Schöneberg (eds.), *Adhesion G Protein-coupled Receptors*, Handbook of Experimental Pharmacology 234,
DOI 10.1007/978-3-319-41523-9_3

In the chapter titled "7TM Domain Structure of Adhesion GPCRs", the sequence of the author name was incorrect. The correct sequence should read as follows:

Saskia Nijmeijer, Steffen Wolf, Oliver P. Ernst, and Chris de Graaf

---

The updated original online version for this chapter can be found at
DOI 10.1007/978-3-319-41523-9_3

S. Nijmeijer • C. de Graaf (✉)
Department of Medicinal Chemistry, Amsterdam Institute for Molecules, Medicines and Systems, Vrije Universiteit Amsterdam, Amsterdam 1081HV, The Netherlands
e-mail: s.nijmeijer@vu.nl; c.de.graaf@vu.nl

S. Wolf
Department of Biophysics, CAS-MPG Partner Institute for Computational Biology, Key Laboratory of Computational Biology, Shanghai Institutes for Biological Sciences, Chinese Academy of Sciences, Shanghai, P.R. China

Institute of Physics, Albert Ludwigs University, Freiburg, Germany
e-mail: swolf@picb.ac.cn

O.P. Ernst
Department of Biochemistry, University of Toronto, 1 King's College Circle, Toronto, ON, Canada, M5S 1A8

Department of Molecular Genetics, University of Toronto, 1 King's College Circle, Toronto, ON, Canada, M5S 1A8
e-mail: oliver.ernst@utoronto.ca

T. Langenhan, T. Schöneberg (eds.), *Adhesion G Protein-coupled Receptors*, Handbook of Experimental Pharmacology 234, DOI 10.1007/978-3-319-41523-9_18

Index

T. Langenhan, T. Schöneberg (eds.), *Adhesion G Protein-coupled Receptors*,
Handbook of Experimental Pharmacology 234, DOI 10.1007/978-3-319-41523-9